NONTHERMAL PLASMA TECHNOLOGY FOR FOOD AND FOOD PRODUCTS

Innovations in Agricultural and Biological Engineering

NONTHERMAL PLASMA TECHNOLOGY FOR FOOD AND FOOD PRODUCTS

R. Mahendran, PhD
Preeti Birwal, PhD
Megh R. Goyal, PhD, PE
Editors

AAP | APPLE ACADEMIC PRESS

First edition published 2025

Apple Academic Press Inc.
1265 Goldenrod Circle, NE,
Palm Bay, FL 32905 USA

760 Laurentian Drive, Unit 19,
Burlington, ON L7N 0A4, Canada

CRC Press
2385 NW Executive Center Drive,
Suite 320, Boca Raton FL 33431

4 Park Square, Milton Park,
Abingdon, Oxon, OX14 4RN UK

Library and Archives Canada Cataloguing in Publication

Title: Nonthermal plasma technology for food and food products / R. Mahendran, PhD, Preeti Birwal, PhD, Megh R. Goyal, PhD, PE, editors.
Names: Mahendran, R. (Professor of food technology), editor. | Birwal, Preeti, editor | Goyal, Megh R., editor
Series: Innovations in agricultural and biological engineering.
Description: First edition. | Series statement: Innovations in agricultural and biological engineering | Includes bibliographical references and index.
Identifiers: Canadiana (print) 20250143259 | Canadiana (ebook) 20250143267 | ISBN 9781774917701 (hardcover) | ISBN 9781774917718 (softcover) | ISBN 9781003543084 (ebook)
Subjects: LCSH: Food industry and trade—Technological innovations. | LCSH: Food—Preservation. | LCSH: Low temperature plasmas.
Classification: LCC TP371.2 .N66 2025 | DDC 664—dc23

Library of Congress Cataloging-in-Publication Data

CIP data on file with US Library of Congress

ISBN: 978-1-77491-770-1 (hbk)
ISBN: 978-1-77491-771-8 (pbk)
ISBN: 978-1-00354-308-4 (ebk)

DEDICATION

To our esteemed mentor

Professor Megh R. Goyal, PhD, P.E.

Senior Technical Editor-in-Chief (Agricultural and Biological Engineering)
Apple Academic Press Inc., (Copartner CRC Press/Taylor and Francis Group)

For your leadership, motivation, and guidance have been instrumental in shaping this book volume.

For your invaluable contributions to the field of science and technology research.

For being a guiding light, providing us with profound impact both personally and academically, throughout our professional careers.

For your mentorship at every stage of our professional and personal development is truly priceless.

For your unwavering support and dedication.

—Global Educational Community—

ABOUT THE BOOK SERIES: INNOVATIONS IN AGRICULTURAL AND BIOLOGICAL ENGINEERING

Under this book series, Apple Academic Press Inc. is publishing book volumes over a span of 8–10 years in the specialty areas defined by the American Society of Agricultural and Biological Engineers (<asabe.org>). Apple Academic Press Inc. aims to be a principal source of books in agricultural and biological engineering. We welcome book proposals from readers in areas of their expertise.

The mission of this series is to provide knowledge and techniques for agricultural and biological engineers (ABEs). The book series offers high-quality reference and academic content on agricultural and biological engineering (ABE) that is accessible to academicians, researchers, scientists, university faculty and university-level students, and professionals around the world.

Agricultural and biological engineers ensure that the world has the necessities of life, including safe and plentiful food, clean air and water, renewable fuel and energy, safe working conditions, and a healthy environment by employing knowledge and expertise of the sciences, both pure and applied, and engineering principles. Biological engineering applies engineering practices to problems and opportunities presented by living things and the natural environment in agriculture.

ABE embraces a variety of the following specialty areas (<asabe.org>): aquaculture engineering, biological engineering, energy, farm machinery and power engineering, food and process engineering, forest engineering, information and electrical technologies, soil and water conservation engineering, natural resources engineering, nursery and greenhouse engineering, safety and health, and structures and environment.

For this book series, we welcome chapters on the following specialty areas (but not limited to):

1. Academia to industry to end-user loop in agricultural engineering
2. Agricultural mechanization
3. Aquaculture engineering
4. Biological engineering in agriculture

5. Biotechnology applications in agricultural engineering
6. Energy source engineering
7. Food and bioprocess engineering
8. Food and water security
9. Forest engineering
10. Hill land agriculture
11. Human factors in engineering
12. Impact of global warming and climatic change on agriculture economy
13. Information and electrical technologies
14. Irrigation and drainage engineering
15. Nanotechnology applications in agricultural engineering
16. Natural resources engineering
17. Nursery and greenhouse engineering
18. Potential of phytochemicals from agricultural and wild plants for human health
19. Power systems and machinery design
20. GPS and remote sensing potential in agricultural engineering
21. Robot engineering and drones in agriculture
22. Simulation and computer modeling
23. Smart engineering applications in agriculture
24. Soil and water engineering
25. Micro irrigation engineering
26. Structures and environment engineering
27. Waste management and recycling
28. Rural electrification
29. Sanitary engineering
30. Circular biosystems in agriculture
31. Farm-to-fork technologies in agriculture
32. Any other focus areas

For more information on this series, readers may contact:

Megh R. Goyal, PhD, PE
Senior Technical Editor-in-Chief:
Book Series: Innovations in Agricultural and Biological Engineering
E-mail: goyalmegh@gmail.com

OTHER BOOKS ON AGRICULTURAL AND BIOLOGICAL ENGINEERING BY APPLE ACADEMIC PRESS, INC.

Management of Drip/Trickle or Micro Irrigation
Megh R. Goyal, PhD, PE, Senior Editor-in-Chief

Evapotranspiration: Principles and Applications for Water Management
Megh R. Goyal, PhD, P.E. and Eric W. Harmsen, Editors

Book Series: Research Advances in Sustainable Micro Irrigation
Senior Editor-in-Chief: Megh R. Goyal, PhD, PE

- **Volume 1:** Sustainable Micro Irrigation: Principles and Practices
- **Volume 2:** Sustainable Practices in Surface and Subsurface Micro Irrigation
- **Volume 3:** Sustainable Micro Irrigation Management for Trees and Vines
- **Volume 4:** Management, Performance, and Applications of Micro Irrigation Systems
- **Volume 5:** Applications of Furrow and Micro Irrigation in Arid and Semi-Arid Regions
- **Volume 6:** Best Management Practices for Drip Irrigated Crops
- **Volume 7:** Closed Circuit Micro Irrigation Design: Theory and Applications
- **Volume 8:** Wastewater Management for Irrigation: Principles and Practices
- **Volume 9:** Water and Fertigation Management in Micro Irrigation
- **Volume 10:** Innovation in Micro Irrigation Technology

Book Series: Innovations and Challenges in Micro Irrigation
Senior Editor-in-Chief: Megh R. Goyal, PhD, PE

- Engineering Interventions in Sustainable Trickle Irrigation: Water Requirements, Uniformity, Fertigation, and Crop Performance
- Management Strategies for Water Use Efficiency and Micro Irrigated Crops: Principles, Practices, and Performance

- Micro Irrigation Engineering for Horticultural Crops: Policy Options, Scheduling, and Design
- Micro Irrigation Management: Technological Advances and Their Applications
- Micro Irrigation Scheduling and Practices
- Performance Evaluation of Micro Irrigation Management: Principles and Practices
- Potential of Solar Energy and Emerging Technologies in Sustainable Micro Irrigation
- Principles and Management of Clogging in Micro Irrigation
- Sustainable Micro Irrigation Design Systems for Agricultural Crops: Methods and Practices

Book Series: Innovations in Agricultural and Biological Engineering
Senior Editor-in-Chief: Megh R. Goyal, PhD, PE

- Advanced Research Methods in Food Processing Technologies
- Advances in Food Process Engineering: Novel Processing, Preservation and Decontamination of Foods
- Advances in Green and Sustainable Nanomaterials: Applications in Energy, Biomedicine, Agriculture, and Environmental Science
- Advances in Sustainable Food Packaging Technology
- Analytical Methods for Milk and Milk Products, 3-volume set:
    - Volume 1: Sampling Methods, Chemical and Compositional Analysis
    - Volume 2: Physicochemical Analysis of Concentrated, Coagulated and Fermented Products
    - Volume 3: Microbiological Analysis, Isolation, and Characterization
- Applications of Supercritical Fluid Extraction in Food Processing
- Biological and Chemical Hazards in Food and Food Products: Prevention, Practices, and Management
- Bioremediation and Phytoremediation Technologies in Sustainable Soil Management, 4-volume set:
    - Volume 1: Fundamental Aspects and Contaminated Sites
    - Volume 2: Microbial Approaches and Recent Trends
    - Volume 3: Inventive Techniques, Research Methods, and Case Studies
    - Volume 4: Degradation of Pesticides and Polychlorinated Biphenyls

- Dairy Engineering: Advanced Technologies and Their Applications
- Developing Technologies in Food Science: Status, Applications, and Challenges
- Emerging Technologies in Agricultural Engineering
- Engineering Interventions in Agricultural Processing
- Engineering Interventions in Foods and Plants
- Engineering Practices for Agricultural Production and Water Conservation: An Interdisciplinary Approach
- Engineering Practices for Management of Soil Salinity: Agricultural, Physiological, and Adaptive Approaches
- Engineering Practices for Milk Products: Dairyceuticals, Novel Technologies, and Quality
- Enzyme Inactivation in Food Processing: Technologies, Materials, and Applications
- Field Practices for Wastewater Use in Agriculture: Future Trends and Use of Biological Systems
- Flood Assessment: Modeling and Parameterization
- Food Engineering: Emerging Issues, Modeling, and Applications
- Food Process Engineering: Emerging Trends in Research and Their Applications
- Food Processing and Preservation Technology: Advances, Methods, and Applications
- Food Technology: Applied Research and Production Techniques
- Functional Dairy Ingredients and Nutraceuticals: Physicochemical, Technological, and Therapeutic Aspects
- Handbook of Research on Food Processing and Preservation Technologies, 5-volume set:
  - Volume 1: Nonthermal and Innovative Food Processing Methods
  - Volume 2: Nonthermal Food Preservation and Novel Processing Strategies
  - Volume 3: Computer-Aided Food Processing and Quality Evaluation Techniques
  - Volume 4: Design and Development of Specific Foods, Packaging Systems, and Food Safety
  - Volume 5: Emerging Techniques for Food Processing, Quality, and Safety Assurance
- Modeling Methods and Practices in Soil and Water Engineering
- Nanotechnology and Nanomaterial Applications in Food, Health, and Biomedical Sciences

- Nanotechnology Applications in Agricultural and Bioprocess Engineering: Farm to Table
- Nanotechnology Applications in Dairy Science: Packaging, Processing, and Preservation
- Nanotechnology Horizons in Food Process Engineering, 3-volume set:
  - Volume 1: Food Preservation, Food Packaging and Sustainable Agriculture
  - Volume 2: Scope, Biomaterials, and Human Health
  - Volume 3: Trends, Nanomaterials, and Food Delivery
- Nonthermal Plasma Technology for Food and Food Products
- Nonthermal Pulsed Electric Field Technology for the Food Industry
- Novel and Alternative Methods in Food Processing: Biotechnological, Physicochemical, and Mathematical Approaches
- Novel Dairy Processing Technologies: Techniques, Management, and Energy Conservation
- Novel Processing Methods for Plant-Based Health Foods: Extraction, Encapsulation and Health Benefits of Bioactive Compounds
- Novel Strategies to Improve Shelf-Life and Quality of Foods: Quality, Safety, and Health Aspects
- Ohmic Heating Technology for Processing of Foods and Food Products: Design, Food Quality, and Applications
  Editors: Preeti Birwal, PhD, Megh R. Goyal, PhD, and Baldev Singh Kalsi, PhD
- Phytochemicals and Medicinal Plants in Food Design: Strategies and Technologies for Improved Healthcare
- Processing of Fruits and Vegetables: From Farm to Fork
- Processing Technologies for Milk and Milk Products: Methods, Applications, and Energy Usage
- Quality Control in Fruit and Vegetable Processing: Methods and Strategies
- Scientific and Technical Terms in Bioengineering and Biological Engineering
- Soil and Water Engineering: Principles and Applications of Modeling
- Soil Salinity Management in Agriculture: Technological Advances and Applications
- State-of-the-Art Technologies in Food Science: Human Health, Emerging Issues and Specialty Topics

- Sustainable and Functional Foods from Plants: Health Impact, Bioactive Compounds, and Production Technologies
- Sustainable Green Nanomaterials: Synthesis, Characterization, and Engineering Applications
- Sustainable Biological Systems for Agriculture: Emerging Issues in Nanotechnology, Biofertilizers, Wastewater, and Farm Machines
- Sustainable Nanomaterials for Biomedical Engineering: Impacts, Challenges, and Future Prospects
- Sustainable Nanomaterials for Biosystems Engineering: Trends in Renewable Energy, Environment, and Agriculture
- Sustainable Use of Wastewater in Agriculture: Applications, Methods, and Technology
- Technological Interventions in Dairy Science: Innovative Approaches in Processing, Preservation, and Analysis of Milk Products
- Technological Interventions in Management of Irrigated Agriculture
- Technological Interventions in the Processing of Fruits and Vegetables
- Technological Processes for Marine Foods, From Water to Fork: Bioactive Compounds, Industrial Applications, and Genomics
- The Chemistry of Milk and Milk Products: Physicochemical Properties, Therapeutic Characteristics, and Processing Methods

ABOUT THE EDITORS

R. Mahendran, PhD
Professor and Head, Centre of Excellence in Nonthermal Processing (CENTP)
National Institute of Food Technology, Entrepreneurship and
Management -Thanjavur (NIFTEM-T), India

R. Mahendran, PhD, is working as a Professor and Head of the Center of Excellence in Nonthermal Processing (CENTP) at the National Institute of Food Technology, Entrepreneurship, and Management–Thanjavur (NIFTEM-T).

Dr. R. Mahendran completed his BTech in Chemical Engineering and MTech in Industrial Biotechnology from SASTRA University, Thanjavur, Tamil Nadu, and PhD in Food Process Engineering from Tamil Nadu Agricultural University, Coimbatore, India. Before his association with NIFTEM-Thanjavur, he worked as a Production Officer in the Fermentation Department of M/s. Krebs Biochemicals and Industries Ltd. Nellore, Andhra Pradesh, and also worked as a Scientist in TIFAC/ Department of Science and Technology, Government of India at New Delhi from 2006 to 2010.

Dr. Mahendran's research focus includes cold plasma applications on agri-food, food structure shape transformation, and novel nonthermal processing on food systems. He has authored over 100 publications in reputed journals, popular articles, and book chapters. He has been awarded the Dr. V. Subrahmanyan Best Scientist Award for the year 2021 for his outstanding research contributions. He is a fellow of the Institute of Food Science and Technology (IFST) and a fellow of the Institution of Engineers (IEI). Dr. Mahendran is a life member of various professional bodies including The Institution of Engineers, AFST(I), ISTE, ISC, IIChE, ISAE, and ISFE. He also serves as a guest editor, editorial board member, and reviewer for reputed food processing journals.

Preeti Birwal, PhD
Scientist (Processing and Food Engineering), Department of Processing and Food Engineering, College of Agricultural Engineering and Technology, Punjab Agricultural University, Ludhiana, Punjab, India

Preeti Birwal, PhD, is working as a Scientist (Processing and Food Engineering) in the Department of Processing and Food Engineering, College of Agricultural Engineering and Technology, Punjab Agricultural University, Ludhiana, Punjab, India.

She holds a BSc (2012) in Dairy Technology from ICAR-National Dairy Research Institute (NDRI), Karnal, an MSc (2014) in Food Process Engineering and Management from NIFTEM, Haryana; and a PhD (Dairy Engineering) from ICAR-NDRI, Bangalore. She is the recipient of MHRD (2008), Nestle India (2009), GATE (2012–14), and UGC-RGN fellowships (2014–18).

She is currently working in the area of nonthermal food preservation, fermented beverages, food packaging, and technology of millet-based beer. She has served Jain Deemed to be University, Bangalore, as a member of the board of examiners and placements. She is advising several MTech. scholars in Food technology. She has participated in several national and international conferences and seminars. She has delivered lectures as a resource person on doubling the farmer's income through dairy technology in a training sponsored by the Directorate of Extension, Ministry of Agriculture and Farmers Welfare, Government of India. She has received an Outstanding Reviewer of the Month award. She has completed AUTOCAD 2D and 3D certification.

She has 18 research papers, one edited book, 5 book chapters, about 28 popular articles, and five conference papers and 56 abstracts, two editorial opinions to her credit. She has successfully guided five postgraduate students in their dissertation work. She is serving as an external examiner for various Indian state agricultural universities. She is also serving as editor and reviewer of several journals. She is a life member of IDEA.

Megh R. Goyal, PhD, PE

Retired Professor in Agricultural and Biomedical Engineering, University of Puerto Rico, Mayaguez Campus; Senior Acquisitions Editor, Biomedical Engineering and Agricultural Science, Apple Academic Press, Inc.

Megh R. Goyal, PhD, PE, is currently a retired professor of agricultural and biomedical engineering from the General Engineering Department at the College of Engineering at the University of Puerto Rico–Mayaguez Campus (UPRM); and Senior Acquisitions Editor and Senior Technical Editor-in-Chief for Agricultural and Biomedical Engineering for Apple Academic Press Inc.

During his professional career, he has worked as a Soil Conservation Inspector; Research Assistant at Haryana Agricultural University and Ohio State University; Research Agricultural Engineer/Professor at the Department of Agricultural Engineering of UPRM; and Professor of Agricultural and Biomedical Engineering in the General Engineering Department of UPRM. He spent a one-year sabbatical leave in 2002–2003 at the Biomedical Engineering Department of Florida International University, Miami, USA.

Dr. Goyal was the first agricultural engineer to receive the professional license in agricultural engineering from the College of Engineers and Surveyors of Puerto Rico. In 2005, he was proclaimed the "Father of Irrigation Engineering in Puerto Rico for the Twentieth Century" by the American Society of Agricultural and Biological Engineers, Puerto Rico Section, for his pioneering work on micro irrigation, evapotranspiration, agroclimatology, and soil and water engineering.

During his professional career of 52 years, he has received many awards, including Scientist of the Year, Membership Grand Prize for the American Society of Agricultural Engineers Campaign, Felix Castro Rodriguez Academic Excellence Award, Man of Drip Irrigation by the Mayor of Municipalities of Mayaguez/Caguas/Ponce and Senate/Secretary of Agriculture of ELA, Puerto Rico, and many others. He has been recognized as one of the experts "who rendered meritorious service for the development of [the] irrigation sector in India" by the Water Technology Centre of Tamil Nadu Agricultural University in Coimbatore, India, and ASABE who bestowed on him the 2018 Netafim Microirrigation Award.

Dr. Goyal has authored more than 200 journal articles and edited more than 100 books. AAP has published many of his books, including *Management of Drip/Trickle or Micro Irrigation; Evapotranspiration: Principles and Applications for Water Management;* 10-volume set on *Research Advances in Sustainable Micro Irrigation.* He has also authored the textbooks *Elements of Agroclimatology* (Spanish) by UNISARC, Colombia, and two *Bibliographies on Drip Irrigation.* Dr. Goyal has also developed several book series with AAP, including Innovations in Agricultural & Biological Engineering (with over 60 titles in the series to date), Innovations and Challenges in Micro Irrigation; and Innovations in Plant Science for Better Health: From Soil to Fork.

Dr. Goyal received his BSc degree in Engineering from Punjab Agricultural University, Ludhiana, India, and his MSc and PhD degrees from the Ohio State University, Columbus, Ohio, USA. He also earned a Master of Divinity degree from the Puerto Rico Evangelical Seminary, Hato Rey, Puerto Rico, USA.

CONTENTS

CONTRIBUTORS

Uday S. Annapure
Professor, Food Chemistry, Department of Food Engineering and Technology, Institute of Chemical Technology, Mumbai, Maharashtra, India; Institute of Chemical Technology, Marathwada Campus, Jalna, Maharashtra, India

Paula Bourke
Professor and Head, School of Biosystems and Food Engineering, University College Dublin, Belfield, Dublin

M. Dharini
PhD Research Scholar, Center of Excellence in Non-Thermal Processing, National Institute of Food Technology, Entrepreneurship, and Management – Thanjavur (formerly Indian Institute of Food Processing Technology (IIFPT)), Thanjavur, Tamil Nadu, India

Gunaseelan Eazhumalai
PhD Research Scholar, Department of Food Engineering and Technology, Institute of Chemical Technology, Mumbai, Maharashtra, India

Gnana Moorthy Eswaran
PhD Research Scholar, Department of Agricultural and Food Engineering, Indian Institute of Technology, Kharagpur, West Bengal, India

Megh R. Goyal
Senior Technical Editor-in-Chief (Agricultural and Biomedical Engineering) for AAP; Retired Professor in Agricultural and Biomedical Engineering, University of Puerto Rico-Mayaguez, USA

T. K. Ranjitha Gracy
PhD Research Scholar, Department of Food Engineering and Technology, Institute of Chemical Technology, Mumbai, Maharashtra, India

Proshanta Guha
Professor, Department of Agricultural and Food Engineering, Indian Institute of Technology, Kharagpur, West Bengal, India

P. Himashree
Project Assistant, Center of Excellence in Non-Thermal Processing, National Institute of Food Technology, Entrepreneurship, and Management (NIFTEM) – Thanjavur (formerly Indian Institute of Food Processing Technology (IIFPT)), Thanjavur, Tamil Nadu, India

Suresh Kumar Kalakandan
Professor and Head, Department of Food Biotechnology, National Institute of Food Technology, Entrepreneurship, and Management – Thanjavur (NIFTEM-T), Tamil Nadu, India

B. Kamalapreetha
PhD Research Scholar, Center of Excellence in Non-Thermal Processing, National Institute of Food Technology, Entrepreneurship, and Management (NIFTEM) – Thanjavur (an Institute of National Importance; formerly Indian Institute of Food Processing Technology (IIFPT)), Thanjavur, Tamil Nadu, India

Sangeetha Karunanithi
PhD Research Scholar, Department of Agricultural and Food Engineering, Indian Institute of Technology, Kharagpur, West Bengal, India

R. Mahendran
Professor and Head, Center of Excellence in Non-Thermal Processing, National Institute of Food Technology, Entrepreneurship, and Management (NIFTEM) – Thanjavur (formerly Indian Institute of Food Processing Technology (IIFPT)), Thanjavur, Tamil Nadu, India

Jeyan A. Moses
Assistant Professor and In-Charge, Computational Modeling and Nanoscale Processing Unit, National Institute of Food Technology, Entrepreneurship, and Management, Thanjavur, Tamil Nadu, India

Pramila Murugesan
National Post-Doctoral Fellow, Computational Modeling and Nanoscale Processing Unit, National Institute of Food Technology, Entrepreneurship, and Management, Thanjavur, Tamil Nadu, India

G. Nagamaniammai
Associate Professor, Department of Food Process Engineering, School of Bioengineering, College of Engineering and Technology – SRM Institute of Science and Technology, Kattankulathur, Tamil Nadu, India

Nisar S. Nikitha
MTech Students, Department of Food Process Engineering, School of Bioengineering, College of Engineering and Technology – SRM Institute of Science and Technology, Kattankulathur, Tamil Nadu, India

S. Niranjan
MTech Students, Department of Food Process Engineering, School of Bioengineering, College of Engineering and Technology – SRM Institute of Science and Technology, Kattankulathur, Tamil Nadu, India

Anjaly Paul
Graduate Research Assistant, Department of Engineering, Faculty of Agriculture, Dalhousie University, Nova Scotia, Canada

G. Sai Preetham
MTech Scholar, Department of Food Product Development, School of Sensory Sciences, National Institute of Food Technology, Entrepreneurship, and Management – Thanjavur (NIFTEM-T), Tamil Nadu, India

Anbarasan Rajan
PhD Research Scholar, Center of Excellence in Non-Thermal Processing, National Institute of Food Technology, Entrepreneurship, and Management (NIFTEM) – Thanjavur (formerly Indian Institute of Food Processing Technology (IIFPT)), Thanjavur, Tamil Nadu, India

Lokeswari Ramireddy
Graduate Research Scholar, University College Dublin, School of Biosystems and Food Engineering, Agriculture and Food Science, Belfield, Dublin

E. J. Rifna
Department of Food Process Engineering, National Institute of Technology Rourkela, Rourkela, Odisha, India

K. Sandhya
PhD Research Scholar, Computational Modeling and Nanoscale Processing Unit, National Institute of Food Technology, Entrepreneurship, and Management, Thanjavur, Tamil Nadu, India

R. Smruthi
MTech Student, Department of Food Process Engineering, School of Bioengineering, College of Engineering and Technology – SRM Institute of Science and Technology, Kattankulathur, Tamil Nadu, India

K. Padma Sree
MTech Scholar, Department of Food Biotechnology, National Institute of Food Technology, Entrepreneurship, and Management – Thanjavur (NIFTEM-T), Tamil Nadu, India

Prem Prakash Srivastav
Professor, Department of Agricultural and Food Engineering, Indian Institute of Technology, Kharagpur, West Bengal, India

Cheeyattil Subith
Research Scholar, National Institute of Food Technology, Entrepreneurship, and Management (NIFTEM) – Thanjavur (formerly Indian Institute of Food Processing Technology (IIFPT)), Thanjavur, Tamil Nadu, India

Aparajhitha Sudarsan
Food and Health Safety Coordinator, Exeter Fresh Produce and Storage, Exeter, Ontario, Canada

PREFACE

The world that is moving toward sustainable food production and processing no longer depends on temperature-assisted processing methods; not just because of the quality concerns associated with the product but because of the consumer preference towards the product. Keeping this in mind, scientists all over the world have decided to explore the possibilities of processing food products without subjecting them to heat treatment, and through this attempt, they have developed processing methods such as high-pressure processing, pulsed light, irradiation, and pulsed electric field, and grouped them as nonthermal processing methods. However, none of these techniques can act as a multifaceted single-point solution as cold plasma technology does.

We are in the era where Industry 4.0 demands sustainability in processing; thus, the food industries also need a sustainable processing method that can be used for multiple applications. If we talk about postharvest processing, we need a technology that can eliminate the microbes, insects, allergenicity, and anti-nutritional factors from food, and also be able to improve germination capacity, stress tolerance level, functional properties (i.e., starch modification), and surface nature. In this regard, cold plasma is the only frontrunner with practically proven study results, and this new understanding made us believe in working on this area to provide comprehensive knowledge to the reader through this book.

The chapters of this book are compiled systematically to provide comprehensive information on cold plasma processing. The book has 16 chapters focusing on four major categories of cold plasma processing including food and non-food applications.

The chapters of the first section deal with the basics of cold plasma technology including its principle, equipment design, and plasma characterization. The second part of this book focuses on the interaction of cold plasma with food systems at different physical states of foods, including their molecular structure. Since plasma application is not limited to food applications, the third section of this book provides an overview of the applications of plasma in biofilms and other food packaging materials. The latter section deals with the possibilities of utilizing plasma technology with other emerging techniques and taking the technology to industrial and space research applications. Like any other technology, cold plasma also has its challenges and limitations, which are listed in three different chapters in the final section of this book to provide an opportunity for future researchers to focus on these areas.

We are grateful to all of the cooperating authors for contributing their knowledge, experience, and ideas to create these highly useful chapters that will make the book a great resource for cold plasma processing of food for many years to come.

Furthermore, we are thankful to all reviewers who offered their time and experience to enhance the chapters. We made sure that the chapters were not overlapping to enhance the reading experience; however, overlapping of some basic information is unavoidable in a book that deals with a different perspective of cold plasma treatment. This, however, will aid readers to visualize the cold plasma basics from different perspectives and understand its mechanism.

We also thank the editorial staff at Apple Academic Press Inc. for their unselfish contribution.

We truly believe that the book will inspire researchers to continue bringing new ideas of utilizing plasma technology in food products to make this technology a sustainable tool for the post-harvest processing of agri-food products.

—Editors

PART I
Basics of Cold Plasma Technology

CHAPTER 1

INTRODUCTION AND PRINCIPLES OF COLD PLASMA TECHNOLOGY

B. KAMALAPREETHA and R. MAHENDRAN

ABSTRACT

Plasma is a non-thermal method that has fascinated many scientists around the world. Cold plasma has numerous applications in the food sector, including the use of active reactive plasma species, radiation of ultraviolet rays (UV), free radicals, and discharged particles. All of these compounds have a massive amount of internal energy. At low or atmospheric pressure, energizing gases or gas mixers are produced by various modes of application. This chapter focused mainly on giving a broad outline of the development of cold plasma in food applications and the principles behind the processing of reactive species. Additionally, it has also included various applications of plasma. Cold Plasma is certainly a valuable tool for the food sector, even though research studies and development are desirable to understand its food matrix interactions. Also, regulatory approval for upscaling of the system to utilize it commercially is required to be a major focus.

1.1 INTRODUCTION

Fresh food products have gained a lot of attention. Fresh commodities including eggs, milk, meat, poultry produce, fruits, and vegetables are expected to augment shelf-life. People get a healthy balanced diet from different fresh produce. Pathogenic microorganisms can, however, be carried by fresh food products. Many food products are abandoned because of poor postharvest handling and foodborne diseases.

Nonthermal Plasma Technology for Food and Food Products.
R. Mahendran, Preeti Birwal, and Megh R. Goyal (Eds.)

Meanwhile, the primary drivers of foodborne pathogens are often the heterogeneity and acquired resistance of microbes/pathogens on fresh produce against the environment. *Salmonella* species, *E. coli, Listeria,* and *Staphylococcus* species (especially *S. aureus*) are the most infectious pathogenic bacteria found in fresh foodstuffs. They can persist in raw as well as frozen products for a longer period, triggering chronic illness and even death. It has the potential to effectively disrupt consumers' health.

Microbes and pathogens are the primary concern of most of the food processing industries. All these organisms have a hostile effect on the health and economy of the consumers. So far, thermal processing methods like canning, pasteurization, and sterilization are employed to overcome the detrimental effect of the spoilage-causing organism. These processing methods have a deleterious and beneficial effect on the spoilage organism but also have an undesirable impact on the quality of the food, especially on its sensory, nutritional, and functional attributes of food.

As a result, there is a demand to develop an appropriate and efficient disinfecting technique for perishable produce that upholds food quality and safety while attempting to minimize adverse changes in the dietary and organoleptic characteristics, as well as the flavor of the food.

This chapter focuses on a general introduction to cold plasma concerning the source of plasma and highlights the significant applications in food matrices, their current advantages, and recent developments.

1.2 EVOLUTION OF NON-THERMAL TECHNOLOGIES

Most of the global goals put forth to encourage and plan sustainable development around the world have a direct or indirect relationship with the global food system, as stated in the millennium development objectives of the United Nations. The major concern in the developing world will be to sustainably nourish the fast-growing population while providing an ample amount of food that is safe, nutritious, tasty, and convenient. Despite recent studies, the current state of the world food system, which is made up of numerous sequential processes from food production to consumption, is associated with a wide variety of holistic vulnerabilities.

According to the investigation, the potential impacts of human activities on agricultural processing account for one-fourth of all greenhouse gas emissions [25]. Furthermore, it emits organic harmful pollutants into waterways while consuming 25% of the total water globally for its production [3].

Most wastage originates in the food supply chain, where 30 to 50% of food is wasted and is primarily caused by fruits, veggies, and cereal grains [61].

The accumulation of food waste in the form of solid, liquid, and gas is attributed to a negative effect on the environment. To develop a sustainable practice, management of resources is the primary focus which can be achieved by managing three aspects: energy, solid waste (upcycle), and water (usage and wastewater treatment) [49]. Other elements that may affect the measures are environmental regulations. Consumer choice must be considered before deploying substitutes for goods or services. Moreover, an increase in energy costs often encourages food industries to invest more in energy management [84].

Considering different aspects of the food sector to ensure food safety, processing techniques are broadly categorized as thermal and non-thermal based on heat application. Thermal treatments include pasteurization, sterilization, evaporation, refrigeration, freezing, drying, ohmic, MW, RF, IR, and inductive heating; while the nonthermal treatments encompass HPP, PEF, high-intensity light, OMF, US, pulsed X-rays, SCFT, membrane technology, pulsed magnetic field, ozonization, and plasma.

The typical concern of all the thermal procedures is ultimately food safety, where all processing conditions are mainly imposed for microbial decontamination and to extend shelf-life. Though it is an extremely energy-intensive technique in the processing of foods.

A significant development in green technologies is required to fulfill the goals of transforming agricultural produce into consumable products, to progress the production efficiency of the food sector with high quality and rich functional properties of goods. The food industry will also need to maintain its competitiveness at a time when governments and consumers demand continuous sustainable development [8]. Moreover, research studies strive to evolve and introduce nonthermal food processing methods and equipment for processing foodstuffs with enhanced food quality and prolonged shelf-life. These techniques are in high demand, providing an alternative to thermal processing.

Considering the limitations, the development of cost-effective techniques is needed to augment food quality. For instance, amidst their ability to minimize pathogenic organisms and extend shelf-life, innovative approaches for natural food matrices have demonstrated deleterious effects on the functional elements present in foods [21, 23]. All these novel technologies will pave the way to satisfy consumers in obtaining healthy food. Among the modern techniques, US, PEF, MW, and HPP processing have been approved by the FDA as safe for human consumption.

The risk of foodborne illness resulting from contaminated food is still a major concern. Food consumed with contamination was the source of 600 million cases with 0.4 million deaths every year [35].

These outbreaks erode trust in the food supply's microbiological safety. Along with worries about eating raw meat and poultry, fresh vegetables have also been linked to human diseases, according to a new assessment. Test experiments have shown that a variety of fresh and freshly-cut vegetable surfaces can support the growth of potentially lethal germs. Despite these studies, there are still considerable knowledge gaps about the best ways to eliminate or inactivate viruses from at-risk produce as well as how to prevent fresh produce from becoming contaminated. Produce does not respond well to conventional postharvest washing and sanitizing procedures, which frequently result in pathogen reductions of less than 2 log units.

Despite extensive research on antimicrobial measures for meat, poultry, and vegetables, there might be an urge for risk-free, unique, novel, cost-effective, and more efficient treatments. One such method of nonthermal treatment is cold plasma. In the material processing industry, plasma plumes are frequently used to treat paper, textiles, glass, electronics, and other materials. Cold plasma, on the other hand, is a relatively novel strategy in terms of enhancing food safety. Nevertheless, food producers, processors, and researchers are not familiar with the technical features of cold plasma.

1.3 INTRODUCTION TO PLASMA TECHNOLOGY

Cold plasma is an advanced novel technique of preservation; hence, the interpretation is continuously evolving. Recent articles have extended the use of the term cool plasma, atmospheric pressure plasma, cold atmospheric gas plasma, and other similar terminology. Other times, the plasma is defined by generative technology, such as arc discharge, homogeneous glow discharge gas plasma, jet type, and DBD. Terminology and research strategies are ascertained to be relatively established as technology evolves. Likewise, the term "cold plasma" pertains to reactive nonequilibrium plasma which is synthesized at or near room temperature and possesses bactericidal effects which don't rely on conventional heating to destroy or inhibit the growth of target organisms.

Nonthermal plasma is also an alternative revolutionary microbial inactivation technique. Although plasma processing is widespread, the history of plasma in sterilization procedures is relatively new. In 1968, for the first time in the history of plasma application, argon plasma exploited the bactericidal effect to inactivate germs on a glass surface.

Although this technique is applicable in the food research sector, the significance of obtained plasma species and their characteristics is yet to be deployed for better understanding.

1.4 DIFFERENCE BETWEEN THERMAL AND NON-THERMAL PLASMA

According to electron temperature (Te) and the method of generation [45], plasma is classified as high electron temperature (Te: 10^6–10^8 K) and low electron temperature (Te: 10^4–10^5 K) plasma. In the former plasma, all the produced ions, plasma species, and electrons stay in thermal equilibrium, whereas the latter plasma is in a non-equilibrium state or quasi-state. Furthermore, it is categorized into thermal (TP) and nonthermal or cold plasma (NTP) [38].

In TP, all temperatures of the gas (T_g), heavy ions (T_i), and electrons (T_e) are equivalent in nature and have the highest ionization degree. In the case of NTP, the temperature of the heavy ions (T_i) is in a much lower range (300–500 K) whereas the T_e can reach up to 10^4 K and T_g is low as room/near the ambient temperature at atmospheric or vacuum condition [36]. In the NTP system, electrifying or applying electromagnetic waves to the gas enables the electrons and gas to emanate higher energy and lower gas temperature, respectively, resulting in the creation of reactive species of oxygen and nitrogen (ROS/RNS). All these reactive gas species are emitted from the varied sources of carrying gas which aids in the application of sterilization and preservation [45].

1.5 PRINCIPLES OF COLD PLASMA

William Crookes invented plasma (a fourth state) in 1879, and in the year 1928, Irving Langmuir [32] coined the term "plasma." Once the molecules heat up, there is a sharp increase in the solid's kinetic energy, which triggers the transition of solids to liquid form. This further tends to raise the energy level of the liquid and ends up turning into gas. Further energizing the gas molecules brings out the ionization of gas due to the disintegration of intermolecular structures [39]. As a result, plasma is produced and implied as the fourth state.

In general, two groups of plasma treatment are thermal plasma and cold plasma. Extreme temperatures activate thermal plasma and are employed to yield high amounts of energy. A non-thermal treatment called cold plasma operates between 25 and 65°C [50]. Free radicals (ions, electrons, etc.,) are created when a gas is ionized. The gas gets ionized, which has a potential effect on the reactive plasma species. To produce plasma often entails gases like argon (Ar), helium (H_e), oxygen (O_2), nitrogen (N_2), and air. Plasma with positive/negative ions and plasma species (ozone – O_3 and singlet oxygen – O) are engendered [28, 50] when subjecting the gaseous molecules to diverse energy supplies including thermal, electrical, magnetic field, etc., and can be operated under low pressure or atmospheric conditions.

Other types of plasma systems are ambient and sub-ambient pressure systems. At atmospheric pressure in an open environment, the produced plasma from the different working gas is with a rich supply of charged ions, molecules, UV photons, and plasma species, causing microbial inactivation. Whereas the low-pressure system does require a costly vacuuming unit and is not simple to work. The plasma's properties have led to a way to use in the realms of electronics, chemical processing, textiles, surface coating, pharmacology, and the food industry. Thirumdas et al. [84] state that plasma is created at or equivalent to ambient temperature which does not involve high temperatures to inactivate microorganisms that are beneficial for handling heat-sensitive foods produce where there is no detrimental thermal damage.

In food processing, cold plasma can be imposed to amend physiochemical features like macronutrients (lipids and proteins), treat food packaging materials, and wastewater, and decontaminate foodstuff or on its surface, sterilize process equipment, inactivate food degenerative enzymes.

1.6 SOURCES OF COLD PLASMA

There are several methods to create cold plasma discharges, some of which have been the research focus since the beginning of addressing electrical phenomena. In food processing applications, plasma is categorized based on generation. The plasma generation systems are classified as follows:

- Barrier discharge (BD) plasma: Dielectric, Resistive
- Plasma jet (PJ)
- Corona discharge (CD)
- Radiofrequency (RF)
- Microwave (MW)

1.6.1 BARRIER DISCHARGE (BD)

- **Dielectric Barrier Discharge (DBD) Plasma:** DBD plasma is the most extensively used system for producing plasma. In this system, plasma is created when the system is supplied with high voltage between the parallelly placed electrodes of similar sizes [46, 91]. These electrodes are layered with a dielectric material like quartz, polymers, glass, or ceramic material [4, 31]. In comparison to other plasma generations, the basic, flexible, and geometric design of the system, its flexibility, construction details, and operating parameters make it more compatible [70, 71]. The main advantage of DBD plasma is that it lessens spark discharge production and creates stream discharge. At low-pressure conditions, the distance between the electrodes and the applied pressure ensures the creation of glow discharge plasma [85] or a plasma plume with legitimate variations in the layers of light emission patterns.

- **Resistive Barrier Discharge Plasma (RBD):** This resistive barrier discharge plasma was developed by Laroussi & Alexeff [34]. It is in the form of DBD plasma wherein the dielectric material is interchanged with a resistive barrier layer (few megaohms. Centimeter – MΩ.cm) over one or both of the electrodes. This resistive barrier limits the current flow through the circuit by utilizing the resistive impedance of the material, which further upholds the transition of discharge from streamer to spark [33, 60]. At low voltage, where the RBD plasma system produces plasma operated at atmospheric pressure at DC (or low-frequency AC 50 Hz) using helium gas.

1.6.2 CORONA DISCHARGE PLASMA (CD)

Basically, Corona discharge plasma depends on the electrode configuration where the discharge plasma is created at the end of the tip/sharp edges of the electrode. It is a type of non-equilibrium plasma. The electric field being applied is non-uniform around the electrode, which accelerates the ionization of the atoms or gaseous molecules around the electrode [82]. Immense energy is delivered at the surfaces with two differing curvature radii – the emitter (sharp edge) and the collector (thin wire) [17]. Glow discharge occurs at the sharp edges of the electrode, which is coined as the corona layer. These CD can be emitted by DC (Direct), AC (alternating), or Pulsed power supply and high frequency. Ozone synthesis now renders more use of CD plasma for various reasons. The applied electric field between the electrodes is kept minimum to lessen the spark discharge.

1.6.3 MICROWAVE DISCHARGE PLASMA (MW)

This type of discharge is coined as electrodeless discharge plasma. The microwave discharge plasma system utilizes the magnetron to produce electromagnetic radiations (300 MHz to 300 GHz frequency). The microwave discharge (plasma) is created as a result of an acceleration of gas molecules/electrons through waveguides [11, 17], which release energies as photons (UV and visible light) [48]. The MW system is operated at low and atmospheric pressure conditions.

1.6.4 RADIOFREQUENCY (RF)

Radiofrequency plasma is classified into ICP, CCP, and helicon wave sources. Among these three types, ICP and CCP are extensively used as radiofrequency-based plasma generation systems for industrial applications [5].

- **CCP:** In a vacuum chamber, two electrodes are placed opposite each other at a certain distance. Here, the RF energy supply is at a frequency of 13.56 MHz, where the OMF causes the ionization of gases resulting in the creation of RF plasma [7, 11].
- **ICP:** This system constitutes two RF sources. One source energizes the coil externally with the plasma parted from the dielectric window and the other source regulates the ionization by positioning the substrate holder [7, 79].

1.6.5 PLASMA JET DISCHARGE (PJ)

The plasma jet discharge system is operated at atmospheric pressure conditions with constant and uniform discharge [38]. The PJ system involves concentric cylinder electrodes in which one electrode is coupled with other plasma sources (such as: DBD, CD, MW, RF) at a higher frequency, in which it ionizes the working noble gas (helium or argon) [26]. At a higher gas flow rate, the generated plasma with highly reactive species from the electrode zone is propelled in the form of a jet into the external environment with a weak glow [41, 51, 73, 81]. Plasma jets are mainly used for direct application, provided that the target surface is not a limitation in this type of discharge. The application of a series of jet configurations can be utilized for the treatment of a large area.

1.7 APPLICATIONS OF COLD PLASMA IN FOODS AND FOOD PRODUCTS

An emerging green technique of NTP, cold plasma is applicable for food, biomedical, and a wide range of applications. This technique has lately gained popularity in agricultural food processing, like thermal techniques such as microwave and ohmic heating. Initially, cold plasma was subjected to non-edible substances like polymers and electronic materials to enrich the binding and printability of the polymer, augment surface modification, also to enhance the properties of glass and paper [10, 22]. As there is a rapid advancement in plasma science, therefore the reactive species obtained from plasma have a potential impact on microbial decontamination in foods, wastewater treatment as well as in packaging material, which results in shelf-life extension. Furthermore, it improves the food safety of agricultural produce. It also aids in the deactivation of enzymes and improves the physicochemical characteristics of fats and lipids in food production. Moreover, the implementation of this technology has modified the barrier characteristics of packaging materials and enhanced the functional attributes of food resources. Some of the advances in the application of cold plasma (Figure 1.1) are discussed in this section. [65].

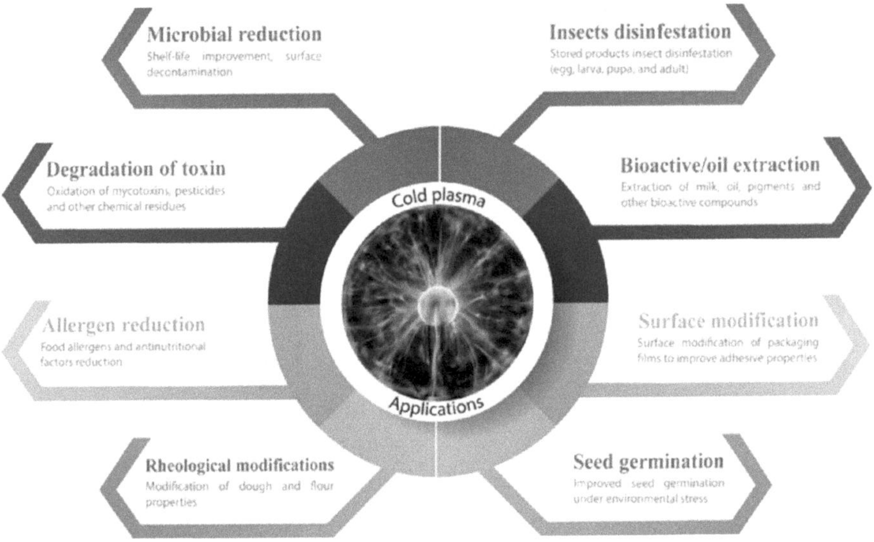

FIGURE 1.1 Applications of cold plasma.

1.7.1 FOOD STERILIZATION AND PRESERVATION

Plasma has been utilized as a sterilizing aid in the medical field for a long time [80], but recent studies have shown that it is also capable of decontaminating food. Conventional techniques of treating germs can be difficult because they may leave residues that could bind with food substances and endure a deadly cause in the food chain. CAP is regarded as a helpful substitute for handling food-borne bacteria deprived of nutritional changes in foods.

The effectiveness of CAP is dependent on the food's surface and topographic characteristics. Similar occurrences have also been observed on the strawberry surface and the stomata of lettuce [90]. These results showed that plasma could act to a modest degree of purification due to inherent characteristics of food surface or waxy cuticle [66].

Dharini et al. [40] investigated the preservation effect on milk using low-pressure plasma. The observed result of microbial reduction (95%) from the study exhibited the potential application of plasma on decontamination without any compromise in the quality of the food. Most of the time, the length of exposure and the gas used for disinfection determine how quickly microorganisms degrade and are inactivated. Excessive exposure levels would increase sterilization, but in some circumstances, high dosages might cause a shift in the bioactive chemicals. Due to the electrostatic force, charged particles are to blame for rupturing the membrane [53]. The disinfection studies using plasma proved to be a suitable future technique of preservation.

1.7.2 DEGRADATION OF TOXINS

The toxic secondary metabolites known as mycotoxins, which are produced by fungal species and infect a variety of food products, can lead to toxicity. Certain mycotoxins have been characterized as genotoxic, mutagenic, and carcinogenic. The lack of total inactivation of mycotoxins by conventional processing and the resistance of mycotoxins to heat processing have sparked ongoing research for alternate processing methods. Mycotoxins present in food products have been shown to degrade significantly in cold plasma [78, 83]. To boost degrading efficiency and gain knowledge of the mechanism, researchers are presently concentrating on process parameter optimization [77, 87]. Even if cold plasma has already been suggested as a potential substitute, more investigations are essential to clarify the chemical reaction of diverse plasma species as well as the toxicity analysis of breakdown products.

1.7.3 PESTICIDE DEGRADATION

Pesticide residue has an adverse effect and it is the major factor that needs to be mitigated in the food industry. Pesticides get ingested into the human body via eyes, skin, oral cavity, and respiration which is toxic to human health. Even without direct ingestion, its residue contaminates the food and water which becomes a human threat. It results in some of the deadly effects [29], such as: carcinogenic, teratogenic, cytotoxic, genotoxic, and immunotoxin. Most pesticides (Chlorpyrifos, Dichlorvos, Cypermethrin, Omethoate) exposure gives rise to genetic changes. Consequently, pesticides hurt human health as well as the environment. Thus, the severity of the pesticide effect needs to be curtailed to prevent harmful effects [15].

Amidst various applications of cold plasma, plasma can be useful for pesticide degradation as a recent research focus. Various research studies proved the potential effect of plasma on pesticide degradation. This plasma effect is aligned with the composition and concentration of formed plasma species [42, 72] and electron energy. Initially, applied plasma breaks the pesticide molecule, further, the generated free radicals (oxidative effect), and reactive species (O_3, OH, H_2O_2) cause the degradation of the pesticide [1, 67]. Other irradiated emissions and UV from plasma gas also have a potential effect on pesticides.

In recent research, Anbarasan et al. [1] studied and confirmed the consequences of low-pressure plasma treatment on the reduction of chlorpyrifos in pesticide-infused soybeans. They observed a 50% reduction at 2.0 kV in 6 minutes due to UV and emitted plasma species, leading to surface modification of the soybean seed coat.

Phan et al. [62] successfully reduced pesticide residues of Chlorpyrifos (74%) and cypermethrin (62.9%) on mangoes using Gliding Arc discharge plasma with Argon gas at 5 L/min. They also noted physicochemical changes. This method could be a successful strategy for ensuring chemical safety in postharvest products. Similarly, Cold plasma has been used on apples and cucumbers to reduce organophosphorus pesticides (diazinon and chlorpyrifos) [47]. This method was found to be more efficient on diazinon than on chlorpyrifos, with apples showing efficient treatment. It was concluded that better treatment effects were achieved with higher voltage and exposure time.

Findings supported by Bai et al. [2] using an ICP – RF with oxygen flow showed a reduction in the amounts of dichlorvos (DDV) and omethoate on maize. Plasma treatment was also applied to strawberries coated with pesticides (azoxystrobin, cyprodinil, fludioxonil, and pyriproxyfen) by Pankaj Misra et al. [56]. After 5 minutes of treatment, the pesticide concentration decreased by 45–71%.

Heo et al. [24] observed a 96% reduction of paraoxon on fresh apples using DBD plasma in a commercial refrigerator. According to previous research on pesticide reduction rates, optimizing cold plasma treatment involves various aspects such as operating gas and its composition, exposure time, discharge gap, applied voltage, power, and frequency.

1.7.4 ALLERGEN REDUCTION

Food allergens are a major concern across the world. Food allergens are water-soluble proteins or glycoproteins with immunogenic compounds of 10–60 kD [20] which initiate an allergic response in the human body. These molecules are highly resistant against hydrolysis, high temperature, pressure treatment, and digestion [89].

Recently, scientific innovations are focused on developing allergen-free food to overcome anaphylaxis (a symptom caused by allergy). So far, conventional processing like heating is followed to overcome the issue, but it comes with the negative effect of nutritional loss. To alleviate allergens in food and prevent nutrition loss necessitates an alternative non-thermal technique. Recent research studies have found that plasma can have a functional application on foodstuff to attenuate food allergenicity while retaining most of the food quality.

1.7.5 FOOD PACKAGING

Food packaging protects the food from external factors during the supply chain. Research studies have proven that plasma can be employed for the decontamination of packing material. For many years, the packaging industry has explored the purpose of cold plasma. It has considerably benefited many applications like surface ablation, surface modification, surface activation, and surface coating, as well as for disinfecting packaging material. There have been numerous reviews written on the usage of cold plasma for food packaging in the past [56].

Recently, it has also been intended to transform the properties of bio-based coating materials in addition to their application to conventional polymers [52, 55, 58]. As per recent studies, cold plasma has been shown to have significant application in the development of active [12, 57] and intelligent [59] packaging materials. Plasma treatment on various packages to enhance their packaging qualities is still a current research trend.

This technology has been acknowledged for improving the functional attributes of the packaging film (antibacterial/antioxidant) [68, 69] and has been used to replace the traditional packaging material in various research studies. It has also improved the surface characteristics of the film [13, 88]. Likewise, cold plasma enhanced the antioxidant property of the film, which prevented oxidation by up to 91.74% when *Momordica charantia* polysaccharide is treated with nitrogen at a rate of 100 cm^3/min [13].

When biopolymers are subjected to cold plasma pretreatment, the physical and structural properties of the packaging film are enhanced to a greater extent, which further coats the bioactive compounds on it. The efficiency of this treatment intensifies the binding attributes and structural arrangement. In recent research Zeng et al. [89], cold plasma-treated zein film loaded with polylactic acid increased the adhesive property of the film with improved characteristics of hydrophobicity, water vapor permeability, and tensile strength. Hence, cold plasma treatment has tremendous application in the packaging industry.

1.7.6 SEED GERMINATION

Cold plasma treatment on seeds can result in early germination as the reactive species penetrate the seed coat and have a direct impact on the interior cells. Changes on the seed surface create ablations that improve the oxygen and moisture transmission to the embryo, which eventually increases the seed germination rates. The plasma species might cause damage to the DNA, fracture the cell wall, provoke signals for growth hormone production, alter the protein structure, influence the enzymatic activities that cause the breakdown of the dormant stage and result in seed germination [19].

Plasma-treated wheat showed a quicker germination rate than untreated samples. The reactive oxygen species and low-energy ions attack the seed and cause ablation, which enhances germination [74]. There was a 6.7% rise in germination rate, which enhanced the plant growth of wheat while using helium plasma [27]. In lab testing and in the field, about 10–20% of the germination rate of legumes was increased. In addition, it also decreased the fungicidal effect by up to 15% and can be presumed to be an advantageous sterilizing agent due to the influence of reactive species, which are the most likely sterilizing agents [18]. Cold plasma was reported as an adaptive technology for surface reform of safflower seeds to enhance their germination rate. The highly active particles caused etching on the seed coat, resulting in germination [16].

Arc plasma was used to enrich the germination of spinach seeds and seed growth, resulting in a 137.2% and 217.6% upsurge in the germination rate and vigor, respectively [75]. Plasma presented a beneficial influence on the germinating seeds and growth rate of soybean. Upon plasma treatment, the seed vigor reached up to 14.66% and germination indices achieved a maximum of 63.33%. The seed's contact angle dropped to 26.19%, which is evident from the result that water uptake has increased by 14.03%. When compared to control (soybean), shoot length (13.77%) and dry weight (21.95%), root length (21.42%), and dry weight (27.51%) of the seeds increased [37].

The utilization of cold plasma was also studied for delaying seed germination by CF4 (carbon tetrafluoride) and ODFD (octadecaflurodecalin) coating. The thickness of the coating material significantly reduced germination. Seed germination was delayed successfully in soy, peas, beans, and corn [86].

Sharanyakanth et al. [54] also proved the effect of plasma on the stimulation of germination on cumin, where it increased from 42 to 74% at 2 kV. To conclude, from all the research studies conducted on seed germination with plasma treatment, there is a positive effect.

1.7.7 OTHER APPLICATIONS

Apart from major research, it is important to recognize the very real prospect of changing food structures and easing different functionalities through plasma treatments. Cold plasma structural modification is already being practiced in the electronics and polymer industries. In food structural and functional modification, the application of plasma is quite a new concept in food science research [28]. The research documented on modification of surface properties specifically the hydrophobic nature or wettability of various substrates for less oil seepage where the application of plasma improved the spreadability of oil on a biscuit surface to a maximum of 50% [44]. Another area of research emerging is the functional characterization of plasma.

Misra et al. [4] have evaluated the influence of atmospheric plasma on the dough functionality of wheat flour. Dough strength was better with the applied voltage. Bußler et al. [6] also found that cold plasma can strengthen the water holding and fat binding capabilities of pea flour, a protein-rich source. Some other areas of research interest were insect disinfestation. It was studied and confirmed that the mortality of Tribolium species is affected by cold plasma, as there is a build-up of applied voltage and exposure time [64]. The effect of cold plasma on mass transfer is one of the areas of growing research. The main principle applied here is the surface etching characteristics of plasma. Higher oil yield was obtained after plasma treatment, which is due to the change in surface properties [30, 63, 76]. It is evident from all the research done on plasma that, regardless of the treatment, it has tremendous applications in various fields of food processing. A large number of research studies are required in the future to explore the potential applications of plasma.

1.8 ADVANTAGES OF COLD PLASMA

Plasma is also being explored as a feasible alternative to novel treatments (HPP, PEF, and irradiation). Plasma methods provide numerous benefits: high reliability at a lower temperature, efficient production of reactive species for appropriate usage, shorter time for manufacturing the sterilizing medium, less effect on the internal structures, residue-free, and resource efficiency. In contrast to conventional treatment, plasma technology is more beneficial as it consumes less energy and operates at a nominal cost. Conversely, this treatment also has the potential to remove organic compounds of high toxicity at low temperatures.

Plasma systems are a novel method of detoxification. This approach offers a lot of benefits, including the capacity to detoxify quickly and at very low temperatures. It is non-hazardous and safe for people, which is another benefit [39].

The primary benefits of plasma technologies are a relatively simple and low-cost design, quick processing periods, lack of toxic effects, and residue generation. The employment of plasmas at normal atmospheric pressure in air and/or different working gas and mixtures causes the degradation of microorganisms, biofilm, and pesticides in heat-sensitive items.

1.9 CURRENT CHALLENGES IN USING COLD PLASMA TECHNOLOGY

Even though the advantages of reactive species of plasma on microbial contamination and other impacts are clear, such intricate chemical interactions make it difficult to evaluate the techniques and obtain regulatory approval. It can be complicated to understand the chemical reactions within biological matrices and the mechanisms of antibacterial activity. Plasma has often been mentioned as a processing technology in food research, although there are significant variations in plasma sources, species, and target reactions. One of the difficulties is setting up standards for plasma discharges in biological applications [28].

Another obstacle is the lack of a quantifiable dose for food goods. The majority of food research describes plasma discharge by identifying spectral lines of neutral atoms and molecular bands using optical emission spectroscopy. It would be possible to standardize the so-called plasma dosage with the availability of absolute standardized emission spectra [14].

Studies on chemical interactions in plasma and the probable toxicity of foods have recently gained attention. Cell lines are susceptible to the toxic effect of plasma-treated liquids like PAW or PAM [9]. Sarangapani et al. [14] used FTIR, NMR, and chromatographic techniques to assess plasma-induced lipid oxidation of dairy and animal fats. Such early investigations suggest a reliable processing technique; however, there is still a knowledge gap in this field.

Nearly every study that has been published so far has been done in the lab, although scale-up initiatives are still in the infant stage. The problems and potential techniques of scaling up technology are examined [14]. Current research is primarily focused on designing plasma operating systems that electrify the atmospheric air or other gases, comprising gas discharges/reactive species and PAW, on a large scale.

1.10 FUTURE SCOPE OF COLD PLASMA

The food sector needs to progress in production, food safety measures, and eco-friendly and efficient practices in response to the increasing need for food in the world. This calls for committed new technology research and development. Lower processing temperature, higher antimicrobial effect, negligible impact on quality, and devoid of residues are all the benefits of cold plasma on foods. The extremely diverse chemistry of plasma, which comprises several reactive species and reactions, is only partially understood by current scientific knowledge. For regulatory endorsement and the progress of effective, large-scale plasma sources, further scientific understanding is essential. It has been noted that research on food using cold plasma has increased rapidly. Although the practice of cold plasma in the food industry is still in its initial stage of development.

For plasma to be recognized, certain constraints might need to be resolved and should abide by the regulatory bodies, industry, and consumers. A few aspects of these obstacles that need to be addressed are inconsistencies in plasma regulatory authorization as a processing tool, technical shortcomings in managing plasma or dosage level, and problems in scaling up for industrial applications.

With the help of collaboration between business, academia, and the government, it is expected that the regulatory and consumer challenges will be resolved by concentrating research efforts and ideologies. It will need a coordinated approach by the plasma community (academic researchers, innovators, marketers, government officials, etc.,) to wholly acknowledge the applicability of the cold plasma technique in the food industry. There is a need for research investigations at all stages: in the lab, then prototype, then pilot plant treatment, and eventually large-scale. To assess the safety of foods treated with cold plasma, researchers and doctors in the sector must continue to collaborate. The ideas addressed in this chapter give a broad outline of the introduction and adoption of plasma research for foods.

1.11 SUMMARY

A novel invention of a non-thermal technique called cold plasma has exhibited promise at several points where the interdisciplinary approaches of physics, plasma chemistry, biology, medicine, and agriculture science coexist. Its applications in the food sector are not only limited to food decontamination but also include food quality improvement, enhancing functional property, removal of toxic compounds, and surface characterization of packaging materials. In comparison to conventional technologies, cold plasma has demonstrated certain distinct advantages that can be used to address current issues in the food sector. But before it can be completely applied at a commercial scale, cold plasma technology still needs to be developed further, focusing on versatile system design, upscaling, toxicity study, and mechanical properties.

KEYWORDS

- **cold plasma**
- **discharge type**
- **food applications**
- **ionized gas**
- **non thermal**
- **reactive species**

REFERENCES

1. Anbarasan, R., Jaspin, S., Bhavadharini, B., Pare, A., Pandiselvam, R., & Mahendran, R. (2022). Chlorpyrifos pesticide reduction in soybean using cold plasma and ozone treatments. *LWT, 159*, Article 113193. https://doi.org/10.1016/j.lwt.2022.113193.
2. Bai, Y., Chen, J., Yang, Y., Guo, L., & Zhang, C. (2010). Degradation of organophosphorus pesticide induced by oxygen plasma: Effects of operating parameters and reaction mechanisms. *Chemosphere, 81*(3), 408–414.
3. Baldwin, C. (2012). Sustainability in the food industry. *Sustainable Food Industry*. Edition 1, pp. 1–280), Publisher: Wiley Blackwell, IFT Press. https://doi.org/10.1002/9781118467589.
4. Becker, K. H., Kogelschatz, U., Schoenbach, K. H., & Barker, R. J. (2004). *Non-Equilibrium Air Plasmas at Atmospheric Pressure*. Edition 1, pp. 1–700, ISBN 9780367864170. CRC Press.
5. Bermudez-Aguirre, D. (2019). *Advances in Cold Plasma Applications for Food Safety and Preservation*. Edition: 1pp. 1–410, ISBN:9780128149225, DOI: https://doi.org/10.1016/C2017-0-02415-1, Publisher: Academic Press.
6. Bußler, S., Steins, V., Ehlbeck, J., & Schlüter, O. (2015). Impact of thermal treatment versus cold atmospheric plasma processing on the techno-functional protein properties from *Pisum sativum* 'Salamanca.' *Journal of Food Engineering, 167*, 166–174.
7. Chabert, P., & Braithwaite, N. (2011). *Physics of Radio-Frequency Plasmas*. Cambridge University Press, pp: 1–385, Edition: 1, ISBN: 9780511974342, DOI: https://doi.org/10.1017/CBO9780511974342.
8. Chemat, F., Rombaut, N., Meullemiestre, A., & Turk, M. (2017). Review of green food processing techniques: Preservation, transformation, and extraction. *Innovative Food Science and Emerging Technologies, 41*, 357–377.
9. Chen, C., Liu, C., Jiang, A., Guan, Q., & Sun, X. (2019). The effects of cold plasma-activated water treatment on the microbial growth and antioxidant properties of fresh-cut pears. *Food and Bioprocess Technology, 12*(11), 1842–1851.
10. Chizoba Ekezie, F. G., Sun, D. -W., & Cheng, J. -H. (2017). A review on recent advances in cold plasma technology for the food industry: Current applications and future trends. *Trends in Food Science & Technology, 69*, 46–58.

11. Chizoba Ekezie, F. G., Sun, D. -W., Han, Z., & Cheng, J. -H. (2017). Microwave-assisted food processing technologies for enhancing product quality and process efficiency: A review of recent developments. *Trends in Food Science & Technology, 67*, 58–69.
12. Clarke, D., Tyuftin, A. A., Cruz-Romero, M. C., Bolton, D., & Fanning, S. (2017). Surface attachment of active antimicrobial coatings onto conventional plastic-based laminates and performance assessment of these materials on the storage life of vacuum packaged beef sub-primals. *Food Microbiology, 62*, 196–201.
13. Cui, H., Yang, X., Abdel-Samie, M. A., & Lin, L. (2020). Cold plasma treated phlorotannin/*Momordica charantia* polysaccharide nanofiber for active food packaging. *Carbohydrate Polymers, 239*, Article 116214. https://doi.org/10.1016/j.carbpol.2020.116214.
14. Cullen, P. J., & Milosavljević, V. (2015). Spectroscopic characterization of a radio-frequency argon plasma jet discharge in ambient air. *Progress of Theoretical and Experimental Physics, 2015*(6), 70–75. https://doi.org/10.1093/ptep/ptv070.
15. Damalas, C. A., & Eleftherohorinos, I. G. (2011). Pesticide exposure, safety issues, and risk assessment indicators. *International Journal of Environmental Research and Public Health, 8*(5), 1402–1419.
16. Dhayal, M., Lee, S. Y., & Park, S. U. (2006). Using low-pressure plasma for *Carthamus tinctorium L.* Seed surface modification. *Vacuum, 80*(5), 499–506.
17. Feizollahi, E., Misra, N. N., & Roopesh, M. S. (2020). Factors influencing the antimicrobial efficacy of dielectric barrier discharge (DBD) atmospheric cold plasma (ACP) in food processing applications. *Critical Reviews in Food Science and Nutrition, 61*(4), 666–689.
18. Filatova, I., Azharonok, V., Kadyrov, M., Beljavsky, V., & Gvozdov, A. (2011). The effect of plasma treatment of seeds of some grain and legumes on their sowing quality and productivity. *Crop Journal, 56*, 139–143.
19. Fridman, A. (2008). *Plasma chemistry*. Cambridge University Press.
20. Fu, L., Cherayil, B. L., H, S., & Y, W. (2009). Allergy/food allergy. *Mucosal Immunology, 2*(1), 31–40.
21. Ganesan, A. R., Tiwari, U., Ezhilarasi, P. N., & Rajauria, G. (2021). Application of cold plasma on food matrices: A review on current and future prospects. *Journal of Food Processing and Preservation, 45*(1), e15070. https://doi.org/10.1111/jfpp.15070.
22. Gavahian, M., Chu, Y. H., Mousavi Khaneghah, A., Barba, F. J., & Misra, N. N. (2018). A critical analysis of the cold plasma induced lipid oxidation in foods. *Trends in Food Science & Technology, 77*, 32–41.
23. Gavahian, M., & Cullen, P. J. (2020). Cold plasma as an emerging technique for mycotoxin-free food: Efficacy, mechanisms, and trends. *Food Reviews International, 36*(2), 193–214.
24. Heo, N. S., Lee, M. K., Kim, G. W., Lee, S. J., Park, J. Y., & Park, T. J. (2014). Microbial inactivation and pesticide removal by remote exposure of atmospheric air plasma in confined environments. *Journal of Bioscience and Bioengineering, 117*(1), 81–85.
25. Smith, P., Bustamante, M., Ahammad, H., Clark, H., Dong, H., Elsiddig, E. A., Haberl, H., Harper, R., House, J., Jafari, M., & Masera, O. (2015). Intergovernmental Panel on Climate Change. Agriculture, Forestry and Other Land Use (AFOLU), *World Meteorological Organization, 52*, 1–43, 811–822.
26. Izadjoo, M., Zack, S., Kim, H., & Skiba, J. (2018). Medical applications of cold atmospheric plasma: State of the science. *Journal of Wound Care, 27*(9), S4–S10.

27. Jiang, J., He, X., Li, L., Li, J., & Shao, H. (2014). Effect of cold plasma treatment on seed germination and growth of wheat. *Plasma Science and Technology, 16*(1), 54–58.
28. Keener, K. M., & Misra, N. N. (2016). Future of cold plasma in food processing. In *Cold Plasma in Food and Agriculture: Fundamentals and Applications* (pp. 343–360). Elsevier Inc.
29. Kim, K. H., Kabir, E., & Jahan, S. A. (2017). Exposure to pesticides and the associated human health effects. *Science of the Total Environment, 575*, 525–535.
30. Kodama, S., Thawatchaipracha, B., & Sekiguchi, H. (2014). Enhancement of essential oil extraction for steam distillation by DBD surface treatment. *Plasma Processes and Polymers, 11*(2), 126–132.
31. Kogelschatz, U. (2003). Dielectric-barrier discharges: Their history, discharge physics, and industrial applications. *Plasma Chemistry and Plasma Processing, 23*(1), 1–46.
32. Langmuir, I. (1928). Oscillations in ionized gases. *Proceedings of the National Academy of Sciences, 14*(8), 627–637.
33. Laroussi, M. (2021). The resistive barrier discharge: A brief review of the device and its biomedical applications. *Plasma, 4*(1), 75–80.
34. Laroussi, M., Alexeff, I., Richardson, J. P., & Dyer, F. F. (2002). The resistive barrier discharge. *IEEE Transactions on Plasma Science, 30*(1), 158–159.
35. Lee, H., & Yoon, Y. (2021). Etiological agents implicated in foodborne illness worldwide. *Food Science of Animal Resources, 41*(1), 1–7.
36. Liao, X., Liu, D., Xiang, Q., Ahn, J., & Chen, S. (2017). Inactivation mechanisms of non-thermal plasma on microbes: A review. *Food Control, 75*, 83–91.
37. Ling, L., Jiafeng, J., Jiangang, L., Minchong, S., & Xin, H. (2015). Effects of cold plasma treatment on seed germination and seedling growth of soybean. *Scientific Reports, 5*, 05859. https://doi.org/10.1038/srep05859
38. Lu, P., Cullen, P. J., & Ostrikov, K. (2016). Atmospheric pressure nonthermal plasma sources. In *Cold Plasma in Food and Agriculture: Fundamentals and Applications* (pp. 83–116). Elsevier.
39. Luo, Q. Z., D'Angelo, N., & Merlino, R. L. (1998). Shock formation in a negative ion plasma. *Physics of Plasmas, 5*(8), 2868–2870.
40. Manoharan, D., Stephen, J., & Radhakrishnan, M. (2021). Study on low-pressure plasma system for continuous decontamination of milk and its quality evaluation. *Journal of Food Processing and Preservation, 45*(2), e15138. https://doi.org/10.1111/jfpp.15138.
41. Matsusaka, S. (2019). Control of particle charge by atmospheric pressure plasma jet (APPJ): A review. *Advanced Powder Technology, 30*(12), 2851–2858.
42. Misra, N. N. (2015). The contribution of non-thermal and advanced oxidation technologies towards dissipation of pesticide residues. *Trends in Food Science & Technology, 45*(2), 229–244.
43. Misra, N. N., Kaur, S., Tiwari, B. K., Kaur, A., Singh, N., & Cullen, P. J. (2015). Atmospheric pressure cold plasma (ACP) treatment of wheat flour. *Food Hydrocolloids, 44*, 115–121.
44. Misra, N. N., Sullivan, C., Pankaj, S. K., Alvarez-Jubete, L., & Cama, R. (2014). Enhancement of oil spread ability of biscuit surface by nonthermal barrier discharge plasma. *Innovative Food Science & Emerging Technologies, 26*, 456–461.
45. Misra, N. N., Tiwari, B. K., Raghavarao, K. S. M. S., & Cullen, P. J. (2011). Nonthermal plasma inactivation of food-borne pathogens. *Food Engineering Reviews, 3*(3–4), 159–170.

46. Misra, N. N., Yepez, X., Xu, L., & Keener, K. (2019). In-package cold plasma technologies. *Journal of Food Engineering, 244*, 21–31.
47. Mousavi, S. M., Imani, S., Dorranian, D., Larijani, K., & Shojaee, M. (2017). Effect of cold plasma on degradation of organophosphorus pesticides used on some agricultural products. *Journal of Plant Protection Research, 57*(1), 25–35.
48. Muhammad, A. I., Liao, X., Cullen, P. J., Liu, D., & Xiang, Q. (2018). Effects of nonthermal plasma technology on functional food components. *Comprehensive Reviews in Food Science and Food Safety, 17*(5), 1379–1394.
49. Murphy, F., McDonnell, K., & Fagan, C. C. (2014). Sustainability and environmental issues in food processing. In *Food Processing: Principles and Applications: Second Edition* (pp. 207–232). John Wiley & Sons, Ltd.
50. Niemira, B. A. (2012). Cold plasma decontamination of foods. *Annual Review of Food Science and Technology, 3*(1), 125–142.
51. Nishime, T. M. C., Borges, A. C., Koga-Ito, C. Y., Machida, M., Hein, L. R. O., & Kostov, K. G. (2017). Non-thermal atmospheric pressure plasma jet applied to inactivation of different microorganisms. *Surface & Coatings Technology, 312*, 19–24.
52. Oh, Y. A., Roh, S. H., & Min, S. C. (2016). Cold plasma treatments for improvement of the applicability of defatted soybean meal-based edible film in food packaging. *Food Hydrocolloids, 58*, 150–159.
53. Olatunde, O. O., Benjakul, S., & Vongkamjan, K. (2020). Shelf-life of refrigerated Asian sea bass slices treated with cold plasma as affected by gas composition in packaging. *International Journal of Food Microbiology, 324*, 108612. https://doi.org/10.1016/j.ijfoodmicro.2020.108612.
54. Paatre Shashikanthalu, S., Ramireddy, L., & Radhakrishnan, M. (2020). Stimulation of the germination and seedling growth of Cuminum cyminum L. seeds by cold plasma. *Journal of Applied Research on Medicinal and Aromatic Plants, 18*, 100259. https://doi.org/10.1016/j.jarmap.2020.100259.
55. Pankaj, S. K., Bueno-Ferrer, C., Misra, N. N., Bourke, P., & Cullen, P. J. (2014). Zein film: Effects of dielectric barrier discharge atmospheric cold plasma. *Journal of Applied Polymer Science, 131*(18), 9541–9546.
56. Pankaj, S. K., Bueno-Ferrer, C., Misra, N. N., Milosavljević, V., & O'Donnell, C. P. (2014). Applications of cold plasma technology in food packaging. *Trends in Food Science & Technology, 35*(1), 5–17.
57. Pankaj, S. K., Bueno-Ferrer, C., Misra, N. N., O'Neill, L., & Jiménez, A. (2014). Surface, thermal and antimicrobial release properties of plasma-treated zein films. *Journal of Renewable Materials, 2*(1), 77–84.
58. Pankaj, S. K., Bueno-Ferrer, C., Misra, N. N., O'Neill, L., & Tiwari, B. K. (2015). Dielectric barrier discharges atmospheric air plasma treatment of high amylose corn starch films. *LWT – Food Science and Technology, 63*(2), 1076–1082.
59. Pankaj, S. K., Kelly, C. A., Bueno-Ferrer, C., Kerry, J. P., & Papkovsky, D. B. (2016). Application of phosphorescent oxygen sensors in in-package dielectric barrier discharge plasma environment. *Innovative Food Science & Emerging Technologies, 33*, 234–239.
60. Pedrow, P., Hua, Z., Xie, S., & Zhu, M. J. (2019). Engineering principles of cold plasma. In *Advances in Cold Plasma Applications for Food Safety and Preservation* (pp. 3–48). Elsevier. https://doi.org/10.1016/B978-0-12-814921-8.00001-3.
61. Peot, E. (2017). American wasteland: How America throws away nearly half of its food (and what we can do about it). *Community Development, 48*(5), 713–714.

62. Phan, K. T. K., Phan, H. T., Boonyawan, D., Intipunya, P., & Brennan, C. S. (2018). Non-thermal plasma for elimination of pesticide residues in mango. *Innovative Food Science & Emerging Technologies, 48*, 164–171.
63. Pragna, C. H., Ranjitha Gracy, T. K., Mahendran, R., & Anandharamakrishnan, C. (2019). Effects of microwave and cold plasma assisted hydrodistillation on lemon peel oil extraction. *International Journal of Food Engineering, 15*(10), 1556–1561. https://doi.org/10.1515/ijfe-2019-0093.
64. Radhakrishnan, M., Ramanan, K. R., Sargunam, R., & Sarumathi, R. (2016). Effect of cold plasma on mortality of *Tribolium castaneum* on refined wheat flour. In *Proceedings of the 10th International Conference on Controlled Atmosphere and Fumigation in Stored Products (CAF 2016)*. pp. 142–146. Publisher: CAF Permanent Committee Secretariat, Winnipeg, Canada.
65. Rajan, A., Boopathy, B., Radhakrishnan, M., Rao, L., Schlüter, O. K., & Tiwari, B. K. (2023). Plasma processing: A sustainable technology in agri-food processing. *Sustainable Food Technology, 1*(1), 9–49.
66. Rana, S., Mehta, D., Bansal, V., Shivhare, U. S., & Yadav, S. K. (2020). Atmospheric cold plasma (ACP) treatment improved in-package shelf-life of strawberry fruit. *Journal of Food Science and Technology, 57*(1), 102–112.
67. Ranjitha Gracy, T. K., Gupta, V., & Mahendran, R. (2019). Influence of low-pressure nonthermal dielectric barrier discharge plasma on chlorpyrifos reduction in tomatoes. *Journal of Food Processing and Engineering, 42*(6), e13242. https://doi.org/10.1111/jfpe.13242.
68. Romani, V. P., Olsen, B., Pinto Collares, M., Meireles Oliveira, J. R., Prentice-Hernández, C., & Guimarães Martins, V. (2019). Improvement of fish protein films properties for food packaging through glow discharge plasma application. *Food Hydrocolloids, 87*, 970–976.
69. Romani, V. P., Olsen, B., Pinto Collares, M., Meireles Oliveira, J. R., Prentice, C., & Guimarães Martins, V. (2019). Plasma technology as a tool to decrease the sensitivity to water of fish protein films for food packaging. *Food Hydrocolloids, 94*, 210–216.
70. Roth, J. R., Dai, X., Rahel, J., & Shermann, D. M. (2005). The physics and phenomenology of paraelectric One Atmosphere Uniform Glow Discharge Plasma (OAUGDPTM) actuators for aerodynamic flow control. In *43rd AIAA Aerospace Sciences Meeting and Exhibit* – Meeting Papers (Vol. 38, No. 4, pp. 14057–14068).
71. Sakudo, A., Misawa, T., & Yagyu, Y. (2019). Equipment design for cold plasma disinfection of food products. In *Advances in Cold Plasma Applications for Food Safety and Preservation* (289–307). Elsevier. https://doi.org/10.1016/B978-0-12-814921-8.00010-4.
72. Sarangapani, C., Patange, A., Bourke, P., Keener, K., & Cullen, P. J. (2018). Recent advances in the application of cold plasma technology in foods. *Annual Review of Food Science and Technology, 9*, 609–629.
73. Segat, A., Misra, N. N., Cullen, P. J., & Innocente, N. (2016). Effect of atmospheric pressure cold plasma (ACP) on activity and structure of alkaline phosphatase. *Food and Bioproducts Processing, 98*, 181–188.
74. Šerá, B., Gajdová, I., Černák, M., Gavril, B., Hnatiuc, E., et al. (2012). How various plasma sources may affect seed germination and growth. In *Proceedings of the International Conference on Optimization of Electrical and Electronic Equipment (OPTIM)* (Vol. 238, pp. 1365–1370).

75. Shao, C., Wang, D., Tang, X., Zhao, L., & Li, Y. (2013). Stimulating effects of magnetized arc plasma of different intensities on the germination of old spinach seeds. *Mathematical and Computer Modelling, 58*(3–4), 814–818.
76. Sharanyakanth, P. S., Lokeswari, R., & Mahendran, R. (2021). Plasma bubbling effect on essential oil yield, extraction efficiency, and flavor compound of Cuminum cyminum L. seeds. *Journal of Food Processing and Engineering, 44*(7), e13750.
77. Shi, H., Ileleji, K., Stroshine, R. L., Keener, K., & Jensen, J. L. (2017). Reduction of aflatoxin in corn by high voltage atmospheric cold plasma. *Food and Bioprocess Technology, 10*(6), 1042–1052.
78. Siciliano, I., Spadaro, D., Prelle, A., Vallauri, D., Cavallero, M., et al. (2016). Use of cold atmospheric plasma to detoxify hazelnuts from aflatoxins. *Toxins, 8*(5), 125. https://doi.org/10.3390/toxins8050125.
79. Sifuentes-Nieves, I., Mendez-Montealvo, G., Flores-Silva, P. C., Nieto-Pérez, M., Neira-Velazquez, G., et al. (2021). Dielectric barrier discharge and radio-frequency plasma effect on structural properties of starches with different amylose content. *Innovative Food Science & Emerging Technologies, 68*, 102630. https://doi.org/10.1016/j.ifset.2021.102630.
80. Soloshenko, I. A., Tsiolko, V. V., Khomich, V. A., Shchedrin, A. I., & Ryabtsev, A. V. (2000). Sterilization of medical products in low-pressure glow discharges. *Plasma Physics Reports, 26*(9), 792–800.
81. Sousa, J. S., Niemi, K., Cox, L. J., Algwari, Q. T., Gans, T., & O'Connell, D. (2011). Cold atmospheric pressure plasma jets as sources of singlet delta oxygen for biomedical applications. *Journal of Applied Physics, 109*(12), e3601347. https://doi.org/10.1063/1.3601347.
82. Surowsky, B., Fröhling, A., Gottschalk, N., Schlüter, O., & Knorr, D. (2014). Impact of cold plasma on *Citrobacter freundii* in apple juice: Inactivation kinetics and mechanisms. *International Journal of Food Microbiology, 174*, 63–71.
83. Ten Bosch, L., Pfohl, K., Avramidis, G., Wieneke, S., Viöl, W., & Karlovsky, P. (2017). Plasma-based degradation of mycotoxins produced by Fusarium, Aspergillus and Alternaria species. *Toxins, 9*(3), e9030097. https://doi.org/10.3390/toxins9030097.
84. Thirumdas, R., Sarangapani, C., & Annapure, U. S. (2015). Cold plasma: A novel non-thermal technology for food processing. *Food Biophysics, 10*(1), 1–11.
85. Turner, M. (2016). Physics of cold plasma. In *Cold Plasma in Food and Agriculture: Fundamentals and Applications* (17–51). Elsevier. https://doi.org/10.1016/B978-0-12-801365-6.00002-0.
86. Volin, J. C., Denes, F. S., Young, R. A., & Park, S. M. T. (2000). Modification of seed germination performance through cold plasma chemistry technology. *Crop Science, 40*(6), 1706–1718.
87. Wang, S. Q., Huang, G. Q., Li, Y. P., Xiao, J. X., Zhang, Y., & Jiang, W. L. (2015). Degradation of aflatoxin B1 by low-temperature radio frequency plasma and degradation product elucidation. *European Food Research and Technology, 241*(1), 103–113.
88. Wu, X., Liu, Q., Luo, Y., Murad, M. S., Zhu, L., & Mu, G. (2020). Improved packing performance and structure-stability of casein edible films by dielectric barrier discharges (DBD) cold plasma. *Food Packaging and Shelf Life, 24*, e100471. https://doi.org/10.1016/j.fpsl.2020.100471.

89. Zeng, Z., Chen, M., Wang, X., Wu, W., & Zheng, Z. (2020). Modeling and optimization for konjac vacuum drying based on response surface methodology (RSM) and artificial neural network (ANN). *Processes, 8*(11), 1–17.
90. Ziuzina, D., Misra, N. N., Han, L., Cullen, P. J., & Moiseev, T. (2020). Investigation of a large gap cold plasma reactor for continuous in-package decontamination of fresh strawberries and spinach. *Innovative Food Science & Emerging Technologies, 59*, e102229. https://doi.org/10.1016/j.ifset.2019.102229.
91. Ziuzina, D., Patil, S., Cullen, P. J., Keener, K. M., & Bourke, P. (2013). Atmospheric cold plasma inactivation of *Escherichia coli* in liquid media inside a sealed package. *Journal of Applied Microbiology, 114*(3), 778–787.

CHAPTER 2

DESIGN AND GENERATION OF COLD PLASMA

P. HIMASHREE and R. MAHENDRAN

ABSTRACT

Cold plasma is a gaseous mixture that is ionized, consisting of reactive oxygen species and reactive nitrogen species, as well as ultraviolet radiation (UV), free radicals, and charged particles. Gas ionization occurs when free electrons collide with gas molecules and produce plasma after electrical energy is delivered to a gas that is present or moving between two electrodes with a significant electrical potential difference. Cold plasma technology shows great promise in various phases of food manufacturing and processing for several applications. This chapter introduces and describes the different types of cold plasma treatment used in food processing applications. The following are the most common generating mechanisms of cold plasma: dielectric barrier discharge, plasma jet, corona discharge, pulsed cold plasma discharge, microwave-assisted discharge, radiofrequency discharge, glow discharge. The design and components for each source of plasma generation, along with their features, are described. Several applications of cold plasma generators in food processing are presented.

2.1 INTRODUCTION

Solids, liquids, and gases are the three states of matter that exist commonly. When external energy is applied, these states can be interchanged from one to another. When sufficiently more energy is given to the gaseous state, then it is transformed into an ionized form known as plasma. Plasma comprises ions, electrons, neutral atoms and molecules, and many other reactive

Nonthermal Plasma Technology for Food and Food Products.
R. Mahendran, Preeti Birwal, and Megh R. Goyal (Eds.)

species. They are electrically neutral as they contain an equal quantity of both positive and negatively charged ions. However, some small fluctuations are there in charges on smaller scales, and hence it is in a quasi-neutral state. When these excited atoms and excited molecules go back to their ground state, energy is released as ultraviolet (UV) radiation.

Thermal and non-thermal plasma are the two types of plasma that are also referred to as equilibrium and non-equilibrium plasma, respectively. Non-thermal plasma is produced at lower gas temperature or room temperature and is also called cold plasma. The capability of plasma to produce highly reactive species, ions, and UV at low temperatures has shown its promise to be used for various disinfection applications. The plasma used for disinfection is mostly non-thermal/cold plasma. Cold plasma treatment of food products is done to inactivate microorganisms that include bacteria and their endospores, molds, and yeasts at temperatures below thermal treatment temperature. Cold plasma, a partially ionized gas, is generated using electric current or microwave or radio-frequency sources. The source of plasma, feed gas, and generation technique influences the plasma's composition that could be used for various applications. Also, the consequence of the plasma technology varies with the conditions of the surface and the microbial load [18]. Cold plasma treatment is an evolving nonthermal technique that provides several possible applications in the food processing and packaging industries that are rapidly advancing. Though cold plasma is rapidly attaining recognition in the field of food processing technology, the effect of long-lasting reactive species produced is still unclear. The basic purpose of cold plasma technology is to protect minimally processed food including fruits and vegetables, meat, poultry, and other fresh produce from microbiological or chemical contaminations. Various products are preserved using cold plasma technology retaining the physical properties and the sensory qualities of food matrices such as color, taste, flavor, appearance, and the composition of food [15].

This chapter discusses the different methods of cold plasma treatment aimed at food disinfection and other applications. The methods of generating cold plasma and the design of cold plasma generating equipment are listed. The generation of plasma changes depending on the availability of resources, need, product, and applications. Its advantages and limitations along with various applications are briefly reviewed.

2.2 METHODS OF COLD PLASMA TREATMENT

The design of plasma treatment intended for application in food processing shall be done in three different methods. This classification is based on the position of the food product in relation to the source of the plasma. This classification is logically drawn from cold plasma chemistry, with outlines of the half-life of reactive oxygen and nitrogen species and the activity of charged and active species present in the generated plasma [46].

The types include:

- Direct cold plasma treatment: product kept within the field of plasma generation
- In-package treatment: inside a package under the cold plasma generator
- Remote cold plasma treatment: at a considerable distance from the cold plasma generation
- Use of plasma-activated medium: Plasma-activated water or plasma-treated solutions act as disinfectants on fresh produce.

2.2.1 DIRECT COLD PLASMA TREATMENT

The food product is exposed directly under active plasma from the plasma-generating source in the direct plasma treatment system. The active plasma consists of short-living and long-living species [15]. The pulse frequencies of the plasma being produced are in hundreds or thousands of time ranges per second, and hence these systems are known to function in pulsed mode. The sample is given exposure to large concentrations of active species since it is kept near the source of cold plasma and treated before the active species reunite or deplete in this system [29].

The electricity is conducted at high voltages through the products with elevated water activity and interior moisture content. To avoid the intense passage of electric current through the product that is treated, the requirements of the systems should be taken carefully. The sources of plasma generation used in direct cold plasma treatment are discussed later in this chapter.

2.2.2 THE IN-PACKAGE COLD PLASMA TREATMENT

More people adopted polymers following their introduction in the 20th century after canning and retorting packaging for processing food in a hermetically sealed condition to prevent contamination and to ensure extended shelf life [38]. Despite the wide applications and usage of thermal processing of foods, researchers explored non-thermal technologies for decontamination and food processing to address many disadvantages of thermal treatments of foods [20]. High-pressure processing has gained success in the commercial industries to prevent post-process decontamination of foods kept inside sealed pouches. Cold plasma technology is having applications in both agriculture and food processing, medicine, and other environmental sectors too [58].

One of the characteristic features of this plasma is its ability to reach foods kept inside sealed packages. This approach is called the "in-package plasma" technology and has an effective treatment capacity as that of canning and high-pressure processing treatments in sealed packages. In-package treatment is an innovative technique that provides a non-thermal treatment for foods that are packed and moreover reduces the post-process contamination possibilities [47]. However, this approach is not considered a sterilization technique for foods since this is only a surface decontamination process. This approach can hence be successfully used for surface decontamination of minimally processed foods or fresh or raw foods to improve shelf life.

The concept of an in-package plasma treatment is to localize disinfectants in a gaseous state inside the package of the food product that has to be decontaminated. The food product that has to be treated with plasma is packed in either plastic or sometimes glass bottles and wrapped in the in-package plasma system. Either atmospheric air or any modified air mixture is filled into the package. When a strong electric field is exposed to the food package or the gas inside it for some time, there occurs a breakdown. The ionized gas, which is the plasma, inactivates the microbes and sterilizes the packaging material without affecting its integrity. Most plasma species are active for a long duration from minutes to even a few hours [41]. The high diffusivity coefficients allow the plasma species to diffuse rapidly around the package and ensure uniform treatment. The unreacted species immediately recombine to form the original gaseous state.

The time taken by the species to return to its original form depends on its energy, pressure, temperature, the type of package and volume, the food present inside, humidity, light, method of generating plasma, and initial concentration of the reactive species. Hence, the process ends with the same gas mixture with which it started before a few hours, simultaneously decontaminating the microbes inside the food package.

Many engineering designs of plasma generation in sealed packages have been created in the last years. Some of the notable designs are indicated in Figure 2.1.

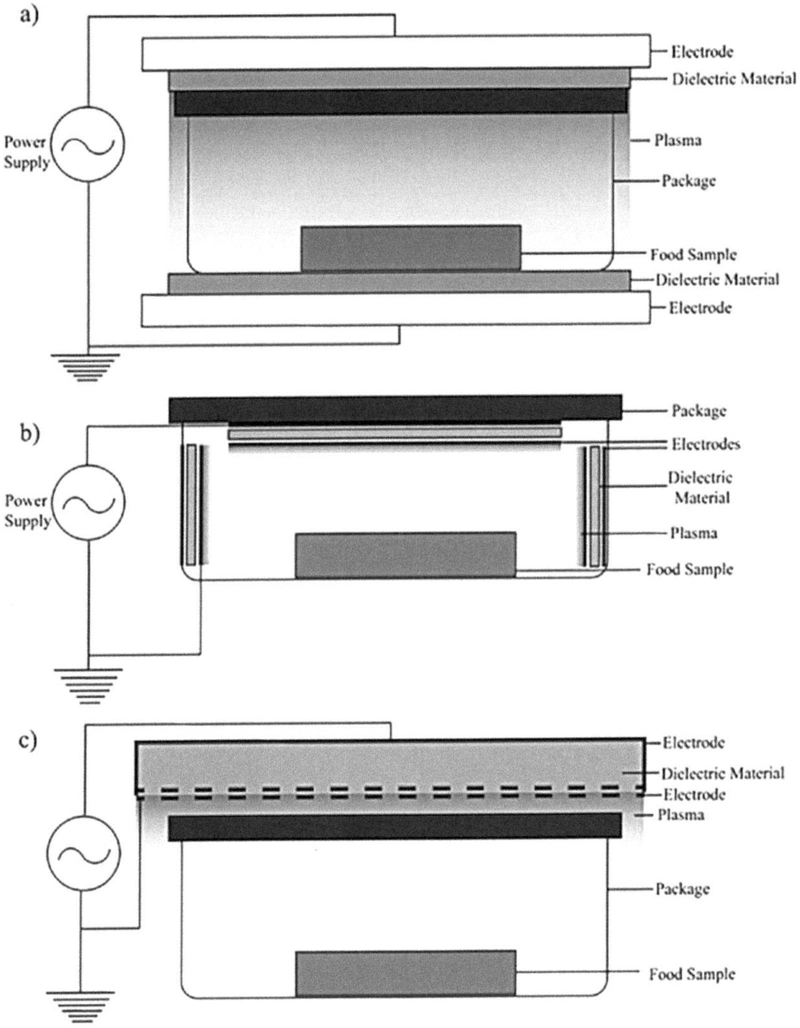

FIGURE 2.1 In-package types of cold plasma treatment: a) volumetric dbd plasma; b) surface dbd plasma inside the package; c) on package dbd plasma.

2.2.2.1 VOLUMETRIC DIELECTRIC BARRIER DISCHARGE SYSTEM

Volumetric DBD works on high voltage applications (about 10 kV to 100 kV) that ionize the gases at continuous frequencies from 50 Hz to 60 Hz throughout the whole package. Hence, there is no need for a frequency generator here.

Volumetric DBD also works even at higher frequencies. This configuration operates best in the package of heights of the range 8–10 cm with air. Though the working principles of the dielectric barrier discharge are the same in all configurations, this design yields high microbial inactivation [24].

2.2.2.2 SURFACE DIELECTRIC BARRIER DISCHARGE PLASMA INSIDE THE PACKAGE

This works at comparatively lower voltages in the range of 1 kV to 10 kV and a frequency of about 10 kHz that restricts the ionization degree and number of reactive plasma species produced. The ionization primarily occurs on the surface of products kept between the electrodes. This causes limited production of the active species and results in the necessity for a lengthier treatment time for the required inactivation of microbes. In this configuration, there is a possibility of the electrodes getting corroded, thus causing contamination. The advantage of this design of in-package plasma treatment has only a mild effect on the package when compared with the volumetric DBD or on-package surface DBD [77].

2.2.2.3 THE ON-PACKAGE SURFACE DIELECTRIC BARRIER DISCHARGE PLASMA

A strong electric field penetrates, causing the headspace air to ionize, thus reactive species are formed when the surface DBD is kept on the food package. This configuration requires a supply of high-frequency power. High frequencies can cause heating of the gases in discharge, and hence reduced treatment times are used. The surface DBD can treat packages of any size, while the dimensions of the package used in volume DBD are restricted by the space allowed for ionized gas. [9].

In-package plasma treatment has several advantages, such as:

- Reduced contamination after processing by external food spoiling and pathogens
- Reduction in microbial load on the food packages
- Ability to freeze directly, refrigerate, or store in ambient conditions
- Utilize varieties of packaging material and modified gas surroundings
- Stimulate the release of antimicrobial or freshness-improving agents at a maintainable rate from the packages.

The polymeric packaging material itself acts as a dielectric material in this method of cold plasma treatment. Several packages like low-density and high-density polyethylene, polystyrene (PS), polypropylene (PP), and polyethylene terephthalate (PET), cryovac pouch, Tygon, etc., were used in different studies. The package is one of the significant aspects of cold plasma discharge and treatment. When the surface of the package is exposed to reactive species, it decontaminates the internal package surfaces too. Plasma is used as a tool to decontaminate the packaging material; however, it is crucial to estimate all the quality parameters and safety aspects of the changes brought in the packaging material due to cold plasma treatment. Properties such as migration limits of the additives, monomers, and oligomers, and also the water vapor permeability and oxygen permeability have to be carefully studied [50]. Recent research showed that natural antimicrobial additives incorporated in packaging material showed better release of the active compound after plasma treatment [51]. Hence, in-package plasma is an efficient and simplified method for decontamination of food products kept in the packages and does not require any process to sterilize the food product and package it separately.

Studies demonstrate cold plasma technology has the possibility to be scaled up and employed for commercial purposes to inactivate microbes and retain the food qualities within sealed packages. The volumetric in-package plasma system has the ability to be upscaled for commercial continuous processing of food products inside closed packages or containers kept in-between electrodes moving on any conveyor belts. The lack of suitable equipment that can be utilized in commercial production rate is the limitation for why cold plasma technology hasn't still been scaled up. This limitation can be overcome by studying the plasma chemistry which is the function of the product, packaging material, and several environmental conditions.

2.3 REMOTE COLD PLASMA TREATMENT

Products that are exposed to plasma for processing sometimes get damaged due to overheating, high electric fields, etching/erosion by the reactive species, or exposure to ultraviolet radiation that may result in undesirable effects. Sometimes large or unevenly shaped products cannot be placed between electrodes or in the plasma generation chamber for treatment. Remote plasma treatment was developed to overcome these limitations and allow larger and more complex three-dimensional products to be treated by cold plasma. Plasmas do not usually cause gas heating, but using a remote source can minimize this effect.

In the remote cold plasma treatment method, the food material is not kept directly under the generated plasma. It is kept at a distance from the plasma production. Here, secondary chemical species are produced from the air or other gases, which can sterilize the food product. Either air, nitrogen, or a noble gas mixture is ionized by an electrical, microwave source, or laser [56]. The generated plasma is driven by either the feed gas flow or can be manipulated using magnetic fields (Figure 2.2).

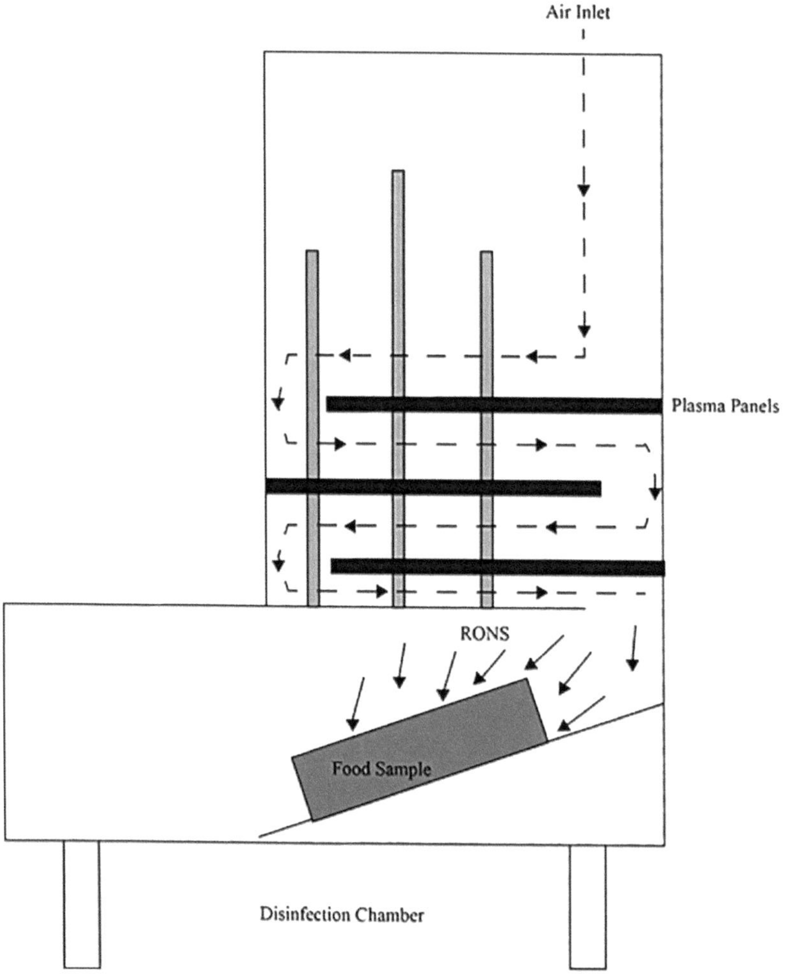

FIGURE 2.2 Remote cold plasma treatment setup.

The benefit of a remote system is that the product to be treated is kept at a location different from the generation area. Thus, the design and operation of this generation system are simplified, and products with various shapes and sizes can be exposed to cold plasma easily. Nevertheless, the most reactive species have the least half-life. The free electrons recombine with other plasma species like heavy ions and atomic species during their travel time to the product. By the time they reach the target, the plasma contains secondary chemical species that have low activity, long-living plasma species formed by recombination within the plasma. The after-glow plasma containing a lower concentration of ions produces UV light, and the chemical species are activated due to their reaction with the target. However, the concentration of ions is lower than that in active plasma [25]. The plasma species formed in the treatment chamber may react with the other reactive species leading to lesser secondary energy and less microbial decontamination, which is a disadvantage of this method [60].

Roth et al. [61] designed a remote mode of cold plasma discharge to inactivate microbes such as *E. coli*. It contains several plasma panels placed in such a way that air has to flow through bending paths in the plasma panels along the active zones. The remote mode of plasma treatment showed similar effects as that of the direct plasma treatment. The reactive air and gas species that sterilize the products are produced on flat panels by glow discharge plasma. The reactive species are sent to a remote exposure chamber by forced air convection where the product that has to be treated is placed. This offers the advantage of treating products of any size or shape without direct plasma contact. Both single time pass and repeated circulation of the reactive species to the remote exposure chamber can be done.

This system consists of two main components: (1) the reactor chamber and (2) the remote exposure reactor chamber. The radiofrequency voltage of several kV and kHz is applied and traps the energized ions and plasma above and below the panels which results in surface plasma in the reactor chamber. If low-frequency RF is applied, the plasma disappears as the ions hit the electrodes or that panel surfaces during half of the cycle itself. If a very high frequency is given, the electrons are trapped causing polarization, filamentation, and quenching of the plasma. Away from this reactor, a centrifugal blower is kept that produces a uniform forced air circulation. The ionized air flows through the plates and then leaves through the remote exposure reactor where the product to be treated is placed. The configuration can be designed in such a way so that the plasma is recirculated through these chambers. This is designed to confine species like ozone and to intensify the concentration of the reactive plasma species for the duration so that it is exposed to the product at least for one recycling loop. In single-pass design systems, the air from the surroundings is taken by a blower and blown to the fume hood vent, thus the long-lived reactive species are not built up in the airflow.

The intensity of ultraviolet radiation at the treated surface is considerably large because there is almost no typical atmosphere prevailing between the cold plasma source and the product that is to be treated. The electrical discharge from a single point focuses the current flow that causes heat damage. Designing and functioning direct cold plasma treatment systems may be more difficult than remote treatment systems. But these systems are supportive in terms of usage for a versatile product shape and size as the plasma emitter allows. However, the probability of conduction of electric current through the product may lead to limited types of commodities that can be treated with a direct cold plasma system [44].

Generation of cold plasma is done using electrodes or without using electrodes; using or not using dielectric barriers over the electrodes, and at a powerful frequency range from direct current to 2.45 GHz; and pressure level from 1.33 to 1.01×10^5 Pa. Atmospheric cold plasma reactors use the maximum atmospheric pressure value for plasma production while low-pressure plasma chambers use a vacuum pump and work at typically 1.33 Pa [53].

2.4 PLASMA-TREATED SOLUTIONS

Plasma-treated water (PTW), or Plasma-activated water (PAW), or Plasma-activated or Plasma-stimulated medium are the names to indicate the plasma-treated solutions that are used for decontaminating fresh produce and also food processing equipment. These solutions are used as washes or sprays or in mists for their applications.

Developing sustainable disinfection technologies is challenging as product quality, cost, environmental impact, and food regulations have to be satisfied. Plasma-treated solutions are an innovative strategy to decontaminate fresh produce while retaining the quality of the products and improving the shelf life. Minimal processing of foods at room temperatures can be achieved by employing cold plasma generated in either gaseous or functionalized liquid delivery mode at atmospheric pressure. The plasma-treated solutions are utilized as the medium to transport the active species and antimicrobial agents to food or any food processing surface to be disinfected.

This solution can be compared to the application and antimicrobial activity of ozonized or chlorine water. The plasma-treated solutions consist of active oxygen and nitrogen species which are responsible for their decontaminating activity along with their low pH.

Plasma-activated mediums are produced by the interactions of plasma species with the liquid medium. Several cold plasma-water reaction systems such as plasma jets, DBDs, pulsed corona, gliding arc, and plasma steamer discharges are designed to generate plasma. The reactive particles produced in the cold plasma interact with the other water molecules and stimulate a flow of chemical reactions to form a distinctive biochemically and extremely reactive mixture called the plasma-activated water. Plasma jets and DBDs are the regularly used designs to create cold plasma. Plasma jets or plasma plume discharges plasma even beyond the area of plasma generation to the surroundings and activates a strong interaction between the plasma and water to produce plasma-activated water. Sterile distilled water, reverse osmosis water, deionized water, and tap water can be utilized to make PAW as stated in different research.

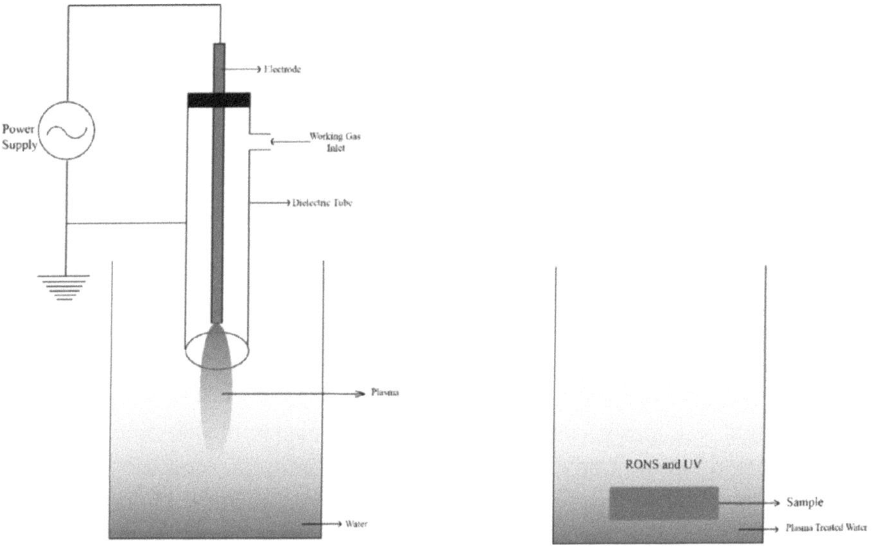

FIGURE 2.3 Disinfection of food samples using plasma treated water.

The power applied and activation time determine the physicochemical characteristics that include pH, oxidation-reduction potential, electrical conductivity, ROS, and RNS of the PAW. The type of working gases used determines the intensity of the plasma species that exist in the PAW. Air, Ar, He, and Ar/O_2 are the various working gases that have been successfully reported to produce PAW. The distance of the water sample from the electrode affects the efficiency of microbial disinfection [17].

Schnabel [68] examined the influence of plasma-activated water on fresh-cut lettuce. Distilled water was treated using air processed with plasma to get plasma-treated water. A microwave-driven device based on a plasma torch was employed to produce plasma at a pressure level equivalent to atmospheric pressure. 1.3 kW power was applied to a 12 slm flow of water at the first stage and then 3 kW power was applied to 60 slm water. Nitronium ion (NO_2^+), Nitrogen monoxide (·NO), and hydroxide ion (OH^-) are predominant components identified in the plasma-treated water. Food quality parameters like texture, color, and nitrate content were retained after processing with plasma-treated water and showed promising results. The study showed plasma-treated water as an alternative to conventional washing techniques in process lines of fresh-cut food products directly.

Plasma-treated solutions can be used as a new technique and sanitizing treatment for shelf-life extension while retaining food quality within regulatory guidelines. However, further sensory and toxicology studies should be done to meet all regulatory requirements. The cost efficiency and marketability of the product life cycle should be examined. The advantages of low-temperature processing, low cost, and simple generation and processing offer potential applications of plasma-treated solutions in food industries (Figure 2.3).

2.5 GENERATION OF ATMOSPHERIC COLD PLASMA

Systems working at one-atmosphere pressure (1 bar or 100 kPa) that generate cold plasma are called atmospheric cold plasma. Vacuum treatment conditions are not suitable for all food products. These do not need any airtight vacuum chambers, making them easier to construct. Products to be treated can be transported through the treatment zone via a conveyor without the need for any vacuum-tight chamber or gaskets. However, due to the challenge of ionization, ambient pressure processing presents significant difficulties in cold plasma systems. When operating at one atmosphere, the voltage between electrodes with the same spacing must be much higher. The higher the ionization voltage, the more difficult it is to establish a uniform plasma field without high-temperature point emissions. This can be particularly significant in cases where the active plasma is directed over the food surface. Plasma can be produced using several methods at atmospheric pressure. The most commonly used designs for cold plasma generation in food applications are DBD, plasma jets, corona discharge, and pulse discharge, which will be detailed in this section [44]. DBD has been used for decontaminating tomatoes [52], fruits' pericarp [55], and meat [62], among others. The application of corona discharge for decontaminating milk [16] and active packaging for apples [11] has been studied. Plasma jet has shown efficient decontamination of almond nuts [45]. Cold plasma has a wide range of applications, such as decontaminating bacteria on artificial packaging materials like glass, paper [40], nitrocellulose [59], and polypropylene [33].

2.5.1 DIELECTRIC BARRIER DISCHARGE (DBD)

The first person to introduce Dielectric Barrier Discharge was Siemens in his experiments in 1857 for ozone production. Then Klemenc studied the physical characteristics of DBD in 1937. Their work explained how several individual small breakdown channels, i.e., micro discharges occur, and they studied their correlation with streamers. This droves many investigations for its applications in various fields for several years. DBD has a wide range of uses, including living tissues, bacteria inactivation, surface treatment, excimer production, etc. [60, 80]. The DBD chamber that works at atmospheric pressure level is one of the most prevalent and commonly employed plasma reactors in the food processing industries.

The generation of strong electric fields by the embedded metals at a dielectric surface starts the discharge in DBD reactors. A self-pulsing operation is responsible for producing plasma by the insulating or dielectric material kept between the electrodes [65]. Materials with low dielectric loss and with high breakdown strength are used, on which coating with metal electrodes is done. Examples of the dielectric materials used include glass, plastic, silicon, quartz, or ceramic plates or tubes. Or the dielectric materials are coated on metal electrodes like enamel-layered steel tubes [26].

The dielectric material placed at the discharge gap halts the passage of electric current and hence spark development is avoided or it suppresses the arc discharge, which can generate thermal plasma. The sparks cause local overheating, local shock waves, and noise. But as there is no spark formation in the DBDs, they are also known as silent discharges. Also, since DBD occurs in a very short period, there is no increase in the temperature of neutral atoms and the ions, and it is hence a non-thermal technology. This technology was originally developed to create ozone gases, but it now has larger agri-food industry applications [15]. The DBDs are used in many industries as they are non-equilibrium discharges running at atmospheric pressure conditions and fairly high power, with no use of any advanced pulsed power supplies.

DBDs can be designed in any geometrical configuration and use any working medium and functioning parameters. The optimized treatment conditions in laboratory research can be easily upscaled for commercial purposes. The high voltage alternative current specifications used in a DBD are indicated in Table 2.1.

TABLE 2.1 Current Requirements for a Dielectric Barrier Discharge

Parameters	Required Range
Clearances	100 μm to several cm
Driving voltage	10 V amplitude
Frequency range	500 Hz to 500 kHz
Power usage	10 to 100 watts

In a DBD reactor, the gas mixture is generally made up of an inert transport gas like Ar or He and a working gas (like dry air, molecular oxygen, or molecular nitrogen). There are neutral and charged chemical species (as well as ultraviolet light) contained in the plasma generated.

DBD has electrically insulated dielectric barriers which keep both electrodes from encountering plasma directly. At times, the dielectric portion covers only a single electrode rather than covering both. The configurations of the electrodes are done in several methods, but the concept underlying all of them is the same. DBDs take both flat and cylindrical-shaped electrode configurations as shown in Figure 2.4.

The working design of the DBD reactor configured in different ways are described below:

- The food product is directly treated by the plasma containing neutral and charged reactive species (including ultraviolet radiation) as the gases flow through the reactor and then out of the exhaust outlet, or the food is exposed to activated neutral chemical species when it is kept under the plasma.
- The food to be treated is kept inside a sealed container where plasma processing occurs within the local gases circulating over it, and the flow fields and gas mixture are predetermined.
- The food is filled in a vented or sealed container and is placed between electrodes.

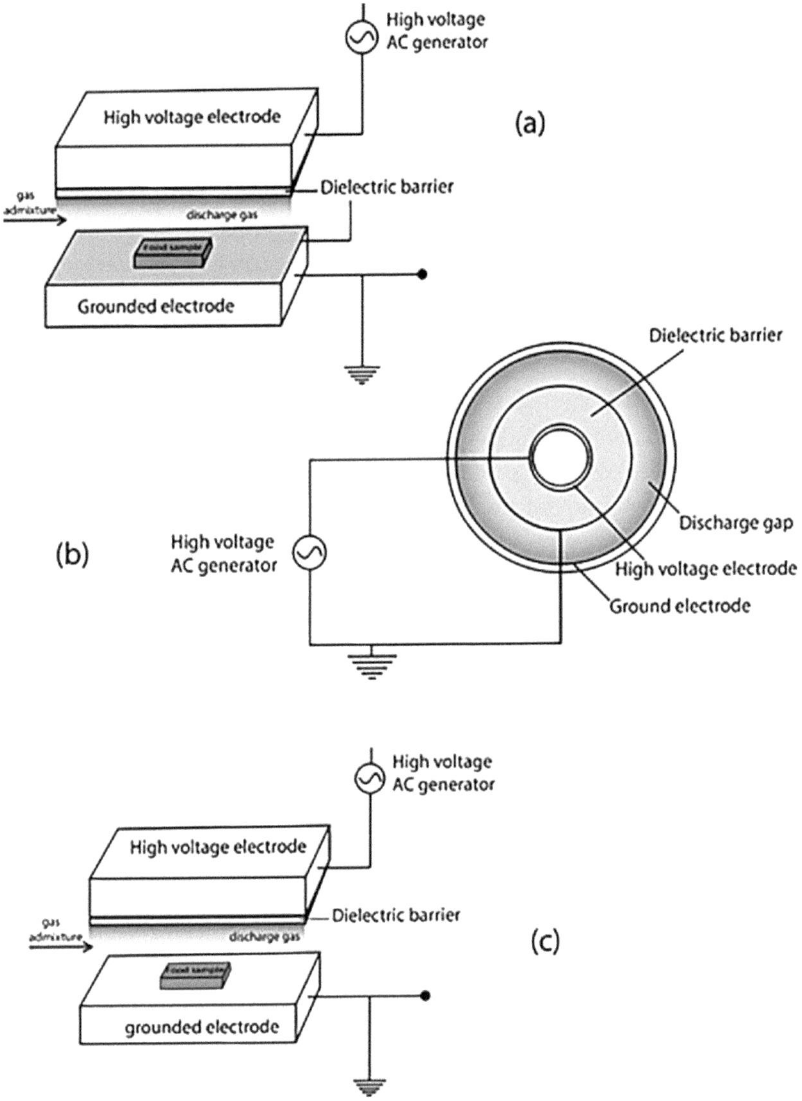

FIGURE 2.4 Configurations of DBD: (a) Planar configuration with dielectric material on one electrode; (b) Cylindrical configuration; (c) Planar configuration with dielectric material on both electrodes.

The chemistry of free radicals is exceptionally quick, allowing for huge production or destruction rates and fast surface treatments at atmospheric pressure. Many independent current filaments or microdischarges occur as breakdown is initiated at atmospheric pressure in most gases. The microdischarge filaments are faintly ionized cold plasma channels with characteristics similar to a transient glow discharge. When the breakdown field is reached, they ignite. When plasma conductivity is reduced due to electron attachment and recombination, they extinguish just below the same breakdown field. At the point of microdischarge, the charge builds up in the dielectric field, causing the microdischarge to diminish in a few nanoseconds after a breakdown, cutting off the flow of current.

Insights about the development and transmission of streamers, and subsequent degradation of plasma channel relate to the physics of these micro discharges. The electrons contained in the plasma of the streamers disperse from the space in just 40 nanoseconds. Whereas the heavy ions and gradually drifting ions stay in that space for more than a few microseconds. Electron deposition on the anode dielectric from the conducting channel causes accumulation of charges and avoids new avalanches and the streamers unless the anode and cathode are inverted (when the voltage given does not exceed the minimum voltage required for breakdown). Operational frequency of DBDs is generally 20 kHz, hence the voltage polarity gets reversed in just 25 microseconds. After the reversal of voltage polarity, the negative charges that were deposited enable the development of new avalanches and streamers at the same point. The duration of micro discharges including electron avalanche, plasma channel, and cathode-directed streamer from atmospheric pressure DBDs with a 1 mm gap is 10 nanoseconds, 1 nanosecond, and 30 nanoseconds, respectively. The charge transferred during this discharge from the electron avalanche was 10^{-9} C, the cathode-directed streamer was 10^{-10} C, and the plasma channel was 10^{-9} C [13].

A cohort of streamers' family is created that can be macroscopically seen as a bright filament localized spatially. From the point of discharge or from dielectric over the cathode or from the cathode itself develops an initial electron that produces secondary electrons developing an electron avalanche by direct ionization. The streamer directed from the cathode is started from the anode section when the avalanche is large enough to meet Meek conditions. The streamer fills the discharge space in just nanoseconds time and creates a weakly ionized channel conducting plasma. Electron-dense current streams over the plasma channel till the resident electric field breaks down. This breakdown is because of the charges deposited on the dielectric surface and the ionic space charge (for the whole length of this current peak, the ions are extremely slower to evacuate the gap). Microdischarge refers to the local processes that are triggered by an avalanche in the gap and continue until the electron current terminates [13].

The amount of the charge discharged is directly related to the space and the dielectric's permittivity but independent of pressure. The intensity of the current is in the range of 100–1,000 A/cm^2 in the micro discharge channel. Since it is a short period, this generally leads to only a very small transient gas heating in the channel remaining. The micro discharge strength is likely to increase because of humidity, whereas irradiation of UV photons on the cathode decreases the micro discharge strength. The function of the dielectric barrier is to restrict the quantity of charge and energy that gets accumulated in micro discharges and spread the micro discharge throughout the whole surface of the electrode. The properties of micro discharges can be customized by optimized operating pressure, gap, and the dielectric barrier [26]. The dielectric barrier insulation permits potential gradient in the reactor's capacity while preventing streamer-to-spark conversions, keeping the plasma cool. During the operation of the reactor, the dielectric barrier's thickness and dielectric strength must be carefully selected to avoid dielectric breakdown. The dielectric barrier discharges either uniform glow plasma or a multitude of small-diameter streamer plasma. By constructing electrodes with greater surface areas, the dielectric barrier discharge plasma reactor may be scaled up for improved processing output. The resistive barrier discharge, a form of DBD, in which the dielectric material is substituted by a resistive barrier that presents enough resistive impedance to restrict current in the electric circuit and avoid streamer-to-spark transition [53].

Characteristic properties of DBD are its simple design and strong thermodynamic non-equilibrium discharge that allow the use of DBD at atmospheric pressure cold plasma for various applications. DBDs can be used for sterilization plasma treatment on targeted pathogens in foods without affecting the food quality. DBD can use various gases and other sterilization methods as hurdle technology to achieve efficient inactivation results. Some applications of the cold plasma generated by dielectric barrier discharge on various foods for inactivating pathogens are discussed in Table 2.2.

2.5.2 CORONA DISCHARGE

Corona is produced at the atmospheric pressure level by the application of low frequency or high voltage pulses on a pair of electrodes of different sizes. When the strength of the electric field is adequately high around a conductor to create a corona region, but not arcing to closer items, a corona discharge is created. The corona discharge method can be effectively used for regulating

TABLE 2.2 Applications of DBD Cold Plasma Generation to Inactivate Microbes in Food Products

Product	Design	Treatment Parameters	Targeted Pathogen	Anti-Microbial Effects	Quality Changes	References
Apple juice	Two aluminum circular plate electrodes of 50 mm diameter; two quartz glass circular dielectric barriers (90 mm upper barrier Dia and 60 mm lower barrier Dia with 1 mm thickness; a fixed 10 kHz frequency power supply and a 14 kV voltage.	30, 40, and 50 W; 0–40 sec	*E. coli*	3.98–4.34 log CFU/mL decrease	A minor change in the Brix value, color value, pH value, titratable acidity value, total phenols, and the antioxidant capacity: However high power and increased treatment time caused significant effects on these qualities and a reduction in the total phenols and its antioxidant capacity.	[34]
Cheese	Encapsulated DBD plasma source; copper electrodes and polytetrafluoroethylene (PTFE) sheet;	15 kHz bipolar square-waveform voltage; 250-W input power; 1–5 mins.	*E. coli, S. Typhimurium,* and *L. monocytogenes*	No viable cells of *E. coli, S. Typhimurium,* or *L.* monocytogenes were found after plasma treatment. More time was essential to inactivate *L. monocytogenes* took more time inactivation than *E. coli* and *S. Typhimurium* by using cold plasma	Increased post-treatment duration improved inactivation efficiency	[76]

TABLE 2.2 *(Continued)*

Product	Design	Treatment Parameters	Targeted Pathogen	Anti-Microbial Effects	Quality Changes	References
Dried blackmouth angler	10 μm thickness silver electrode and 1.8 mm thick substrate glass; SiO_2 of thickness 100 μm was employed as a dielectric material.	N_2 feed gas with a 1.5 L per minute flow rate; 3 mm distance; 1–30 mins.	*S. aureus* and *B. cereus*	0.10–1.03 log_{10} CFU/g decrease	1.5–1.6-fold growth in the thiobarbituric acid content in blackmouth angler under DBD plasma treatment for 20 and 30 minutes; Decrease in textural properties values on the samples treated for over 20 mins; No significant differences in L a b color values, pH value, and sensory attributes.	[5]
Fresh oyster	Silver electrode; SiO_2 dielectric.	1.1 kV; 43 kHz; N_2 feed gas at 1.5 L/min; 3 mm distance from the sample; 10–60 mins.	*E. coli* O157:H7 and Non-pathogenic *E. coli*	>1 log decrease of *E. coli* O157:H7, and >2 log decrease of non-pathogenic *E. coli*	No significant changes were observed in glycogen content and the overall texture of oysters on 1 hr. treatment	[6]
Romaine lettuce	0 to 76 kV voltage; 0 to 2,400 Hz frequency	Treated at 34.8 kV, 1 A; 5 min.	*E. coli, Salmonella,* and *L. monocytogenes;* Tulane virus	*Salmonella* had the minimum sensitivity to the plasma; 1.1 log CFU/g decrease in *E. coli*; 1.0 log CFU/g decrease in L. monocytogenes, and 1.3 log PFU/g decrease in Tulane virus		[37]

TABLE 2.2 *(Continued)*

Product	Design	Treatment Parameters	Targeted Pathogen	Anti-Microbial Effects	Quality Changes	References
Tiger nut milk	Aluminum 50 mm diameter electrodes; glass dielectric material having 90 mm diameter.	30 V; 6 mm distance; 0–12 mins treatment.	Bacteria and molds	Treatment up to 12 mins resulted in microflora detection below 1.0 log CFU/mL)	Significant changes were observed in the physico-chemical properties; (only slight alterations in the Brix value and fat content); Peroxidase activity inactivation; Lipid oxidation was increased.	[42]

the plasma emission and breakdown by improving the strength of the electric field using sharp electrodes. A local non-uniform discharge can be formed by improvising the local electric field, and a corona is begun. They can be formed using either direct or alternating current or pulsed voltage systems but are usually operated by low-frequency alternating current. The plasma generated from this system contains a sequence of minor lightning-type discharges. The discharge is non-uniform and has extreme local energy levels [74].

The weak luminance that forms after a thin corona section looks similar to a crown and hence it is named the "corona." Sharp edges are used as one or both electrodes of this discharge system and hence the shape of the electrode has a vital role in this system of discharge. Generally, asymmetric electrodes like a point or plane are used to create corona discharge cold plasma. The process of ionization is constricted to the confined region around the high-voltage corona discharge electrode using these configurations because of the non-uniformity of the electric field. At atmospheric pressure, corona discharge is very effective to generate reactive plasma and offers many applications in food processing. They can produce huge quantities of oxygen species either in air or gases comprising oxygen that are used for having antimicrobial effects or surface altering effects. Single point or pin, multiple points, coaxial, ducts and packed beds are the different configurations that can be used to generate corona discharges. Each of these designs can be configured according to its applications and target. For example, a pin-to-plate configuration of corona discharge delivers a bigger plasma volume and discharge gap (Figure 2.5) [75].

Corona-based plasma reactor does not have a dielectric material but is similar to a dielectric barrier reactor in appearance. The corona reactor produces cold plasma utilizing electron avalanches, glows, and streamers between two metal electrodes that have no dielectric material. The neutral gas is induced to ionize because of the intense electric field at this point, which generates the corona discharge plasma due to local dielectric breakdown [65]. Either one or both of the electrodes have a small radius of curvature like an array of needles or wires with a small diameter. The avalanches, glows, and streamers generated at the sharp electrodes by huge electric fields are together called the corona discharge. The streamer-to-spark transitions are prevented as the electric field in the center space is low. However, the food placed in the high voltage gap supports the creation of a spark. Therefore, the food sample is kept downstream of the plasma area [22].

Here, the activity of ions or electrons is crucial in creating the plasma. Identities of atoms generated, on the other hand, are determined by the

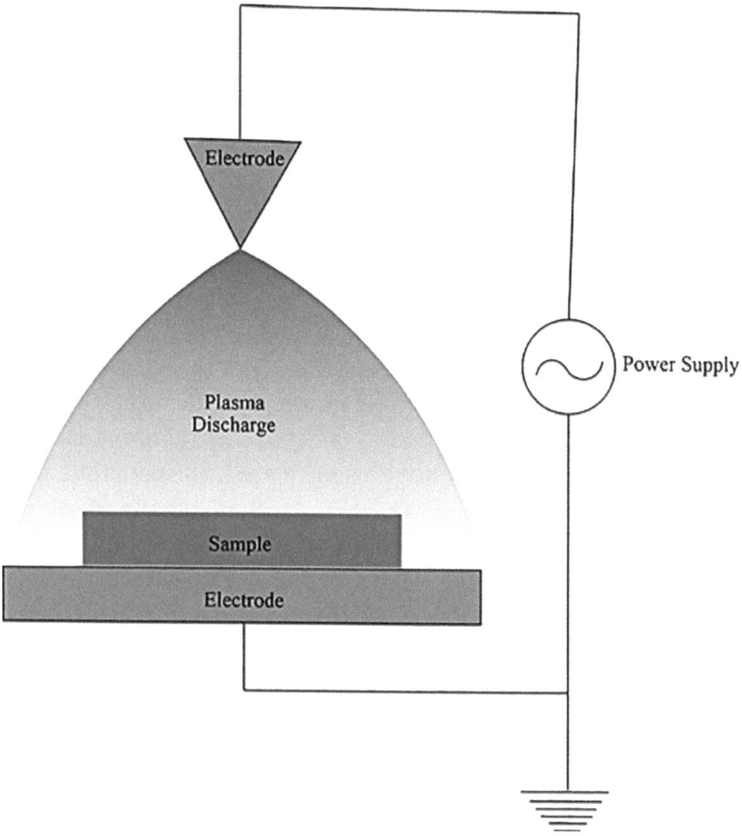

FIGURE 2.5 Single point to plate corona discharge.

properties of the gaseous mixture and discharge's polarity, while energies of the electrons are determined by the corona generation method as well as the gas properties [2]. Corona discharge is a weak luminous and non-uniform discharge at atmospheric pressure levels with positive and negative currents [14]. The weak luminous discharge occurs when the electric field is quite huge near sharp ends or thin wires. The electrode near the circuit pulls the charged particles towards itself by its weak electric fields. The breakdown circuit is shut by displacement current instead of charged particle transport in the early phases of the corona discharge. The corona is also found in the air surrounding the transmission lines of high voltage, lightning rods, and ship masts, which are named 'Saint Elmo's fire.' It is ignited by a high voltage around the region of one of the electrodes. The remaining discharge gap breaks when the voltage is increased, and a spark is formed from the corona.

The electrode's polarity where there are high electric fields influences the maintenance of the continuous ionization level in a corona discharge. When a high electric field surrounds the cathode, it is called a negative corona, whereas it is known as a positive corona when it surrounds the anode. The cathode's secondary emission is brought about by an effect of ions and offers continuous electric current flow from the cathode to the plasma. The negative corona is ignited by the mechanism that Townsend breakdown works on, and the possible electron attachment process is expressed as follows (Eqn. 1):

$$\int_0^{x_{max}} [\alpha(x) - \beta(x)]dx = \ln\left(1 + \frac{1}{\tilde{a}}\right) \quad (1)$$

Where, $\alpha(x)$, $\beta(x)$ and γ are the 1st, 2nd, and the 3rd Townsend coefficients. respectively relating ionization, attachment with an electron, and the cathode's secondary electron emission; x_{max} represents the space from the cathode when the electric field is low. $\alpha(x_{max}) = \beta(x_{max})$ depicts that extra electron multiplication do not occur.

This links to the electric field breakdown (E_{break}) in the electronegative gases. x_{max} is chosen to owe to an exponential decrease of $\alpha(x)$,while the integration equation will not be limited if $\beta = 0$ i.e., the gas is not electronegative. The critical distance x_{max} governs the ionization, electronic excitation zone, and plasma luminosity zone. This allows x_{max} to be taken as the visible zone of the plasma.

The cathode phenomenon does not provide ionization in the positive corona region as the field is low at the cathode. Cathode-directed streamer formation is associated with the ionization processes. The conditions for igniting a positive corona can be expressed using the cathode-directed streamer formation criteria. Meek's breakdown criteria can be considered by taking the non-uniformity and electron attachments also into account:

$$\int_0^{X_{max}} \left[\alpha(x) - \beta(x)\right] dx \approx 18 - 20 \quad (2)$$

Since $\ln(\frac{1}{\gamma}) \approx 6 - 8$,the minimum amplification coefficient values are 2–3 times lesser to give the criteria for negative corona ignition. Though positive and negative coronas are associated with different breakdown mechanisms, they still have close critical electric field values. The electric field critical value for the start of the negative corona is not influenced by the composition of the electrode according to the Townsend breakdown. This is attributed to the robust exponential reliance of the coefficients of amplification on the values of the electric field. The indirect ionization phenomenon, like metastable–metastable collisions, contributes to amplification coefficients that can be related to this explanation.

The ignition criteria for the positive and negative coronas are defined by the maximum electric field values surrounding the electrode region during the initiation of the discharge. When coaxial electrodes are used, the critical igniting electric field values are determined by the empirical Peek formula (Eqn. 3):

$$E_{cr}\left(\frac{kV}{cm}\right) = 31\delta\left(1+\frac{0.308}{\sqrt{\delta r\ (cm)}}\right) \quad (3)$$

Where, δ represents the air density and standard value ratio and r represent the internal electrode's radius.

This formula can be used for 0.1–10 atm pressures, smooth internal electrodes of radius ≈ 0.01–1 cm, having up to 1 kHz of DC and AC with frequencies. The electric field reduces up to 20% when the electrodes are rough.

The Peek formula can also be used for corona discharges with configurations other than coaxial electrodes using a slightly different coefficient value. Using an empirical formula (Eqn. 4), the corona discharge initiating from electric fields at two parallel wires is calculated as follows:

$$E_{cr}\left(\frac{kV}{cm}\right) = 30\delta\left(1+\frac{0.301}{\sqrt{\delta r\ (cm)}}\right) \quad (4)$$

The given two equations agree with the simplified empirical formulae of α, the Townsend coefficient at the air when in lowered electric fields $E/p <$ 150 Vcm^{-1} $Torr^{-1}$:

$$\alpha\left(cm^{-1}\right) = 0.14\times\delta\left[\left(\frac{E(kV\ cm^{-1})}{31\delta}\right)^{2} - 1\right] \quad (5)$$

The region around the electrode at which the electric field is adequately high enough for the ionization of charged particles to happen is known as the active corona volume. The volume of the active corona region is very crucial in the corona discharge for all the applications of the plasma as the majority of excitation and reaction phenomena occur in this region. The electric field value agreeing to the Ebreak value, the active volume boundary is used to calculate the active corona volume's external radius.

The lowest voltage necessary to ignite the corona between the wire electrode having a radius = 0.1 cm and a coaxial external electrode having a radius = 10 cm in the air is around 30 kV. Simultaneously, the electric field nearby the external electrode is too low and is insufficient to ionize. The electric field breakdown approximately equal to $E_{break} \sim 25$ kV cm^{-1} is necessary for the effective multiplication of charges. This is used to calculate the active corona region case's external radius as $rAC = rE_{cr} / E_{break} \approx 0.25$ cm. Therefore, the active corona region covers the thin wire with a cylinder layer of 0.1 cm < x < 0.25 cm. The active corona region surrounding a thin wire has an external radius of r_{AC} that is calculated as follows:

$$r_{AC} = \frac{V}{E_{break} \ln\left(R/r\right)} \tag{6}$$

where V is the applied voltage to withstand the corona discharge. It is clear from the above equation that the region of the active corona rises with the applied voltage.

The charged species are generated in the active corona region only around the electrode. Electric current away from the active corona region is given by the drifting of charged particles produced at the active corona region. The positive corona has positive ions as drifting particles, while the negative ions or electrons (when a non-electronegative gas mixture generates corona) are the drifting particles in the negative corona.

The continuous corona discharge supplies low power, and an increase in voltage and current also causes the conversion of the corona to sparks. These limitations can be resolved by converting this to the pulse-periodic mode of discharge. The total power in the corona when a very long wire is used becomes noteworthy even though the corona power is comparatively lower for unit wire length.

2.5.3 PLASMA JET

Many designs utilize a dielectric barrier or pointed edges at corona discharge for plasma generation to prevent spark transitions and use gas at reduced temperature. However, these designs do not produce plasma with uniform composition throughout its discharge. To overcome this limitation, plasma jets were designed. These jets have two concentric electrodes that contain a mixture of gases flowing through them. The discharge is ignited in the gas when voltage is given to the inner electrode. A plasma jet consists of ionized gas flushing out of the nozzle to the product that has to be treated and kept below a few millimeters. Generally, the velocity of the gas emitted from the jet is about 12 m/s and is at 150°C. In this system, gas is allowed to rush into the discharge volume and given exposure to plasma excitation. Along with the effluent, long-lived radicals created are also carried in the jet so that they are applied to the products to be treated (Figure 2.6).

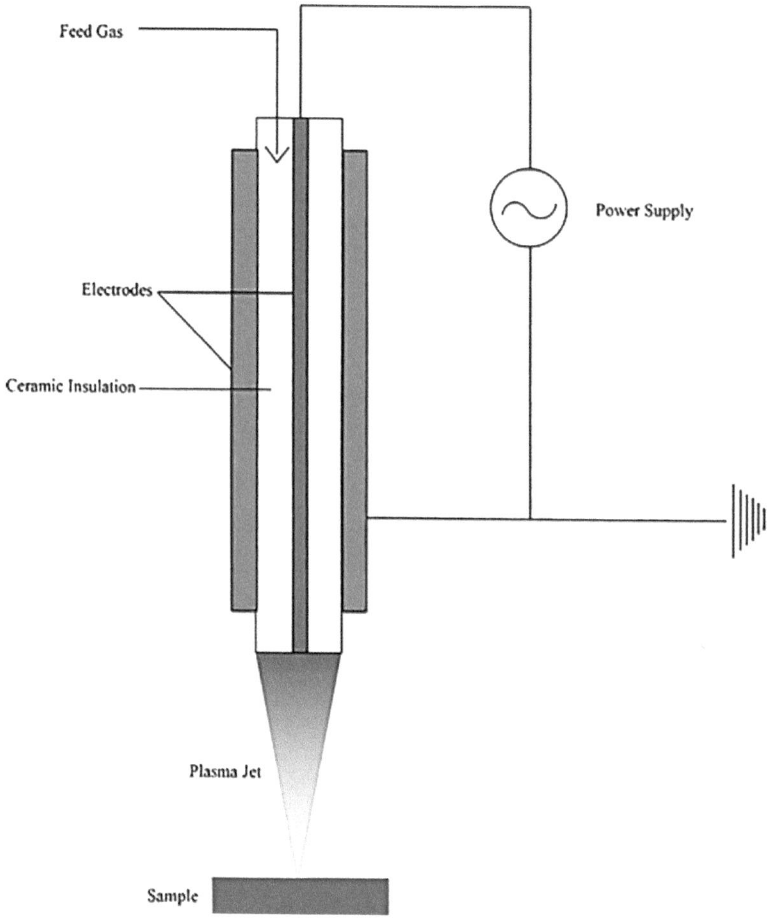

FIGURE 2.6 Design of a plasma jet.

The plasma jet, also known as an atmospheric pressure plasma torch, discharges several kHz and kV voltage using high voltage alternating current [72]. The feed gas is a mixture of oxygen, helium, and other gases with a high flow rate between two coaxial in the atmospheric pressure plasma jet [19]. Ceramic tube, copper tape, and stainless-steel mesh are the materials that can be used for the static electrode. This is a batch-type system [67]. It might seem that the plasma jets utilize only inert carrier gases, but they use the atmosphere air as their working gas. Air is conveyed by convection and diffusion in the plasma jet zone. Stabilization of gas discharge and temperature control of the gas are the major challenges for researchers in other designs of generating plasma for its use. Atmospheric pressure plasma jets were developed as a good option due to their cost-effectiveness and convenience compared to low-pressure plasma generators. The high voltage applied on small gaps (0.5 cm–3.0 cm) filled with air produces discharge currents with sharp edges in just a few nanoseconds.

Plasma characterization of plasma jets shows that they have similar properties as a low-pressure plasma discharge. The breakdown occurs at about 130 V when the current is 0.01 A, which is comparatively lower than the parallel plate discharge. On increasing the current, there will be slight changes in the voltage above spark breakdown. The plasma jet shows a voltage-current characteristics curve similar to that exhibited by the low-pressure DC discharge [69]. But it showed three- or four-times reduced voltage. The concentrations of the charged species are in the order of 1×10^{10} cm^{-3} at the cold plasma jet discharge. Nevertheless, the electron density is higher inside the jet. The range of the plasma jet's electron temperature is 1–2 eV. The range of the ozone concentration was from $2–5 \times 10^{15}$ cm^{-3} [23]. Table 2.3 displays the different characterizations of the cold plasma generated from different sources [10].

TABLE 2.3 Comparison of Different Plasma Discharges

Plasma Source		Dielectric Barrier Discharge	Corona Discharge	Plasma Jet
Breakdown voltage (kV)		5–25	10–50	0.05–0.2
Electron Temperature (eV)		0.9	8.6	1–2
Flow rate (slm)		0–1.7	0–0.5	8–92
Frequency		2–22 kHz	DC, Pulsed DC	1 kHz–27.12 kHz
Gases used		Ar/ air/ humidity	Ar/ air	O_2/He/Ar/ air
Plasma Density		$10^{12}–10^{15}$	$10^{9}–10^{13}$	$10^{11}–10^{12}$
Power (W)		1.8–360	50	15–250
Temperature (K)		~300	~300	300–430
Voltage (kV)		1.5–10.4	1–14	1–5
Oxygen species density	O^+, O^{2+}, O^-	10^{10}	10^{10}	10^{12}
	O	10^{12}	10^{12}	10^{16}
	O_3	10^{18}	10^{18}	10^{16}

A plasma jet is a type of source that can also be made in combination with any other plasma generating system. It may be dielectric barrier discharge type, corona discharge, or glow discharges. A plasma jet can be generated from a dielectric tube when the gas is forced out at a high flow rate. Generally, from a few kHz up to 27 MHz electrical power coupling occurs with from a few W up to kW power ranges. A dielectric barrier is kept in between the needle-shaped electrode and a ring-shaped electrode that is grounded and has a high voltage. The electrodes that are needed for power coupling are mostly kept in the ring outside the tube at a distance conditional to the experimental setup. A plasma jet is an open system, i.e., the air in the room is circulated by convection and diffuses into the plasma zone. But a larger gas flow rate of the admixture decreases the entrapment of the room airflow into the plasma zone. This design has a limitation as many jets have to be coupled to get a higher area throughput. Also, system cost and the operating cost have to be considered as it consumes a higher volume of air [47].

A jet of plasma containing charged and also neutral activated species encroaches on the food surface exposed to it. Protection should be provided for the jet operator and also for the food surface to be treated under a high-voltage electrode. The plasma jet should be designed in such a way that there is no spark created between the main electrode and the product being treated. The streamer-to-sparker transitions are prevented by the dielectric material. Classification of jets is done based on the temporal shape, and the frequency of the voltage applied ranges from square-shouldered pulsed voltage to microwave excitation. The voltages for the pulsed plasma jets with a DC power source are generally in the range of kV, and the pulse widths range from a few nanoseconds to microseconds at kHz repetition rates [53].

Corona discharge plasma jet electrodes are steel needles having a diameter range of 0.6 to 1 mm and radii of 20 to 50 μm. Electrodes that are grounded can either be in a flat or cylindrical configuration. Polytetrafluoroethylene plates were used to divide the high and low potentials. In an atmospheric plasma jet, the feed gas flows in between two coaxial electrodes at a high rate. The feed gas includes a mixture of oxygen, helium, and a few other gases. One of the electrodes is the grounded electrode, whereas RF power is given to the other electrode which is the fundamental electrode that causes the discharge. The active species are generated and leave at a very high velocity from the nozzle and reach the product that has to be treated.

Plasma jets have various applications including etching and deposition at reduced temperatures. Plasma jet equipment is simple in design and can be operated easily to produce a cold plasma that has a great capacity to be employed in online microbial decontamination of food substances. In a study conducted by Choi et al. [7], a corona plasma jet was utilized as a tool to decontaminate dried pollock shreds from *Staphylococcus aureus* by treating for 0–3 mins. The plasma was produced at 1.5 A current and a 25 mm long gap was kept between the tip of the electrode and the product that was treated. The corona discharge plasma jet inactivated all microbes satisfactorily in the pollock shreds. The plasma jet used in this research satisfactorily showed >1 log reduction of all identified microbes. The plasma did not show any significant difference effect in most of the physicochemical properties of the pollock shreds.

A plasma jet may be a so-called remote source because the plasma that is excited by power electronics is from a location a little away from the processing area. It is designed in this way to prevent direct exposure of plasma to fragile products. Even though nonthermal plasma does not heat the gas much, still usage of the remote source is preferred for minimizing gas heating. Since jet sources are expensive, it is questionable to be favored in food processing when cost consideration is taken.

2.3.4 PULSED DISCHARGE

The continuous use of plasma discharge is restricted to using very small currents and, as a result, very low discharge power, which results in a poor rate of plasma treatment for the product. To overcome this limitation, a pulsed-periodic voltage can be used at higher voltage and power without the formation of sparks. The direct treatment type of cold plasma, where the plasma is directly exposed to the product to be treated, can be operated in a pulsed mode in which the plasma is produced hundreds or thousands of times per second at pulse frequencies. Currently, pulsed plasma discharge is one of the promising atmospheric pressure plasma discharges used for several applications. Food products with high water activity and moisture conduct electricity when sufficient voltage is applied. Either the meat or plant tissues conduct electricity through them, causing the development of localized heat that can damage sensory attributes like aroma, texture, and appearance or cause burn marks or protein coagulation in the food. Hence, the design of direct treatment cold plasma generators should be carefully made in such a way that concentrated conduction of electricity through the food product must be prevented. For this reason, pulsed mode generation of cold plasma was developed, which can be advantageous to prevent undesirable changes in food products [44].

Pulsed plasma discharge is produced when a series of short pulses of discharges are applied that result in a sharp increase in voltage of 100 V/ns or more. Unlike electrons, the heavily charged species do not accelerate in these conditions. A stable plasma discharge can be formed easily at atmospheric pressure itself using this method. The arc discharge is prevented as the voltage is applied in very short times. Power sources of high voltage producing short pulses and highly recurring plasma discharges have recently been developed for various applications. A pulse power source using the static induction thyristor was developed to create nitrogen gas plasma [66].

The velocity of the streamer is 108 cm/s, which is 10 times more than the usual velocity of the electron drift in an avalanche. The total time taken for avalanche development, the transition from avalanche to the streamer, and propagation of the streamer between electrodes kept at about 1–3 cm distance will be about 100–300 ns. This depicts that the voltage pulses can sustain streamers in these periods and efficiently transmit power to cold plasma without formations of sparks. It is necessary to design a pulsed power supply for pulsed plasma discharges that can generate adequate short voltage pulses having a steep front and less time for rising.

A plasma pencil that is a miniature form of a plasma jet was designed by Laroussi. Two disk electrodes of 2.5 cm diameter were inserted into a cylindrical dielectric tube of equal diameter. The distance between the two electrodes can be altered from 0.3 to 1 cm. The cylinder also contained one thin copper ring which was joined to the dielectric disk. A DC supply that has variable voltage output supplies high voltage to the pulse generator. This pulse generator is employed to supply the electrodes with electrical power of high voltage. High voltage was applied as pulses in sub-microseconds between the electrodes where gas was allowed to flow to generate plasma. Once the discharge is produced, a plasma plume was ignited into the air through the outer electrode's hole. The generated plasma plume is up to 5 cm long and is at low temperature (about 290 K). The plasma pencil was effectively used for treating *E. coli, Leukemia cells, P. gingivalis*, and many other biomedical applications too [1, 30, 36]. Few methods of generating pulsed plasma and the parameters for their discharges are discussed in the following sections.

2.3.4.1 FLASHING CORONA

The corona discharge can run in periodic pulses at a constant voltage. Frequencies go up to 10^4 Hz for positive corona and up to 10^6 Hz for negative corona. Corona-to-spark transitions still occur in this type of self-organized pulsed corona, and hence it is obvious that this method cannot overcome the limitations of the continuous corona. Nevertheless, this is still a crucial method for developing non-steady-state coronas having high current, voltage, and power. Flashing corona occurs due to the consequence of the positive space charge, which is formed as the electrons in streamers reduce faster at the anode, and the positive ions that are slow stay in the discharge opening. The increasing positive space charge at the discharge gap led to the reduction in an electric field close to the anode and does not allow new streamers to generate. Unless the positive space charge moves towards the cathode and makes clear the discharge opening, the positive corona current remains suppressed. Then ignition of a new corona occurs, and the cycle repeats. This flashing corona is not formed at intermediate voltages where the active corona region's outer electric field is adequately elevated enough to give an efficient steady-state positive ions clearance but not very high, which causes rigorous ionization. During the flashing corona, the electric current does not go down to zero in between pulses but remains at some constant value that is present. An increase in the power of the corona occurs due to the pulse-periodic regime. Still, the increase in power is limited by the formation of sparks as the voltage applied is continuous.

2.3.4.2 TRICHEL PULSES

The negative corona discharge held by unceasing voltage also functions in a pulse-periodic administration at lower voltage values approximately equal to the ignition value. The period of the pulses in the negative corona region is roughly equal to 100 ns. The peak current value in each pulse goes up to 10 mA when the mean corona current is about 20 μA. At higher voltages, pulses disappear while the positive steady-state corona discharge lasts until a spark forms.

The negative corona's pulse-periodic regime is generally called Trichel pulses. It is similar to flashing corona but with a few differences. The avalanche evolution from the cathode causes the formation of a double layer: (1) an internal positively charged layer, and (2) an external negatively charged layer (consisting of negative ions in the air or the electronegative gases or the electropositive gases' electrons). Only electronegative gases are more suitable to generate Trichel pulses. Negative ions create a negative space charge surrounding the cathode that is not to be compensated with the positive ions' narrow layer. Hence, the space charge of these ions represses the electric field close to the cathode region, resulting in the suppression of the corona current. The negative corona gets re-ignited as the ions move out from the discharge opening and neutralize on electrodes, and the cycle restarts again.

2.3.4.3 PULSED CORONA DISCHARGES MAINTAINED BY NANOSECOND PULSED POWER SUPPLIES

The pulses that are generated in short durations like 100–300 ns to prevent corona spark transition are known as pulsed corona discharges maintained by power supplies with nanosecond pulses. The power supplied should be adequately high to increase the rate of voltage (0.5 kV/ns–3 kV/ns) that generates high corona ignition voltage and high power. This also results in more efficient plasma-chemical processes that require high electron energy. Higher values of the discharge energy are required to reduce a part of discharge power getting to the vibrationally excited state of molecules in the plasma-chemical processes. This is also needed to ionize and for electronic excitation and dissociate the molecules.

Hence, there exist various designs of pulsed discharges that are powerful and luminous, having good applicability and scope in the food sector.

2.4 GENERATION OF LOW-PRESSURE COLD PLASMA

Low-pressure plasma generators involve a vacuum pump and are kept closed from the atmosphere. Foods that can withstand low gas pressures and that do not get mechanically damaged are treated using these low-pressure cold plasma reactors (e.g., nuts and spices are not susceptible to damage due to low-pressure gas). Food treated in these reactors should be resistant to mechanical injury induced by low pressures. Foods that are exposed to these low-pressure systems might outgas significantly. The user of these systems must be concerned about the impacts of the complex chemistry of the plasma that can lead to food derivative molecules flowing into the cold plasma and portions of plasma species molecules might come back as dense condensates to the food [43].

Low-pressure cold plasma is little exploited compared to atmospheric pressure cold plasma as it is relatively expensive. The low-pressure cold plasma generating systems require extra manpower and expertise to handle the vacuum systems, which add to the challenges related to the output rates in food processing. Low or vacuum pressure causes important implications beyond potential mechanical damage to the food product that is exposed to it. As the pressure drops in plasma-to-solid interfaces, sheaths of plasma are formed, similar to the plasma-to-food interface to treat food directly under low-pressure plasma. In a few low-pressure systems, downstream processing must be taken into account, in which electrodes with electrically-biased screens are able to interact with activated neutral chemical species other than the electric field and charged molecules. Low-pressure plasma systems possess high plasma density, do not get contaminated by the metals from electrodes, and cause a smaller amount of plasma sheath voltages [53].

2.4.1 MICROWAVE-ASSISTED DISCHARGE

Low-pressure cold plasma is generally produced by microwaves. Gases are contained at low pressure in a plasma reactor chamber and when the energy is supplied to gases via an antenna, it creates plasma. The most common gases employed to produce plasma are O_2, N_2O, and Ar. Rather than employing electrodes having varying potentials, microwave cold plasma discharges use an electromagnetic field of high frequency produced by a magnetron. Microwave energy oscillates electrons, which collide with gas atoms and molecules to deliver ions with a high-level ionization degree in microwave plasma [15].

Microwave discharges are generally produced using electromagnetic radiation of frequency in a range of 300 MHz and 300 GHz. Electrodes are not required for this design of plasma generation. The plasmas are produced using antennas or waveguides and shielding is necessary here [65]. Microwave plasma discharges produce plasma with high electron density and have higher reactive species generating efficiency. Under low pressure, a large diameter plasma can be produced using this method as a contamination-free system for plasma generation [4].

An electric field is coupled with electromagnetic waves that are highly sufficient to cause an electron avalanche and a plasma glow. The food samples are either placed in the cold plasma area directly or downstream of the cold plasma region. Depending on the design of the system, DC magnetic fields can be included in the reactor that can cause electron cyclotron resonance discharge plasma [35]. The collision of ions and electrons with the food surface is minimized in this system as the plasma species travel only short distances before changing their directions (movement of ions is halted because of the high frequency and the large mass of ions) [54].

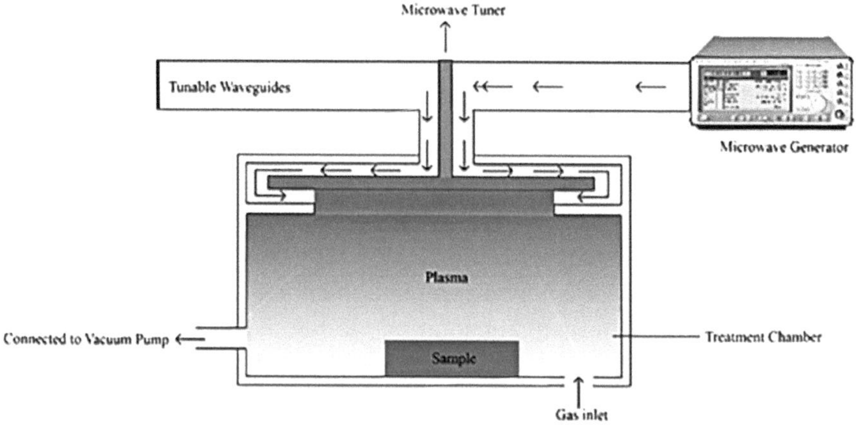

FIGURE 2.7 Low-pressure microwave plasma discharge system.

The basic components of the microwave plasma generator are as follows (Figure 2.7) [32]:

- Cooling system
- Gas mass flow rate controller
- IR Thermometer
- Microwave generator – magnetron
- N_2 supply
- Parameter controller
- The-O_2 supply
- Treatment chamber
- Vacuum

Microwave plasma can successfully be used to decontaminate food products by inactivating the proteins and damaging the DNA of the microbes. Microwave plasma discharges at low pressure show efficient effects of decontamination because of the etching effects by oxygen and argon species and also the UV radiation produced [21]. Microwave system cold plasma has similar advantages to conventional microwave ovens. These systems provide a more rapid technology development by utilizing microwaves for gas ionization present in the sample treatment chamber [73].

Advantages of low-pressure plasma generation using microwaves for the applications of food decontamination or sterilization are low-temperature treatment, time-saving processing, and non-toxic methods. The inactivation effect of microwaves is directly proportional to the power density applied. Microwave plasma UV lamps can generate UV at 245 nm, which is a germicidal region, and at 185 nm, in which ozone forms at the same time. A microwave plasma UV lamp is more advantageous than the electrode configuration lamp as there are no limitations on its power and shape [49]. A microwave plasma UV lamp is a potential alternative to commercial UV irradiation systems [48].

In a microwave plasma generator, the electric field of the electromagnetic waves is so high that it can produce an electron avalanche and plasma glow. DC magnetic fields added in a microwave cold plasma generator create electron cyclotron resonance discharge [35]. Electron cyclotron resonance cold plasma (ECR) is due to the circular motion of the free electrons in a uniform and static magnetic field. This has many advantages such as low electron temperature (5–10 eV), high ionization, and flexible working pressure and temperature. Bombardment of ions and electrons on the food surface that is exposed is reduced as the electrons in the plasma zone navigate only for a short distance prior to shifting directions (the ions are highly immobilized because of high frequency and a huge mass of the ions) [54].

Soleiman Abbasi et al. [71] designed a low-pressure microwave plasma discharge system consisting of a gas pressure monitor, a gas flow monitor, a microwave, an afterglow chamber where the products to be treated are placed, and a vacuum pumping system. The gas inlet flow was controlled by a needle-shaped valve and monitored using a flowmeter. A vacuum meter monitored the internal pressure in the plasma reactor. A household microwave oven (1,200 W and 2.45 GHz) was used as the energy source to generate plasma in the glow discharge tube. The glow discharge tube was connected to the afterglow chamber, which was covered with aluminum foil and positioned outside the microwave chamber. The products to be exposed to plasma treatment were placed in the afterglow chamber using holders. One end was connected to a vacuum pump while the other end was linked to the discharge tube. The vacuum pump regulated the outlet and the flow rate of the inlet gas to maintain the required pressure. The low-pressure microwave plasma generator designed by Soleiman Abbasi, capable of producing afterglow cold plasma, was successfully used for sterilization under optimized conditions. It was employed to decontaminate *Salmonella* Spp. from eggs and aluminum foils used as packaging materials effectively.

2.4.2 RADIO FREQUENCY DISCHARGE

Radio Frequency (RF) discharge is used to produce cold plasma by employing pulsed electricity within the electric coil's center. Capacitively coupled, inductively coupled, and helicon wave sources (HWSs) low-pressure cold plasma reactors generally utilize radio frequency excitation at 13.56 MHz frequency. RF discharges can be used at low pressure, intermediate pressure level, and atmospheric pressures using gases such as argon, nitrogen, and air. Figure 2.8 shows the discharge system contains a parallel plate arrangement that has two disc-shaped electrodes with a minor space in between them, the upper electrode is powered by radiofrequency whereas the lower electrode is grounded. The hole drilled in the lower electrode acts as a nozzle. A ceramic or quartz tube of cylindrical shape surrounds the discharge area and acts as the chamber for the plasma source. The feed gases (argon, nitrogen, or air) flow through this discharge chamber, cross the gap, and expel out from the outlet to the outer space. The source of plasma is kept inside an expansion chamber that is propelled out using the vacuum system. Plasma discharge is maintained in a tiny slit between the RF-powered electrode and a nozzle. Nevertheless, it can also be transported through the grounded aperture's expansion chamber.

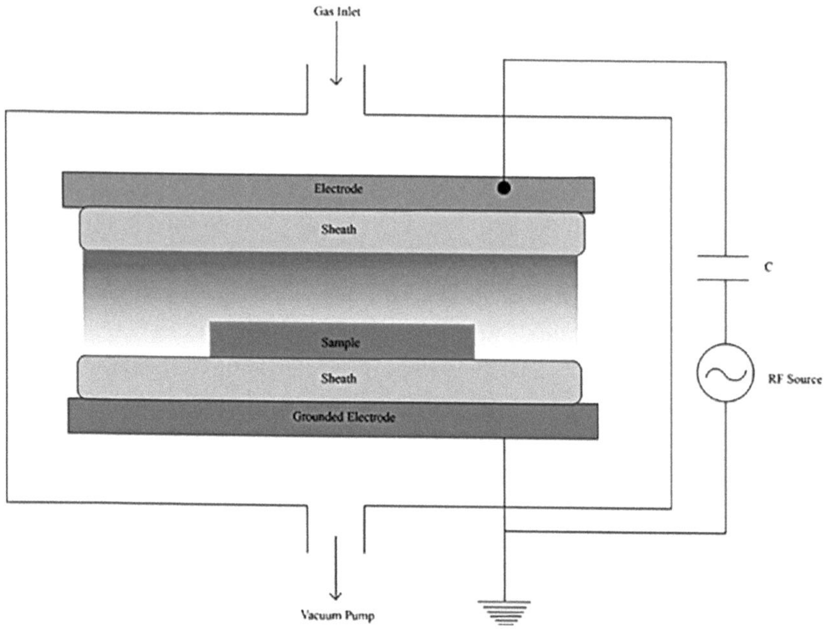

FIGURE 2.8 Radio-frequency plasma discharge system.

Radiofrequency plasma ionizes the gases by using rapid cyclic electrical impulses, working at different voltages and powers. Lassen et al. employed a radiofrequency plasma discharge at 13.56 MHz frequency to inactivate *B.*

stearothermophilus spores. Different mixtures of gases including O_2, Ar, H_2, and CF_4 at different powers from 100 W to 400 W were used for the study [31].

Radiofrequency plasma discharge can be made in various configurations. The food can either be placed in between the electrodes making it a contact treatment system. Or, the electrodes are placed above the product to which the plasma exposure is applied. An example of this kind of system is an RF-driven plasma jet.

Plasma generated using radiofrequency is not only easily ignitable but also can work at lower current and power, and with less instability, when compared with DC discharges. The breakdown voltage for the plasma discharge is dependent on the frequency. Hence, Radiofrequency-based plasma discharges are frequently used for research. Many different configurations were designed such as planar geometry, needle design, multi-electrode assembly, and cylindrical geometry based on their applications.

Hollow cathode expanding cold-plasma jet systems are radiofrequency-based plasma discharging sources. The radiofrequency is applied to the hollow electrode, generally a metal tube, that may or may not contain any dielectric material inside the tube. Gas is flushed into the cathode, and the jet arises from the radiofrequency-applied electrode. For low-pressure plasma discharges, the RF electrode is kept inside a vacuum chamber [12, 28].

Arcing and filamentation increase with pressure in RF plasma discharges. This is due to thermal instability caused by gas and electrode heating. As a result, effective cooling options were considered, with the most common being the usage of chilled electrodes, higher gas flow rates, or, if possible, high thermal conductivity gases (like helium). Due to transport processes, the created plasma is directed in the presence of gas flow. Collisional mechanisms that lead to recombination and loss of excitation have lower rates at low pressure, and the plasma spreads over long distances [8].

Another method of radiofrequency plasma generation is by circulating radiofrequency current in coils or antennas placed inside the plasma or away from the plasma, separated by a dielectric window. The radiofrequency excitation design governs the efficiency of power coupling into charged particles and the uniformity of the plasma [3].

Koinuma et al. [27] designed one of the earliest radiofrequency plasma jets in 1992. A tungsten or stainless-steel needle electrode of 1 mm diameter was used as the cathode. This cathode is coupled with the RF source of 13.56 MHz frequency. The cathode is kept inside a quartz tube and a grounded anode is used. Either helium or argon was mixed with other gases depending on the usage.

Sharma et al. [70] used the open-air hollow plasma system having an afterglow plume for inactivating *E. coli,* and *Bacillus atrophaeus* and spores. The electrodes in this system were powered with 13.56 MHz radiofrequency to ignite Ar (flow rate = 32 L/min) and O_2 (flow rate = 6 standard cm^3/min) gas mixture by a capacitively coupled discharge. 177 W power and 1.5 A current were supplied to the system and plasma formed at 203 V of RF voltage.

Radiofrequency discharge can be divided based on the shape of the electrode or antenna. The radiofrequency power is given between the facing electrodes in capacitively coupled plasma, while coiled shaped antenna with radiofrequency power is given in inductively coupled plasma. The high-frequency plasma was generated inside this coil. A huge quantity of dense cold plasma can be produced under this low pressure with minimal heating. Disinfection of materials using this method is advantageous as it does not damage them during the treatment. The choice of plasma produced should be made convenient for the product's dimensions that are to be treated.

2.4.3 GLOW DISCHARGE

In Glow discharge, a high voltage input is given between two electrodes to generate plasma by the alternating current that is passed through a gas. The products that must be treated are positioned in a vacuum chamber, and the cold plasma is produced when the vacuum chamber pressure is stabilized at 2 Pa [63].

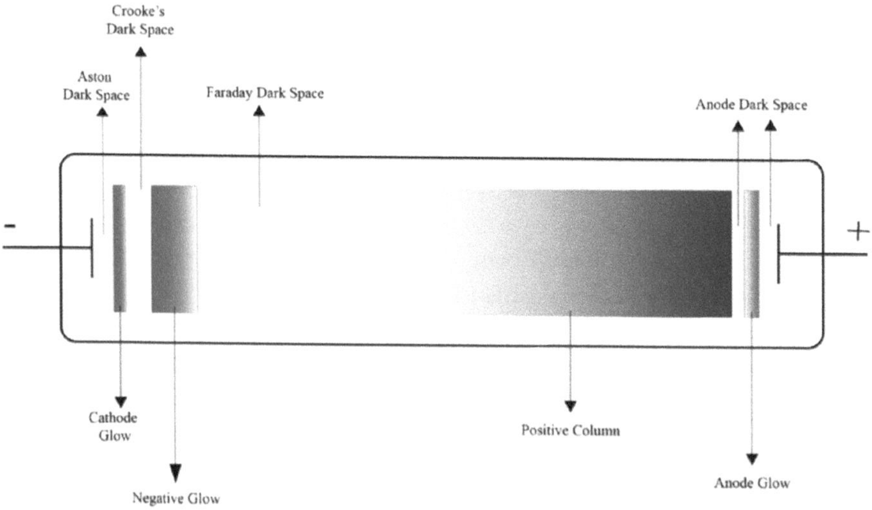

FIGURE 2.9 Regions of glow discharge.

Glow discharge is produced when a voltage from 100 V to many kV is applied to gas at low pressure in the range of 1–1,000 Pa. The electrons possess higher energy than the neutral gases and ions in the glow discharge. This discharge is spatially extremely uniform and can produce a plasma that is sustainable at low temperatures in huge volumes [65]. Glow discharges have electrodes separated by space in between that are either in partial vacuum or filled with some gas.

The current in glow discharge is in the range of a few mA and is upheld by secondary electron emission due to the bombardment of positive ions from the cathode. This system is the most extensively used gas discharge, with applications ranging from sources of light to cold plasma reactors for processing materials [35, 64].

Glow discharges are usually operated at low pressure and are recognized for their distinctive glowing appearance, which is contrary to the Townsend dark discharges. The light emission of the low discharge is divided into many layers and shows a stratified structure (Figure 2.9). Each layer is given a separate name and relates to some portion of the discharge. The first layer starts from the cathode side and is known as the Aston dark space, occasionally covered by the cathode glow, from where the electron is emitted at low energy from the cathode. The cathode glow is the next thin layer where the electrons excite the atoms. The electron multiplies by impact ionization inside the cathode dark space, which is next to the cathode glow. The next bright zone is the glow region in the negative section that is distinctly divided from the cathode dark layer. This layer progressively becomes dim, where it is known as Faraday dark space as it moves towards the anode. The positive column brightens up with a uniform pattern; nevertheless, it is not brighter than the negative glow.

The Aston dark space, a cathode glow, a cathode dark space, and the negative glow are components of the cathode layer. The positive column is transmitted to the dark space of the anode and lastly into the anode glow. The color of the glow changes with the gas present, reflecting the gas spectrum. The gas pressure and gap distance influence the glow pattern. The effect of space charge and the secondary electron emission process is more prevalent in this discharge system. The cathode layer is a unique feature of the glow discharge as it is crucial for maintaining the discharge. Cathode fall is the slow movement of the positive ion space of the cathode layer's dark space that makes a potential drop in between the negative glow of about 100 V–400 V and the cathode. This important parameter defines the secondary electron emission of the cathode and subsequent discharge lighting. The cathode fall is a key mark to differentiate subnormal, abnormal, and normal glow discharges. The current density is constant at the cathode in a normal glow discharge. The current of the glow discharge covers the cathode surface partially and is directly related to the area of glow in the cathode. The density of the current on the cathode surface, cathode fall, and thickness of the cathode layer are fixed in every normal glow discharge for the specified composition of gases, pressure, and the cathode material. Von Engel-Steenbeek's theory's minimal power concept for achieving the smallest feasible cathode fall explains this mechanism in the normal discharge [64].

Higher power causes abnormal glow discharge, which may even result in a transition to arc discharge. The current density of glow discharge is generally 100 $\mu A/cm^2$ at 1 torr. The cathode layer is about 0.5 cm thick at this pressure. The cathode potential drop is independent of pressure and temperature and is around 200 V. A long and uniform positive column is visible between the Faraday dark space and the anode when the separation between electrodes is adequately large. Electron impact ionization maintains the generation of electrons in a constant field. An electrical connection of a cathode and anode forms the positive column. It is generally a faintly ionized equilibrium cold plasma maintained by the electric field. The diffuse positive column has a range of 10^8–10^{12} electron density [39].

Photoionization can efficiently replace the lost primary electrons and cause steady glow discharges. The secondary electron emission from the cathode due to the positive ions' flux is crucial as it replaces the lost primary electrons to maintain the glow discharge.

2.5 SUMMARY

Foodborne illness related to contaminated food continues to be a serious issue. Plasma technology has proven potential on a lab-scale as a consistent, eco-friendly technique that is used for manufacturing safe food products. Industries will definitely adopt this technology in the near future since this gives a pollution-free working process. They are used widely for food decontamination and other applications on food products. There are several methods and conditions used to generate cold plasma with its strengths and limitations relevant to specific applications. Investigations in this line are greatly invited to commercialize this technology. The plausible future development should be focused on designing generating techniques of cold plasma based on its applications to realize bulk process and production.

Investigations on accurately mapping the characteristics of species in plasma depending on plasma generator and process parameters are very essential for process optimization. Designing and optimizing cold plasma generators are indeed one of the future developments which the food industries demand. Optimizing the composition of the working gas to have an equilibrium between UV and the reactive species-mediated approaches has to be done depending on the requirements of the products. Economic modeling including scaling-up factors, investment, and operational costs, consumables expenses, commercial-scale throughput, and others are to be calculated. Necessities for designing cold plasma applications in refrigerated conditions and high-moisture surroundings should also be made.

KEYWORDS

- **atmospheric-pressure plasma**
- **design of plasma generator**
- **low-pressure plasma**
- **plasma discharge**
- **plasma-activated medium**

REFERENCES

1. Barekzi, N., & Laroussi, M. (2012). Dose-dependent killing of leukemia cells by low-temperature plasma. *Journal of Physics D: Applied Physics, 45*(42), e422002. https://doi.org/10.1088/0022-3727/45/42/422002.
2. Bussiahn, R., Brandenburg, R., Gerling, T., Kindel, E., & Lange, H. (2010). The hairline plasma: an intermittent negative dc-corona discharge at atmospheric pressure for plasma medical applications. *Applied Physics Letters, 96*(14), e3380811. https://doi.org/10.1063/1.3380811.
3. Chabert, P., & Braithwaite, N. (2011). Physics of radio-frequency plasmas. Cambridge University Press Edition : 1, pp. 1–395, ISBN: 9780511974342, DOI: https://doi.org/10.1017/CBO9780511974342.
4. Chizoba Ekezie, F. G., Sun, D. W., Han, Z., & Cheng, J. H. (2017). Microwave-assisted food processing technologies for enhancing product quality and process efficiency: A review of recent developments. *Trends in Food Science & Technology, 67*, 58–69.
5. Choi, M. S., Jeon, E. B., Kim, J. Y., Choi, E. H., & Lim, J. S. (2020). Impact of non-thermal dielectric barrier discharge plasma on *Staphylococcus aureus* and *Bacillus*

cereus and quality of dried blackmouth angler (*Lophiomus setigerus*). *Journal of Food Engineering, 278*, e109952. https://doi.org/10.1016/j.jfoodeng.2020.109952.
6. Choi, M. S., Jeon, E. B., Kim, J. Y., Choi, E. H., & Lim, J. S. (2022). Application of dielectric barrier discharge plasma for the reduction of non-pathogenic *Escherichia coli* and *E. coli* O157 and the quality stability of fresh oysters (*Crassostrea gigas*). *LWT, 154*, e112698. https://doi.org/10.1016/j.lwt.2021.112698.
7. Choi, S., Puligundla, P., & Mok, C. (2016). Microbial decontamination of dried alaska pollock shreds using corona discharge plasma jet: Effects on physicochemical and sensory characteristics. *Journal of Food Science, 81*(4), M952–M957.
8. Dinescu, G., & Ionita, E. R. (2008). Radio frequency expanding plasmas at low, intermediate, and atmospheric pressure and their applications. *Pure and Applied Chemistry, 80*(9), 1919–1930.
9. Diver, D. A., & Potts, H. (2015). Plasma generation and use of plasma generation apparatus, US Patent No: 8980190B2, URL: https://patentimages.storage.googleapis.com/72/bb/bd/1b613d36af9d1d/US8980190.pdf, pp. 1–17.
10. Ehlbeck, J., Schnabel, U., Polak, M., Winter, J., & Von Woedtke, T. (2011). Low temperature atmospheric pressure plasma sources for microbial decontamination. *Journal of Physics D: Applied Physics, 44*(1), e013002. https://doi.org/10.1088/0022-3727/44/1/013002.
11. Fernández-Gutierrez, S. A., Pedrow, P. D., Pitts, M. J., & Powers, J. (2010). Cold atmospheric-pressure plasmas applied to active packaging of apples. *IEEE Transactions on Plasma Science, 38*(4), 957–965.
12. Foest, R., Kindel, E., Lange, H., Ohl, A., Stieber, M., & Weltmann, K. D. (2007). RF capillary jet – A tool for localized surface treatment. *Contributions to Plasma Physics, 47*(1–2), 119–128.
13. Fridman, A., Chirokov, A., & Gutsol, A. (2005). Non-thermal atmospheric pressure discharges. *Journal of Physics D: Applied Physics, 38*(2), e3727. https://doi.org/10.1088/0022-3727/38/2/R01.
14. Fridman, A., & Kennedy, L. A. (2010). Nonequilibrium cold atmospheric pressure plasmas. *Plasma Physics and Engineering, 1*, 561–611.
15. Ganesan, A. R., Tiwari, U., Ezhilarasi, P. N., & Rajauria, G. (2021). Application of cold plasma on food matrices: A review on current and future prospects. *Journal of Food Processing and Preservation, 45*(1), e15070. https://doi.org/10.1111/jfpp.15070.
16. Gurol, C., Ekinci, F. Y., Aslan, N., & Korachi, M. (2012). Low temperature plasma for decontamination of E. Coli in milk. *International Journal of Food Microbiology, 157*(1), 1–5.
17. Herianto, S., Hou, C. Y., Lin, C. M., & Chen, H. L. (2021). Nonthermal plasma-activated water: A comprehensive review of this new tool for enhanced food safety and quality. *Comprehensive Reviews in Food Science and Food Safety, 20*(1), 583–626.
18. Hertwig, C., Reineke, K., Ehlbeck, J., Erdoɫdu, B., Rauh, C., & Schlüter, O. (2015). Impact of remote plasma treatment on natural microbial load and quality parameters of selected herbs and spices. *Journal of Food Engineering, 167*, 12–17.
19. Hoffmann, C., Berganza, C., & Zhang, J. (2013). Cold atmospheric plasma: Methods of production and application in dentistry and oncology. *Medical Gas Research, 3*(1), 1–15.
20. Holdsworth, S. D., & Simpson, R. (2007). Thermal processing of packaged foods, https://doi.org/10.1007/978-0-387-72250-4, pp. 1–407, Edition: 2, Publisher: Springer New York, NY, Series Title: *Food Engineering Series*. Vol. 284. Springer, p. 466.

21. Hueso, J. L., Rico, V. J., Frías, J. E., Cotrino, J., & González-Elipe, A. R. (2008). Ar + no microwave plasmas for *Escherichia coli* sterilization. *Journal of Physics D: Applied Physics, 41*(9), e092002. https://doi.org/10.1088/0022-3727/41/9/092002.
22. Islam, R., Pedrow, P. D., & Englund, K. R. (2017). Phenomenology of corona discharge in helium admixtures inside a point-to-point electrode geometry. *IEEE Transactions on Plasma Science, 45*(10), 2848–2856.
23. Jeong, J. Y., Babayan, S. E., Schütze, A., Tu, V. J., & Park, J. (1999). Etching polyimide with a nonequilibrium atmospheric-pressure plasma jet. *Journal of Vacuum Science & Technology A: Vacuum, Surfaces, and Films, 17*(5), 2581–2585.
24. Keener, K. M., & Klockow, P. A. (2013). Method and system for treating packaged products. US Patent No: 20130189156A1. pp. 1–17, URL: https://patentimages.storage.googleapis.com/1f/67/70/a7ce69337dfb98/US20130189156A1.pdf.
25. Kelly-Wintenberg, K., Montie, T. C., Brickman, C., Roth, J. R., & Carr, A. K. (1998). Room temperature sterilization of surfaces and fabrics with a one atmosphere uniform glow discharge plasma. *Journal of Industrial Microbiology and Biotechnology, 20*(1), 69–74.
26. Kogelschatz, U., Eliasson, B., & Egli, W. (1997). Dielectric-barrier discharges: Principle and applications. *Journal de Physique IV, 7*(4), 47–66.
27. Koinuma, H., Ohkubo, H., Hashimoto, T., Inomata, K., & Shiraishi, T. (1992). Development and application of a microbeam plasma generator. *Applied Physics Letters, 60*(7), 816–817.
28. Korzec, D., Nithammer, D., Engemann, J., Ikeda, T., Aoki, T., & Hatanaka, Y. (2001). Radio frequency hollow cathode jet: Optical emission study. *Surface and Coatings Technology, 142–144*, 21–27.
29. Laroussi, M., & Lu, X. (2005). Room-temperature atmospheric pressure plasma plume for biomedical applications. *Applied Physics Letters, 87*(11), e2045549. https://doi.org/10.1063/1.2045549.
30. Laroussi, M., Tendero, C., Lu, X., Alla, S., & Hynes, W. L. (2006). Inactivation of bacteria by the plasma pencil. *Plasma Processes and Polymers, 3*(6–7), 470–473.
31. Lassen, K. S., Nordby, B., & Grün, R. (2005). The dependence of the sporicidal effects on the power and pressure of rf-generated plasma processes. *Journal of Biomedical Materials Research Part B: Applied Biomaterials, 74*(1), 553–559.
32. Lee, H., Kim, J. E., Chung, M. S., & Min, S. C. (2015). Cold plasma treatment for the microbiological safety of cabbage, lettuce, and dried figs. *Food Microbiology, 51*, 74–80.
33. Lee, K., Paek, K. H., Ju, W. T., & Lee, Y. (2006). Sterilization of bacteria, yeast, and bacterial endospores by atmospheric-pressure cold plasma using helium and oxygen. *Journal of Microbiology, 44*(3), 269–275.
34. Liao, X., Li, J., Muhammad, A. I., Suo, Y., & Chen, S. (2018). Application of a dielectric barrier discharge atmospheric cold plasma (dbd-acp) for *Escherichia coli* inactivation in apple juice. *Journal of Food Science, 83*(2), 401–408.
35. Lieberman, M. A., & Lichtenberg, A. J. (2005). Principles of plasma discharges and materials processing: Second edition. John Wiley & Sons, p. 757.
36. Mahasneh, A., Darby, M., Tolle, S. L., Hynes, W., Laroussi, M., & Karakas, E. (2011). Inactivation of *Porphyromonas gingivalis* by low-temperature atmospheric pressure plasma. *Plasma Medicine, 1*(3–4), 191–204.

37. Min, S. C., Roh, S. H., Niemira, B. A., Sites, J. E., Boyd, G., & Lacombe, A. (2016). Dielectric barrier discharge atmospheric cold plasma inhibits *Escherichia coli* O157, Salmonella, *Listeria monocytogenes*, and tulane virus in romaine lettuce. *International Journal of Food Microbiology, 237*, 114–20.
38. Misra, N. N., Koubaa, M., Roohinejad, S., Juliano, P., & Alpas, H. (2017). Landmarks in the historical development of twenty first century food processing technologies. *Food Research International, 97*, 318–339.
39. Misra, N. N., Schlüter, O., & Cullen, P. J. (2016). Cold plasma in food and agriculture: Fundamentals and applications. Edition: 1, ISBN:9780128014899, pp. 1–380, https://doi.org/10.1016/C2014-0-00009-3. Academic Press.
40. Moisan, M., Barbeau, J., Crevier, M. C., Pelletier, J., Philip, N., & Saoudi, B. (2002). Plasma sterilization: Methods and mechanisms. *Pure and Applied Chemistry, 74*(3), 349–358.
41. Moiseev, T., Misra, N. N., Patil, S., Cullen, P. J., & Bourke, P. (2014). Post-discharge gas composition of a large-gap DBD in humid air by UV-Vis absorption spectroscopy. *Plasma Sources Science and Technology, 23*(6), e065033. https://doi.org/10.1088/0963-0252/23/6/065033.
42. Muhammad, A. I., Li, Y., Liao, X., Liu, D., & Ye, X. (2019). Effect of dielectric barrier discharge plasma on background microflora and physicochemical properties of tiger nut milk. *Food Control, 96*, 119–127.
43. Muhammad, A. I., Xiang, Q., Liao, X., Liu, D., & Ding, T. (2018). Understanding the impact of nonthermal plasma on food constituents and microstructure—A review. *Food and Bioprocess Technology, 11*(3), 463–486.
44. Niemira, B. A. (2012). Cold plasma decontamination of foods. *Annual Review of Food Science and Technology, 3*(1), 125–142.
45. Niemira, B. A. (2012). Cold plasma reduction of Salmonella and *Escherichia coli* O157 on almonds using ambient pressure gases. *Journal of Food Science, 77*(3), M171–M175.
46. Niemira, B. A., & Gutsol, A. (2011). Nonthermal plasma as a novel food processing technology. In *Nonthermal Process. Technol. Food* (Vol. 10, Issue 2). https://doi.org/10.1002/9780470958360.ch20.
47. Ojha, S., Frohling, A., Durek, J., Ehlbeck, J., & Tiwari, B. K. (2020). Principles and application of cold plasma in food processing. *Innovative Food Processing Technologies: A Comprehensive Review, 1857*, 519–540.
48. Ortoneda, M., O'Keeffe, S., Cullen, J. D., Al-Shamma'a, A. I., & Phipps, D. A. (2008). Experimental investigations of microwave plasma UV lamp for food applications. *Journal of Microwave Power and Electromagnetic Energy, 42*(4), 13–23.
49. Pandithas, I., Brown, K., Al-Shamma'a, A. I., Lucas, J., & Lowke, J. J. (2003). Biological applications of a low-pressure microwave plasma UV lamp. *Proceedings of the IEEE International Conference on Plasma Science, 2,* 1112–1115. ISBN 0-7803-7915-2.
50. Pankaj, S. K., Bueno-Ferrer, C., Misra, N. N., Milosavljević, V., & O'Donnell, C. P. (2014). Applications of cold plasma technology in food packaging. *Trends in Food Science & Technology, 35*(1), 5–17.
51. Pankaj, S. K., Bueno-Ferrer, C., Misra, N. N., O'Neill, L., Bourke, P., & Cullen, P. J. (2017). Effects of cold plasma on surface, thermal and antimicrobial release properties of chitosan film. *Journal of Renewable Materials, 5*(1), 14–20.

52. Pankaj, S. K., Misra, N. N., & Cullen, P. J. (2013). Kinetics of tomato peroxidase inactivation by atmospheric pressure cold plasma based on dielectric barrier discharge. *Innovative Food Science & Emerging Technologies, 19*, 153–157.
53. Pedrow, P., Hua, Z., Xie, S., & Zhu, M. J. (2019). Engineering principles of cold plasma. In *Advances in Cold Plasma Applications for Food Safety and Preservation*, pp. 3–48, Elsevier; https://doi.org/10.1016/B978-0-12-814921-8.00001-3.
54. Pedrow, P., Hua, Z., Xie, S., & Zhu, M. J. (2019). Engineering principles of cold plasma. *Advances in Cold Plasma Applications for Food Safety and Preservation* (pp. 3–48). ISBN:978-0-12-814921-8, Publisher: Academic Press. https://doi.org/10.1016/B978-0-12-814921-8.00001-3.
55. Perni, S., Shama, G., & Kong, M. G. (2008). Cold atmospheric plasma disinfection of cut fruit surfaces contaminated with migrating microorganisms. *Journal of Food Protection, 71*(8), 1619–1625.
56. Pinela, J., & Ferreira, I. C. F. R. (2017). Nonthermal physical technologies to decontaminate and extend the shelf-life of fruits and vegetables: Trends aiming at quality and safety. *Critical Reviews in Food Science and Nutrition, 57*(10), 2095–2111.
57. Priya Arjunan, K., & Morss Clyne, A. (2011). Hydroxyl radical and hydrogen peroxide are primarily responsible for dielectric barrier discharge plasma-induced angiogenesis. *Plasma Processes and Polymers, 8*(12), 1154–1164.
58. Puač, N., Gherardi, M., & Shiratani, M. (2018). Plasma agriculture: A rapidly emerging field. *Plasma Processes and Polymers, 15*(2), e1700174. https://doi.org/10.1002/ppap.201700174.
59. Purevdorj, D., Igura, N., Hayakawa, I., & Ariyada, O. (2002). Inactivation of *Escherichia coli* by microwave-induced low-temperature argon plasma treatments. *Journal of Food Engineering, 53*(4), 341–346.
60. Raviteja, T., Dayam, S. K., & Yashwanth, J. (2019). A study on cold plasma for food preservation. *Journal of Scientific Research and Reports, 23*(4), 1–14.
61. Reece Roth, J., Sherman, D. M., Gadri, R. Ben, Karakaya, F., & Chen, Z. (2000). A remote exposure reactor (RER) for plasma processing and sterilization by plasma active species at one atmosphere. *IEEE Transactions on Plasma Science, 28*(1), 56–63.
62. Rød, S. K., Hansen, F., Leipold, F., & Knøchel, S. (2012). Cold atmospheric pressure plasma treatment of ready-to-eat meat: Inactivation of *Listeria innocua* and changes in product quality. *Food Microbiology, 30*(1), 233–238.
63. Romani, V. P., Olsen, B., Pinto Collares, M., Meireles Oliveira, J. R., Prentice-Hernández, C., & Guimarães Martins, V. (2019). Improvement of fish protein films properties for food packaging through glow discharge plasma application. *Food Hydrocolloids, 87*, 970–976.
64. Rycroft, M. J. (1993). Gas discharge physics. *Journal of Atmospheric and Terrestrial Physics, 55*(10), e90114. https://doi.org/10.1016/0021-9169(93)90114-E.
65. Sakudo, A., Misawa, T., & Yagyu, Y. (2019). Equipment design for cold plasma disinfection of food products. *Advances in Cold Plasma Applications for Food Safety and Preservation, 10*, 289–307. https://doi.org/10.1016/B978-0-12-814921-8.00010-4.
66. Sakudo, A., Shimizu, N., Imanishi, Y., & Ikuta, K. (2013). N2 gas plasma inactivates influenza virus by inducing changes in viral surface morphology, protein, and genomic RNA. *Biomedical Research International, 2013*, e694269. https://doi.org/10.1155/2013/694269.

67. Sakudo, A., Yagyu, Y., & Onodera, T. (2019). Disinfection and sterilization using plasma technology: Fundamentals and future perspectives for biological applications. *International Journal of Molecular Sciences, 20*(20), e05216. https://doi.org/10.3390/ijms20205216.
68. Schnabel, U., Handorf, O., Winter, H., Weihe, T., & Weit, C. (2021). The effect of plasma-treated water unit processes on the food quality characteristics of fresh-cut endive. *Frontiers in Nutrition, 7*, e627483. https://doi.org/10.3389/fnut.2020.627483.
69. Schütze, A., Jeong, J. Y., Babayan, S. E., Park, J., Selwyn, G. S., & Hicks, R. F. (1998). The atmospheric-pressure plasma jet: A review and comparison to other plasma sources. *IEEE Transactions on Plasma Science, 26*(6), 1685–1694.
70. Sharma, A., Pruden, A., Neisan, N., Yu, Z., & Collins, G. J. (2004). Bacterial deactivation in open-air by the afterglow plume emitted from a grounded hollow slot electrode. *IEEE International Conference on Plasma Science, 39*(1), e1339615. https://doi.org/10.1109/plasma.2004.1339615.
71. Soleiman Abbasi, & Harouni, A. (2017). Designing a microwave-assisted low pressure cold plasma (LPCP) generator: A case study on Salmonella decontamination. *Journal of Food Science and Technology, 2*(2), 197–204.
72. Teschke, M., Kedzierski, J., Finantu-Dinu, E. G., Korzec, D., & Engemann, J. (2005). High-speed photographs of a dielectric barrier atmospheric pressure plasma jet. *IEEE Transactions on Plasma Science, 33*(2), 310–311.
73. Tran, N., Amidi, M., & Sanguansri, P. (2008). Cool plasma for large scale chemical-free microbial inactivation of surfaces. *Food Australia, 60*(8), 345–347.
74. Vaideki, K. (2016). Plasma technology for antimicrobial textiles. In *Antimicrobial Textiles*. Elsevier. https://doi.org/10.1016/B978-0-08-100576-7.00005-5.
75. Wei, S., Chelliah, R., Oh, D., & Liu, S. (2022). Applications of cold plasma in food safety. Springer. https://doi.org/10.1007/978-981-16-1827-7.
76. Yong, H. I., Kim, H. J., Park, S., Alahakoon, A. U., & Kim, K. (2015). Evaluation of pathogen inactivation on sliced cheese induced by encapsulated atmospheric pressure dielectric barrier discharge plasma. *Food Microbiology, 46*, 46–50.
77. Yong, H. I., Kim, H. J., Park, S., Kim, K., & Choe, W. (2015). Pathogen inactivation and quality changes in sliced cheddar cheese treated using flexible thin-layer dielectric barrier discharge plasma. *Food Research International, 69*, 57–63.

CHAPTER 3

INTERACTION OF REACTIVE SPECIES WITH FOOD MOLECULES

UDAY S. ANNAPURE, T. K. RANJITHA GRACY, and
GUNASEELAN EAZHUMALAIE

ABSTRACT

This chapter describes detailed information about the generation and numerous reaction mechanisms involved during the interaction of reactive species with food components such as carbohydrates, proteins, lipids, and other functional food components. The reactive species are dynamic and unstable molecules produced from various sources and processes, which tend to affect the functional and nutritional attributes of the food upon interaction. Cold plasma is a novel budding technique widely applied in the field of food processing for various applications including microbial decontamination, ingredient modification, pesticide degradation, enzyme inactivation, allergen and toxin reduction. The fundamental principle of cold plasma technology lies in the generation of reactive species which propagate and produce different kinds of energetic species that cause the targeted effect on food molecules upon interaction. Furthermore, these reactive species induce multiple reactions involving the abstraction and subtraction of radicals from the structure, producing different new products and crosslinks related to the food component substrate. These mechanisms are significant in understanding the modification and tailoring of food ingredients according to the application requirements. The chapter also emphasizes the undesirable effects of cold plasma on different food components, which are essential to understand the limitations and challenges associated with this emerging technology.

Nonthermal Plasma Technology for Food and Food Products.
R. Mahendran, Preeti Birwal, and Megh R. Goyal (Eds.)

3.1 INTRODUCTION

Food is an intricate system composed of numerous constituents complexed with one another to provide a unique and specific nutrition and functionality. Based on the proportionate existence and energy contribution, there are major and minor constituents such as carbohydrates, proteins, lipids, vitamins, and minerals, respectively [39]. The chemical structure and interaction of these macro and micro molecules impart the distinct nutritional and quality attributes of every food commodity. Thus, any modification in their structure would alter the characteristics of the food significantly. Understanding this fact, food is processed and preserved to obtain palatability, safety, and convenience through different thermal, non-thermal, chemical, enzymatic, and biological technologies wherein the constituents are operated to yield varied end products, so as the attributes [77]. For instance, in the fundamental food process of cooking, a hydro or oleo thermal operation of food each constituent is transformed, namely, the hydrophilic interactions followed by the breaking of carbohydrate chains, denaturation of proteins, pyrolysis of fatty acids, and oxidation of phenols [49]. These interactions would change according to the composition of the food and the technology involved in the processing.

In the past two decades, the insistence on unprocessed or fresh food products has increased, which introduced safety-ensuring yet quality-retaining technologies like non-thermal processing techniques including gamma irradiation, ozonation, ultra-violet radiation, ultrasonication, and cold plasma technology [76]. Though these technologies comprise different mechanisms of action for preserving and processing the food, the intervention of reactive species renders a substantial effect [51]. This necessitates the comprehension of such reactive species with the constituents of food. Generally, the reactive species produced from various external sources like smoke, alcohol, pollution, and heavy metals were considered to be causative agents for cell toxicity through affecting the balance of antioxidants and the reactive species [117]. But the equivalent reactions would contribute to the cell metabolic pathways due to their high energy and reactivity [32]. In this regard, the consequences of reactive species interaction might either be beneficial or detrimental according to the event of occurrence and the requisites of processing. With the aim that, having an extensive acquaintance with the possible interactions of reactive species would be effective.

The present chapter focuses on discussing the physically generated exogenous reactive species that interact with the individual constituents (such as carbohydrates, proteins, lipids), and the antioxidative elements including phenolics and flavonoids. Previously, the significance of comprehending the interaction of reactive species with each element of the food composition was explained. The existing investigations on the interaction of reactive species produced from the mentioned non-thermal technologies were taken into principal consideration to explain the principal reaction mechanism, implications on the paramount functionalities, and the application prospects.

3.2 CHEMISTRY OF REACTIVE SPECIES

3.2.1 MECHANISM OF FREE RADICAL GENERATION

Reactive species are radical and non-radical species that are energetic, unstable, and capable of reacting instantaneously with the constituents coming in contact with them to attain stability. Reactive species radicals are elements with at least one uncoupled electron in their farthest orbit responsible for their unstable nature and high reactivity (superoxide anion (O_2^{-}), hydroxyl (OH.), peroxyl (RO_2.), hydroperoxyl (OH_2.), nitric oxide (NO.), nitrogen dioxide (NO_2.)), whereas non-radical species are stable and efficient stimulants to induce radical generation (hydrogen peroxide (H_2O_2), oxygen singlet state ($^1O_2$), hypochlorous acid (HOCl), ozone (O_3), nitrous acid (HNO_2), nitroxyl anion (NO^-), peroxynitrite (ONOO–), alkyl peroxynitrite (ONOOH), peroxy nitrous acid (ROONO), nitrosyl cation (NO^+). Both the radical and non-radical reactive species are together named as reactive oxygen and nitrogen species (RONS) based on their composition and as oxidants/pro-oxidants based on their mechanism of interaction [119]. When the free radicals encounter a stable molecule, they abstract or donate electrons to the molecule to attain stability, making the molecule itself an unstable radical. This cascade of reactions incessantly proceeds until two free radicals collide and react with each other to terminate the reaction by forming stable molecules. This sequence of interactions is called initiation (Eqns. (1) & (2)), propagation (Eqns. (3) & (4)), and termination (Eqns. (5) & (6)) reactions of free radicals as follows.

$$Initiator \rightarrow Initiator \quad (1)$$

$$Initiator. + RH \rightarrow R. + Initiator\ H \quad (2)$$

$$R. + X_2 \rightarrow RX_2 \quad (3)$$

$$RX_2. + RH \rightarrow R. + RX_2H \quad (4)$$

$$R. + R. \rightarrow R - R \tag{5}$$

$$X. + X. \rightarrow X_2 \tag{6}$$

where, RH represents the target alkyl compound; R = alkyl radical; X_2 = molecular oxygen; RX_2 = peroxy radical [100].

3.2.2 SOURCES

Generally, the abundant existence of biradical oxygen tends to initiate the free radical reaction through endogenous as well exogenous sources. Endogenous sources comprise the intracellular components such as mitochondria, endoplasmic reticulum, and peroxisomes involved in the metabolic reactions [98]. Whilst, the exogenous sources are of prime interest as they induce the accumulation of free radicals to yield desirable as well as undesirable metabolites depending on the target molecules. Exogenous or external factors causing the reactive species generation are heavy and transition metals, pesticides, tobacco smoke, alcohol, pesticides, polluted air, contaminated water, and radiations [106]. Apart from the negative impacts of reactive species via oxidative and nitrosative stress, these implications could be positively employed in varying aspects of food applications. To illustrate, the oxidative stress phenomenon is beneficial in food as the instances of microbial destructions enhances the storage stability. Similarly, the oxidative degradations of chemical contaminants, pollutants, and pesticides by reactive species aid in improving the chemical safety of the food. Concerning this, several physical technologies composing the reactive species generation were introduced as an alternative to thermal processing technologies of food to preserve the thermolabile nutrients and the quality attributes associated with them. Such technologies list as gamma irradiation, ozonation, ultraviolet radiation, ultrasonication, and cold plasma.

Gamma irradiation is a process of subjecting the food to the exposure of high energy γ rays which would energize the air molecules in and around the product and dissociate them to form free radicals and reactive species [99]. Whereas in the ozonation process, the food matrix is directly or indirectly exposed to one of the reactive oxygen species, ozone (O_3) which would persuade primary and secondary oxidative species with the treated product [26].

Ultra-violet radiation (UV) is also similar to the gamma irradiation process where the ionizing UV rays provoke photocatalytic production of reactive species. The penetrability of ultraviolet radiation over food is considerably lesser among the listed technologies, yet the impact is majorly contributed by the secondarily produced reactive species on collision with the primary reactive species excited by the UV radiation [27].

Differing from the above technologies, ultrasonication involves the application of elated frequency (> 20 kHz) sound waves mediated through a homogeneous solvent usually water. The high-frequency sound waves give rise to the formation of bubbles that collapse with belligerent energy release which causes the water molecules in the air-water interface of the bubble to excite and bring in hydroxyl radicals. The produced radicals from cavitation subsequently form furthermore reactive species [107]. These emerging technologies play a vital role in the recent food processing operations due to their green, non-thermal, and efficient processing capabilities. Thus, it becomes important to look into the mechanism of radical's reaction with the constituents of food.

3.2.3 REACTIVE SPECIES PRODUCED IN COLD PLASMA

Cold plasma is considered the fourth physical state of matter as it is produced by energizing the third physical state of matter, i.e., the gas. The energy imparted by the plasma reactor through either microwave, radiofrequency or electric discharges disassociates the gas molecules present in the reactor space to yield the radical and non-radical species along with free electrons and atoms [101]. These dynamic plasma species in reaction with the moisture present in the atmosphere and the food, further instigate certain stable non-radical acidic species. The chemistry of the reactive species generated will be as follows (Eqns. (7)–(16) [40, 136]:

$$e^{\cdot} + O_2 \rightarrow O^{\cdot} + O^{\cdot} + e \tag{7}$$

$$O^{\cdot} + O_2 \rightarrow O_3 \tag{8}$$

$$O^{\cdot} + H_2O \rightarrow OH^{\cdot} + OH^{\cdot} \tag{9}$$

$$e^{\cdot} + O_2 \rightarrow \mathrm{O_2 .\text{-}+e} \tag{10}$$

$$O_2^{\cdot} + O_2^{\cdot} + 2\ OH^{\cdot} \rightarrow H_2O_2 + 2O_2 \tag{11}$$

$$H_2O_2 + O_2^{\cdot} \rightarrow O_2 + 2OH^{\cdot} \tag{12}$$

$$e^{\cdot} + O_2 + N_2 \rightarrow 2NO^{\cdot} + e \tag{13}$$

$$O^{\cdot} + NO^{\cdot} \rightarrow NO_2 \tag{14}$$

$$NO_2 + O^{\cdot} \rightarrow ONOO^{\cdot} \tag{15}$$

$$O_2 + N_2 \rightarrow N_2O + O \tag{16}$$

Distinct plasma process parameters including the operating gas, operating pressure, discharge power, reactor configuration, moisture content, the humidity of the atmosphere, and the exposure or interaction time significantly influence the reactive species profile, intensity, stability, and their energy to dissociate the bonds in the targeted molecules. Particularly the various operating gases used in the plasma generation listing atmospheric air, O_2, N2, Ar and He are providing varied reactive species profiles and intensity. The existing studies unveiled that the spectroscopic peak intensity of reactive species is higher among the operating gases in the following order [21]:

air > (air + oxygen) > oxygen > nitrogen > (argon + oxygen) > (helium + oxygen) > argon > helium.

The existence of O_2 and N_2 in the atmospheric air makes the formation of reactive species more facile than compared to that of the noble gases, however, to incur controlled reactions, specific noble gases are used over the air [138] for plasma generation. Additionally, the moisture of the food system and the humidity of the reaction atmosphere remarkedly effect the reactive species concentration as the water molecules promotes hydroxyl radicals and hydrogen peroxide production [94]. The operating pressure and the reactor configuration determine the energy required to overcome the critical electric potential to excite and disassociate the stable molecules [130].

The discharge power is linked to the concentration and intensity of reactive species i.e., the higher discharge power naturally intensifies the reactive species production [105]. Over and above all the process variables, the exposure time can change the mechanism the reactive species interaction on the food constituents since the reactive species are exceptionally transient and their limited half-life manifests discrete dependence on the treatment time [16]. The involvement of numerous process and product variables regulating the reactive species characteristics demonstrate the feasibility of the cold plasma processing to be tailored as per the required modifications.

3.3 INTERACTION OF REACTIVE SPECIES WITH CARBOHYDRATES

3.3.1 POLYSACCHARIDES

The most abundantly occurring storage polysaccharide in nature is starch. Being the primary source of energy, this macronutrient plays a vital role in the human diet globally. Various classes of food are rich in starch viz, cereals (corn, rice, wheat, maize), tubers (potato, cassava, arrowroot, sweet potato, yam), millets (proso millet, kodo millet, pearl millet), certain pulses (pea, mung bean, and non-conventionally used fruit seeds (mango seed kernel, litchi seed, avocado seed, jackfruit seed) and serve as the nutritional and functional starch reserves. Processing of these foods in the attempt to decontaminate, disinfect, add value, and enhance palatability through numerous thermal (drying, blanching, frying), non-thermal (irradiation, cold plasma, UV exposure, ozonation), and biological methods (fermentation) generate the possibility of interaction with the reactive species. Apart from this, employing starch as a functional ingredient in food, textile, pharmaceutical, paper, and biopolymers industries as a thickener, emulsifier, fat replacer, binder, encapsulant, and plasticizer requires modification of starch to provide desirable hydration, rheological and structural characteristics.

The chemical modification methods include esterification, etherification, oxidation, grafting, and cross-linking are widely practiced due to their high feasibility. However, the concern for environmental health and safety demands green technologies with the null discharge of chemical residues. Thus recently, a plethora of studies investigate the potential of physical techniques such as irradiation, cold plasma, ozonation, ultrasound, and pulsed electric field. These techniques involve intense energy for the process of engendering reactive species which would be responsible for the modification. Hence. it is necessitated to investigate the effects of reactive species on starch to comprehend and control the reaction as required for the application.

Chemically, starch is a homopolymer of glucose units constructed by two distinct structures *viz.*, the linear amylose with α-1,4 linkages and the branched amylopectin linked with α-1,6 linkages [91]. This polymer of glycosidic bonds is capable to undergo modification easily due to the presence of numerous hydroxyls in the structure. The high reactivity of radical and non-radical reactive species tends to incur three different mechanisms of interaction namely, depolymerization/ degradation, crosslinking/ aggregation, and oxidation.

3.3.1.1 DEPOLYMERIZATION

When the energetic and unstable reactive species encounter a starch chain, they result in the random cleavage of the covalent bonds (C–O–C) between the glycosides, resulting in a reduced chain length. This would render a decrease in the number of monomers in the cleaved starch giving to a significant decrement in the degree of polymerization. Such a phenomenon is called 'depolymerization' (Figure 3.1). Depolymerization of starch would facilitate the formation of starch fragments of lesser saccharides units including maltose, maltotriose, and several oligomers of low molecular weight [53]. The C–O–C linkage at the C 1,6 position of the glycoside is predominantly susceptible to dissociation compared to the other linkages [29]. Thus, the branched amylopectin side chains are prone to be detached from the polymeric structure, leaving out the amylose freely. This further contributes to the rise in hydroxyl content which would enhance the solubility of starch by promoting hydrophilic affinity at relatively lesser temperatures [50].

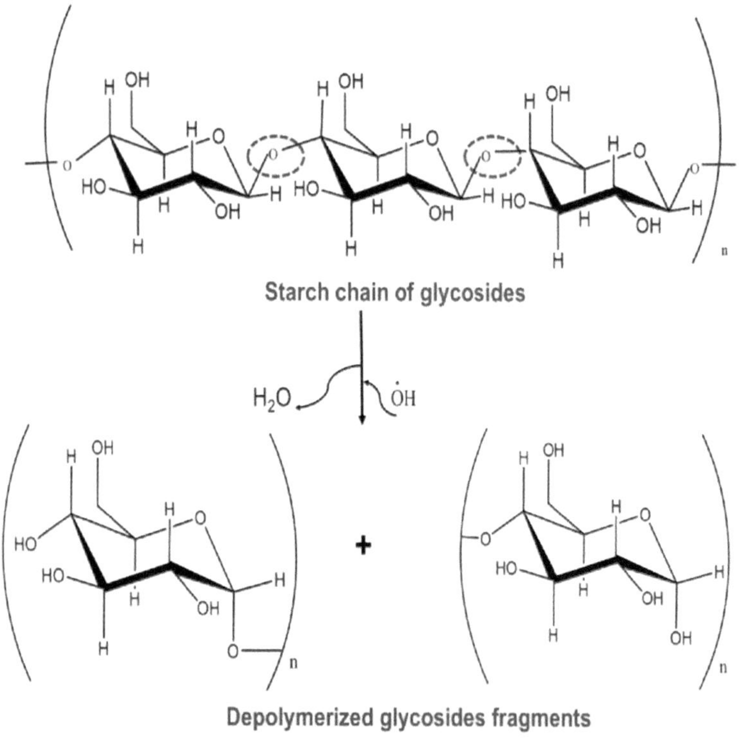

FIGURE 3.1 Depolymerization of starch.

The reduced degree of polymerization declines the rheological strength of the starch on swelling due to the weak structure represented by the reduced viscosity values at higher concentrations. This is reasoned out of the incapability of fragmented starch to form entanglements with the neighboring remnants. Several physical technologies involving reactive species such as ultrasound [140], gamma irradiation [2, 15], cold plasma [50], UV [62], and ozonation [66] demonstrated the occurrence of depolymerization in starch irrespective of the variations in sources. Additionally, depolymerization alters the morphology and crystallinity of starch which would influence the functional, thermal, hydration, and gelation characteristics.

The structurally deformed starch was observed with higher polar functional groups including hydroxyls that were exposed while uncoiling the polymer providing improved hydration properties such as water solubility, water absorption, and swelling power. The increased hydro-affinity induces the formation of clear gels and dispersions regardless of the weak structure of depolymerized starch with reduced gel strength [35]. Besides, nutritionally the resistant starch content of the degraded starch was observed to be increased as the cleavage of amylopectin side chains tends to cause amylose-amylopectin; amylopectin-amylopectin, or amylose-amylose interactions to produce aggregated structures difficult to be hydrolyzed by the digestive enzymes [121]. Also, this would aid in the increased yield of starch fermentative products with less intervention of enzymes in a relatively shorter time making the process cost and time-efficient [18].

3.3.1.2 CROSS-LINKING

Another possible reaction leads to cross-linking of the polymer. The reactive species like unstable OH– radicals detach a hydrogen from the hydroxyl at the C–2 position (C–OH) and condense the adjacent chains to crosslink via C–O–C linkages with the formation of a water molecule (Eqns. (17) [127].

$$Starch + 2OH^{\cdot} + Starch \rightarrow Starch - O - Starch + H_2O \tag{17}$$

Compared to the primary (C–6) and secondary (C2, C3) hydroxyls at the glycosidic units, the C2 position is found to be more susceptible to cross-linking [150]. The molecular weight of the cross-linked starch increases notably corresponding to the native starch demonstrating the polymer strength. These intra-cross-linked starches swell and form a compact dense structure on gelation which is associated with the increased viscosity and gel strength. The viscosity of crosslinked starches deviates among themselves based on the level of cross-linking through the phenomenon of swelling. Highly cross-linked starches retard the intervention of hydro groups and thereby object to the swelling process yielding low hot paste viscosity on gelation while the lesser cross-linked starches enlarge and form a dense structure providing higher viscous gels compared to the native starches [114]. Besides, the cross-linked starch gels exhibit a thermostable behavior preventing the viscosity loss during excessive heating as the additional glycosidic bonds unveil resistance towards thermal disruption [25]. Similarly, the enzymatic hydrolysis is also retarded remarkedly in the cross-linked starches [128]. The enhanced stability, strength, and resistance to hydration of the cross-linked starches provide the possibility of using these in bio-degradable packaging films [78].

The functional properties rely on the level of cross-linking which can be tailored by controlling the process variables. In cold plasma, the medium for plasma generation, discharge power, exposure time, the pressure of the system, and the presence of a cross-linking agent would alter the mechanism of modification. When the intensity of the process in terms of the concentration and energy distribution of reactive species and the duration of their interaction is scarce to outdo the bond disassociation energy of C–O–C linkages, the occurrence of cross-linking prevails [115]. The duration of interaction plays a prominent role as observed in the existing studies, the shorter exposure to reactive species results in the cross-linking of polymers whereas the extended reactions facilitate disruption or depolymerization of the structure [45, 83]. Utilizing this efficient cross-linking of starches with reactive species interaction, reactive species assisted grafting of the starch with auxiliary components like ethylene [149], citric acid [55], dichloro-dimethyl-silane [69], etc.

3.3.1.3 OXIDATION

Oxidation of starch is another prominent reaction anticipated on interaction with the reactive species, as the innate nature of the reactive species is to produce oxidative stress to the subjected system. Hydroxyl radicals attack the hydrogen atom involved in the methyl end of the glycoside ($-CH_2OH$) at the C–6 position and condense to form a water molecule (Eqn. (18)). In the subsequent reaction successive encountering with OH– radicals oxidize the –CHO further to form –COOH functionalities which in turn increase the carboxyl content of the starch (Eqns. (19)–(20)).

$$R\text{-}CH_2OH\ +.OH \rightarrow R\text{-}CHOH.\ +H_2O \quad (18)$$

$$R\text{-}CHOH\ +.OH \rightarrow R\text{-}CHO.\ +H_2O \quad (19)$$

$$R\text{-}CHO.+2OH \rightarrow R\text{-}COOH.\ +H_2O \quad (20)$$

Since the carboxyl and carbonyl functionalities are introduced as a result of oxidation, the surge in this content defines the degree of oxidation of the starch. It is because of this reason, that oxidized starch (OS) dispersions with the cold plasma treatment were found with the low pH values corresponding to the carboxylic groups' addition [8]. On further proceeding of oxidation, the reactive species tend to depolymerize the glycosidic linkage leaving behind glucose monomers and oligomers which would also be oxidized to yield formic acid and erithronic acid. respectively [92]. However, these reactions are less anticipated as the bond dissociation energy of glycosidic bonds involved in the starch biopolymer is higher to be degraded with limited reactive species interactions as observed in the novel non-thermal technologies like irradiation, ozonation, ultra-violet, and cold plasma. Particularly, in a dense solid food matrix, the restricted penetration power of reactive specifies and their superficial reaction constrain the degradation of the polymer [141].

Looking into the functionality of oxidized starches, as the introduced carboxyl and carbonyl functionalities are inherently hydrophilic, the ability to absorb, retain and solubilize in water increases in the oxidized starches [135], which is useful to develop cold-water-soluble starches. This would also facilitate better freeze-thaw stability preventing the syneresis of water on thawing, making it suitable for frozen food applications [82]. In addition, the carboxyl functionalities prevent the functioning of amylose and amylopectin chains on retrogradation via electrostatic repulsive forces [64]. This is responsible for the high light transmittance of the OS gels which is notably desirable from the consumer perspective as it affects the product appeal. It is because of the same fact of electrostatic repulsion, that the oxidized starches exhibit lower gel viscosity and gel strength even at higher concentrations [4]. However, the oxidized starches were investigated to form better complexes due to their improved binding ability which provides the potential application as binders and fortification carriers [56].

Furthermore, the mechanical strength of the films made with OS [118] and the tensile strength of the fabric containing OS were superior to the native starches [81]. Owing to these attributes of OS, conventional chemical-induced oxidation by sodium hypochlorite [148], or potassium permanganate [43] was employed. To overcome the ecological constraint of their inorganic residues, hydrogen peroxide-induced starch oxidation has emerged [147]. The discussed physical technologies (irradiation, ozonation, ultra-violet, and cold plasma) engender an amalgamation of different reactive species as per the medium of reaction including hydrogen peroxide, these can be an efficient green substitute for the chemical oxidative methods [87].

Another important class of polysaccharides includes functional and structural polysaccharides. Functional polysaccharides are hydrocolloids that would form a three-dimensional gel network on hydration via absorbing and trapping the water molecules such as xanthan gum, gum Arabic, Konjac glucomannan, and some other novel gums [137]. Due to their ability to yield high viscous gels with minimum concentrations, they are used as a functional ingredient in strengthening the gels, films, stabilizers, and also as fat replacers [10]. Whereas the structural polysaccharides comprise pectin, cellulose, and chitin which are the structural part of the fruits, vegetables, and crustaceans, respectively [126]. Owing to their characteristics these structural polysaccharides as well isolated and utilized as functional ingredients. Besides, these non-starch polysaccharides are feasible and cost-effective sources of soluble dietary fiber widely employed in fiber-fortified products [36]. Exposing these polysaccharides alone as an isolate or in a compound with the food or food product to reactive species would incur the discussed interactions such as depolymerization, cross-linking, and oxidation. Nevertheless, the end product of the interaction would vary as per the monomeric profile of the polymer. To illustrate, depolymerization of a polymeric glucomannan might increase free glucose and mannose content and the same is the case with another polysaccharide [65].

The reduced gel strength of depolymerized polysaccharides would cause a reduction in the hardness of vegetables and fruits as reported in the existing investigations [105]. Non-thermal technologies like cold plasma processing were preferred to microbially and chemically decontaminate the fruits and vegetables with less impact on the nutrients and organoleptic properties compared to the thermal treatments [51]. However, the modification of the commodity by the surface-acting reactive species alters the structural polysaccharides which are present in the exocarp region thereby affecting the texture. Alternatively, the cross-linking mechanism of the polysaccharides along with enzyme inactivation might aid in preserving the texture of the fruits and vegetables from softening [86].

An increase in the complexing ability of the oxidized polysaccharides may also facilitate the retention of bioactive components of the food and their stability would assist in the anchored availability for absorption [133].

It was also found that the reactive species interacted Gum Arabic exhibited enhanced interfacial properties yielding better emulsification therefore can be used as fat reducers and replacers in fat-based food emulsions [3]. The overall mechanism of reactive species interaction with the polysaccharides, their effect on the physicochemical, functional and rheological properties and their applications are summarized in Table 3.1.

3.4 SIMPLE CARBOHYDRATES

Simple carbohydrates comprise the mono-, di- and oligosaccharides playing important role in providing the sweetness of the food, particularly in fruits and honey. Several investigations demonstrated the latent ability of plasma for microbial decontamination, pesticide degradation, and enzyme inactivation of the whole, minimally processed, and other fruit-based products like juices [73]. This shows the importance of understanding the mechanism of plasma-induced reactive species with the mono, di- and oligosaccharides which tend to influence the organoleptic properties of the products. Though there is a scarcity of literature on the cold plasma impact on the simple saccharides which are intact in the food systems, the mechanism of interaction with the non-radical reactive species hydrogen peroxide was well established. Hydrogen peroxide tends to produce hydroxyl radicals on further free radical initiation reactions [31].

Oxidation of the monosaccharides including Aldo and keto sugars occurs in reaction with the hydroxyl radicals. The unstable radical [47] attack the carbon at the C–1 position and cleave the aldehyde group from the pyranose structure forming a subsequent monosaccharide with a lower number of carbons and formic acid (Figure 3.2). This might reduce the total glucose content in the processed juices as observed in cashew apple juice after plasma treatment [108]. On further oxidation of low carbon sugars, the formation of hydroxymethyl furfural and furfurals is a critical concern to be investigated and addressed [146]. Nevertheless, the superficial effect of reactive species and the compounded existence of the sugars in the fibrous matrix preclude such negative consequences from occurring.

In matrices where the simple sugars exist in complexation with other constituents such as lipids or ascorbic acid, the oxidation of those constituents is prevented due to the sugars-reactive species interaction [30, 63]. Another aspect of plasma-induced oxidation was employed for the reduction of hydroxymethyl furfural, furfurals, and formic acid in honey [85] and

TABLE 3.1 Reactive Species Interaction with Carbohydrate Polymers: Mechanism, Effect and Applications

Food Components	Mechanism	Physical Properties	Rheological Properties	Chemical Properties	Applications
Polysaccharides	Depolymerization	• Molecular weight ↓ • Solubility ↑ • Clarity of paste ↑ • Syneresis ↑ • Water absorption & swelling ↑ • Crystallinity ↑ • Least gelation concentration ↓ • Foaming capacity ↓ • Emulsion capacity ↑	• Peak viscosity ↓ • Apparent viscosity ↓ • Structuring velocity ↑ • Gel hardness ↓ • Gel strength ↓ • Thermal stability of gels ↓	• Hydroxyl groups ↑ • Free monosaccharides content ↑	• Soluble starch formulations • Fat replacers • Substrate for enzymatic and fermentative hydrolysis • Binders in meat analogs • Emulsifiers
	Crosslinking	• Molecular weight ↑ • Solubility ↓ • Clarity of paste ↓ • Syneresis ↓ • Water absorption & swelling ↓ • Crystallinity ↓ • Least gelation concentration ↑	• Peak viscosity ↑ • Apparent viscosity ↑ • Gel hardness ↑ • Gel strength ↑ • Thermal stability of gels ↓↑	• Hydroxyl groups ↓ • Free monosaccharides content ↓	• Thickeners • Stabilizers • Drug delivery carriers • Frozen products
	Oxidation	• Molecular weight ↓ • Solubility ↑ • Clarity of paste ↑ • Syneresis ↓ • Water absorption & swelling ↑ • Crystallinity ↑ • Emulsion capacity ↑	• Peak viscosity ↓ • Apparent viscosity ↓ • Gel hardness ↓ • Gel strength ↓ • Thermo plasticity of gels ↑	• Carboxyl and carbonyl groups ↑ • Free monosaccharides content ↑ (in excessive oxidation) • pH ↓	• Thermoplastic starch formulations • Biodegradable polymeric films • Dairy analogs • Binders in textile and paper industry

bagasse hydrolysate from sugarcane [67]. This is evident that plasma-induced reactive species oxidation is highly customizable as per the requirement by controlling the process variables involved in the treatment.

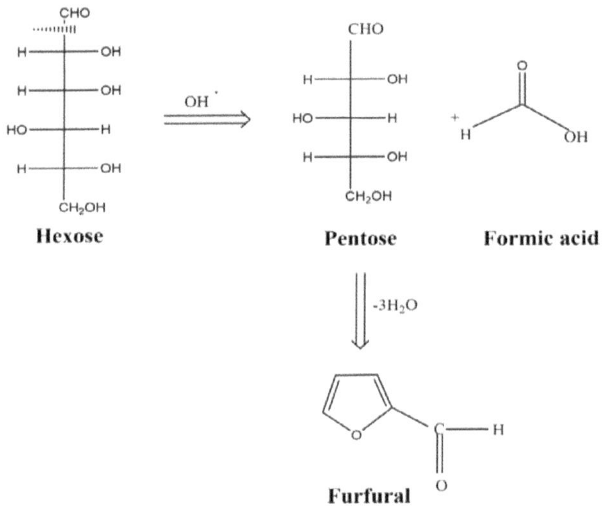

FIGURE 3.2 Oxidation of monosaccharides.

Source: Isbell, Frush, & Martin [47]. Need reprint permission.

In the case of oligosaccharides, the reactive species incur hydrolysis of the saccharides chain and engender products with a low degree of polymerization. Oligosaccharides depolymerization would also yield disaccharides like sucrose, lactose, and maltose depending on the structure of oligosaccharides [23] thereby increasing the disaccharide content of the product while decreasing the oligosaccharide content [6]. The degradation of the oligosaccharides into soluble di- and monosaccharides might also cause the rise in total soluble solids of the products containing oligosaccharides after they interacted with the reactive species [5, 28].

3.5 INTERACTION OF PLASMA REACTIVE SPECIES WITH PROTEINS

Protein is one of the major constituents of plant and animal food matrices contributing to structural development. Apart from the nutritional benefits, it also provides several functional properties including gelation, emulsification, foaming, and structuring [103]. It also binds with other food constituents such as carbohydrates, lipids, polyphenols, and vitamins via different hydrophilic and hydrophobic interactions. The functional properties of a protein are influenced by the cumulative effect of chemical characteristics of amino acids with specific functional groups and its structural conformation. Thus, the alteration or damage caused in the amino acid sequence and protein conformation causes changes in protein functional properties as amino acids are an integral part of the protein structure [33].

In this case, the plasma-produced reactive species and radicals are highly energetic to cause alteration in the structural conformation of protein molecules via various reaction mechanisms [95]. These alterations affect the peptide bonds as well as the structural conformation of the protein. As the amino acids differ in structure and function due to their chemical and side chain groups, there are multiple mechanisms involved during the protein-radical interaction in the plasma treatment [136]. Among several plasma-activated reaction mechanisms, the oxidation of protein and its amino acids is most commonly caused by the reactive oxygen species (ROS) and hydroxyl radicals [97]. Hydroxyl radical, the highly produced plasma reactive species, reacts with the polypeptide chain to cause either polypeptide chain cleavage or polypeptide chain cross-linking. The unstable radical initiates the reaction by abstracting a hydrogen atom of the alpha-carbon from one of the peptide units of the protein-polymer, producing a peptide radical which may proceed to cross-link with other stable polypeptide chains or it might get oxidized further to yield alkoxy peptide radicals [145]. The alkoxy radical [120] would terminate the reaction by causing a cleavage in the polypeptide chain either via di-amide or α-amidation pathways, yielding di-amide & isocyanate derivatives and amide & α-keto acyl derivatives, respectively (Figure 3.3). Because these reactions tend to increase the peptide fragments with carbonyl functionalities such as aldehydes and ketones in the peptide unit, the carbonyl content is discerned as the protein oxidative marker.

The process of increasing the carbonyl content in the peptide fragments through oxidative degradation is called protein carbonylation. Carbonylation of proteins can also occur through other reactions such as glycation and glycoxidation in the presence of simple carbohydrates, which may also be expected in the case of food matrices like sprouts rich in protein and available sugars [71]. Another important oxidative reaction is sulfoxidation, in which sulfur-containing amino acids such as cysteine and methionine undergo oxidation by reactive species to produce cysteine and methionine sulfoxide through disulfide bond formation [109]. This would alongside reduce the sulfhydryl (S–H) content of the amino acid with the increase in di-sulfide (S–S) content. Thus, the changes in these functionalities (S–H; S–S) would also be employed as the oxidative markers to estimate the level of protein oxidation [113]. Whereas, the oxidation of phenylalanine and tyrosine engenders the cross-linking of their radicals to form a di-tyrosine compound, which can also indicate the oxidative degradation of protein-polymer [46].

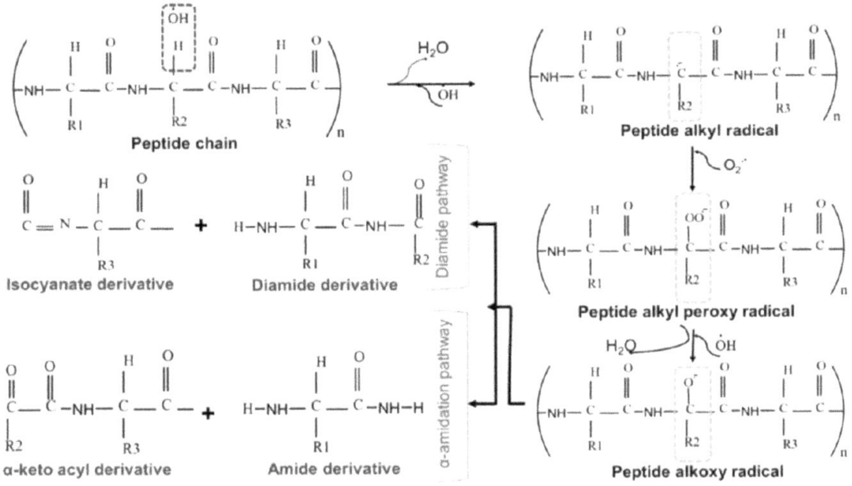

FIGURE 3.3 Oxidative polypeptide cleavage on reactive species interaction.

Source: Adapted from Ref. [120].

All the amino acids are affected by ROS and free radicals. The most vulnerable effects are observed with amino acids containing sulfur (cysteine, methionine), amine (arginine, lysine, and histidine), and side chain groups (phenylalanine, tryptophan, and leucine) [109]. Also, the oxidative products are different depending on the amino acid. For example, lysine gives γ-glutamyl semialdehyde, tryptophan oxidizes to N-formyl kynurenine, and arginine and proline result in α-amino adipyl semialdehyde. In the case of amino acids like arginine, histidine, lysine, proline, threonine, and tryptophan, the oxidative reaction produces carbonyl groups through the carbonylation process. These amino acids are highly prone to oxidation in the presence of oxidized lipids due to the presence of side chains such as sulfoxide, methionine, amino group, and imidazole [139].

The conformational changes can be elucidated by estimating the secondary structures of the protein. The nature of hydrogen bonding between the functional groups and peptides confers these secondary structures. Thus, amino acid oxidation would affect the secondary structures as well. It was reported that the proteins attain a more ordered structure after oxidation with the surge in α-helix content and with the decrement in the unordered random coil structure [72, 75, 111]. Typically, the secondary structures are entangled to construct the tertiary random coiled protein structures. The peptide cleavage induced by the reactive species unveils the embedded secondary structural elements from the unordered structure to bring a more ordered nature to the protein, which yields higher stability to unfolding [19]. However, in certain instances, the aggregation of peptide fragments brings in the rise in unordered structure with the depletion of alpha-helix content [123].

The enzymes, which are the proteinaceous catalysts of multiple desirable and degradative reactions, would also undergo a similar oxidation process as aforementioned. The oxidation of the individual amino acids in the polypeptide chains causes the degradation of C–H and N–H bonds, producing metabolites such as carbon dioxide, water, and nitric oxide. This implicates a loss in the secondary structure of the protein, thereby inactivating or retarding the enzyme activity [41]. Owing to this, cold plasma and other reactive species producing non-thermal technologies are employed for the inactivation of several detrimental enzymes such as polyphenol oxidase [124], peroxidase [88], alkaline phosphatase [112], lipase, lipoxygenase [20], pectin methyl esterase [125], and α-amylase [90].

The oxidation of proteins changes the functional properties, such as solubility, interfacial tension affecting foaming and emulsifying properties, gelation, and stability due to peptide cleavage, peptide chain cross-linking, and structural changes like aggregation or unfolding of proteins [111]. The cleavage and cross-linking of peptide units facilitate unfolding and aggregation of the protein fragments, respectively, which intend to alter the particle size, hydrophobicity, ionic behavior, and thus the solubility [72]. The unfolding of the polypeptide chains via oxidative cleavage facilitates enhanced polar and non-polar adsorption properties. This could efficiently reduce the interfacial tension of the biphasic layers in colloidal systems such as emulsions and foams [144]. The decrease in the size of the protein after reactive species may also assist in the formation of better and stable complexes useful for encapsulating sensitive bioactive compounds [19].

3.6 INTERACTION OF REACTIVE SPECIES WITH LIPIDS

Lipids are one of the important components of food as they provide high calorific value and play a vital function in plant and animal systems. Lipids are mainly triglycerides made up of chains of fatty acids and glycerol. The length of a fatty acid is classified into 3 categories (short (1–6), medium

(7–12), and long (13–22)) based on the number of carbons in the fatty acid chain. The fatty acids could be saturated with no double bond or unsaturated with one or more double bonds present in them. Oxidation is a well-known degrading process of lipids triggered by food processing operations such as cooking, heating, radiation, etc. Catalysts like light, heat, and metals favor and support the oxidation process of food lipids [14]. This results in the generation of off-flavors due to the formation of secondary oxidative metabolites, thus producing a rancid flavor [44].

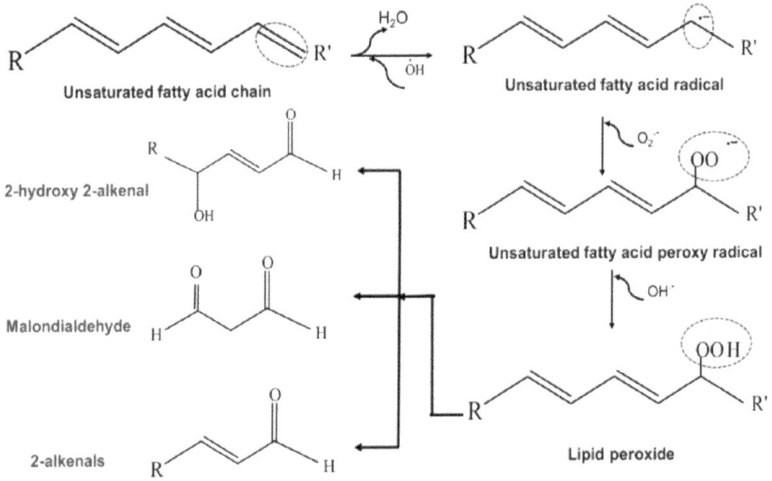

FIGURE 3.4 Oxidative degradation of lipids.

Source: Adapted from Ref. [60].

The rancidity could be either oxidative rancidity or hydrolytic rancidity, where the former is caused by free radicals produced during food processing and storage, and the latter is by lipolytic enzymes (lipase). The lipid oxidation can be detected subjectively through sensory evaluation of rancid off-flavors or through measuring the oxidative by-products with analytical techniques including nuclear magnetic resonance spectroscopy (NMR), Fourier-transform infrared spectroscopy (FTIR), or by chemical methods like peroxide value and Thiobarbituric acid reactive substance assay (TBARS) [11].

Lipid oxidation is a chain of reactions that occurs in three stages: initiation, propagation, and termination. During the initiation step, reactive oxygen species (ROS) attack the lipid moieties (methyl groups) in the unsaturated fatty acid chain, which are attached through double bonds (requiring less energy for hydrogen abstraction), thus producing alkyl free radicals. The oxidation initiation is followed by the propagation step where the alkyl free radical ($R^{\cdot}$) reacts with oxygen (O_2) to produce a lipid peroxide radical ($ROO^{\cdot}$). This further reacts with unsaturated fatty acid (RH) resulting in lipid peroxides and a lipid radical ($R^{\cdot}$), which then react with various oxygen molecules to propagate more reactions, producing several radicals. Finally, during the termination step, the generated reactive species such as $ROO^{\cdot}$, $RO^{\cdot}$, and $R^{\cdot}$ combine among themselves to form non-radical secondary metabolic components (ROOR, ROR, R–R) (Figure 3.4). These secondary metabolites produced from the lipid oxidation process contribute to the unpleasant rancid flavor of food [60, 68]. The oxidation of lipids is a deteriorating process that affects the sensory and nutritional profile of food. It also produces toxic components and free radicals that can cause serious health effects, including cardiovascular diseases. Carcinogenic components like 4-hydroxynonenal, epoxides, and hydrogen peroxide by-products are produced in the lipid oxidation process [24].

In cold plasma interaction with food components, the plasma generated ROS and RNS play a vital role in the oxidation process. The reactive species of various categories produced during the plasma treatment differ based on the atmosphere (argon, oxygen, helium, nitrogen) maintained between the electrodes, which can be altered and controlled in vacuum and jet plasma treatment [52]. In atmospheric cold plasma treatment, the most commonly produced species are ROS and RNS. Especially, ozone and singlet oxygen are the most common sources of lipid deterioration through the oxidation process [136]. Following the discussed mechanism of the reaction, the ROS of cold plasma reacts with the unsaturated fatty acids to produce lipid hydroperoxides ($ROO^{\cdot}$), which are converted further to various secondary products such as ketones, aldehydes, and alcohols through different reactions.

The oxidation process is more probable in foods with more unsaturated fatty acids (linoleic acid, linolenic acid, etc.) [34]. In another case, the ozonide is produced by a reaction of unsaturated fatty acid with ozone, and the ozonide further decomposes and forms various oxidative cyclic components [110]. Other effects such as an increase in aldehyde content, production of short-chain fatty acid by fatty acid cleavage [57], and hydrogenation of fat [142] have also been observed in past studies.

Apart from the degradative reactions, the cold plasma treatment of lipids and oil in the presence of hydrogen reactive species might aid in fractional hydrogenation. The plasma treatment, in which the feed gas is hydrogen or its combination with inert gas, could help in the partial hydrogenation of unsaturated fatty acids. In a study conducted by [42], hydrogen plasma has been successfully used to generate hydrogen reactive species which hydrogenate toluene in a plasma-driven catalytic reaction. The hydrogenation of palm oil and soybean oil was also reported with H_2 as the feed gas [102]. The degree of unsaturation of fatty acids could be understood from a decrement in the iodine value of oils. The iodine value was found to be decreasing while treating soybean oil at a higher voltage; a corresponding observation was made in the plasma-treated soybean oil after 4 hours of treatment [134, 142]. Cold plasma was also reported to generate trans-fatty acids associated with the saturation of elaidic acid at a higher voltage [131].

Though the reactive species of cold plasma are effectively employed for hydrocolloid modification [17], microbial decontamination [28], starch modification [50], flour modification, etc., [48], it has few negative deteriorating effects on lipids causing oxidation. The cold plasma treatment has shown lipid oxidation in various food products such as milk [54], meat, plant seed oil [132], cereals [7], seafood [96], cheese [143], and plant-based milk analogs [80]. Processed food with considerable lipid content always tends to exhibit lipid damage over storage. Since the lipid oxidation process is influenced by parameters like process parameters, time, treatment methods, packaging, and the temperature and humidity of the storage atmosphere [104].

Lipid oxidation is also triggered by the presence of pro-oxidants in the foods [116]. Thus, lipid oxidation is inevitable during the long-term storage of processed foods. To inhibit or slow down the oxidative changes, the addition of a natural antioxidant, and improved storage and packaging conditions can be employed [61]. The cold plasma treatment of high-fat products has a higher chance of developing lipid oxidation over storage, due to the free radical-triggered mechanisms leading to lipid oxidation. Thus, merging other hurdles along with cold plasma such as non-thermal and novel methods along with good storage practices including controlled (less oxygen) and modified (inert gas) packaging or storage could effectively reduce the deteriorating outcomes of cold plasma application on lipids [12, 74].

3.7 INTERACTION OF REACTIVE SPECIES WITH FUNCTIONAL CONSTITUENTS

The functional food components offer functional properties like antioxidant properties, radical scavenging properties, immunity-boosting, vitamin precursors, and anti-cancerous properties along with providing regular nutritional benefits. The functional components present in plants are flavonoids, polyphenols, minerals, vitamins, phytosterols, tannins, etc. [13]. Though the animals are sourced for functional components like ω-3 fatty acids, and bioactive peptides, the potential of fruit and vegetables to provide functional components which benefit human health is massive [84].

The traditional and modern food processing methods including thermal processing and chemical treatments impact the loss or reduction of these valuable food components. Thus, giving way for a surge in exploration studies in non-thermal technology like cold plasma processing for various technological processing benefits over conventional methods. The influence of reactive species on functional molecules is poorly specified owing to various structural and functional differences which lead to multiple reactions and products. It is also because of numerous plasma processing conditions as well as the varied nature of reactive species, that the mechanism of action needs to be revealed to understand its impact.

The possible consequences of reactive species interaction with polyphenols are the increase in carbonyl and carboxylic groups and the addition of hydroxyl groups in phenolic rings of polyphenols [1, 37]. The phenolic degradation (rutin and quercetin) is induced by the interaction of various reactive species produced on cold plasma application [37].

There can be two possible pathways of reactive species interacting with the phenolic compounds, such as the hydroxyl radical-induced and the ozone-induced oxidation. The hydroxyl radicals abstract a hydrogen atom from the C–3 position of quercetin and convert it into an unstable quercetin radical, which would further degrade into low molecular weight phenols through substitution and hydrogen inhibition reactions [38, 70]. In the ozone-induced reactions, the formation of ozonide, which is the cyclic compound attached to ozone, occurs. Subsequently, the produced ozonide gets oxidatively metabolized to carboxylic, carbonyl, and ketone compounds. These reactions result in the reduction of the antioxidative ability of the phenolic compounds.

In the case of plasma interaction with vitamins, the increase in vitamin C was observed in many products, which might be due to cell breaking, size reduction of food matrices, and reaction induced by ROS. Besides, few reports mentioned loss which might be associated with increased treatment time [79]. An important antioxidant in fruits is vitamin C (ascorbic acid) which gets oxidized on interaction with oxidative reactive species. Oxidation of ascorbic acid generates dehydroascorbic acid which would further be transformed into 2,3-diketo-gluonic acid [122]. Due to its susceptibility to oxidation, 2,3-diketo-gluonic acid tends to be hydrolyzed to give away the end products, such as threosone, oxalic acid, and erythrulose [93]. This reaction would result in the reduction of vitamin C content and the antioxidant capacity of the subjected food matrix. However, the superficial reactions of cold plasma less often result in such negative consequences [22, 129].

The effect of cold plasma on bioactive components depends on the power and exposure time of treatment when it comes to the treatment of food products like juices and beverages. It is also based on the sample size, thickness, and treatment mode. Polyphenols are important bioactive components which include isoflavones, flavones, anthocyanins, quercetins, catechins that provide vital functional and physiological benefits. The plasma treatment on fruit and vegetables showed both increases in the extraction yield of polyphenols by breaking the cell wall through surface itching mechanism and also modified functionality and structure of polyphenols.

The effect of reactive species on bioactive components primarily depends on the reaction time and their stability to reactive species. These species might increase the content by releasing flavonoids [89], anthocyanins [59], and ellagic acid [9] by breaking the mesophyll and epidermal cells of the products, though it also showed a decrease in bioactive content in certain treatments [58]. The antioxidant activity of these bioactive components is also affected in both positive and negative manners and concluded that short time exposure was found to be increasing the antioxidant capacity, whereas reduction was observed in prolonged exposure. It is also observed to be difficult to measure the impact of cold plasma on bioactives and functional activities via the combined effect of reactive species, which leads to multiple reaction pathways [79].

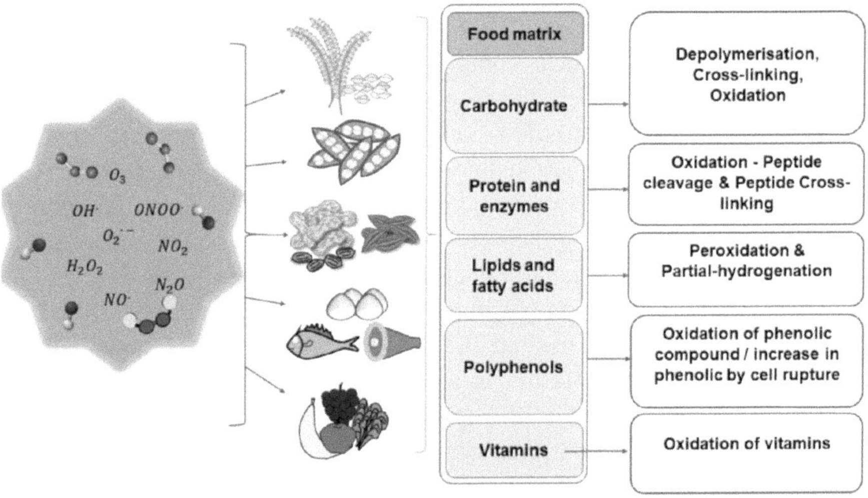

FIGURE 3.5 Summary of reactive species interaction with food constituents.

Reactive species interactions with food constituents are summarized in Figure 3.5.

3.8 SUMMARY

Free radicals are high-energy species that can initiate various reaction pathways in the food components and be the prime reason for all the changes observed in treated food. The free radicals containing non-equilibrium plasma are generated by different kinds of systems where process parameters like feed gas, system configuration, input voltage, and exposure time are constituted in an optimum way to produce a high concentration plasma cloud.

These plasma radicals, when made to interact with food components like carbohydrates, proteins, lipids, and other functional components, trigger and propagate various reaction mechanisms and pathways. These pathways produce different types of end-products based on the substrate of plasma exposure and alter the structural and functional properties of the components as well as the food system. In the case of carbohydrates, the free radical interaction results in cross-linking or degradation of the polymeric chain through oxidation mechanisms. Whereas the protein interaction leads to numerous reaction pathways like oxidation, hydroxylation, nitration, cross-linking, and peptide cleavage. Similarly, the exposure of enzymes might lead to its activity loss as the free radical affects its structural conformations. The plasma radicals initiate lipid oxidation in food containing high-fat or high unsaturated fatty acids. It might also partially hydrogenate the unsaturated fatty acids in the presence of hydrogen radicals. Concerning the plasma treatment of food products containing functional food components, the plasma process parameters like exposure time and their chemical stability influence the impacts. It is also understood that the cold plasma acts in a dual manner with the functional components, either to increase their availability in the food matrix by breaking the cell wall and tissue layers or to degrade them by affecting their chemical structure. Thus, the consequences of plasma-generated reactive species' interaction with the complex food matrices highly depend on plasma generation and exposure parameters to attain the controlled and desired effect on food ingredients' functionality.

Nevertheless, cold plasma is an evolving novel process explored in various food applications, with a lesser impact on vital components of food compared to conventional processing methods. It is certainly advantageous over thermal processing methods and chemical modification techniques, but a clear explanation of the interaction of cold plasma reactive species needs to be further explored to formulate and control the underlying mechanisms of action, which would aid in better application and industrial scale-up. It is also associated with a few undesired effects in food products, especially in lipid oxidation. However, these issues could be resolved by incorporating natural antioxidants, better storage practices, and by combining cold plasma with other novel technologies. Though various studies are being carried out globally to improve or mitigate the undesirable effects, the challenges involved in the repeatability, reproducibility, and consistency of reactive species generation are yet to be addressed to support transferring the cold plasma technology for commercialization.

KEYWORDS

- **cold plasma**
- **free radical**
- **functional properties**
- **mechanism**
- **reactive species**

REFERENCES

1. Aadil, R. M., Zeng, X. A., Han, Z., & Sun, D. W. (2013). Effects of ultrasound treatments on quality of grapefruit juice. *Food Chemistry, 141*(3), 3201–3206.
2. Aaliya, B., Sunooj, K. V., Navaf, M., Akhila, P. P., Sudheesh, C., Sabu, S., & Khaneghah, A. M. (2021). Effect of low dose γ-irradiation on the structural and functional properties, and in vitro digestibility of ultrasonicated stem starch from *Corypha umbraculifera* L. *Applied Food Research, 1*(2), 1–11. E-article ID 100013. https://doi.org/10.1016/j.afres.2021.100013.
3. Abbasi, E., Sarteshnizi, R. A., Gavlighi, H. A., Nikoo, M., Azizi, M. H., & Sadeghinejad, N. (2019). Physicochemical, texture, oxidative stability, and sensory property of reduced fat emulsion type sausage. *Meat Science, 147*, 135–143.
4. Ai, Y., & Jane, J. L. (2015). Gelatinization and rheological properties of starch. *Starch-Stärke, 67*(3–4), 213–224.
5. Ali, M., Cheng, J. H., & Sun, D. W. (2021). Effects of dielectric barrier discharge cold plasma treatments on degradation of anilazine fungicide and quality of tomato (*Lycopersicon esculentum* Mill) juice. *International Journal of Food Science & Technology, 56*(1), 69–75.

6. Almeida, F. D., Cavalcante, R. S., Cullen, P. J., Frias, J. M., Bourke, P., Fernandes, F. A., & Rodrigues, S. (2015). Effects of atmospheric cold plasma and ozone on prebiotic orange juice. *Innovative Food Science & Emerging Technologies, 32*, 127–135.
7. Bahrami, N., Bayliss, D., Chope, G., Penson, S., Perehinec, T., & Fisk, I. D. (2016). Cold plasma: A new technology to modify wheat flour functionality. *Food Chemistry, 202*, 247–253.
8. Banura, S., Thirumdas, R., Kaur, A., Deshmukh, R. R., & Annapure, U. S. (2018). Modification of starch using low pressure radio frequency air plasma. *LWT, 89*, 719–724.
9. Bao, T., Hao, X., Shishir, M. R., Karim, N., & Chen, W. (2021). Cold plasma: an emerging pretreatment technology for the drying of jujube slices. *Food Chemistry, 337*, E-article ID 127783. https://doi.org/10.1016/j.foodchem.2020.127783.
10. Barak, S., Mudgil, D., & Taneja, S. (2020). Exudate gums: chemistry, properties, and food applications–a review. *Journal of the Science of Food and Agriculture, 100*(7), 2828–2835.
11. Barriuso, B., Astiasarán, I., & Ansorena, D. (2013). A review of analytical methods measuring lipid oxidation status in foods: A challenging task. *European Food Research and Technology, 236*(1), 1–15.
12. Bauer, A., Ni, Y., Bauer, S., Paulsen, P., Modic, M., Walsh, J. L., & Smulders, F. J. (2017). The effects of atmospheric pressure cold plasma treatment on microbiological, physical-chemical and sensory characteristics of vacuum packaged beef loin. *Meat Science, 128*, 77–87.
13. Behl, T., Rocchetti, G., Chadha, S., Zengin, G., Bungau, S., Kumar, A., & Montesano, D. (2021). Phytochemicals from plant foods as potential source of antiviral agents: An overview. *Pharmaceuticals, 14*(4), 381–387.
14. Bekhit, A. E., Hopkins, D. L., Fahri, F. T., & Ponnampalam, E. N. (2013). Oxidative processes in muscle systems and fresh meat: Sources, markers, and remedies. *Comprehensive Reviews in Food Science and Food Safety, 12*(5), 565–597.
15. Bertolini, A. C., Mestres, C., Colonna, P., & Raffi, J. (2001). Free radical formation in UV-and gamma-irradiated cassava starch. *Carbohydrate Polymers, 44*(3), 269–271.
16. Bourke, P., Ziuzina, D., Han, L., Cullen, P. J., & Gilmore, B. F. (2017). Microbiological interactions with cold plasma. *Journal of Applied Microbiology, 123*(2), 308–324.
17. Bulbul, V. J., Bhushette, P. R., Zambare, R. S., Deshmukh, R. R., & Annapure, U. S. (2019). Effect of cold plasma treatment on Xanthan gum properties. *Polymer Testing, 79*, E-article ID 106056. https://doi.org/10.1016/j.polymertesting.2019.106056.
18. Chatterjee, S., & Mohan, S. V. (2021). Refining of vegetable waste to renewable sugars for ethanol production: Depolymerization and fermentation optimization. *Bioresource Technology, 340*, E-article ID 125650. https://doi.org/10.1016/j.biortech.2021.125650.
19. Chen, G., Dong, S., Chen, Y., Gao, Y., Zhang, Z., Li, S., & Chen, Y. (2020). Complex coacervation of zein-chitosan via atmospheric cold plasma treatment: Improvement of encapsulation efficiency and dispersion stability. *Food Hydrocolloids, 107*, E-article ID 105943. https://doi.org/10.1016/j.foodhyd.2020.105943.
20. Chen, H. H., Hung, C. L., Lin, S. Y., & Liou, G. J. (2014). Effect of low-pressure plasma exposure on the storage characteristics of brown rice. *Food and Bioprocess Technology, 8*(2), 471–477.
21. Choi, H. S., Kim, K. N., You, E. M., Choi, E. H., Kim, Y. H., & Kim, K. M. (2013). Tooth whitening effects by atmospheric pressure cold plasmas with different gases. *Japanese*

Journal of Applied Physics, 52(11S), E-article ID NF02. https://doi.org/10.7567/JJAP.52.11NF02.
22. Criegee, R. (1975). Mechanism of ozonolysis. *Angewandte Chemie International Edition in English, 14*(11), 745–752.
23. Crittenden, R. A., & Playne, M. (1996). Production, properties, and applications of food-grade oligosaccharides. *Trends in Food Science & Technology, 7*(11), 353–361.
24. Csala, M., Kardon, T., Legeza, B., Lizák, B., Mandl, J., Margittai, É., & Bánhegyi, G. (2015). On the role of 4-hydroxynonenal in health and disease. *Biochimica et Biophysica Acta (BBA) – Molecular Basis of Disease, 1852*(5), 826–838.
25. Cui, S. W. (Ed.). (2005). *Food Carbohydrates: Chemistry, Physical Properties, and Applications*. Edition: 1, ISBN 9780849315749, pp. 1– 432, https://doi.org/10.1201/9780203485286. CRC Press.
26. Cullen, P. J., Valdramidis, V. P., Tiwari, B. K., Patil, S., Bourke, P., & O'Donnell, C. P. (2010). Ozone processing for food preservation: an overview on fruit juice treatments. *Ozone: Science & Engineering, 32*(3), 166–179.
27. De Jager, T. L., Cockrell, A. E., & Du Plessis, S. S. (2017). Ultraviolet light induced generation of reactive oxygen species. *Ultraviolet Light in Human Health, Diseases and Environment, 2017*, 15–23.
28. Eazhumalai, G., Ranjitha Gracy, T. K., Mishra, A., & Annapure, U. S. (2021). Atmospheric pressure nonthermal pin to plate plasma system for the microbial decontamination of oat milk. *Journal of Food Processing and Preservation, 46*(10), e16181. E-article ID 16181. https://doi.org/10.1111/jfpp.16181.
29. Erlander, S. R., & Tobin, R. (1968). The stability of the helix of amylose and amylopectin in DMSO and water solutions. *Die Makromolekulare Chemie (Macromolecular Chemistry and Physics), 11*(1), 194–211.
30. Faraji, H., & Lindsay, R. C. (2004). Characterization of the antioxidant activity of sugars and polyhydric alcohols in fish oil emulsions. *Journal of Agricultural and Food Chemistry, 52*(23), 7164–7171.
31. Florence, T. M. (1984). The production of hydroxyl radical from hydrogen peroxide. *Journal of Inorganic Biochemistry, 22*(4), 221–230.
32. Franchina, D. G., Dostert, C., & Brenner, D. (2018). Reactive oxygen species: involvement in T cell signaling and metabolism. *Trends in Immunology, 39*(6), 489–502.
33. Ganesan, A. R., & Rajauria, G. (2021). Application of cold plasma on food matrices: A review on current and future prospects. *Journal of Food Processing and Preservation, 45*(1), E-article ID 15070; https://doi.org/10.1111/jfpp.15070.
34. Gavahian, M., Chu, Y. H., Khaneghah, A. M., Barba, F. J., & Misra, N. N. (2018). A critical analysis of the cold plasma induced lipid oxidation in foods. *Trends in Food Science & Technology, 77*, 32–41.
35. Ge, X., Shen, H., Su, C., Zhang, B., Zhang, Q., Jiang, H., & Li, W. (2021). The improving effects of cold plasma on multi-scale structure, physicochemical and digestive properties of dry heated red adzuki bean starch. *Food Chemistry, 349*, E-article ID 129159; https://doi.org/10.1016/j.foodchem.2021.129159.
36. Goff, H. D., Repin, N., Fabek, H., El Khoury, D., & Gidley, M. J. (2018). Dietary fiber for glycaemia control: Towards a mechanistic understanding. *Bioactive Carbohydrates and Dietary Fibre, 14*, 39–53.

37. Grzegorzewski, F., Rohn, S., Kroh, L. W., Geyer, M., & Schlüter, O. (2010). Surface morphology and chemical composition of lamb's lettuce (*Valerianella locusta*) after exposure to a low-pressure oxygen plasma. *Food Chemistry, 122*(4), 1145–1152.
38. Grzegorzewski, F., Schluter, O., Ehlbeck, J., Weltmann, K. D., Geyer, M., Kroh, L. W., & Rohn, S. (2009). Plasma-oxidative degradation of polyphenolics – Influence of non-thermal gas discharges with respect to fresh produce processing. *Czech Journal of Food Science, 27*, S35–S39.
39. Gulati, P., Brahma, S., & Rose, D. J. (2020). Impacts of extrusion processing on nutritional components in cereals and legumes: Carbohydrates, proteins, lipids, vitamins, and minerals. In *Extrusion Cooking* (pp. 415–443). London: Woodhead Publishing.
40. Halliwell, B. (1993). The chemistry of free radicals. *Toxicology and Industrial Health, 9*(1–2), 1–21.
41. Han, Y., Cheng, J. H., & Sun, D. W. (2019). Activities and conformation changes of food enzymes induced by cold plasma: A review. *Critical Reviews in Food Science and Nutrition, 59*(5), 794–811.
42. Hao, H., Wu, B. S., Yang, J., Guo, Q., Yang, Y., & Li, Y. W. (2015). Non-thermal plasma enhanced heavy oil upgrading. *Fuel, 149*, 162–173.
43. Hebeish, A., El-Rafie, M. H., El-Sisi, F., Hafiz, S. A., & Abdel-Rahman, A. A. (1994). Oxidation of maize and rice starches using potassium permanganate with various reductants. *Polymer Degradation and Stability, 43*(3), 363–371.
44. Heiniö, R. L., Lehtinen, P., Oksman-Caldentey, K. M., & Poutanen, K. (2002). Differences between sensory profiles and development of rancidity during long-term storage of native and processed oat. *Cereal Chemistry, 79*(3), 367–375.
45. Hu, J., Li, X., Cheng, Z., Fan, X., Ma, Z., Hu, X., & Xing, Y. (2022). Modified tartary buckwheat (*Fagopyrum tataricum* Gaertn.) starch by gaseous ozone: Structural, physicochemical and in vitro digestible properties. *Food Hydrocolloids, 125*, E-article ID 107365. https://doi.org/10.1016/j.foodhyd.2021.107365.
46. Huggins, T. G., Wells-Knecht, M. C., Detorie, N. A., Baynes, J. W., & Thorpe, S. R. (1993). Formation of o-tyrosine and dityrosine in proteins during radiolytic and metal-catalyzed oxidation. *Journal of Biological Chemistry, 268*(17), 12341–12347.
47. Isbell, H. S., Frush, H. L., & Martin, E. T. (1973). Reactions of carbohydrates with hydroperoxides: Part I. Oxidation of aldoses with sodium peroxide. *Carbohydrate Research, 26*(2), 287–295.
48. Johnsy, K. J., Kalaivendan, R. G., Eazhumalai, G., Khakar, S. P., & Annapure, U. S. (2022). Effect of pin-to-plate atmospheric cold plasma on jackfruit seed flour functionality modification. *Innovative Food Science & Emerging Technologies*, E-article ID 103009. https://doi.org/10.1016/j.ifset.2022.103009.
49. Joseph, J. P., Keri, L. C., Brenda, S. K., & Mark, A. W. (Eds.). (2016). *The Science of Cooking: Understanding the Biology and Chemistry Behind Food and Cooking*. pp. 1–544, ISBN: 9781118674208, DOI:10.1002/9781118910771. John Wiley & Sons.
50. Kalaivendan, R. G., Mishra, A., Eazhumalai, G., & Annapure, U. S. (2022). Effect of atmospheric pressure non-thermal pin to plate plasma on the functional, rheological, thermal, and morphological properties of mango seed kernel starch. *International Journal of Biological Macromolecules, 196*, 63–71.
51. Kalaivendan, T. R., Shashikanthalu, P. S., & Radhakrishnan, M. (2022). Non-thermal technologies: Solution for hazardous pesticides reduction in fruits and vegetables. *Critical Reviews in Food Science and Nutrition, 62*(7), 1782–1799.

52. Khan, M. S., Lee, N. R., Ahn, J., Kim, J. Y., Kim, J. H., Kwon, K. H., & Kim, Y. J. (2021). Degradation of different pesticides in water by microplasma: Roles of individual radicals and degradation pathways. *Environmental Science and Pollution Research*, 2021, 8296–8309.
53. Khlestkin, V. K., Peltek, S. E., & Kolchanov, N. A. (2018). Review of direct chemical and biochemical transformations of starch. *Carbohydrate Polymers, 181*, 460–476.
54. Kim, H. J., Yong, H. I., Park, S., Kim, K., Choe, W., & Jo, C. (2015). Microbial safety and quality attributes of milk following treatment with atmospheric pressure encapsulated dielectric barrier discharge plasma. *Food Control, 45*, 451–456.
55. Kim, H. S., & Min, S. C. (2017). Effects of microwave-discharged cold plasma on synthesis and characteristics of citrate derivatives of corn starch granules. *Food Science and Biotechnology, 26*(3), 697–706.
56. Komulainen, S., Pursiainen, J., Perämäki, P., & Lajunen, M. (2013). Complexation of Fe (III) with water-soluble oxidized starch. *Starch-Stärke, 65*(3–4), 338–345.
57. Korachi, M., Ozen, F., Aslan, N., Vannini, L., Guerzoni, M. E., Gottardi, D., & Ekinci, F. Y. (2015). Biochemical changes to milk following treatment by a novel, cold atmospheric plasma system. *International Dairy Journal, 42*, 64–69.
58. Kovačević, D. B., Kljusurić, J. G., Putnik, P., Vukušić, T., Herceg, Z., & Dragović-Uzelac, V. (2016). Stability of polyphenols in chokeberry juice treated with gas phase plasma. *Food Chemistry, 212*, 323–331.
59. Kovačević, D. B., Putnik, P., Dragović-Uzelac, V., Pedisić, S., Jambrak, A. R., & Herceg, Z. (2016). Effects of cold atmospheric gas phase plasma on anthocyanins and color in pomegranate juice. *Food Chemistry, 2016*, 317–323.
60. Kubow, S. (1992). Routes of formation and toxic consequences of lipid oxidation products in foods. *Free Radical Biology and Medicine, 12*(1), 63–81.
61. Kumar, V., Chatli, M. K., Wagh, R. V., Mehta, N., & Kumar, P. (2015). Effect of the combination of natural antioxidants and packaging methods on quality of pork patties during storage. *Journal of Food Science and Technology, 52*(10), 6230–6241.
62. Kurdziel, M., Łabanowska, M., Pietrzyk, S., Pająk, P., Królikowska, K., & Szwengiel, A. (2022). The effect of UV-B irradiation on structural and functional properties of corn and potato starches and their components. *Carbohydrate Polymers*, Vol. 289. E-article ID 119439. https://doi.org/10.1016/j.carbpol.2022.119439.
63. Kyzlink, V., & Čurda, D. (1970). Influence of sucrose concentration and oxygen access on the course of L-ascorbic acid oxidation in liquid medium. *European Food Research and Technology, 143*(4), 263–273.
64. Lawal, O. S. (2004). Composition, physicochemical properties, and retrogradation characteristics of native, oxidized, acetylated, and acid-thinned new cocoyam (*Xanthosoma sagittifolium*) starch. *Food Chemistry, 87*, 205–218.
65. Li, F., Sun, X., Yu, W., Shi, C., Zhang, X., Yu, H., & Ma, F. (2021). Enhanced konjac glucomannan hydrolysis by lytic polysaccharide monooxygenases and generating prebiotic oligosaccharides. *Carbohydrate Polymers, 253*, E-article ID 117241. https://doi.org/10.1016/j.carbpol.2020.117241.
66. Lima, D. C., Castanha, N., Maniglia, B. C., Matta Junior, M. D., La Fuente, C. I., & Augusto, P. E. (2021). Ozone processing of cassava starch. *Ozone: Science & Engineering, 43*(1), 60–77.
67. Lin, S. P., Kuo, T. C., Wang, H. T., Ting, Y., Hsieh, C. W., Chen, Y. K., & Cheng, K. C. (2020). Enhanced bioethanol production using atmospheric cold plasma-assisted

detoxification of sugarcane bagasse hydrolysate. *Bioresource Technology, 313*, E-article ID 123704. https://doi.org/10.1016/j.biortech.2020.123704.
68. Lorenzo, J. M., Pateiro, M., Domínguez, R., Barba, F. J., Putnik, P., Kovačević, D. B., & Franco, D. (2018). Berries extracts as natural antioxidants in meat products: A review. *Food Research International, 106*, 1095–1104.
69. Ma, Y. C., Manolache, S., Sarmadi, M., & Denes, F. S. (2004). Synthesis of starch copolymers by silicon tetrachloride plasma-induced graft polymerization. *Starch-Stärke, 56*(2), 47–57.
70. Makris, D. P., & Rossiter, J. T. (2002). Hydroxyl free radical-mediated oxidative degradation of quercetin and morin: a preliminary investigation. *Journal of Food Composition and Analysis, 15*(1), 103–113.
71. Matamoros, M. A., Kim, A., Peñuelas, M., Ihling, C., Griesser, E., Hoffmann, R., & Becana, M. (2018). Protein carbonylation and glycation in legume nodules. *Plant Physiology, 177*(4), 1510–1528.
72. Mehr, H. M., & Koocheki, A. (2020). Effect of atmospheric cold plasma on structure, interfacial and emulsifying properties of grass pea (*Lathyrus sativus* L.) protein isolate. *Food Hydrocolloids, 106*, E-article ID 105899. https://doi.org/10.1016/j.foodhyd.2020.105899.
73. Mir, S. A., Siddiqui, M. W., Dar, B. N., Shah, M. A., Wani, M. H., Roohinejad, S., & Ali, A. (2020). Promising applications of cold plasma for microbial safety, chemical decontamination and quality enhancement in fruits. *Journal of Applied Microbiology, 129*(3), 474–485.
74. Misra, N. N., & Jo, C. (2017). Applications of cold plasma technology for microbiological safety in meat industry. *Trends in Food Science & Technology, 64*, 74–86.
75. Misra, N. N., Kaur, S., Tiwari, B. K., Kaur, A., Singh, N., & Cullen, P. J. (2015). Atmospheric pressure cold plasma (ACP) treatment of wheat flour. *Food Hydrocolloids, 44*, 115–121.
76. Monteiro, C. A. (2009). Nutrition and health: Issue is not food, nor nutrients, so much as processing. *Public Health Nutrition, 12*(5), 729–731.
77. Moubarac, J. C., Parra, D. C., Cannon, G., & Monteiro, C. A. (2014). Food classification systems based on food processing: significance and implications for policies and actions: a systematic literature review and assessment. *Current Obesity Reports, 3*(2), 256–272.
78. Mu, B., Xu, L., & Yang, Y. (2021). Flexible and wet stable starch films crosslinked with sugar-based aldehydes. *Industrial Crops and Products, 173*, E-article ID 114109. https://doi.org/10.1016/j.indcrop.2021.114109.
79. Muhammad, A. I., Liao, X., Cullen, P. J., Liu, D., Xiang, Q., Wang, J., & Ding, T. (2018). Effects of nonthermal plasma technology on functional food components. *Comprehensive Reviews in Food Science and Food Safety, 17*(5), 1378–1394.
80. Muhammad, A. I., Lv, R., Liao, X., Chen, W., Liu, D., Xiang, Y., & Ding, T. (2019). Modeling the inactivation of *Bacillus cereus* in tiger nut milk treated with cold atmospheric pressure plasma. *Journal of Food Protection, 82*(11), 1828–1836, https://doi.org/10.4315/0362-028X.JFP-18-586.
81. Nachtergaele, W. (1989). The benefits of cationic starches for the paper industry. *Starch-Stärke, 41*(1), 27–31.
82. Naknaen, P., Tobkaew, W., & Chaichaleom, S. (2017). Properties of jackfruit seed starch oxidized with different levels of sodium hypochlorite. *International Journal of Food Properties, 20*(5), 979–996.

83. Nawapat, D., & Thawien, W. (2013). Effect of UV-treatment on the properties of biodegradable rice starch films. *International Food Research Journal, 20*(3), 1313–1322.
84. Nouman, S. M., Shehzad, A., Butt, M. S., Khan, M. I., & Tanveer, M. (2015). Phytochemical profiling of curry (*Murraya koenigii*) leaves and its health benefits. *Pak. J. Food Sci., 25*(4), 204–215.
85. Onal-Ulusoy, B. (2021). Effects of cold atmospheric gliding arc discharge plasma, non-thermal ultrasound, and low-temperature oven treatments on quality parameters of Turkish blossom honey. *Food Bioprocess Technology, 14*, 1763–1771.
86. Pan, Y. W., Cheng, J. H., & Sun, D. W. (2021). Inhibition of fruit softening by cold plasma treatments: affecting factors and applications. *Critical Reviews in Food Science and Nutrition, 61*(12), 1935–1946.
87. Pandiselvam, R., Manikantan, M. R., Divya, V., Ashokkumar, C., Kaavya, R., Kothakota, A., & Ramesh, S. V. (2019). Ozone: An advanced oxidation technology for starch modification. *Ozone: Science & Engineering, 41*(6), 491–507.
88. Pankaj, S. K., Misra, N. N., & Cullen, P. J. (2013). Kinetics of tomato peroxidase inactivation by atmospheric pressure cold plasma based on dielectric barrier discharge. *Innovative Food Science & Emerging Technologies, 19*, 153–157.
89. Pankaj, S. K., Wan, Z., Colonna, W., & Keener, K. M. (2017). Effect of high voltage atmospheric cold plasma on white grape juice quality. *Journal of the Science of Food and Agriculture, 97*(12), 4016–4021.
90. Park, H., Puligundla, P., & Mok, C. (2020). Cold plasma decontamination of brown rice grains: Impact on biochemical and sensory qualities of their corresponding seedlings and aqueous tea infusions. *LWT, 131*, E-article ID 109508. https://doi.org/10.1016/j.lwt.2020.109508.
91. Parker, R., & Ring, S. G. (2001). Aspects of the physical chemistry of starch. *Journal of Cereal Science, 34*(1), 1–17. https://doi.org/10.1006/jcrs.2000.0402.
92. Parovuori, P., Hamunen, A., Forssell, P., Autio, K., & Poutanen, K. (1995). Oxidation of potato starch by hydrogen peroxide. *Starch-Stärke, 47*(1), 19–23.
93. Parsons, H. T., & Fry, S. C. (2012). Oxidation of dehydroascorbic acid and 2, 3-diketogulonate under plant apo-plastic conditions. *Phytochemistry, 75*, 41–49.
94. Patil, S., Moiseev, T., Misra, N. N., Cullen, P. J., Mosnier, J. P., Keener, K. M., & Bourke, P. (2014). Influence of high voltage atmospheric cold plasma process parameters and role of relative humidity on inactivation of *Bacillus atrophaeus* spores inside a sealed package. *Journal of Hospital Infection, 88*(3), 162–169.
95. Pérez-Andrés, J. M. (2019). Effect of cold plasma on the techno-functional properties of animal protein food ingredients. *Innovative Food Science & Emerging Technologies, 58*, E-article ID 102205. https://doi.org/10.1016/j.ifset.2019.102205.
96. Pérez-Andrés, J. M., de Alba, M., Harrison, S. M., P, B. N., Cullen, P. J., & Tiwari, B. K. (2020). Effects of cold atmospheric plasma on mackerel lipid and protein oxidation during storage. *LWT, 118*, E-article ID 108697. https://doi.org/10.1016/j.lwt.2019.108697.
97. Perinban, S., Orsat, V., & Raghavan, V. (2019). Nonthermal plasma–liquid interactions in food processing: A review. *Comprehensive Reviews in Food Science and Food Safety, 18*(9), 1985–2008.
98. Phaniendra, A., Jestadi, D. B., & Periyasamy, L. (2015). Free radicals: properties, sources, targets, and their implication in various diseases. *Indian Journal of Clinical Biochemistry, 30*(1), 11–26.

99. Piri, I., Babayan, M., Tavassoli, A., & Javaheri, M. (2011). The use of gamma irradiation in agriculture. *African Journal of Microbiology Research, 5*(32), 5806–5811.
100. Porter, N. A., Caldwell, S. E., & Mills, K. A. (1995). Mechanisms of free radical oxidation of unsaturated lipids. *Lipids, 30*(4), 277–290.
101. Pragna, C. H., Kalaivendan, T. R., Mahendran, R., & Anandharamakrishnan, C. (2019). Effects of microwave and cold plasma assisted hydro distillation on lemon peel oil extraction. *International Journal of Food Engineering, 15*(10), 1–10. online; https://doi.org/10.1515/ijfe-2019-0093.
102. Puprasit, K., Wongsawaeng, D., Ngaosuwan, K., Kiatkittipong, W., & Assabumrungrat, S. (2020). Non-thermal dielectric barrier discharge plasma hydrogenation for production of margarine with low trans-fatty acid formation. *Innovative Food Science & Emerging Technologies, 66*, E-article ID 102511. https://doi.org/10.1016/j.ifset.2020.102511.
103. Queirós, R. P., Saraiva, J. A., & da Silva, J. A. (2018). Tailoring structure and technological properties of plant proteins using high hydrostatic pressure. *Critical Reviews in Food Science and Nutrition, 58*(9), 1538–1556, DOI: 10.1080/10408398.2016.1271770.
104. Raisi, M., Ghorbani, M., Mahoonak, A. S., Kashaninejad, M., & Hosseini, H. (2015). Effect of storage atmosphere and temperature on the oxidative stability of almond kernels during long term storage. *Journal of Stored Products Research, 62*, 16–21.
105. Ranjitha-Gracy, T. K., Gupta, V., & Mahendran, R. (2019). Influence of low-pressure nonthermal dielectric barrier discharge plasma on chlorpyrifos reduction in tomatoes. *Journal of Food Process Engineering, 42*(6), E-article ID 13242. https://doi.org/10.1111/jfpe.13242.
106. Rao, P. S., Kalva, S., Yerramilli, A., & Mamidi, S. (2011). Free radicals and tissue damage: Role of antioxidants. *Free Radicals and Antioxidants, 1*(4), 2–7.
107. Riesz, P., & Kondo, T. (1992). Free radical formation induced by ultrasound and its biological implications. *Free Radical Biology and Medicine, 13*(3), 247–270.
108. Rodríguez, Ó., Gomes, W. F., Rodrigues, S., & Fernandes, F. A. (2017). Effect of indirect cold plasma treatment on cashew apple juice (*Anacardium occidentale* L.). *LWT, 84*, 457–463.
109. Saremnezhad, S., Soltani, M., Faraji, A., & Hayaloglu, A. A. (2021). Chemical changes of food constituents during cold plasma processing: A review. *Food Research International, 147*, E-article ID 110552. https://doi.org/10.1016/j.foodres.2021.110552.
110. Sega, A., Zanardi, I., Chiasserini, L., Gabbrielli, A., Bocci, V., & Travagli, V. (2010). Properties of sesame oil by detailed 1H and 13C NMR assignments before and after ozonation and their correlation with iodine value, peroxide value, and viscosity measurements. *Chemistry and Physics of Lipids*, 1632, 148–156.
111. Segat, A., Misra, N. N., Cullen, P. J., & Innocente, N. (2015). Atmospheric pressure cold plasma (ACP) treatment of whey protein isolate model solution. *Innovative Food Science & Emerging Technologies, 29*, 247–254.
112. Segat, A., Misra, N. N., Cullen, P. J., & Innocente, N. (2016). Effect of atmospheric pressure cold plasma (ACP) on activity and structure of alkaline phosphatase. *Food and Bioproducts Processing, 98*, 181–188.
113. Shacter, E. (2000). Quantification and significance of protein oxidation in biological samples. *Drug Metabolism Reviews, 32*(3–4), 307–326.
114. Shah, N., Mewada, R. K., & Mehta, T. (2016). Crosslinking of starch and its effect on viscosity behavior. *Reviews in Chemical Engineering, 32*(2), 265–270.

115. Shen, H., Guo, Y., Zhao, J., Zhao, J., Ge, X., Zhang, Q., & Yan, W. (2021). The multi-scale structure and physicochemical properties of mung bean starch modified by ultrasound combined with plasma treatment. *International Journal of Biological Macromolecules, 191*, 821–831.
116. Shin, J., Song, M. H., Oh, J. W., Keum, Y. S., & Saini, R. K. (2020). Pro-oxidant actions of carotenoids in triggering apoptosis of cancer cells: A review of emerging evidence. *Antioxidants, 9*(6), 532–539.
117. Sies, H., & De Groot, H. (1992). Role of reactive oxygen species in cell toxicity. *Toxicology Letters, 64*, 547–551.
118. Sifuentes-Nieves, I., Flores-Silva, P. C., Gallardo-Vega, C., Hernández-Hernández, E., Neira-Velázquez, G., Mendez-Montealvo, G., & Velazquez, G. (2020). Films made from plasma-modified corn starch: Chemical, mechanical and barrier properties. *Carbohydrate Polymers, 237*, 1–7. E-article ID 116103. https://doi.org/10.1016/j.carbpol.2020.116103.
119. Slater, T. F. (1988). Free radical mechanisms in tissue injury. *Cell Function and Disease*, pp. 209–218, ISBN 978-1-4612-8095-8, https://doi.org/10.1007/978-1-4613-0813-3.
120. Stadtman, E. R., & Levine, R. L. (2003). Free radical-mediated oxidation of free amino acids and amino acid residues in proteins. *Amino Acids, 25*(3), 207–218.
121. Sun, X., Saleh, A. S., Sun, Z., Ge, X., Shen, H., Zhang, Q., & Li, W. (2022). Modification of multi-scale structure, physicochemical properties, and digestibility of rice starch via microwave and cold plasma treatments. *LWT, 153*, E-article ID 112483. https://doi.org/10.1016/j.ijbiomac.2022.05.128.
122. Surowsky, B., Bußler, S., & Schlüter, O. K. (2016). Cold plasma interactions with food constituents in liquid and solid food matrices. In: *Cold Plasma in Food and Agriculture*. pp. 179–203, Academic Press.
123. Surowsky, B., Fischer, A., Schlueter, O., & Knorr, D. (2013). Cold plasma effects on enzyme activity in a model food system. *Innovative Food Science & Emerging Technologies, 19*, 146–152.
124. Tappi, S., Berardinelli, A., Ragni, L., Dalla Rosa, M., Guarnieri, A., & Rocculi, P. (2014). Atmospheric gas plasma treatment of fresh-cut apples. *Innovative Food Science & Emerging Technologies, 21*, 114–122.
125. Tappi, S., Gozzi, G., Vannini, L., Berardinelli, A., Romani, S., Ragni, L., & Rocculi, P. (2016). Cold plasma treatment for fresh-cut melon stabilization. *Innovative Food Science & Emerging Technologies, 33*, 225–233.
126. Teramoto, Y. (2015). Functional thermoplastic materials from derivatives of cellulose and related structural polysaccharides. *Molecules, 20*(4), 5487–5527.
127. Thirumdas, R., Kadam, D., & Annapure, U. S. (2017). Cold plasma: An alternative technology for the starch modification. *Food Biophysics, 12*(1), 129–139.
128. Thirumdas, R., Trimukhe, A., Deshmukh, R. R., & Annapure, U. S. (2017). Functional and rheological properties of cold plasma treated rice starch. *Carbohydrate Polymers, 157*, 1723–1731.
129. Tiwari, B. K., O'Donnell, C. P., & Cullen, P. J. (2009). Effect of non-thermal processing technologies on the anthocyanin content of fruit juices. *Trends in Food Science & Technology, 20*(3–4), 137–145.
130. Turner, M. (2016). Physics of cold plasma. In: *Cold Plasma in Food and Agriculture*. Academic Press, pp. 17–51.

131. Upadhyay, R., Thirumdas, R., Deshmukh, R. R., Annapure, U., & Misra, N. N. (2019). An exploration of the effects of low-pressure plasma discharge on the physicochemical properties of chia (*Salvia hispanica* L.) flour. *Journal of Engineering & Processing Management, 11*(2), 73–80.
132. Van Durme, J., & Vandamme, J. (2016). Non-thermal plasma as preparative technique to evaluate olive oil adulteration. *Food Chemistry, 208*, 185–191.
133. Wang, P., Luo, Z. G., & Xiao, Z. G. (2021). Preparation, physicochemical characterization and in vitro release behavior of resveratrol-loaded oxidized gellan gum/resistant starch hydrogel beads. *Carbohydrate Polymers, 260*, E-article ID 117794. https://doi.org/10.1016/j.carbpol.2021.117794.
134. Wang, T., Zhang, X., Wang, H., Yuan, T., Yu, D., Wang, L., & Jiang, L. (2019). Study on the electrochemical hydrogenation of soybean oil under H2 conditions. *Journal of Oleo Science, 68*(4), 311–320.
135. Wang, Y. J., & Wang, L. (2003). Physicochemical properties of common and waxy corn starches oxidized by different levels of sodium hypochlorite. *Carbohydrate Polymers, 52*(3), 207–217.
136. Wende, K., von Woedtke, T., Weltmann, K. D., & Bekeschus, S. (2019). Chemistry and biochemistry of cold physical plasma derived reactive species in liquids. *Biological Chemistry, 400*(1), 19–38.
137. Williams, P. A., & Phillips, G. O. (2021). Introduction to food hydrocolloids. In: *Handbook of Hydrocolloids*. London: Woodhead Publishing, pp. 209–218, ISBN 978-1-4612-8095-8, https://doi.org/10.1007/978-1-4613-0813-3, Edition Number: 1, 3–26. ISBN 978-0-12-820104-6. https://doi.org/10.1016/B978-0-12-820104-6.00017-6.
138. Wu, Y., Liang, Y., Wei, K., Li, W., Yao, M., Zhang, J., & Grinshpun, S. A. (2015). MS2 virus inactivation by atmospheric-pressure cold plasma using different gas carriers and power levels. *Applied and Environmental Microbiology, 81*(3), 996–1002.
139. Xiong, Y. L., & Guo, A. (2020). Animal and plant protein oxidation: Chemical and functional property significance. *Foods, 10*(1), 40–48.
140. Xu, B., Ren, A., Chen, J., Li, H., Wei, B., Wang, J., & Ma, H. (2021). Effect of multi-mode dual-frequency ultrasound irradiation on the degradation of waxy corn starch in a gelatinized state. *Food Hydrocolloids, 113*, E-article ID 106440. https://doi.org/10.1016/j.foodhyd.2020.106440.
141. Xu, L., Yepez, X., Applegate, B., Keener, K. M., Tao, B., & Garner, A. L. (2020). Penetration and microbial inactivation by high voltage atmospheric cold plasma in semi-solid material. *Food and Bioprocess Technology, 13*(10), 1688–1702.
142. Yepez, X. V., & Keener, K. M. (2016). High-voltage atmospheric cold plasma (HVACP) hydrogenation of soybean oil without trans-fatty acids. *Innovative Food Science & Emerging Technologies, 38*, 169–174.
143. Yong, H. I., Kim, H. J., Park, S., Kim, K., Choe, W., Yoo, S. J., & Jo, C. (2015). Pathogen inactivation and quality changes in sliced cheddar cheese treated using flexible thin-layer dielectric barrier discharge plasma. *Food Research International, 69*, 57–63.
144. Yu, X., Huang, S., Nie, C., Deng, Q., Zhai, Y., & Shen, R. (2020). Effects of atmospheric pressure plasma jet on the physicochemical, functional, and antioxidant properties of flaxseed protein. *Journal of Food Science, 85*(7), 2010–2019.
145. Zhang, K., Perussello, C. A., Milosavljević, V., Cullen, P. J., Sun, D. W., & Tiwari, B. K. (2019). Diagnostics of plasma reactive species and induced chemistry of plasma treated foods. *Critical Reviews in Food Science and Nutrition, 59*(5), 812–825.

146. Zhang, L., Xi, G., Chen, Z., Jiang, D., Yu, H., & Wang, X. (2017). Highly selective conversion of glucose into furfural over modified zeolites. *Chemical Engineering Journal, 307*, 868–876.
147. Zhang, S. D., Zhang, Y. R., Wang, X. L., & Wang, Y. Z. (2009). High carbonyl content oxidized starch prepared by hydrogen peroxide and its thermoplastic application. *Starch-Stärke, 61*(11), 646–655.
148. Zhang, Y. R., Wang, X. L., Zhao, G. M., & Wang, Y. Z. (2012). Preparation and properties of oxidized starch with high degree of oxidation. *Carbohydrate Polymers, 87*(4), 2554–2562.
149. Zhu, F. (2017). Plasma modification of starch. *Food Chemistry, 232*, 476–486.
150. Zou, J. J., Liu, C. J., & Eliasson, B. (2004). Modification of starch by glow discharge plasma. *Carbohydrate Polymers, 55*(1), 23–26.

PART II

Interaction of Cold Plasma with Food Systems and Its Applications

CHAPTER 4

CHARACTERIZATION METHODS FOR COLD PLASMA TECHNOLOGY

P. HIMASHREE and R. MAHENDRAN

ABSTRACT

Cold plasma has made substantial advances in food and agriculture processing. This method enables the production of reactive species with high efficiency and minimal power consumption. Understanding the interactions of plasma with food matrices requires knowledge of the densities and fluxes of plasma species generated by the sources. Plasma diagnosis may provide some useful information about the role of each reactive species for modification in food products to provide insights into its potential applications. This chapter discusses the various characterization techniques for analyzing the reactive oxygen species and the reactive nitrogen species generated from the cold plasma. The benefits and limitations of each approach of characterizing reactive species in cold plasma are compared. The applicability of each characterization method in various food processing methods is outlined.

4.1 INTRODUCTION

Atmospheric plasma is greatly utilized in surface disinfection and offers a great breakthrough in food and agri-material processing [33]. Dry air is a preferred working gas for food processing as it is readily available and can produce reactive species through the utilization of a cold plasma reactor. This technique generates bactericidal species effectively at room temperatures with fewer power requirements. The feed gas stream can be

Nonthermal Plasma Technology for Food and Food Products.
R. Mahendran, Preeti Birwal, and Megh R. Goyal (Eds.)

pure or an admixture of the gas stream that can be chosen according to the processing requirements for the food product [32]. Water molecules (acting as precursors) can be added slowly to the input gas stream carefully to create desired chemical radicals. The applications of plasma for processing and preserving food materials are influenced by the activities of the reactive species on the food components including DNA, protein, lipid, and carbohydrates. Hence, it is essential to diagnose the plasma species involved for understanding the mechanisms behind the plasma processing of foods in a better way [51].

The research studies and development of cold plasma have progressed dramatically during the last few decades. The high chemical reactivity of the plasma at low temperatures of gases even at room temperature is one of their most essential features. Interaction of plasma with biological systems affects the quality which has to be considered. Knowledge of discharge species fluxes and related elementary processes is crucial in applications such as food processing. Small cavities can be accessed by the common plasma jets, but it is difficult to monitor and compute the fluxes of the reactive oxygen and nitrogen species. Plasma's substantially non-equilibrium condition renders many usual diagnostic procedures useless. Plasma's strong non-equilibrium condition renders many standard diagnostic techniques and associated evaluation procedures invalid, as they are based on equilibrium state assumptions. Invasive techniques having a low spatial resolution do not suit plasma jets because of their small dimensions and severe non-uniformity, since they produce inaccurate mean values and drastically alter plasma characteristics [39].

For the diagnosis of plasma jets, non-invasive optical characterization techniques are preferred. These approaches include: Spectroscopic methods (such as: optical emission, optical absorption), methods using lasers such as laser-induced fluorescence, scattering techniques such as Rayleigh, Raman, and Thomson scattering, and measurements [40]. Other frequent chemical analysis techniques include mass spectrometry, electron paramagnetic resonance spectroscopy, and gas chromatography, among others.

Plasma chemistry for food processing is described in this chapter. Optical plasma diagnosing techniques including optical emission spectroscopy, optical absorption spectroscopy, and fluorimetry techniques are also described. Other non-optical plasma characterization approaches in this chapter include mass spectroscopy, scattering techniques, and electron paramagnetic resonance spectrometry. Applications, advantages, and disadvantages of the listed characterization techniques are summarized.

4.2 REACTIVE COLD PLASMA SPECIES

The source of plasma, process parameters like the gas employed, gas flow rate, frequency, pressure, and ionization degree influence the reactivity of the plasma generated [42]. The reactive species and free radicals produced in the plasma cause changes in functional properties, modifications in surface exposure, and postharvest preservation [36]. The feed gases commonly used are air, oxygen, nitrogen, hydrogen, and argon. Different gases can be utilized as feed gas for generating plasma to produce several antimicrobial and disinfection effects and the extent of changes [4].

The reactive species have an important role in producing antimicrobial effects. Radicals like hydroxyl (OH), superoxide (O_2), hydroperoxyl (HO_2), carbon dioxide radical (CO_2^-), carbonate (CO_3^-), Singlet ($^1O_2$), Peroxyl (RO_2), alkoxyl (RO) and non-radicals like H_2O_2, ozone (O_3), Singlet oxygen ($^1O_2$ Dg), Hypochlorous acid (HOCl), Hypobromous acid (HOBr), Hypoiodous acid (HOI), organic peroxide (ROOH), carbon monoxide (CO) are examples of reactive oxygen species. The reactive nitrogen species include NO (Nitric oxide), NO_2 (Nitrogen dioxide), and NO_3 (Nitrate), and non-radicals including HNO_2 (Nitrous acid), NO^- (Nitroxyl anion), NO^+ (Nitrosyl cation), N_2O_3 (Dinitrogen trioxide), N_2O_4 (Dinitrogen tetroxide), N_2O_5 (Dinitrogen pentoxide), NO_2Cl (Nitryl chloride), Alkyl peroxynitrites (ROONO), Alkyl peroxy nitrates (RO_2ONO) [17]. It is hard to underestimate the importance of these reactive species in biological and other food applications. Radicals were suspected to be harmful for decades, but later it has been recognized to have beneficial effects since they can decrease the concentration in the area where they are exposed [19]. The two-edged sword nature of the RONS makes it essential to characterize all species generated in the plasma for various applications.

4.3 OPTICAL TECHNIQUES FOR CHARACTERIZATION OF COLD PLASMA

Optical spectroscopy (Figure 4.1) is a non-invasive technique to study how molecules or atomic reactive substances react with electromagnetic radiation. Spectroscopy reports data on the excited and ionized states, and more process parameters [42]. UV-vis spectrophotometry is a non-destructive method that visualizes many RONS. The analyzed compounds are stable. However, this is an indirect technique for the characterization of plasma. It measures the intensity of the improvised color or the reduced color due to the reactions of RONS produced in the plasma with the chemical probes.

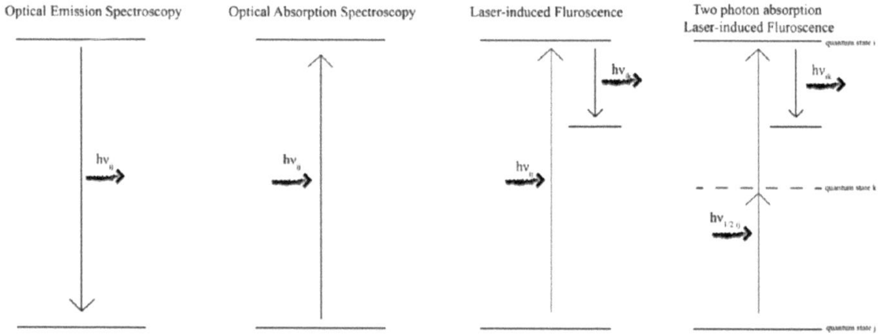

FIGURE 4.1 Schematic representation of optical characterization methods of cold plasma.

4.3.1 OPTICAL EMISSION SPECTROSCOPY (OES)

For years, Optical Emission Spectroscopy has been employed for characterizing and diagnosing the reactive species that are excited during plasma generation [25]. This is the most widely used analytical technique to detect short-lived RONS. It measures the radiations from excited atoms or molecules that occur in the plasma and only investigates the excited levels. The optical emission spectra interpretation is possible for the plasmas at low pressure and in thermal equilibrium. Whereas it may be difficult to interpret the spectra measured in strong non-equilibrium plasma such as plasma jets to get absolute values. Nonetheless, OES provides more information on atmospheric pressure itself that is determined by the absolute and relative intensity of lines, widths of the lines, the presence of particular lines ascribed to prohibited states, or emission continuum radiation. There is a transfer of energy due to collision and radiative processes that cause emission profiles in plasmas. Electrons get excited to higher states due to this energy transfer which later falls back due to relaxation transition to a lower or steady state. Photons are emitted during relaxation transition to follow the law of conservation of energy. The photons produce light energy having a certain wavelength. The energy emitted produces different wavelengths which vary for different atoms and molecules. Hence, this principle is utilized for the identification of reactive particles generated in plasma using OES [26].

The density of the investigated energy state generated by the plasma is correlated to the lower energy states using the Saha equation with Boltzmann distribution. Data attained from the OES is analyzed later to get the absolute emission intensity of every spectral line detected. Every spectral line has an individual constant related to its spontaneous emission called the Einstein coefficient. There are many spectral bands or lines representing a specific chemical composition. The representative spectral line for N is 760.9 nm and for O is 844.6 nm. Whereas, the representative spectral lines for Ar are at six different levels from 696.5 nm to 912.3 nm [8].

The spectral breadth of the optical emission line reveals a lot of data relevant to the reactive species that emit light. Several processes influencing the line width make analyzing difficult. Because certain processes produce the Lorentzian line profile while others produce the Gaussian profile. Profile type analysis for the lines generated or the Lorentzian vs. Gaussian component may already give us predictions related to the processes involved. The line broadening part that impacts every element that appears in the examined section determines this which is called the homogeneous line broadening. The average effect which differs for every particle in the examined section known as the inhomogeneous broadening also influences the analysis. The major line broadening mechanisms in the OES are as follows:

- Natural broadening is the homogeneous mechanism of line broadening which is because of the state of excitation of emitting species' finite lifetime τ. This is the result of the Heisenberg uncertainty where the time uncertainty causes a corresponding energy uncertainty; therefore, the transition of energy is smeared over an array of energies that contribute. A Lorentzian shape of the line is the result of homogeneous broadening. Mostly, it is ignored in the study of cold plasma jets since the line broadening is in orders of 10^{-4}.
- Doppler line broadening is due to the Doppler effect. Here, there is a shift of the emitted frequency from the particle that is in movement during a transition influenced by the radiating particle's velocity with respect to the radiation propagation. The increase of the full width due to the Doppler line broadening at half the Gaussian line's maximum is $\Delta v D_{opp} = 2vo\sqrt{\frac{2\ln 2k_BT}{mc^2}}$ where m represents the moving particle's mass and t represents the temperature.
- Most processes in atmospheric pressure plasma generation are dominated by collisions. A collision of an excited particle decreases the lifetime of excitement, causing an emission line broadening similar to the finite natural lifetime broadening. Collisional line broadening increases with an increase in operating pressure. The particles get de-excited in a short time due to the higher collision frequencies. The

broadening coefficient for an N-APPJ which has a low gas temperature is expressed as $\frac{\Delta v}{N} = K(J_G, J_R)\frac{r_e}{\delta}\sqrt{\frac{g_G}{g_R}\frac{f_R}{v_R}}$, where f_R and v_R represent the strength of an oscillator and the resonant transition wavenumber. gG represents statistical weight in ground state and g_R refers to the resonant state and re represents the electron radius. The density of ground state is denoted by N and K (J_G, J_R) is the numerical coefficient depending on the total angular momentum J_G and the resonant state J_R. Line broadening is also expressed as $\Delta\lambda = 9.3492 \times 10^{-14}\sqrt{\frac{g_G}{g_R}}\lambda_R f_R N$. The λ represents the line wavelengths under observation line and λ_R represents the line wavelengths under resonant line, then Δλ is expressed in centimeters.

- Diagnostics-based modeling combined with emission spectroscopy was used to examine the radiofrequency-based cold plasma jet to find the absolute density values of the reactive species. This modeling technique is complex and requires laser beam access and observation optics. It combines the kinetic treatment of the electrons with a 1-D simulation based on fluid equations.

The method is actinometry-based, where the intensities of two optical emission lines are related. One line originates from the actinometer species whose density is known (already measured), whereas the other line is from the species that needs to be diagnosed. It is presumed that the two emission lines are excited due to electron excitation from their ground state. Photo-emission is assumed to cause primary decay. Electron excitation impacts the cross-sections of the two transitions similarly in shape. The ratio of the intensity of the lines is directly related to the ratio of the density of the two species in this case. However, these are only precise assumptions at atmospheric pressure. Therefore, Spatio-temporal properties obtained from the modeling are used to obtain space and time-resolved electron dynamics. Excitation rate coefficients of transitions observed are deduced from this and additionally used to obtain the atomic density of the species by considering related processes such as dissociative attachment [46].

Friedl & Fantz [14] found that the spectral lines' absolute intensity is affected by operational parameters of plasma including the electron temperature, the temperature of the gas, and the density of electrons. In low-temperature plasma under atmospheric pressure, the effective values of T_e and T_g are reported in a very limited range of temperatures 300–600 K and

1–2 eV, respectively. As a result, it is assumed that the population density (Np) is affected by changes in ne than T_e and T_g. As a result, any variations in Np are traced to the electron density changes.

Chung & Ra Kang [9] coupled OES with the Langmuir probe RF-compensated to explore the physical characteristics of a plasma mixture inductively coupled with Ar, O_2, and Ar/O_2. On increasing power and pressure, OES models detected an electron temperature reduction. The temperature also decreased on increasing Argon levels. Dissociation fraction of oxygen in the Ar/ O_2 discharge was also estimated using OES. On increasing the Ar levels and power, the dissociation fraction increased.

Walsh et al. [47] studied the properties of the active species in sub-microsecond pulsed plasma discharge using atmospheric pressure glow discharge with He-O_2. In comparison with a glow discharge, the sub-microsecond pulsed plasma discharge reported low-intensity peaks which correspond to excited O (5p) and O (3p) at 777 nm and 856 nm, respectively, revealing a lower concentration of excited atoms. The stronger emission intensity of reactive nitrogen species was detected below 400 nm, showing a higher concentration of nitrogen in air plasma.

Cherry tomatoes treated under dielectric barrier cold plasma discharge system were evaluated for their physical properties and respiration rates. The cold plasma generated numerous gaseous reactive species which were detected using the optical emission spectral properties of the cold plasma. A spectrometer controlled by a computer was used for optical emission spectroscopy (OES). An optical fiber captured the light emitted from the generated plasma inside the empty package that had high species density at the center of the space between electrodes. The evaluation was done in empty packages as it is hard to maintain a constant arrangement of optical fibers in the packages containing the product. The spectrometer had a 40 mm radius of curvature diffraction grating with 590 grooves every one mm and a 25 mm wide entrance slit. The resolution of the spectrometer is 1.5 nm and it worked in the 190 nm–850 nm spectrum. The fiber with a 0.22 aperture was suitable to be used in the ultraviolet region and the visible part of the spectrum [26].

Another study that evaluated the quality of milk treated with low-pressure plasma was characterized using optical emission spectroscopy. This operated at 0.61–0.72 nm resolution coupled with thin and black charged image sensing equipment that registers the intensity within the spectral range (200–1,100 nm). An optical fiber was used to connect the plasma discharge obtained from the viewport of the glass that was 10 cm distant from the source of plasma to the spectrometer [24]. The excited atomic species were identified from the spectra collected using a NIST library, and the *Identification of Molecular Spectra* book was referred to for the molecular species [10]. The experimental simplicity of OES to characterize the reactive species discharged in plasma makes it an advantageous diagnostic technique for wide usage. However, it is tough to infer the spectra observed to obtain absolute values of the cold plasma [16].

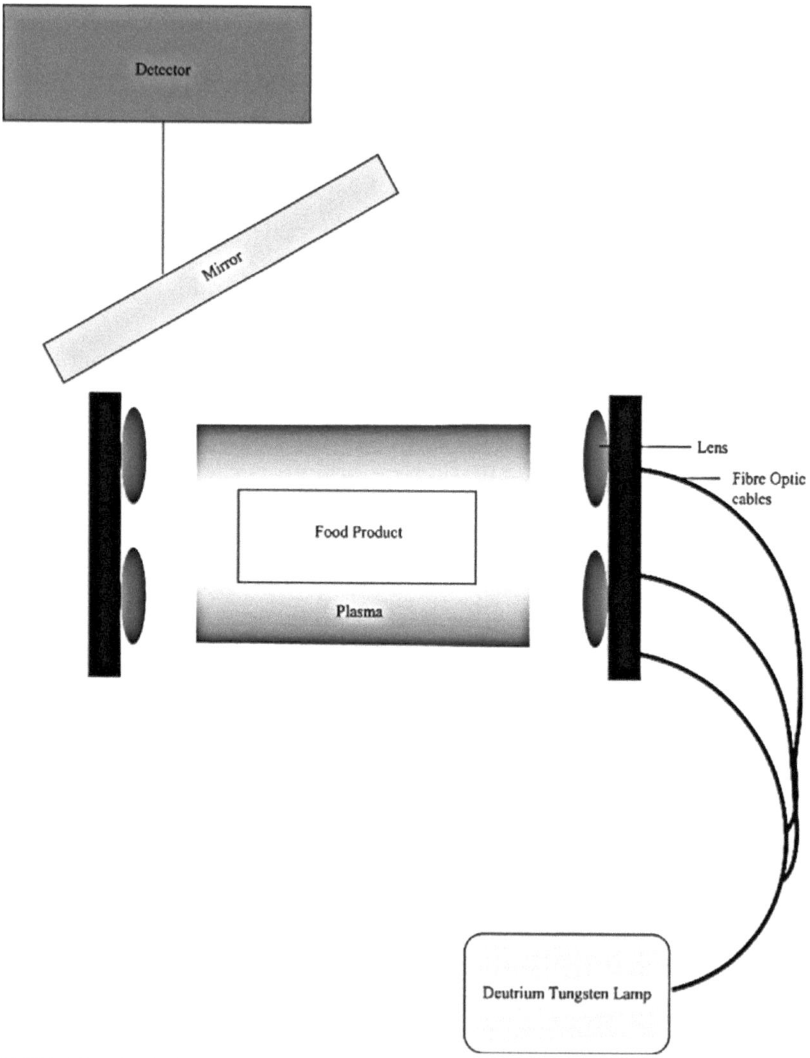

FIGURE 4.2 Optical absorption spectroscopy.

4.3.2 OPTICAL ABSORPTION SPECTROSCOPY (OAS)

Optical absorption spectroscopy (Figure 4.2) is the simplest and a direct approach to measuring absolute densities of reactive plasma species that does not require calibration. However, this supplies only the particles' densities in line of sight. The sensitivity of this spectrometer is quite poor. Nonetheless, absorption spectroscopy may provide a lot of information about the investigated species and its surroundings. Absorption spectroscopy is becoming more useful in the diagnosis of cold plasma jets as sensitivity and light sources become more stable [34].

When there is no or less spectral radiation, employing optical absorption spectroscopy is better than an OES. The plasma species are treated as absorbers of the electromagnetic waves that are conveyed by them or that get reflected at the boundaries of the plasma so that the reactive species can be detected and quantified in the plasma. OAS naturally uses fiber optic cables and one spectrometer similar to that of an OES, along with the Deuterium-Tungsten lamp as a light source.

Plasma diagnostics by absorption spectroscopy can be done from the vacuum ultraviolet region (VUV) to the mid-infrared range (MIR). The oxygen atoms, nitrogen atoms, and the singlet delta oxygen show absorption features in the VUV spectrum. However, VUV absorption radiation will generally function in a vacuum for the regulation of the ambient gas as it has a good absorption capacity with the particles in humid air. The UV absorption spectroscopic technique has more sensitivity for the detection of ozone and hydroxyl radicals, but the measurements should be done to prevent misinterpretations of the species density. Argon and helium metastables can be measured in the visible region and the near-infrared range, respectively.

The Beer-Lambert law explains the light absorption by an absorbent material, which provides the basis for absorption spectroscopy. The absorbing species density is distributed uniformly throughout the light route of length L is linked to the wavelength-dependent intensity I_0 of light of incidence as follows:

$$I(\lambda) = \mathrm{I0}(\lambda)e^{-N\sigma(\lambda)L} = \mathrm{I0}(\lambda)e^{-k(\lambda)L} \tag{1}$$

In Eqn. (1), the transmitted radiation is represented as $I(\lambda)$, while absorption cross-section is referred by $\sigma(\lambda)$, and absorption coefficient as $k(\lambda)$ that are wavelength-dependent.

The absorber density may be determined using the absorbance A as follows:

$$A(\lambda) = -\ln\left(\frac{I}{Io}\right) = N\sigma(\lambda)L \tag{2}$$

OAS was used to determine the nitrogen dioxide and ozone concentration using their absorbance between 385–415 nm and 250–160 nm, respectively in the plasma discharge inside the food packages in a study [31]. The absorbance bands revealed the existence of N_2O_5, H_2O_2, and HNO_4 in the 220 nm to 240 nm wavelength range. The absorbance bands showed the presence of OH and water-bonded OH compounds at a wavelength range of 290 nm to 340 nm. Owing to the quick reaction times, the peak of absorption for the existence of OH radicals was not observed at 308 nm. UV-vis absorbance was used to quantify the gas composition after the plasma discharge was subjected to the food package for a short period of 15–20 seconds [27]. The studies revealed that nitrogen oxides and ozone concentrations increased with increasing relative humidity in the enclosed gas. However, in the closed container, the levels of both ozone and nitrogen oxides started to decrease when the relative humidity increased. OAS was employed to detect the plasma species produced at high voltage from 70 kV to 88 kV. Reactive species concentrations are calculated with relation to the voltage and the gas humidity of the plasma.

Schmidt-Bleker, Reuter, and Weltmann's simple modeling method in combination with absorption spectroscopy investigated the long-living plasma species at various levels of humidity that were present in the mixture of O_2 and N_2 plasma jet [37]. The outcome revealed that clear absorptions were found in the O_3 and NO_2 zones, even though the absorption of nitric acid (HNO_3) and hydrogen peroxide (H_2O_2) were discovered. The quantitative assessments are impossible because of poor signals. The O_3 density lowered dramatically with increased humidity up to 1.75 $\times$ 10^{14} cm^{-3} humidity in dry gas and up to 1.25 $\times$ 10^4 cm^{-3} at a relative humidity of 36%, however, it increases with the oxygen concentration of the gas mixture on a continual basis. Nitrogen dioxide had a density approximately equal to 8 $\times$ 10^{12} cm^{-3} with 18% relative humidity. The density was maximum when it was a gas mixture with 30% oxygen. Other gases' densities were analyzed using an OES which were 5 $\times$ 10^{12} cm^{-3} for H_2O_2, 3 $\times$ 10^{12} cm^{-3} for nitric acid, and 10^{14} cm^{-3} for OH produced in plasma jets [48].

Meike Muller investigated microbial inactivation using a surface microdischarge plasma treatment. UV absorption spectroscopic technique was employed to evaluate the ozone concentration, the dominant species generated in the cold plasma afterglow circulation equipment. UV light having an initial intensity of I_0 enters the volume which reduces on entering the volume containing the absorbing gas in the length of the optical path. The intensity of the residual UV light I (t) is further determined. A spectrometer and a mercury or argon UV lamp were utilized for this measurement. Optical fibers containing collimating lenses are fitted for this purpose in the treatment chamber's side walls center or the measurement box. One fiber is connected to the light source whereas another fiber to a spectrometer. The small measurement box aids in analyzing the concentration of ozone at various positions within the apparatus. The UV measurement box was kept hot using warm heating oil at 50°C to prevent condensation over the lenses in the spectrometer. The concentration of ozone is monitored using the UV absorption at a wavelength of around 254 nm. The intensity that was measured is integrated from 253.5 nm to 254.5 nm. Signal integration time is altered between 50 and 200 ms and data points are recorded for every 500 ms [28].

The gas composition after the exposure of plasma was evaluated using an OAS by Patil et al [31]. The light source used was a UV-visible deuterium-hydrogen lamp and an Ocean Optics spectrometer having a resolution of 1 nm and UV-Vis range 200–1,100 nm was used to analyze the light transmitted. Inside the enclosed container used for all tests, the optical probes that insulated the UV-Vis collimators are arranged. Between the probes, the length of the optical path was 2.4 cm. The optical assembly between the probes, lamp, and the spectrometer was established using optical fibers having a range within UV-Vis, a numerical aperture of 0.22, 600 μm core, and 30 cm length. The concentrations of NO_2 and O_3 are determined using the Beer-Lambert equation, which included averaging concentrations throughout a wavelength range in which every species had a maximum absorption cross-section, else had values significantly greater than other absorbent species. The MPI database provided the absorption cross-sections that were used for this study. A direct deconvolution was employed to detect the concentration of every species by analyzing the NO_2 in the wavelength range from 385 nm till 415 nm and ozone levels on the basis of their absorbance from 250 nm till 260 nm wavelength range. Measurements are made by the OAS for the treatments done by generating plasma using ozone at various relative humidity levels in the air at the same plasma treatment time. Before and after plasma exposure, the intensity of the light that was transmitted was measured. Food applications of cold plasma using optical techniques for plasma characterizing are shown in Table 4.1.

TABLE 4.1 Food Applications of Cold Plasma using Optical Techniques for Plasma Characterizing

Treatment Objective	Cold Plasma Discharge Type	Characterizing Technique of Cold Plasma	Analyzed Reactive Species	References
Evaluation of physicochemical properties and microbial load in Coconut Neera	Atmospheric air cold plasma system	Electron paramagnetic resonance	OH. Radicals	[1]
Regulating aflatoxin M1 in skimmed milk	High Voltage Atmospheric pressure cold plasma	Optical absorbance spectroscopy	Ozone, NO_2, NO_3, N_2O_4, (N_2O_5 + H_2O_2)	[29]
To inactivate *Listeria monocytogenes* occurring in Queso Fresco cheese	DBD of High Voltage Atmospheric pressure cold plasma	Optical absorbance spectroscopy	O_3, NO_2, NO_3, N_2O_4, N_2O_5	[30]
To inactivate *Salmonella enterica* and *E. coli* in baby spinach leaves	High Voltage Atmospheric pressure cold plasma (DBD type)	Optical emission spectroscopy	N_2^+, NO and excited states of hydrogen	[41]
To inactivate *Salmonella* in pet foods freeze-dried	In-package atmospheric pressure cold plasma	Optical emission spectroscopy	Excited nitrogen species, hydroxyl (OH) and oxygen species; Gas temperature of O_3, NO^+NO_2, and H_2O_2 were also measured	[49]
Treating liquid media kept inside sealed packages to inactivate *E. Coli*	Dielectric barrier discharge	Optical Emission Spectroscopy	N_2 and N_2^+ excited species	[52]

4.3.3 FLUORIMETRY

Laser-induced fluorescence technique is a widely used plasma analytical technology that can quantify plasma density and the temperature of the ground-state plasma reactive particles with high spatial and temporal resolutions. Laser-induced fluorescence works on the principle that uses laser radiation whose wavelength is tunable and is utilized to excite atoms or molecules to a higher energy state. Light is emitted by a material to which a dye laser has provided energy. The energy is absorbed by the fluorophore at a specific wavelength from light (in the ultraviolet or visible ranges) and subsequently releases energy as light with a higher wavelength. This fluorescence radiation provides information on the population density of energy states when the parameters of laser beam excitation and the impacts of the quenched substances and the energy dissipating processes are identified. As a result, it can offer useful data on the density of atoms and molecules in the ground or excited states. Although Laser-induced Fluorescence detection systems have a high level of experimental complexity, they offer great sensitivity with good spatial and temporal resolutions. The distribution of the density and temporal characteristics of oxygen species like hydroxide and O radicals created using helium cold plasma jets has been effectively investigated using Laser-induced fluorescence [50].

In a fluorimetry system induced by a single-photon laser, only one photon is used for every excitation of the investigated atom or molecule. It causes difficulty since fluorescence photons may get re-absorbed in optically dense plasmas, obscuring measurements. Moreover, using single-photon transitions restricts the energy transitions that are possible. Light atoms, like H_2, O_2, and N_2 have transition energies at a ground state greater than 6.5 eV, necessitating the usage of vacuum ultraviolet laser (VUV) if laser-induced fluorescence is the preferred approach. The experimental hurdles of producing and transporting the radiation are substantial, and the VUV radiation may be less appropriate for N-APPJ diagnosis when this was done for Oxygen atoms and Hydrogen atoms [12]. For this case, two-photon laser fluorescence technique can be employed. The single-photon technique has the advantage of being simple experimentally since the fluorescence emitted depends linearly only on the intensity of the laser, whereas two-photon laser-induced fluorescence has a quadratic dependence. Hence data evaluation is easy in the case of the single-photon technique. However, data evaluation is complicated as laser-induced fluorescence is mostly employed for molecule detection. The measurements are obscured by vibrational energy and rotational energy dissipation and also collisional quenching. Laser scattering noise might also cause concealed measurements.

The electronic transitions of light atoms to their ground state are above the range of 6.5 eV, hence laser-induced fluorescence technique needs high-energy VUV photons. The media used is not always transparent for the VUV photons, which makes the volume to be investigated optically thicker. It is also challenging to deliver an environment free of oxygen to allow VUV by the path of the laser to reach the plasma. Hence, the two-photon laser-induced fluorescence technique can be used for simpler excitation schemes instead of the one-photon technique. Two-photon techniques can detect light atoms using high laser intensity. This is a non-linear process that makes the interpretation and quantification of signals complex when compared to the single-photon technique. The main disadvantage of the two-photon technique is that the excitation and detection systems cannot be calibrated using Rayleigh scattering. A linear relation between laser energy and Rayleigh scattering energy cannot be assumed for multi-photon processes like single-photon laser-induced fluorescence. Hence, other methodologies should be used to calibrate two-photon techniques. The signal from the fluorescence induced by a two-photon laser is compared with the density of any one of the known species. A reference plasma discharge can be used for this, whose density is determined by other methods. But it may be problematic to detect the transient species and the intermediate species densities. Another method is measuring the collisional quenching for the pure oxygen molecule at an excited state due to the laser that is high but has a variable dissociation. As oxygen has a lower quenching rate when compared to an oxygen molecule, the measurements can be related to the density of an oxygen atom [11]. The general method to calibrate a two-photon absorption fluorescence signal is by utilizing densities of noble gases that are already known. Noble gases are used as their concentration doesn't decay, and hence the density is evaluated by the ideal gas law [15].

Fluorescence induced by a two-photon laser was employed to determine the absolute density of atomic oxygen and atomic hydrogen from the plasma discharges. The utilization of the two-photon technique is to get the target atom's absolute density by comparing it with the laser, same as the calibration gas whose density is known. This was employed to determine the density of atomic oxygen in the discharge of an Ar or O_2 atmospheric level pressure cold plasma jet that was calibrated using xenon gas [35]. The $2P^4\ {}^3P_2$ state, which is oxygen atoms' ground state, is excited to the $3P^3P_J$ state using two photons (225.65 nm), and the ensuing fluorescence radiation is measured in the infrared region at 844.87 nm. The density of atomic oxygen is found to be around 3.75×10^{15} cm^{-1} near the opening of the plasma jet and decreases as the distance from the opening increases.

Dichlorodihydrofluorescein diacetate is extensively used as a fluorescence probe for detecting hydrogen peroxide and oxidants in cellular plasma diagnostics. Hydroethidine probe is commonly used for measuring O_2^- and dihydrorhodamine probe is generally used for the assessment of $ONOO^-$ [22]. Laser-induced fluorescence can be successfully used to measure the population density and energy state of reactive species. It is also employed in high spatial and time resolution interactions with biological cells. However, laser-induced fluorescence is based on a few model assumptions and lacks accuracy in testing the temperature of the gas and electron characteristics [23].

4.4 NON-OPTICAL TECHNIQUES FOR CHARACTERIZATION OF COLD PLASMA

The multi-phase systems of interest in plasma-processed foods need other characterization methods that do not depend on optical spectroscopic methods. The study of plasma interaction with liquids requires different approaches to characterization. Non-optical techniques including mass spectroscopy, scattering techniques, and electron paramagnetic resonance spectroscopy are alternatives to optical techniques for the characterization of cold plasma. With these techniques, highly soluble particles like ions, and highly reactive particles like radicals, and their interaction with other states of matter can be evaluated.

4.4.1 MASS SPECTROSCOPY

Molecular beam mass spectrometry is one characterization technique used for detecting neutral and charged atomic particles and molecular species generated by plasma jets. The examined particles are typically sampled into a three-step differential pump through a tiny aperture. Skimmers link the second and third stages, determining which particle beam will arrive in the ionizing compartment and the detector. The objective is to create a collisionless particle ray that mass spectrometry techniques measure. However, the ratio of the investigated species composition is disturbed by many processes that should be considered. The opening should be kept small so the plasma does not enter the orifice and produce more reactive species into the mass spectrometer. The plasma sheath thickness determines the size of the opening, which must be in the range of tens of micrometers at atmospheric pressure. The species size that is investigated is extended by suction using vacuum pumps. Collisions inside the orifice with other particles or with walls also disturb the composition of the particles. This accelerates the particles and forms a Mach cone within the orifice. Therefore, the first skimmer must be kept in the silent zone to detect the required collision-less particle ray. The substance's mean free path when the background pressure is low is long to facilitate the second stage to directly catch collision-free particles. For every species, the particle diffusion and particle velocity vary based on their specific weight within the mass spectrometer. For example, the atomic weight of helium is different from the nitrogen or oxygen molecular species.

Hence, the molecular beam mass spectrometer does not provide absolute or relative densities of the species without calibration. Absolute concentrations are measured when it is calibrated by keeping a compartment with the species whose composition is known before the opening. Neutral species can be sampled by gas ionizing varying the electron energy. 70 eV electron energy is generally noticed at the maximum ionization. The high electron energy ionizes as well as dissociates the gas, hence influencing the composition of the particle. As a result of dissociative ionization, each ion is accompanied by more ions. This cracking pattern is specific to each electron energy and species. The original composition of the species can be determined as a superposition of each detected species' cracking pattern at the known cracking pattern. Therefore, the molecular beam mass spectrometer detects ions more easily. Ions produced by plasma certainly do not require any further ionization in the spectrometer and can be diagnosed directly [23].

Molecular beam mass spectrometry is an important method for diagnosing plasma to provide information about the absolute densities of the reactive oxygen species. A molecular beam mass spectrometer contains two or three pumping stages that are interconnected by small, grounded orifices. On the line of sight is a gas ionizer placed with the sampling orifice in the mass spectrometer. A mixture of gases is permitted to flow via small orifices in the high-pressure compartment while in operation, which forms the molecular beam containing atoms, molecules, ions, and also free radicals. This molecular beam is analyzed by the mass spectrometer to diagnose and quantify plasma using differences in their mass. Through this, the reactive species involved in the process can be assessed. An orifice of about 100 μm in diameter was placed on the plasma jet's axis to capture the mixture of gases for further investigation [13].

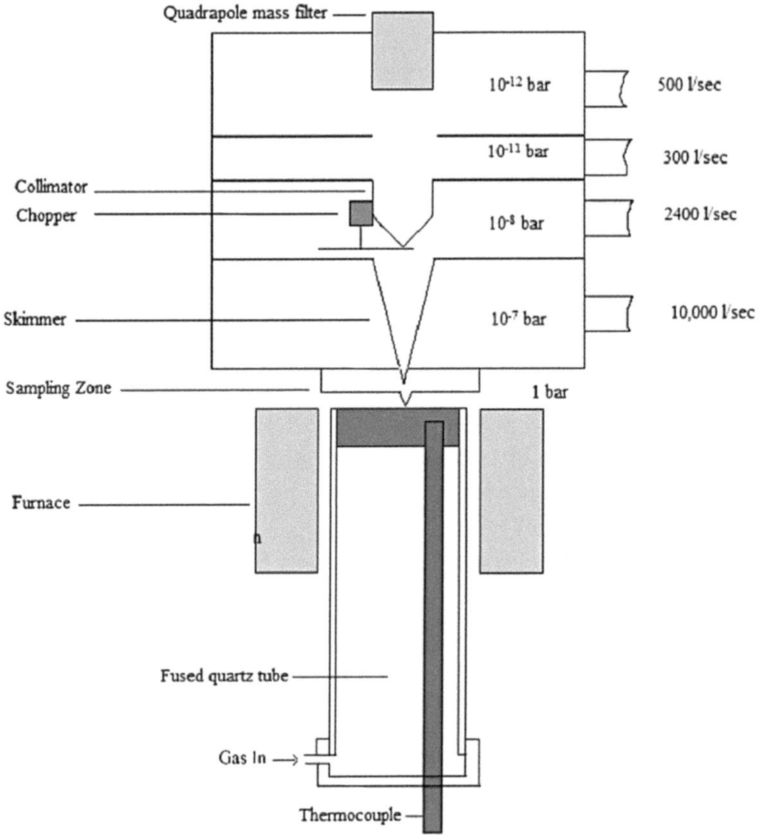

FIGURE 4.3 Schematic diagram of molecular beam mass spectroscopy.

The gas flow through the primary orifice is less than the flow of gas at the plasma jet, which causes stagnation. Pulsed gas flow identified by the rotating skimmer is only allowable via this system. It is a disk made of stainless steel consisting of fine embedded skimmers. When the skimmer aligns to the opening or is blocked in the blocking position by the disk, the molecular beam penetrates to the second stage. The density molecular beam mass spectroscopy detects the plasma species density as a combination of the density of the species present in the molecular ray and background density associated with the pressure level in the pumping system in the mass spectroscopy [6]. Mass spectrometers can diagnose neutral and charged atomic and molecular active species but do not give absolute and relative densities of these species without calibration.

Using molecular beam mass spectrometry (Figure 4.3), the discharge from a microscale plasma jet using helium that had a lower proportion of oxygen was measured [13]. The density of oxygen was recorded as 4.7×10^{15} cm^{-3} in the effluent, and the depletion of oxygen up to 10% was also measured. With a higher input of oxygen proportion, a continuous increase in ozone density was observed. However, the ozone density decreased as the distance from the jet nozzle increased up to 40 mm. The absolute density of atomic nitrogen (N) was also assessed using mass spectrometry in the effluent of a microscale cold plasma jet operating in helium with a slight admixture of nitrogen gas [38]. As the applied voltage increased, the density of nitrogen was found to increase linearly. The maximum density of nitrogen was around 1.5×10^{14} cm^{-3} when the proportion of nitrogen gas was varied, which was about 25% of the nitrogen gas admixture. Molecular beam mass spectrometry was also used to quantify the nitric acid and ozone concentration produced by an RF-driven atmospheric pressure cold plasma jet [45].

Another detection approach involves using threshold ionization spectroscopy, in which the electron energy that ionizes rises from zero to its highest value, and a spectrum is obtained. A signal can be expected at the species' threshold ionization energy identified. The ground-state particles are ionized by the electrons, and the excited species are present in the plasma. In threshold ionization spectroscopy, the species are identified as they appear in the spectrum before the species in their ground state [18].

4.4.2 SCATTERING TECHNIQUES

Fluorimetry and emission techniques for measurements are often non-destructive and provide excellent spatial and temporal resolution. They do, however, rely on a number of modeled assumptions and are not precise, as well as they cannot investigate parameters relevant to the electrons or measure gas temperatures directly. Scattering methods are especially beneficial to probe these characteristics (Figure 4.4). Elastic scattering techniques such as Rayleigh from heavy particles, while Thomson from electrons, whereas inelastic Raman scattering technique from molecules are examples of scattering methods used for the characterization of plasma. Having a high spatial resolution with an exact calibration, all these scattering signals can reveal the temperature and densities of the electron, including the gas temperature, partial pressure of the air, and the ratio of N_2 to O_2 in the atmospheric-pressure plasmas [44].

Rayleigh scattering signal is elastic laser radiation of atoms, molecules, or any other particles whose wavelength is less than the radiation. It is utilized to analyze the gas temperature or density. Rayleigh scattering is caused by the interaction of a light wave's fluctuating electric field with the investigated particles' localized charges. The charges shift with the incident light frequency due to the ability of scattering particles to polarize, resulting in the particle's dipole radiation, that is perceived as the scattered light signal.

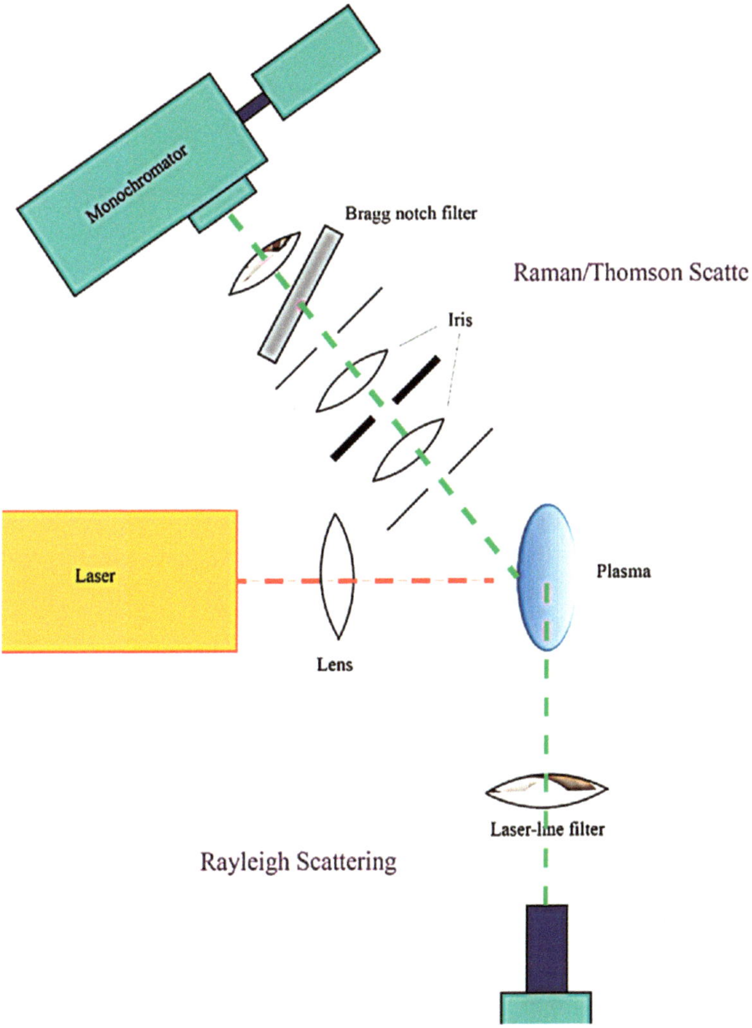

FIGURE 4.4 Scattering techniques used for cold plasma characterization.

The ideal gas law states that the signal is directly proportionate to the "n" number of particles scattered,

$$n = \frac{p}{k_B T_G} \tag{3}$$

where, k_B represents Boltzmann constant, and the gas temperature represented by T_G at pressure p. Inference of the temperature of the gas from the Rayleigh scattering technique is done by taking a known gas temperature as the reference temperature.

It is evaluated using the equation $T_{ref}\, I_{plas} = T_{ref}\, I_{ref}$, where, I_{plas} is the Rayleigh scattering signal's intensity on the species when plasma is turned on while the Iref is the scattering intensity on the species when plasma is turned off.

Gas temperature can be deduced assuming room temperature for calculations when plasma is switched off. The Rayleigh cross-section differs for different gases. Hence for the reference spectrum, it must be performed optimally at the same gaseous composition. The density of the gas can be evaluated from the Rayleigh using:

$$n_g = \frac{S_{Pl+st} - S_{st}}{S_{ref+st} - S_{st}} n_{ref} \tag{4}$$

where, n_{ref} is the gas density that is known and injected into the plasma chamber at room temperature at a fixed pressure. S_{Pl} is the Rayleigh signal of the plasma; S_{St} is the stray light evaluated at zero measurements with a reference measurement, and S_{ref} is the reference gas signal at a particular temperature and pressure.

The density and temperature of the electron are the least evaluable but important parameters in an atmospheric pressure plasma discharge. Langmuir probes can be employed to measure these parameters at low-pressure plasma discharges. However, they haven't given accurate results for small-size collision-dominated atmospheric pressure plasma jet measurements as there are no applicable theories. The Thomson scattering technique is one of the methods that can be used to evaluate the electron energy and density accurately. But scattering signal in the Thomson technique is extremely weak and hence has a less efficient scattering signal. In a study, where a 20 W laser having a flux of photons in several 10^{19} photons/seconds was used, only a few 10^5 photons/seconds were Thomson scattered. Only 10^3 photons/sec were collected from these by the optical device of 60 cm focal length and 10 cm diameter. Considering the probability of the transmission of the detection system with the lenses, the spectrograph, and the detector, only 10 photons/seconds are detected out of the initial 10^{19} photons/seconds [7].

The 18-order magnitude difference is problematic in the Thomson scattering signal. The photons scattered inelastically are noticed at a somewhat different wavelength when compared to the wavelength of the laser. The laser light was stronger than the measured signal due to the laser's higher flux. Thus, complicated spectroscopic techniques intended to filter the central laser wavelengths with steeper filter edges and high optical density are used to quantify Thomson scattering. Triple grating spectrographs can be utilized, where the negative apertures filter the laser radiation, and the walls and apertures within spectrographic devices restrict other dispersed light. Another option is to employ a vapor cell that contains a gas which absorbs the laser energy's wavelength.

The scattered laser light power is measured as plasma per wavelength. P_λ is computed to calibrate the scattering signal by:

$$P_\lambda = n\varepsilon\,\Delta\Omega\,P_i \frac{d\sigma}{d\Omega} S_\lambda(\lambda) \tag{5}$$

where, n refers to the scattering signal's density, Pi represents incident laser power; S_λ is the spectral distribution, ε is a constant representing optics and detection efficiency, and also the detection volume's length; ΔΩ refers to the detection solid angle and $\frac{d\sigma}{d\Omega}$ is the differential cross-section.

Thomson scattering signals have high-velocity electrons and hence are Doppler broadened. From the scattering signal's line broadening profile, the electron temperature is evaluated:

$$T_e = \frac{m_e c^2}{4k_B}\left(\frac{\Delta\lambda}{\lambda_1}\right)^2 \tag{6}$$

where, Δλ refers to the Thomson scattering profile's half-maximum, λ_1 is the wavelength of the laser, m_e represents the electron mass, c represents the light speed.

Electron density is computed from the Eqn. (7):

$$n_e = n_g \frac{P_\lambda^T}{P^R} \frac{d\sigma^R / d\Omega}{d\sigma^T / d\Omega} \tag{7}$$

where, PT and PR represents the scattered Thomson and Raman signals power, respectively, and ng is the density of the gas, $d\sigma^T/d\Omega$ and $d\sigma^R/d\Omega$ refers to the Thomson and Raman scattering cross sections, respectively.

4.4.3 ELECTRON PARAMAGNETIC RESONANCE (EPR) SPECTROSCOPY

The electron paramagnetic resonance spectroscopic technique (EPR) is also known as electron spin resonance spectroscopy (ESR). It can be employed to directly identify highly active oxygen species, as well as organic and inorganic species that are paramagnetic in food systems after applying novel processing technologies. This spectroscopy detects the degenerate energy levels splitting into many sublevels inside the magnetic field as a result of the Zeeman effect. The spectra obtained are every radical's distinguishing feature. The magnetic moment of the electron m_s aligns itself in a parallel direction or in an anti-parallel direction to the magnetic field.

The magnetic moment $m_s = ½$ for parallel alignment and $m_s = – ½$ for anti-parallel moment each having an energy of: $E = m_s g_e \mu_B B_0$, where: g_e is called the Lande factor that links the electron's magnetic moment with angular spinning momentum quantum number, which is 2.0023 (for a free electron), $\mu_B = eh/ 2m_e$ representing Bohr magnetron and B_0 refers to the magnetic field.

Energy difference between two energy states of parallel alignment and the electron's magnetic anti-parallel alignment moment is directly related to the magnetic field applied by the relation $\Delta E = g_e \mu_B B_0$. The magnetic field agrees with the resonance energy criteria in EPR; then an absorption line can be seen. EPR is a familiar surface diagnosing technique and is also employed for the detection of water-based short-living reactive radicals like O_2^-, NO, OH, and. OOH. The detection of unpaired electrons, i.e., free radicals and transition metal ions using this approach, is widely used in quantum physics and structural biology. 5,5-dimethyl-pyrroline N-oxide, 5,5-dimethyl-pyrroline N-oxide, and alpha-phenyl N-tertiary-butyl nitrone are the most often utilized spin traps for plasma diagnostics. Furthermore, non-optical methods (having low visualization capacity) and scattering techniques (Rayleigh scattering signal, Thomson scattering signal, and Raman scattering signal) are alternatives. The electron spin trapping technique necessitates extra caution in selecting the suitable spin trap and liquid, both of which might affect measurement results [51]. Detection of transient species in liquids needs extra care because OH radicals have a short life, and *in situ* diagnostics can be conducted rarely, but it needs liquid handling steps until measurements are initialized.

Spin trapping technique was developed, where nitroso compounds or nitrone compounds form a stable spin adduct that is paramagnetic and hence electron paramagnetic resonance active. Here, radicals with microsecond lifetime form a spin adduct with a lifetime ranging from a few minutes to a few hours, which is stabilized enough to conduct *ex-situ* EPR studies [21]. The spin trap must be added before the cold plasma treatment or to the plasma-activated medium. The plasma-treated medium must then be placed in a cuvette, which is introduced into the resonator in the EPR spectrometer. Choosing a suitable spin trap is crucial for better measurement quality as it significantly reduces background noise and false signals. The choice of liquid used also strongly affects the measurement results [43]. The possible impact of the liquid medium is more directly associated with the kinetic process of the ongoing reaction rather than EPR excitation. EPR spectroscopy can also detect the effect of plasma jets producing VUV radiation, generating radicals in a small enclosed liquid volume since this is a remote technique [20].

High sensitivity, specificity, small sample quantity requirement, and commercial availability of several types of EPR are the main advantages of using this technique. ESR was employed to diagnose the short-living radical species like hydroxyl radicals with a lifetime <100 ms, peroxynitrous lifetime ~ 1 ms, and a lifetime of superoxide radicals and hydroperoxyl radicals <10 ms in liquids [3]. It has great potential to detect the penetration depth of cold plasma by diagnosing the formations of free radicals generated by the cold plasma at the sub-surface layers for proper decontamination assurance. It can also be employed to evaluate the nature and concentration of the free radicals which affect the cold plasma treated food quality [5].

4.5 SUMMARY AND FUTURE PROSPECTIVES

Cold plasma is one promising technology that is studied as a processing technique for several food products. It is mostly employed for the destruction of microbes but also used for toxin degradation, inactivating enzymes, degrading residual pesticides, and functionalization of food systems. The majority of the present cold plasma literature in the realm of food science and technology is the knowledge concerning its chemical effects on food and is centered on microbial inactivation efficacy. This chapter focuses on characterizing the plasma's chemical interactions for an improved understanding of these effects.

To successfully adopt cold plasma treatment in the food processing industries, it is important to elucidate the interactions of the plasma species with the food systems. The reliability and safety of plasma processing technology in food commodities increase the confidence level in the utilization of efficient atmospheric pressure-level cold plasma approaches and tools. Various approaches such as optical and non-optical techniques employed in various studies relating to cold plasma applications in foods were described. Development, detection, and simulation of plasmas generated and a scale-up process having potential in commercial applications have to be done in the future. Research on the plasma effects on the gas dynamics in the plasma jets should be done further. Kinetic and photoionization models for different plasma compositions need to be developed. Characterizing the plasma also helps in quantifying the effects of multiple plasma discharge pulses on foods further.

KEYWORDS

- **cold plasma**
- **diagnostics**
- **non-optical characterization methods**
- **optical characterization methods**
- **reactive nitrogen species**
- **reactive oxygen species**

REFERENCES

1. Aparajhitha, S., & Mahendran, R. (2019). Effect of plasma bubbling on free radical production and its subsequent effect on the microbial and physicochemical properties of coconut neera. *Innovative Food Science & Emerging Technologies, 58*, Article 102230. https://doi.org/10.1016/j.ifset.2019.102230.
2. Aranda Gonzalvo, Y., Whitmore, T. D., Rees, J. A., Seymour, D. L., & Stoffels, E. (2006). Atmospheric pressure plasma analysis by modulated molecular beam mass spectrometry. *Journal of Vacuum Science & Technology A: Vacuum, Surfaces, and Films, 24*(3), 550–553.

3. Attri, P., Kim, Y. H., Park, D. H., Park, J. H., & Hong, Y. J. (2015). Generation mechanism of hydroxyl radical species and its lifetime prediction during the plasma-initiated ultraviolet (UV) photolysis. *Scientific Reports, 5*(1), 1–8.
4. Attri, P., Kumar, N., Park, J. H., Yadav, D. K., & Choi, S. (2015). Influence of reactive species on the modification of biomolecules generated from the soft plasma. *Scientific Reports, 5*(1), 1–12.
5. Barba, F. J., Roohinejad, S., Ishikawa, K., Leong, S. Y., El-Din, A., & Bekhit, A. (2020). Electron spin resonance as a tool to monitor the influence of novel processing technologies on food properties. *Trends in Food Science & Technology, 100*, 77–87.
6. Benedikt, J., Ellerweg, D., & Von Keudell, A. (2009). Molecular beam sampling system with very high beam-to-background ratio: The rotating skimmer concept. *Review of Scientific Instruments, 80*(5), Article 3133804. https://doi.org/10.1063/1.3133804.
7. Carbone, E., & Nijdam, S. (2015). Thomson scattering on non-equilibrium low density plasmas: Principles, practice and challenges. *Plasma Physics and Controlled Fusion, 57*(1), Article 014026. https://doi.org/10.1088/0741-3335/57/1/014026.
8. Chiper, A. S., Chen, W., Mejlholm, O., Dalgaard, P., & Stamate, E. (2011). Atmospheric pressure plasma produced inside a closed package by a dielectric barrier discharge in Ar/CO_2 for bacterial inactivation of biological samples. *Plasma Sources Science and Technology, 20*(2), Article 025008. https://doi.org/10.1088/0963-0252/20/2/025008
9. Chung, T. H., Ra Kang, H., & Bae, M. K. (2012). Optical emission diagnostics with electric probe measurements of inductively coupled Ar/O_2/Ar-O_2 plasmas. *Physics of Plasmas, 19*(11), Article 4765357. https://doi.org/10.1063/1.4765357.
10. Dharini, M., Jaspin, S., & Mahendran, R. (2022). Optical emission spectra characterization of plasma discharge during the processing of milk and evaluation of its quality attributes. *IEEE Transactions on Plasma Science, 50*(4), 1128–1136.
11. Dilecce, G., Vigliotti, M., & De Benedictis, S. (2000). A TALIF calibration method for quantitative oxygen atom density measurement in plasma jets. *Journal of Physics D: Applied Physics, 33*(6), 53–56.
12. Durant, J. L. (1994). Product branching fractions in the reaction of NH (ND) (3Σ-) with NO. *The Journal of Physical Chemistry, 98*(2), 518–521.
13. Ellerweg, D., Benedikt, J., Von Keudell, A., Knake, N., & Schulz-Von Der Gathen, V. (2010). Characterization of the effluent of a He/O_2 microscale atmospheric pressure plasma jet by quantitative molecular beam mass spectrometry. *New Journal of Physics, 12*(1), Article 013021. https://doi.org/10.1088/1367-2630/12/1/013021.
14. Friedl, R., & Fantz, U. (2012). Spectral intensity of the N2 emission in argon low-pressure arc discharges for lighting purposes. *New Journal of Physics, 14*(4), Article 043016. https://doi.org/10.1088/1367-2630/14/4/043016.
15. Goehlich, A., Kawetzki, T., & Döbele, H. F. (1998). On absolute calibration with xenon of laser diagnostic methods based on two-photon absorption. *The Journal of Chemical Physics, 108*(22), 9362–9370.
16. Granier, A., Vervloet, M., Aumaille, K., & Vallée, C. (2003). Optical emission spectra of TEOS and HMDSO derived plasmas used for thin film deposition. *Plasma Sources Science and Technology, 12*(1), 89–96.
17. Graves, D. B. (2012). The emerging role of reactive oxygen and nitrogen species in redox biology and some implications for plasma applications to medicine and biology. *Journal of Physics D: Applied Physics, 45*(26), Article 263001. https://doi.org/10.1088/0022-3727/45/26/263001.

18. Große-Kreul, S., Hübner, S., Schneider, S., Ellerweg, D., & Von Keudell, A. (2015). Mass spectrometry of atmospheric pressure plasmas. *Plasma Sources Science and Technology, 24*(4), Article 044008. https://doi.org/10.1088/0963-0252/24/4/044008.
19. Halliwell, B. (2006). Reactive species and antioxidants. Redox biology is a fundamental theme of aerobic life. *Plant Physiology, 141*(2), 312–322.
20. Jablonowski, H., Bussiahn, R., Hammer, M. U., Weltmann, K. D., Von Woedtke, T., & Reuter, S. (2015). Impact of plasma jet vacuum ultraviolet radiation on reactive oxygen species generation in bio-relevant liquids. *Physics of Plasmas, 22*(12), Article 4934989. https://doi.org/10.1063/1.4934989.
21. Janzen, E. G., & Blackburn, B. J. (1969). Detection and identification of short-lived free radicals by electron spin resonance trapping techniques (Spin Trapping). Photolysis of organolead, -tin, and -mercury compounds. *Journal of the American Chemical Society, 91*(16), 4481–4490.
22. Kalyanaraman, B., Darley-Usmar, V., Davies, K. J. A., Dennery, P. A., & Forman, H. J. (2012). Measuring reactive oxygen and nitrogen species with fluorescent probes: Challenges and limitations. *Free Radical Biology and Medicine, 52*(1), 1–6.
23. Lu, X., Naidis, G. V., Laroussi, M., Reuter, S., Graves, D. B., & Ostrikov, K. (2016). Reactive species in non-equilibrium atmospheric-pressure plasmas: Generation, transport, and biological effects. *Physics Reports, 630*, 1–84.
24. Manoharan, D., Stephen, J., & Radhakrishnan, M. (2021). Study on low-pressure plasma system for continuous decontamination of milk and its quality evaluation. *Journal of Food Processing and Preservation, 45*(2), Article e15138. https://doi.org/10.1111/jfpp.15138.
25. Misra, N. N., Kaur, S., Tiwari, B. K., Kaur, A., Singh, N., & Cullen, P. J. (2015). Atmospheric pressure cold plasma (ACP) treatment of wheat flour. *Food Hydrocolloids, 44*, 115–121.
26. Misra, N. N., Keener, K. M., Bourke, P., Mosnier, J. P., & Cullen, P. J. (2014). In-package atmospheric pressure cold plasma treatment of cherry tomatoes. *Journal of Bioscience and Bioengineering, 118*(2), 177–182.
27. Moiseev, T., Misra, N. N., Patil, S., Cullen, P. J., & Bourke, P. (2014). Post-discharge gas composition of a large-gap DBD in humid air by UV-Vis absorption spectroscopy. *Plasma Sources Science and Technology, 23*(6), Article 065033. https://doi.org/10.1088/0963-0252/23/6/065033.
28. Muller, M. (2019). Characterisation of cold atmospheric plasma afterglow for decontamination. *School of Electronics and Computer Science, 1*, 1–186. https://doi.org/10.5282/edoc.24909.
29. Nguyen, T., Palmer, J., Phan, N., Shi, H., Keener, K., & Flint, S. (2022). Control of aflatoxin M1 in skim milk by high voltage atmospheric cold plasma. *Food Chemistry, 386*, Article 132814. https://doi.org/10.1016/j.foodchem.2022.132814.
30. Ott, L. C., Jochum, J., Burrough, L., Clark, S., Keener, K., & Mellata, M. (2022). High voltage atmospheric cold plasma inactivation of Listeria monocytogenes in fresh queso fresco cheese. *Food Microbiology, 105*, Article 104007. https://doi.org/10.1016/j.fm.2022.104007.
31. Patil, S., Moiseev, T., Misra, N. N., Cullen, P. J., & Mosnier, J. P. (2014). Influence of high voltage atmospheric cold plasma process parameters and role of relative humidity on inactivation of Bacillus atrophaeus spores inside a sealed package. *Journal of Hospital Infection, 88*(3), 162–169.

32. Pedrow, P., Hua, Z., Xie, S., & Zhu, M. J. (2019). Engineering principles of cold plasma. In *Cold Plasma in Food and Agriculture* (pp. 3–48). Elsevier.
33. Rajan, A., Boopathy, B., Radhakrishnan, M., Rao, L., Schlüter, O. K., & Tiwari, B. K. (2023). Plasma processing: A sustainable technology in agri-food processing. *Sustainable Food Technology, 1*(1), 9–49.
34. Reuter, S., Sousa, J. S., Stancu, G. D., & Van Helden, J. P. H. (2015). Review on VUV to MIR absorption spectroscopy of atmospheric pressure plasma jets. *Plasma Sources Science and Technology, 24*(5), Article 054001. https://doi.org/10.1088/0963-0252/24/5/054001.
35. Reuter, S., Winter, J., Schmidt-Bleker, A., Schroeder, D., & Lange, H. (2012). Atomic oxygen in a cold argon plasma jet: TALIF spectroscopy in ambient air with modelling and measurements of ambient species diffusion. *Plasma Sources Science and Technology, 21*(2), Article 024005. https://doi.org/10.1088/0963-0252/21/2/024005.
36. Sarangapani, C., Ryan Keogh, D., Dunne, J., Bourke, P., & Cullen, P. J. (2017). Characterisation of cold plasma treated beef and dairy lipids using spectroscopic and chromatographic methods. *Food Chemistry, 235*, 324–333.
37. Schmidt-Bleker, A., Reuter, S., & Weltmann, K. D. (2014). Non-dispersive path mapping approximation for the analysis of ambient species diffusion in laminar jets. *Physics of Fluids, 26*(8), Article 4893573. https://doi.org/10.1063/1.4893573.
38. Schneider, S., Dünnbier, M., Hübner, S., Reuter, S., & Benedikt, J. (2014). Atomic nitrogen: A parameter study of a micro-scale atmospheric pressure plasma jet by means of molecular beam mass spectrometry. *Journal of Physics D: Applied Physics, 47*(50), Article 505203. https://doi.org/10.1088/0022-3727/47/50/505203.
39. Schulz-Von Der Gathen, V., Schaper, L., Knake, N., Reuter, S., & Niemi, K. (2008). Spatially resolved diagnostics on a microscale atmospheric pressure plasma jet. *Journal of Physics D: Applied Physics, 41*(19), Article 194004. https://doi.org/10.1088/0022-3727/41/19/194004.
40. Stancu, G. D., Kaddouri, F., Lacoste, D. A., & Laux, C. O. (2010). Atmospheric pressure plasma diagnostics by OES, CRDS and TALIF. *Journal of Physics D: Applied Physics, 43*(12), Article 124002. https://doi.org/10.1088/0022-3727/43/12/124002.
41. Sudarsan, A., & Keener, K. (2022). Inactivation of spoilage organisms on baby spinach leaves using high voltage atmospheric cold plasma (HVACP) and assessment of quality. *Innovative Food Science & Emerging Technologies, 79*, Article 103023. https://doi.org/10.1016/j.ifset.2022.103023.
42. Surowsky, B., Schlüter, O., & Knorr, D. (2015). Interactions of non-thermal atmospheric pressure plasma with solid and liquid food systems: A review. *Food Engineering Reviews, 7*(2), 82–108.
43. Tresp, H., Hammer, M. U., Weltmann, K. D., & Reuter, S. (2013). Effects of atmosphere composition and liquid type on plasma-generated reactive species in biologically relevant solutions. *Plasma Medicine, 3*(1–2), 45–55.
44. Van Gessel, A. F. H., Carbone, E. A. D., Bruggeman, P. J., & Van Der Mullen, J. J. A. M. (2012). Laser scattering on an atmospheric pressure plasma jet: Disentangling Rayleigh, Raman, and Thomson scattering. *Plasma Sources Science and Technology, 21*(1), Article 015003. https://doi.org/10.1088/0963-0252/21/1/015003.
45. Van Ham, B. T. J., Hofmann, S., Brandenburg, R., & Bruggeman, P. J. (2014). In situ absolute air, O3, and NO densities in the effluent of a cold RF argon atmospheric pressure plasma jet obtained by molecular beam mass spectrometry. *Journal of Physics D: Applied Physics, 47*(22), Article 224013. https://doi.org/10.1088/0022-3727/47/22/224013.

46. Walkup, R. E., Saenger, K. L., & Selwyn, G. S. (1985). Studies of atomic oxygen in O2+CF4 RF discharges by two-photon laser-induced fluorescence and optical emission spectroscopy. *Journal of Chemical Physics, 84*(5), 2668–2674.
47. Walsh, J. L., Liu, D. X., Iza, F., Rong, M. Z., & Kong, M. G. (2010). Contrasting characteristics of sub-microsecond pulsed atmospheric air and atmospheric pressure helium-oxygen glow discharges. *Journal of Physics D: Applied Physics, 43*(3), Article 032001. https://doi.org/10.1088/0022-3727/43/3/032001.
48. Winter, J., Wende, K., Masur, K., Iseni, S., & Dünnbier, M. (2013). Feed gas humidity: A vital parameter affecting a cold atmospheric-pressure plasma jet and plasma-treated human skin cells. *Journal of Physics D: Applied Physics, 46*(29), Article 295401. https://doi.org/10.1088/0022-3727/46/29/295401.
49. Yadav, B., & Roopesh, M. S. (2020). In-package atmospheric cold plasma inactivation of Salmonella in freeze-dried pet foods: Effect of inoculum population, water activity, and storage. *Innovative Food Science & Emerging Technologies, 66*, Article 102543. https://doi.org/10.1016/j.ifset.2020.102543.
50. Yonemori, S., & Ono, R. (2014). Flux of OH and O radicals onto a surface by an atmospheric-pressure helium plasma jet measured by laser-induced fluorescence. *Journal of Physics D: Applied Physics, 47*(12), Article 125401. https://doi.org/10.1088/0022-3727/47/12/125401.
51. Zhang, K., Perussello, C. A., Milosavljević, V., Cullen, P. J., Sun, D. W., & Tiwari, B. K. (2019). Diagnostics of plasma reactive species and induced chemistry of plasma-treated foods. *Critical Reviews in Food Science and Nutrition, 59*(5), 812–825.
52. Ziuzina, D., Patil, S., Cullen, P. J., Keener, K. M., & Bourke, P. (2013). Atmospheric cold plasma inactivation of *Escherichia coli* in liquid media inside a sealed package. *Journal of Applied Microbiology, 114*(3), 778–787.

CHAPTER 5

EFFECT OF COLD PLASMA ON THE MODIFICATION OF STARCH IN FOODS

K. PADMA SREE, G. SAI PREETHAM, and
SURESH KUMAR KALAKANDAN

ABSTRACT

Plasma is an unbound medium that contains both positive and negative particles with a net electrical charge of roughly zero. Nonthermal plasma processing has been a breakthrough, eco-sustainable technology in the food processing sector that has the ability to improve the quality and microbiological safety of a wide variety of food items. For industrial purposes, natural starch is modified to improve its characteristics. The impact of cold plasma on starches is determined by the type of input gas, the voltage applied, and the treatment period. In this chapter, the effect of starch modification in cold plasma focuses on the generation of plasma and its sources, types, mechanism of modifications of starch by cold plasma treatment, the effect of this treatment on characteristics of starch, and the uses of starch modified by cold plasma treatment.

5.1 INTRODUCTION

Nature is the primary source of food for mankind, including plants and animals. Different food components are responsible for the essential functioning of the human body and leading a healthy life. Carbohydrates are a crucial necessary dietary component that serves as one of

Nonthermal Plasma Technology for Food and Food Products.
R. Mahendran, Preeti Birwal, and Megh R. Goyal (Eds.)

the energy sources for the human body. Starch is the principal source of carbohydrates derived from plant sections such as stems, tubers, roots, leaves, fruit, seeds, etc., synthesized by chloroplasts and amyloplasts via photosynthesis [73]. Starch is often present in the form of granules, differentiated by two major polymers known as amylose and amylopectin. Amylose is mostly organized as a linear chain polymer of 840–22,000 units of (1→4) D-glucopyranosyl linkages, accounting for 20–30% of starch. Amylopectin consists of flattened disks of α-(1,4)-glucan chains joined by α-(1→6)-D-glucopyranose, constituting about 70–80% of total starch. Native starch, due to its relatively complex structure and sensitivity to pressure, light, temperature, pH, and radiation, is not digestible without thermal treatment [95]. Therefore, native starch cannot be used directly in the food industry. So, starch needs modification to improve its functional properties and digestibility for use in food, pharmaceutical, and biomedical industries. The starch, after modification, can be used as a thickening and binding agent, coatings, production of edible films, and conditioner and can bring consistency and smoothness.

Physical, chemical, and enzymatic modification are the three forms of modification. Physical modification can be easily commercialized in industry [116]. Cold plasma is among the most recent areas of interest for starch modification that does not produce hazardous compounds [34]. Over the conventional ones, cold plasma treatment doesn't need any water or chemicals, which is usually a dry and clean process [24]. In recent days, research has been conducted on cereal and tuby waxy starches [77, 78], corn and tapioca starches [1], rice starch [110], and other starches. Cold plasma modification of starch for industrial scale-up was intensively reviewed by several investigators [78].

This chapter briefly explains plasma classification, generating techniques, cold plasma sources, the mechanism by which starch is modified by treating it with cold plasma, and its influence on various characteristics of starch.

5.2 PLASMA AND CLASSIFICATION

Plasma, often known as the fourth state of matter, is a gaseous substance that may be perceived as a brilliant fluorescent spark or light and is composed of a large number of tiny interacting species of charged particles which include photons, free radicals, excited molecules, electrons, and both positive and negative ions [60, 86]. Plasma is generated when the atoms in the gas dissociate to release charged particles due to the supply of ample energy. Plasma is a quasi-neutral substance home [20] to unbound positive and negative particles, indicating that its total charge is almost zero. Plasma is generally categorized into two based on the method of generation.

When a gas is sufficiently heated at a high temperature, thermal or equilibrium plasma is generated (20,000 K). In this type, all the reactive species are in equilibrium condition w.r.t temperature. The former is the low-temperature plasma, which is further classified into two based on the temperature [94]. Quasi-equilibrium plasma has a temperature range of 373.15 to 423.15 K, indicating that there is a local thermodynamic equilibrium present between the species, while non-equilibrium plasma has a lower temperature range of (333.15 K) [66]. In this state, the collision frequency is relatively low at low pressure. The transfer of kinetic energy is inefficient owing to the huge mass difference between heavier particles and electrons, and this is the reason plasma gas is at a lower temperature [62]. Non-equilibrium plasma can be called nonthermal plasma (NTP) or cold plasma [85].

5.2.1 MECHANISM OF PLASMA GENERATION

Cold plasma is generated and classified based on the pressure applied. The classification includes atmospheric non-thermal plasma (ANTP) and low-pressure plasma with pressure ranging from 10^{-2} to 10^{-3} mbar or nearly vacuum by vacuum pumps [89]. Cold plasma is generated by exciting specific feed gases like argon, helium, oxygen, and nitrogen by applying different ranges of frequencies, voltage, heat, electricity (either AC or DC with the application of both static and pulsed fields), UV light, X-rays, and gamma radiation [29]. Various power discharges include Inductively Coupled Plasma (ICP) and Capacitive Coupled Plasma (CCP) [26].

Townsend's theory explains the generation of cold plasma [110], which states that when sufficient energy is supplied for the gas in between electrodes, the kinetic energy of the gas molecules increases, causing electrons to be released from the cathode and move towards the anode, where elastic collisions occur. Even yet, the electron energy is inadequate or low enough to ionize or excite more molecules. Collisions continue until there is enough energy to dissociate more atoms and molecules in the gas, resulting in the production of reactive species. Due to the presence of an electric field, the collision of electrons increases, producing more and more electrons, positive ions, and an avalanche [50].

Further motion of electrons in an electric field was studied by [42]. Electrons have less mass and lower speed in ranges of 10^{-5} to 10^{-6} m/s, and due to the presence of an electric field, they move towards the avalanche head and positive ions with higher speeds ranging from 50 to 500 m/s, which make up the tail. Due to the low mass of electrons, they acquire higher energies from electromagnetic fields, and electron energies are higher than the ions due to this non-equilibrium state that exists in cold plasma. The coefficients for the probability of electrons generated from the cathode (α), which collide along the distance (d) and the formation of secondary electrons due to the impact of positive ions moving towards the cathode (γ) are given by Townsend. For steady-state ionization, the balanced equation is given in Eqn. (1).

$$\lambda\left[\exp(\alpha \times d) - 1\right] = 1 \tag{1}$$

The formation of secondary electrons is vital for ionization and steady-state current [2]. The cold plasma generally contains hot electrons with an average energy between 0 and 10 eV [5, 62]. For gas to ionize, there is a minimum voltage required to break down gas particles. As ionization of gases produces cold plasma, the breakdown voltage (V_b) is the minimum voltage necessary for the ionization of gas-forming particles and for plasma production, which is given by Paschen's law [23]. The voltage required to ignite a plasma between two electrodes for a given gas is calculated by Paschen's law, which is based on the product pressure (p) and electrode gap distance (d) provided by Eqn. (2).

$$V_b = p.d \tag{2}$$

The breakdown voltage is different for different gasses. The breakdown voltages for different gasses can be obtained from Paschen's curve, which shows the relationship between breakdown voltage and gap (= p.d) [113]. The Paschen curve has a concave upward shape for all gases, suggesting that there is always an optimal value for the product of variables, pressure, and gap (p.d) for the lowest V_b [81].

The significant properties for the generation of cold plasma include density and energies of charged particles, which are varied by the amount and type of discharge, type of gas (flow rate, pressure), power supply (input voltage, frequency, pulse), etc. [2, 50].

5.3 SOURCES OF COLD PLASMA

Diverse technologies are being quickly developed to produce cold plasma (Figure 5.1). These may function at ambient pressure or in a partial vacuum.

Cold plasma systems may be categorized into three groups depending on the location of food inside the system [75]: (1) Food placed at a significant distance from cold plasma; (2) Food placed within the plasma field; and (3) Food placed close to the generation point. The diversity of cold plasma generating sources allows for the creation of one-of-a-kind designs that are compatible with current food industry equipment.

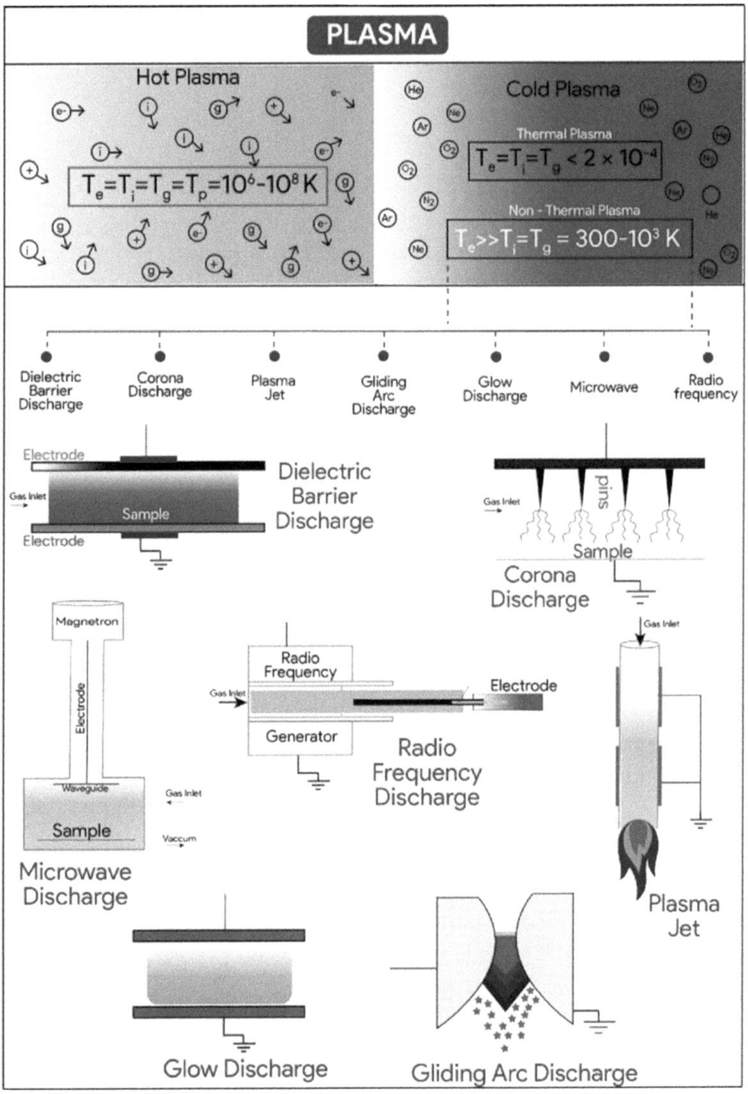

FIGURE 5.1 Overview of cold plasma and mechanism of generation.

The major sources for the generation of cold plasma include dielectric barrier discharge (DBD), corona discharge (CD), plasma jet (PJ), gliding arc discharges (GAD), Glow Discharge (GD), microwave (MW), and radio frequency (RF) [17, 18, 37, 50]. Because of their simplicity, convenience, and affordability, plasma jet designs were the focus of early research on atmospheric plasmas for biological disinfection. DBD is a nonthermal atmospheric gas discharge source widely used due to the ease of operation, operating parameters, versatility, and industrial use. For numerous biological applications, plasma jet is more acceptable and attractive than solid food preparation [80].

5.3.1 DIELECTRIC BARRIER DISCHARGE (DBD)

DBD is the easiest way of producing cold plasma due to its configuration, convenience, and equipment flexibility. It can be quickly developed and offer low cost at an industrial scale. In DBD, cold plasma is generated between the two electrodes. When energy is applied, ionization of gas occurs between the electrodes, one of which is grounded. Dielectric materials can be applied to one or two electrodes. The main reason for covering electrodes is to avoid the production of arc or spark. The different configurations of DBD were studied by [35]. DBD can generate plasma at temperatures ranging from 300 K to 2,500 K so that it can be used in a wider range of applications. The different materials for dielectric barriers include ceramic, Teflon, quartz, polymer, glass layers, etc. [37, 50, 111]. The metals used for electrodes include copper, stainless steel, silver, brass, aluminum, etc. [46, 80]. The operational parameters for the operation of DBD are i) gas pressure in the range of 10^4–10^5 Pa, Voltage is applied between 1 and 100 kV and current in the form of AC or pulsed DC with frequencies ranging from 50 Hz to 10 MHz maintaining an electrode distance between 0.1 mm to a few centimeters [25]. The different aspects determining the performance of DBD include the type of gas, flow rate, voltage, frequency, current, relative humidity, temperature, treatment duration, product type, structure, and composition [25]. Different patterns of DBD, such as filamentary, glow, and resistive, can be well suited to food applications. The reactor design may be planar or tubular based on the application [18]. One of the disadvantages of the system is the use of higher voltages to ignite the air. Utilization of DBD to increase food shelf-life, thereby decreasing food spoilage, and to inactivate germs in meat, was investigated by Nasiru et al. [72] and Roy et al. [90].

5.3.2 CORONA DISCHARGE

When a sufficiently high electric field is applied to a pin-type electrode with a sharp edge, a thin wire, or an electrode with curvature at the end, a corona discharge occurs. The anode is the HV electrode in the positive corona, while vice versa in negative corona. The discharge only occurs when the field is highly non-uniform [57]. However, there are various modes of discharge that occur as a result of an electrode's polarity shift [13]. One of the crucial features of corona discharge is the generation of an "electric wind." Because the charged particle flow in atmospheric density coronas is viscous (dominated by collisions), practically all momentum and energy absorbed from the electric field by the ions is transferred to the neutral gas molecule [33]. To increase the surface area of the treatment, a multiple electrode configuration is used for food applications [50]. The strong electric field is concentrated towards the end of the electrode surface due to the high voltage input. The dielectric breakdown ionizes the neutral gas around the electrode, resulting in cold plasma [91]. In many cases, the operating parameters for cold plasma generation include voltage ranging from 60–100 kV, 10–30 mm for the single pin and up to 70 mm in multi-pin configuration [18, 50]. Coronas are very weak discharges with very low concentrations of electrons and ions, that are often used in ozone generators and particle precipitators [26].

5.3.3 PLASMA JET

This system produces cold plasma using radiofrequency as an excitation source [75]. This system is stabilized by passing either helium or argon gases [26]. However, efficacy is dependent on the kind of gas injected. The system comprises spherical hollow or tubular electrodes with gas flowing between them. The plasma is expelled from the reactor zone towards the target material to an open environment [18]. The plasma coming out of this type is called 'plasma brush' [51]. Apart from noble gases, a small concentration of reactive gases such as oxygen and nitrogen is introduced to make the ignition process more accessible and produce reactive oxygen species for enhanced inactivation. Typical operating parameters include voltage in the range of 100–250 V to the inner electrode, RF power with 13.56 MHz which excites one electrode while the other is grounded [65]. The electrons present accumulate within the RF field, which collides with gas molecules to generate plasma [72], which is relatively stable, uniform, and homogeneous even at atmospheric pressure [50]. The foremost vital issues in plasma jet are that the target material's temperatures may exceed 60°C, which needs special emphasis. The different types of plasma jets, such as noble gas plasma jet, di-electrode free electrode jet, DBD and DBD-like jets, single electrode jets, N_2 plasma jet, and air plasma jet, were reviewed by Lu et al. [58]. However, a cooling system has to be provided for the mechanism in cases when the temperature exceeds.

5.3.4 GLIDING ARC DISCHARGE (GAD)

GAD is an efficient and inexpensive source that exhibits characteristics of both thermal and nonthermal plasma at atmospheric pressure with humid air [44]. The air flows between two metal knife-shaped electrodes with a pulsed potential difference, causing the arc discharge to transition into a bulk plasma flash [105]. The discharge is triggered by a high voltage at a point when the distance between divergent electrodes is the shortest (in the mm range). For elevated plasma flow rates, the electrodes may be coupled or two pairs of electrodes can be joined. The discharge expands in volume and length as the gas flow increases. The gas (generally humid air) is flushed from the top of the electrode where the pressure breaks the plume, and a new plume is created at the top of the electrode [71]. GAD produces high electron energies with high electron density and temperature at atmospheric pressures with a gas temperature of 350 K [45]. This technique is efficient in surface and liquid treatment for inactivation and degradation [17].

The major inactivation principle is involved near the liquid-gas interface system in most cases. GAD does not necessitate the use of costly noble gases, especially helium or argon, or any other gases. Substrate gas flow or supply water or oil causes a cooling effect [19]. Operating voltage differs (9 kV or higher) based on the application used with a general frequency ranging from 50–120 Hz. GAD can deposit cold plasma onto the surface more forcefully than corona discharges [63].

5.3.5 GLOW DISCHARGE (GD)

A glow discharge is a micro discharge created between the electrode gap with a filamentary appearance [2]. It consists of 2 electrodes with some gap of separation facilitating the gas to pass through. Nanosecond pulse discharges are used to produce stable glow discharges, which are highly energy-efficient than other systems [83]. RF, also as a power source, can be used in this mechanism. The operating parameters include voltage ranging from a few volts to 100 kV, currents in the order of 1 mA, and operating pressures in the range of 1–1,000 Pa. The density of charged particles ranged from 10^8–10^{13} cm^3. The electrode distance is generally maintained around 13 cm [29].

A minor concentration of oxygen and nitrogen is used along with other noble gases to decrease breakdown voltage, making the system efficient. Majorly atmospheric air is generally forced through the system apart from noble gases. GD is highly selective but is limited to lower power and pressures [88]. Optimum glow can be achieved by adjusting voltage or RF frequency. The significant advantage of the system is that it can be easily configured and used with DBD to produce plasmas of various parameters for different purposes according to the need. In the case of GD, UV photons play a less important role as they are absorbed by gas molecules and atoms [67]. The electron mean free path permits the light between the electrodes if vacuum is introduced [82].

5.3.6 MICROWAVE (MW)

This system employs a magnetron as a power source to produce microwave waves, as seen in a microwave oven. In this case, electrodes are not used as a source of energy in electromagnetic waves, so a waveguide is used to direct the waves towards the specific gas, which gets heated and ionized, due to which inelastic collisions occur and plasma is generated [17]. The advantages of the system include more electron density and greater efficiency in generating reactive species [50]. The disadvantages include the requirement of high power and restriction of size due to the design of the waveguide [2]. The gas is passed through a metallic nozzle to stabilize the plasma, and the gas is also passed tangentially [53]. The general operating parameters include a microwave frequency of 915 MHz or 2.45 GHz and operating pressures below 1,000 Pa, with power in the range of 50–1,000 W [37, 50, 68]. The input power of the microwave source is more important than the applied voltage.

5.3.7 RADIO FREQUENCY (RF)

Cold plasma is generated between 2 electrodes separated by a few centimeters, usually operated in a vacuum chamber, with RF as a power source that typically produces oscillating electromagnetic fields that ionize the gas. The typical charge densities range from 10^{15} to 10^{16} m^{-3} of electrons and ions. The temperature of cold plasma increases with an increase in electron density. The amount of power necessary to keep the plasma going is determined by the device's design, the plasma's environment, and the flow velocity [6]. Inductively coupled plasma (ICP), capacitively coupled plasma (CCP), and helicon wave generators are examples of RF plasmas [50].

RF plasma can be operated with various noble gases with some concentrations of oxygen and nitrogen and a combination of their mixtures. The general operating parameters include RF in the range of 1 kHz–500 MHz but usually operated at 13.56 MHz with operating voltage ranging between 1 to a few hundred kV. RF-driven systems require characteristic impedance between the power source and the plasma to maximize the power deposited in the plasma while minimizing reflected power.

5.4 TYPES OF STARCH MODIFICATION

Starch in its natural condition has various limitations owing to its poor functional, physical, and colloidal properties. Native starches have various drawbacks, including insoluble nature, inertness, enzyme resistance, and inability to withstand high temperatures [108]. Modified starch is utilized to alleviate these drawbacks. There are three different approaches for the modification of starch which include physical modification, chemical modification, and enzymatic modification (Figure 5.2).

5.5 MECHANISM OF STARCH MODIFICATION BY COLD PLASMA TREATMENT

5.5.1 DEPOLYMERIZATION

Depolymerization is the scientific process where a polymer is converted to a monomer or a mixture of monomers. Starch depolymerization is the reduction of the molecular size of starch to produce smaller molecules. Smaller molecules indicate sugars. After depolymerization, starch converts into lower sugars. Three fundamental processes can achieve depolymerization.

- Acid-catalyzed
- Enzyme catalyzed
- Oxidation followed by alkaline pH

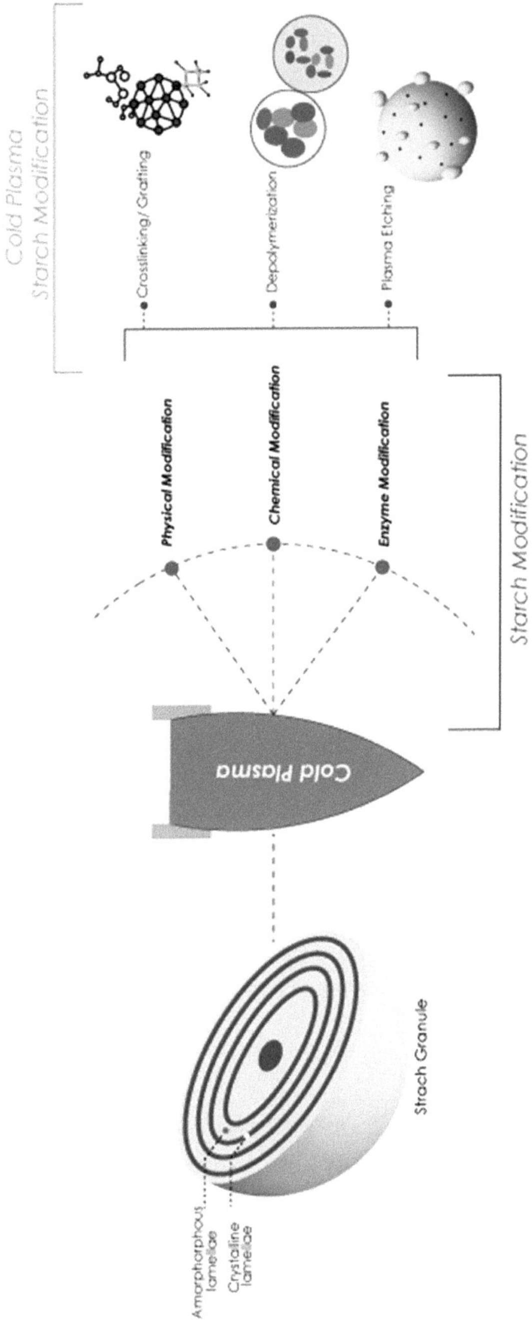

FIGURE 5.2 Pictorial representation of starch modification by cold plasma.

The oldest known method for depolymerization of starch is only by acid hydrolysis. Another alternative to depolymerization is heating, extrusion, and spray drying. Still, depolymerization occurs partially below the glass transition temperature [90], and this depolymerization is caused by heat-sheared treatment [28]. The starch, after these processes, generally becomes thin and is called "thinned starch." Due to these processes, the granule structure loosens without loss of birefringence, making starch useful in many food and non-food applications. Hypochlorite and ammonium persulfate are used to induce oxidation in starches [7].

Recent developments in using the potential of cold plasma for the modification of starch by depolymerization have been discussed [99]. It is well known that cold plasma produces ROS and RNS species, which can cause oxidation. Therefore, depolymerization occurs as a consequence of the oxidation of starch by ROS. When starch is oxidized, it produces carbonyl and carboxyl groups, decreasing recrystallization. The depolymerization of amylose chains in starch is induced by an increase in treatment duration and power levels. The extent to which starch can be depolymerized depends on the origin and type of starch, gas flow, and type of system employed. During the depolymerization of starch, endothermic enthalpy and crystallinity can be observed [15].

The increase in the amylose content of starches can be observed after depolymerization. By-products, also known as simple sugars, can be obtained, such as glucose and maltose, and they leach out during cooking and other operations. A direct correlation can be made for the gelatinization properties of starch [1]. This depolymerization process generally occurs in different steps. Starch hydroxyl groups are initially converted to carboxyl groups, followed by depolymerization through the breakage of α-(1,4)-glycosidic bonds. The ionization process, which is primarily responsible for oxidation, influences the degree of depolymerization [85].

5.5.2 PLASMA ETCHING

Plasma etching is defined as any process of removal of part of material when subjected to a plasma gas mixture. As discussed in the previous section, plasma generates the reactive species which are also known as "etchers." These etchers generally modify the surface due to their high energy charges by bombarding the surface, which causes the surface to break down forming pinholes and other surface modifications [112]. The plasma etching process boomed into the semiconductor chip industry in the early 1970s [22]. Plasma etching is also known as "Bosch etching" in industrial terms [61], "dry etching" because this process occurs without the use of the liquid phase, and "reactive ion etching" which involves the reactive species of plasma, the main important characteristic for plasma to cause etching [92]. Plasma etching occurs either by chemical means or physical sputtering, resulting in the increase of surface energies of the substrate. This part of plasma etching is also called "volatilization." However, food materials can be influenced by both ion bombardment and free radical bombardment, or by both, but the influencing of characteristics of each individual is least studied in the case of starch modification. The effectiveness of etching depends on ion bombardment; increased ion density leads to increased ion bombardment with increased power levels. In the case of semiconductors, the major purpose of plasma etching is to remove oxidation layers, and a similar kind of oxidation was found in the case of starches, generating distinct changes in amylose and amylopectin, as well as different surface modifications on the surface of starch granules.

The plasma etching relies on the kind of gas employed, plasma dynamics, and composition of the exposed substrate. The increased effect can be observed with increased voltage and discharge power. The type of gas also alters the modification induced. So, oxygen gas, as it causes oxidation, is more effective than nitrogen gas. Plasma etching mainly causes surface modifications. The hydrophilicity of starch can be increased, and amylose molecules will leach out faster than normal starch during gelatinization. Surface modification with non-penetrative damage was reported in the instance of banana starch [115, 117], while an increase in surface area was found in the case of tapioca starch [1], fissure and particle aggregation in the case of aria starch [11], surface granule damage in the case of mung bean starch by combined treatment of ultrasound and plasma [98], changes in physical properties of corn starch [114], surface etching on normal and waxy maize starch [121]. This non-penetrative damage leads to surface corrosion of starch granules. Liu et al. [55] observed superficial damage in Chinese milled rice and aggregation of small particles on the surface due to the Van der Waal's and electrostatic forces due to plasma treatment. Hence, it is evident that plasma etching causes surface modification altering the different properties of starch.

5.5.3 CROSSLINKING OR GRAFTING

Crosslinking of starch is most widely used to convert the starch in its native form to the required usable form. Crosslinking is defined as replacing the

OH groups of starch with any other functional groups by using different chemical reagents such as phosphoryl chloride, sodium trimetaphosphate (STMP), and mixtures of STMP, adipic acetic mixed anhydride, and sodium tripolyphosphates (STPP). The replaced functional groups generally include phosphoryl and ether groups. There are primary and secondary hydroxy groups that are being replaced by these functional groups [96]. The mechanism of starch crosslinking by cold plasma can be observed in following equation (Eqn. (3)) [123]:

$$Starch - OH \ + OH - Starch \rightarrow Starch - O - Starch \tag{3}$$

Crosslinking typically takes place in the amylose region of starch granules, which occurs by the release of a water molecule between two molecules, and the crystalline structure is not affected. Crosslinking is mainly pH-dependent, and an increase in molecular size can be observed. The induced changes in functional groups can be confirmed with FTIR spectroscopy. The replacement of OH groups forms C-O-C bonds, and its content is directly correlated to the degree of crosslinking [122]. The stabilization of starch with the phenomena of crosslinking can be observed during plasma treatment and restricts degradation. As a result of crosslinking, the hydrophobic nature of starch is increased, affecting starch swelling and solubility activities, exhibiting high peak viscosity. Braşoveanu et al. [10] observed that crosslinking caused dissociation of chains, which limited the granular swelling, due to which light cannot pass through granules and has lower clarity than native starch. The author also observed that sample loading affects crosslinking.

5.6 THE EFFECT OF COLD PLASMA ON THE PROPERTIES OF STARCH

5.6.1 CRYSTALLINITY

Starch is generally stored in the form of granules that exhibit crystallinity. Moisture content plays an important factor in starch's crystallinity, which can be observed through X-ray diffraction analysis [108]. The crystalline region of starch contains double helices of amylopectin and the percentage of crystallinity depends on structural ordering [20]. The size and shape variation may be noticed depending on the origin and composition that are generated in nature. Based on the crystalline structure, starches are divided into A-type native to cereals, B-type native to tubers, fruits, and stems, and

C-type starches native to legumes. Native starches have crystallinity in the range of 1445%. During retrogradation of starch from a disordered state to a more crystalline state, the crystallization kinetics can be described by Avrami models, which better predict the degree of crystallization based on two parameters: the "rate constant," which is based on growth and nucleation, and the Avrami exponent, which is based on crystal dimension [27]. In earlier sections, we have witnessed the creation of free radicals and reactive species during cold plasma generation. When exposed to cold plasma, the water molecules in starch are accessible and break down in the same way as electrolysis does, resulting in the generation of hydroxyl ions and a loss in crystallinity. So, the regions which contain more water molecules are more susceptible to free radicals, thereby altering crystallinity. Another important aspect is based on the structure and interruption of hydrogen bonds by species due to treatment leading to a reduction in crystallinity [101].

The melting of microcrystals, as seen in the case of thermal degradation, was observed the same when the fresh wet noodles subjected to cold plasma treatment decreased their crystallinity as the double-helical structure of amylose was depolymerized, but fluctuations were observed with an increase in the intensity of treatment, making it evident that cold plasma can alter the degree of crystallinity but doesn't change crystal shape [16]. Starch surface modification and disintegration of chains, which created the loose organization, was found in the instance of banana starch when treated with DBD plasma. This loose rearrangement has led to changes occurring in the solubility and pasting behavior [117]. In the context of potato starch, the combined action of twin-screw extrusion and cold plasma resulted in improved crystallinity even after depolymerization [104]. Crystallinity also affects the enthalpy of starch where it is directly proportional. Cold plasma can only modify surface characteristics because it can only penetrate to very minute depths, so many of the changes can only be observed in amorphous regions rather than in crystalline regions [1]. The atmospheric pressure plasma jet had no effect on the change in crystallization type in waxy and normal maize starch; however, it did lower crystallinity as treatment time increased [121]. Microcrystalline starch, when prepared with native potato starch, destroyed its structure when treated with atmospheric plasma jet, which was made evident from XRD and FTIR analysis [118]. The increase in crystallinity can also be observed in some starches, which may be due to the breaking of amylose chains and the formation of low-branching polysaccharides. These polysaccharides can again reconstitute to form more crystalline regions [94]. The increase in relative crystallinity was observed in the case of tartary buckwheat grain starch, and the type was unchanged due to the treatment by DBD [30].

5.6.2 LAMELLAR STRUCTURE

It is well known that starch microstructures contain lamellar domains in the range of (7–11) nm thick, where one lamella contains about 100 double-stranded helices of amylopectin, where each helix consists of about 20 glucose molecules. Lamellae are closely packed with high regularity. The lamellar structure is one of the factors determining enzyme binding and catalysis during starch digestion in humans. Correlating with curves acquired from small-angle X-ray scattering may be employed to identify the conformation and changes in the lamellar structure of starch. Both amylose and amylopectin are stacked together to form lamellar structures, which are only visible when granules are humidified [8]. However, the formation of the lamellar structure by amylose and amylopectin is less understood. Li & Gong [54] concluded that the formation of the lamellar structure depends on chain length distribution but does not depend on the molecular size of starch after experimenting with 16 rice starches. Retrogradation also depends on the lamellar structure. The higher the order of the lamellar structure, the greater the tendency towards retrogradation [49]. The reactive species present in the cold plasma impact the free water molecules in the starch, transforming them from lamellar to non-lamellar structure [38]. The average repeat distance between the lamellae can be calculated by Bragg's equation. When potato starch was treated with glow plasma supplied by helium and nitrogen, the lamellae repeat distance increased [120]. Maize starch subjected to RF plasma with high amylopectin content (30%) showed that amorphous lamellae are less ordered and more affected by cold plasma [100]. When treated with DBD plasma, there was a reduction in intensity ratio to approximately 995/1,022 cm^{-1} with an increase in treatment time due to the decrement of the water molecule in crystalline lamellae of potato starch [36].

5.6.3 GRANULES MORPHOLOGY

The amylose and amylopectin in the starch are stored as water-insoluble granules of definite size, shape, and morphology based on the type and origin [41]. The different shapes of starch include oval, polygonal, disk, spherical, dome, and elongated rod shapes. Different types of starches based on crystallinity were described in the previous section. A-type starches exhibited pinholes on the surface and serpentine-like channels inside the granules, and the structure was also loosely packed [40]. This unique morphology has led to opening a way to modify A-type starch in cold plasma, making a pathway for reactive species to modify easily. Scientists have been able to better understand the granule structure of different starches by using confocal scanning laser microscopy and scanning electron microscopy. Heating starch with too much water destroys its crystalline structure and induces swelling attributable to repulsion between phosphate groups and neighboring chains, which increases its hydration capacity. Lipids also play an important role in swelling, where higher lipid content suppresses the swelling of granules.

However, cold plasma can significantly change the grain morphology with treatment power and time. The size, shape, and regularity can also affect the morphology during plasma treatment. The pinholes or channels existent by nature or formed due to plasma treatment by etching enable simple ingress of reactive species into the granules, causing amylose to leach out and modifying the characteristics of starch [109]. The major changes occurring in the starch granules include irregular deposition, formation of pores, loss of rigidity, melted and interlinked structure based on type and origin after cold plasma treatment [108]. The native corn starch subjected to cold plasma along with cross-linking treatment for producing modified starch has increased its swelling power after treatment [106]. The destruction of granules caused by obverted cracks while maintaining the integrity on the surface was observed in the case of red adzuki bean starch when subjected to cold plasma treatment [31]. Starch citrate prepared from corn starch exhibited dents distributed on the surface of granules [47]. Wu et al. [115] reported that roughness and shrinkage of granule size and irregularity were observed in the case of banana starch native to Taiwan, and the possible reason was due to the passage of electric current by high-intensity plasma. Crosslinking during plasma treatment is also one of the reasons for changes in grain morphology. The plasma composition can also alter the changes in the grain morphology by the use of different gases. Helium was known to create surface erosions, whereas nitrogen had no effect [122].

Color changes may also be induced after cold plasma treatment due to different reactions occurring by reactive species. Increased yellowness and redness with no significant variations in lightness were noticed in the case of cassava starch [12]. The surface morphology of parboiled rice was changed by cold plasma, resulting in shorter cooking times and a greater water uptake ratio with a smaller contact angle [93]. Helium-powered RF cold plasma, when treated with rice kernels, exhibited the production of a scraggly rough surface followed by cavities and cracks and the formation of aggregated piles with increasing power levels [55].

5.6.4 GELATINIZATION TEMPERATURE

In various food processing operations, starch undergoes gelatinization to bring out textural and rheological properties suitable for final consumption. This process of gelatinization plays an important role, especially in ready-to-eat foods where final product characteristics are important. When starch is heated in excess water, the gelatinization process occurs, where starch molecules change from an ordered to a disordered state. During this stage, the starch granules swell, lose their birefringence, and later burst open leaching out substances leading to the formation of colloidal dispersion [87]. This starch gelatinization occurs at a particular temperature (onset temperature) where crystal melting starts and reaches the final constant temperature (completion temperature). It depends on the type and origin of starch, where it generally starts as low as 35–40°C. It can also be defined as a temperature at which all the starch molecules lose birefringence.

As cross-linking is an important parameter that occurs during plasma treatment, it is the key factor affecting the gelatinization temperature. Crosslinking can lead to a decrease in the mobility of amorphous chains and form intermolecular bridges, causing the gelatinization temperature to rise. However, no proper studies have been found on the reason for a decrease in gelatinization temperature due to crosslinking. The reduction in starch granules also increases the gelatinization temperature. Reactive oxygen species (ROS) cause oxidation on starch granules in the presence of water, changing the proportions of amylose and amylopectin and influencing the gelatinization temperature. The etching phenomena on starch granules lead to more water and heat being absorbed into the structure, which is also a reason for the decrease in gelatinization temperature.

A decrease in the gelatinization temperature was observed in the case of aria starch with increased voltage [11], waxy maize and potato starch when treated with RF cold plasma [77], quinoa flour [119], basmati rice flour [107], and bamboo flour [84], but an increase was observed in the case of red adzuki bean starch [31] and wheat flour starch [15]. Authors reported that crosslinking was the reason for the increase in gelatinization temperature, and other factors include process parameters and the type of gas used. However, no substantial change in milled rice flour was found when treated with DBD type plasma [55].

5.6.5 CHANGE IN PH

It is well known that cold plasma produces different reactive species of different functional groups. When subjected to these species, oxidative and reduction reactions occur in food complexes based on the composition of both plasma and food, which leads to changes in pH after treatment. The different functional groups include carboxyl, peroxide, and carbonyl groups, which are acidic. FTIR studies indicated the conversion of carbonyl groups to carboxyl groups due to oxidation [14]. The main cause of pH changes in starch is mainly the oxidation of starches caused by cold plasma, which in the majority of cases led to decreases. In cases when nitrogen gas is used, the formation of nitric acid (HNO_3) and nitrous acid (HNO_2) due to reactive nitrogen species (RNS) can lead to a decrease in pH. The different reactive species responsible include NOx, O, O_3 [52], O_2^-, O_2^+, H_3O^+ [21]. The formation of new compounds when plasma is exposed to starch may also cause an acidic medium [122]. The reduction in pH was observed in Aria starch [11], wheat gluten [4], and mango starch [43], etc.

5.6.6 MOLECULAR WEIGHT

When starch is exposed to cold plasma, it undergoes depolymerization, resulting in the formation of low molecular weight sugars and changes in molecular weight can be observed. The major sugars that are formed after treatment include maltose, maltotetraose, and maltotriose. The molecular weights of starch can be found from gel-permeation chromatography. Cross-linking and polymerization result in an increase in molecular weight. The molecular weight of starch may be increased or decreased by cold plasma, depending on the kind of starch, process parameters, feed gas, and reactor type. The compositions of amylose and amylopectin and their behavior towards cold plasma, which differ in oxidation and reduction properties, also lead to changes in molecular weight. The change in the radius of gyration of molecules alters the molecular weight [122]. After 5 minutes of treatment with DBD plasma, the molecular weight of corn starch rose from 14.80% to 85.93%. But in the case of potato starch, the molecular weights were increased [120].

The decrease in molecular weights in the case of wheat, sorghum, maize, quinoa, sweet potato, and barley starch subjected to DBD plasma [30, 32, 97], Kithul starch when subjected to glow discharge air plasma [103], high methoxylated pectin when subjected to glow discharge plasma powered by nitrogen [70]. So, it can be concluded that plasma can increase the molecular weight of polymers due to crosslinking and polymerization, whereas the decrease in molecular weight is due to depolymerization.

5.6.7 THERMAL PROPERTIES

Starch gelatinization is related to starch thermal properties. The different thermal properties of starch include thermal conductivity (k), specific heat (c_p), thermal diffusivity (α), onset temperature (T_o), peak temperature (T_p), completion temperature (T_c), enthalpy of gelatinization (ΔHgel) and peak height index. Peak height index (PHI) can be calculated by Eqn. (4).

$$PHI = \frac{\Delta H_{gel}}{T_p - T_o} \quad (4)$$

Where, T_p= peak temperature; T_o= onset temperature

Onset temperature is defined as the initial start of the temperature of gelatinization, and peak temperature is where the crystal melting occurs and the process of gelatinization is occurring at its maximum. Thermal conductivity plays an important role, especially during processing, which helps us better understand how evenly the heat is flowing throughout the starch. Enthalpy of gelatinization is the energy needed for starch gelatinization. Differential scanning calorimetry (DSC) is an instrument used for measuring the thermal properties of starch. DSC is equipment that determines the temperature and heat flow associated with material transitions as a property of change in the function of temperature and time. The thermal conductivity can be calculated from Eqn. (5).

$$k = \frac{Q}{4\pi S} \quad (5)$$

where, Q= power consumed (W/m); S= slope of log(time) vs temperature.

However, different prediction models have been developed for a better understanding of thermal properties. Many changes start to occur when starch is heated, especially structural changes leading to changes in physical properties like density and viscosity. These physical properties affect the transfer of heat during different food processes. The percentage of retrogradation can also be evaluated by the use of thermal properties.

Cold plasma tended to show both corrosion on the surface and crosslinking. Starches, when treated with cold plasma, tend to show lower thermal stability than native starch due to structural modifications like corrosion and degradation. While crosslinking increases thermal degradation, overcoming the degradation. However, breakdown viscosity is an important parameter in deciding the thermal stability of starch [76]. Enthalpy also depends on the starch origin and process parameters. All three temperatures, including enthalpy, tend to decrease in case of aria starch and kithul starch, but increase in enthalpy of both corn and cassava starch [1, 11, 103].

Waxy maize and potato starch exhibited minimal reduction in onset temperature, while waxy potato starch demonstrated a considerable drop in peak temperature without influencing the completion temperature. The enthalpy tended to increase in the case of both due to molecular species inducing the amylopectin side chains showing a stabilizing effect [78]. It can be concluded that there might be an increase or decrease in enthalpy due to the effect of cold plasma. Even though thermal conductivity and specific heat as effects of cold plasma were not widely studied.

5.6.8 RHEOLOGICAL PROPERTIES

Generally, starch-based suspensions are categorized as non-Newtonian fluids. As these fluids exhibit complex behavior during heating and other operations, changes in rheological parameters are observed for different conditions. Changes or modifications in starch structure lead to changes in deformation and flow corresponding to applied stress. The measurement of these properties is called the rheological properties of starch.

The general rheological parameters for starch measured include viscosity, shearing stress (τ), shearing time (s), consistency index (K), and flow behavior index (n). Dynamic rheological parameters include loss modulus, storage modulus (G′), loss modulus (G″), loss tangent (δ) [59], gel strength, the tendency of retrogradation, etc. G′ represents the elastic behavior, i.e., amount of deformation recovered per cycle, and G″ represents the viscous behavior, i.e., measure of the amount of energy dissipated as heat per cycle. However, different models have been developed for these terms to understand the rheological behavior of starch suspensions, such as the power-law model (Eqn. 6), Weltman model [3] (Eqn. 7), and Herschel-Bulkley model (Eqn. 8) [48].

$$\tau = K \times (\gamma)^n \tag{6}$$

$$\tau = A + B * \ln(t) \tag{7}$$

$$\tau = \tau_{\circ} + (\gamma)^n \quad (8)$$

$$\tan \delta = \frac{G''}{G'} \quad (9)$$

Plasma contains reactive oxygen species. The starch undergoes oxidation primarily on the surface, and due to the granular structure, pinholes present facilitate oxidation in the interior of the granule, leading to the degradation of the ordered structure and molecular chains due to high oxidation activity. Degradation of size and number of pinholes due to molecular fragmentation of corn starch was observed [9]. The formation of holes makes the amylose leach out easily, which alters the solubility [102]. Due to this, changes in the water holding capacity can be observed, leading to a change in flow behavior. Rheological properties depend on the population, type, and energies of reactive species, type and method of generation of cold plasma, sample distance, plasma temperature, power levels, and frequency. Retrogradation tendency is a function of changes in storage and loss modulus during heating and cooling.

The rice starch, when treated at power level 60 W, showed a significant decrease in both moduli. However, when treated for 5 minutes at the same power level, an increase in moduli was observed [109]. The author reported that this is due to the formation of more crosslinked networks and rigid gel structures. An increase in both moduli and shear stress was observed in cassava starch when treated with low argon plasma [12]. At relatively high concentrations, plasma lowers viscosity, and cold plasma may be used to change the viscosity of starches [9]. However, in terms of viscosity, it did not show a similar trend and was dependent on the nature of the source of starch. Potato starch, when subjected to corona discharge, increased apparent viscosity with decreased consistency [74]. A significant reduction in viscosity of potato starch was found when treated with glow plasma, and an increase in viscosity of rice starch was observed [109] [120].

5.6.9 RETROGRADATION

Gelatinization occurs during heating, which is a non-equilibrium phase, and associated changes were discussed in previous sections. Once the starch tends to cool down, it reorganizes back to form its crystalline structure. During retrogradation, amylose and amylopectin play very different roles.

Amylose retrogradation is a 2-stage process, first involving the reformation of helices and the second involving crystal formation. Amylopectin retrogrades slower. Many conformation changes and theories are involved in the evidence of amylopectin retrogradation. Co-crystallization of both polysaccharides may also help regain the stable starch. They both tend to form lipid complexes, which play an important role in retrogradation [79]. The botanical origin and water molecules at the molecular level impact retrogradation. The influence of short-range molecular order on starch retrogradation demonstrates that ideal short-range molecular order produces more efficient starch retrogradation [39].

Process parameters, especially gas flow rate and temperature, showed a significant change in retrogradation. Mitsugi et al. [69] investigated the impact of gas flow rate on potassium iodide starch and observed that an increased gas flow rate decreased surface temperature and caused retrogradation. Starch retrogradation can be better described by measuring rheological parameters like loss and storage modulus. The loss and storage modulus depend on amylose content, which may be ascribed to a change in retrogradation. The formation of newly formed functional groups, oxidation, and cross-linking by water molecules in starch alter these properties. An increase in G' and a decrease in G" was observed in wheat flours [64].

A decrease in both moduli was observed in the case of rice starch, indicating a decreased tendency towards retrogradation when treated with DBD plasma, mango kernel starch [43], wheat flour [15], pearl millet flour [56], but an increase in the case of aria starch [11]. The increasing tendency is typically not an appropriate property since it might modify the final qualities of the food product. The tendency to withstand retrogradation was observed in the case of waxy maize and corn starch [77]. It can be concluded that increased setback viscosity indicates increased retrogradation and a lesser tendency to retrograde towards the treatment.

5.7 APPLICATIONS OF STARCHES IN FOOD PROCESSES

The modification of starch changes its functionality and reactivity during processing. The altered properties include changes in crystallinity, lamellar structure, granule morphology, gelatinization temperature, molecular weight, and retrogradation capability. These starches can be used in fruit product manufacturing for viscosity building and give a better fruit taste than chemically modified starches, maintaining food stability for canned food even at higher temperatures. Products show better textural characteristics when incorporated during freeze-thawing, etc. The cold plasma-modified starches have clear appearances, which can be combined with coloring agents directly. The new possible application includes the incorporation of cold plasma-modified starches for the production of ice cream and its wafers, low-fat mayonnaise, dips, savories, and sauces, etc.

5.8 SUMMARY

This chapter provides an overview of cold plasma with various generating techniques, and the potential of cold plasma to change different kinds of starch was explored in this chapter. The major systems for the generation of cold plasma include dielectric barrier discharge (DBD), corona discharge (CD), plasma jet (PJ), gliding arc discharges (GAD), Glow Discharge (GD), microwave (MW), and radiofrequency (RF). The true potential of cold plasma in modifying the starch makes it commercially viable for food industries for its use. Native starch cannot be used directly in food industries due to its complex nature, lack of tolerance to process conditions during manufacturing due to acid and shear, and undesirable texture in food. So, there is a need for starch modification to provide desirable characteristics in final food products. To generate distinctive and varied textures or cost-effective solutions, modified starches are frequently advised for tough process conditions (pH, temperature, and shear) and prolonged shelf stability. Until now, starches are modified physically, chemically, and enzymatically, based on the type and origin of starch and ease of production. As cold plasma contains many highly reactive species such as ROS and RNS, they bombard the starch molecules during processing and cause modifications in starch. Modifications occur through the processes of crosslinking/grafting, plasma etching, and depolymerization. However, different process parameters affect the modifications, bringing about new changes. Even though the principles involved are the same as the modifications involved previously, the resulting starch obtained shows better characteristics. Because no toxic constituents remain after the reaction, cold plasma can be considered as "green technology" for the modification of starch, opening the way for the manufacture of food that is both safe and high in quality.

KEYWORDS

- **amylopectin**
- **amylose**
- **cold plasma**
- **modification**
- **retrogradation**
- **starch**

REFERENCES

1. Banura, S., Thirumdas, R., Kaur, A., Deshmukh, R. R., & Annapure, U. S. (2018). Modification of starch using low pressure radio frequency air plasma. *LWT, 89*, 719–724.
2. Bárdos, L., & Baránková, H. (2010). Cold atmospheric plasma: Sources, processes, and applications. *Thin Solid Films, 518*(23), 6705–6713.
3. Barrera, G. N., Bustos, M. C., Iturriaga, L., Flores, S. K., León, A. E., & Ribotta, P. D. (2013). Effect of damaged starch on the rheological properties of wheat starch suspensions. *Journal of Food Engineering, 116*(1), 233–239.
4. Barros, J. H. T., Montenegro, F. M., & Steel, C. J. (2022). Characterization and regeneration potential of vital wheat gluten treated with non-thermal plasma. *Journal of Cereal Science, 104*, Article 103402. https://doi.org/10.1016/j.jcs.2021.103402.
5. Basner, R., Schmidt, M., Becker, K., & Deutsch, H. (2000). Electron impact ionization of organic silicon compounds. In *Advances in Atomic, Molecular and Optical Physics* (Vol. 43, pp. 147–185).
6. Becker, K. H., Kogelschatz, U., Schoenbach, K. H., & Barker, R. J. (2004). *Non-Equilibrium Air Plasmas at Atmospheric Pressure.* pp. 1–700, ISBN 9780367864170. CRC Press.
7. BeMiller, J. N. (2012). Starch-based gums. In *Industrial Gums: Polysaccharides and Their Derivatives: Third Edition*, pp. 579–600. Elsevier. https://doi.org/10.1016/B978-0-08-092654-4.50025-5.
8. Bertoft, E. (2017). Understanding starch structure: Recent progress. *Agronomy, 7*(3), 1–56.
9. Bie, P., Pu, H., Zhang, B., Su, J., Chen, L., & Li, X. (2016). Structural characteristics and rheological properties of plasma-treated starch. *Innovative Food Science & Emerging Technologies, 34*, 196–204.
10. Braşoveanu, M., Nemţanu, M. R., & Ticoş, D. (2021). Influence of the sample loading on the contribution of competitive effects for granular starch exposed to radio-frequency plasma. *Innovative Food Science & Emerging Technologies, 72*, Article 102740. https://doi.org/10.1016/j.ifset.2021.102740.
11. Carvalho, A. P. M. G., Barros, D. R., da Silva, L. S., Sanches, E. A., & Pinto, C. da C. (2021). Dielectric barrier atmospheric cold plasma applied to the modification of Ariá

(*Goeppertia allouia*) starch: Effect of plasma generation voltage. *International Journal of Biological Macromolecules, 182*, 1618–1627.
12. Chaiwat, W., Wongsagonsup, R., Tangpanichyanon, N., Jariyaporn, T., & Deeyai, P. (2016). Argon plasma treatment of tapioca starch using a semi-continuous downer reactor. *Food and Bioprocess Technology, 9*(7), 1125–1134.
13. Chang, J. S., Lawless, P. A., & Yamamoto, T. (1991). Corona discharge processes. *IEEE Transactions on Plasma Science, 19*(6), 1152–1166.
14. Chang, R., Ji, N., Li, M., Qiu, L., & Sun, C. (2019). Green preparation and characterization of starch nanoparticles using a vacuum cold plasma process combined with ultrasonication treatment. *Ultrasonics Sonochemistry, 58*, Article 104660. https://doi.org/10.1016/j.ultsonch.2019.104660.
15. Chaple, S., Sarangapani, C., Jones, J., Carey, E., & Causeret, L. (2020). Effect of atmospheric cold plasma on the functional properties of whole wheat (*Triticum aestivum* L.) grain and wheat flour. *Innovative Food Science & Emerging Technologies, 66*, Article 102529. https://doi.org/10.1016/j.ifset.2020.102529.
16. Chen, Y., Zhang, Y., Jiang, L., Chen, G., & Yu, J. (2020). Moisture molecule migration and quality changes of fresh wet noodles dehydrated by cold plasma treatment. *Food Chemistry, 328*, Article 127053. https://doi.org/10.1016/j.foodchem.2020.127053.
17. Chizoba Ekezie, F. -G., Sun, D. -W., & Cheng, J.-H. (2017). A review on recent advances in cold plasma technology for the food industry: Current applications and future trends. *Trends in Food Science & Technology, 69*, 46–58.
18. Cullen, P. J. (2022). Systems for generation of cold plasma. In *Applications of Cold Plasma in Food Safety* (Springer). https://doi.org/10.1007/978-981-16-1827-7_2.
19. Dasan, B. G., Onal-Ulusoy, B., Pawlat, J., Diatczyk, J., Sen, Y., & Mutlu, M. (2017). A new and simple approach for decontamination of food contact surfaces with gliding arc discharge atmospheric non-thermal plasma. *Food and Bioprocess Technology, 10*(4), 650–661.
20. Dome, K., Podgorbunskikh, E., Bychkov, A., & Lomovsky, O. (2020). Changes in the crystallinity degree of starch having different types of crystal structure after mechanical pretreatment. *Polymers, 12*(3), Article 12030641. https://doi.org/10.3390/polym12030641.
21. Dong, S., Gao, A., Zhao, Y., Li, Y., & Chen, Y. (2017). Characterization of physicochemical and structural properties of atmospheric cold plasma (ACP) modified zein. *Food and Bioproducts Processing, 106*, 65–74.
22. Donnelly, V. M., & Kornblit, A. (2013). Plasma etching: Yesterday, today, and tomorrow. *Journal of Vacuum Science & Technology A: Vacuum, Surfaces, and Films, 31*(5), Article 4819316. https://doi.org/10.1116/1.4819316.
23. Esmaeili, Z., Hosseinzadeh Samani, B., Nemati, A., Nazari, F., & Rostami, S. (2021). Development of novel green pesticide system by using cold plasma to control *Plodia interpunctella* in pistachio. *Journal of Food Processing and Preservation, 45*(7), Article 15621. https://doi.org/10.1111/jfpp.15621.
24. Fazeli, M., Florez, J. P., & Simão, R. A. (2019). Improvement in adhesion of cellulose fibers to the thermoplastic starch matrix by plasma treatment modification. *Composites Part B: Engineering, 163*, 207–216.
25. Feizollahi, E., Misra, N. N., & Roopesh, M. S. (2020). Factors influencing the antimicrobial efficacy of dielectric barrier discharge (DBD) atmospheric cold plasma in food processing applications. *Critical Reviews in Food Science and Nutrition, 61*(4), 666–689.

26. Firdaus Zainal, M. N., Redzuan, N., & Misnal, M. F. I. (2015). Brief review: Cold plasma. *Jurnal Teknologi, 74*(10), Article 11113. https://doi.org/10.11113/jt.v74.4834.
27. Fu, Z. Q., Wang, L. J., Li, D., Zhou, Y. G., & Adhikari, B. (2013). The effect of partial gelatinization of corn starch on its retrogradation. *Carbohydrate Polymers, 97*(2), 512–517.
28. Fujio, Y., Igura, N., & Hayakawa, I. (1995). Depolymerization of molten-moisturized-starch molecules by shearing-force under high temperature. *Starch – Stärke, 47*(4), 143–145.
29. Ganesan, A. R., Tiwari, U., Ezhilarasi, P. N., & Rajauria, G. (2021). Application of cold plasma on food matrices: A review on current and future prospects. *Journal of Food Processing and Preservation, 45*(1), Article 15070. https://doi.org/10.1111/jfpp.15070.
30. Gao, S., Liu, H., Sun, L., Liu, N., & Wang, J. (2019). The effects of dielectric barrier discharge plasma on physicochemical and digestion properties of starch. *International Journal of Biological Macromolecules, 138*, 819–830.
31. Ge, X., Shen, H., Su, C., Zhang, B., & Zhang, Q. (2021). The improving effects of cold plasma on multi-scale structure, physicochemical and digestive properties of dry heated red adzuki bean starch. *Food Chemistry, 349*, Article 129–159.
32. Ge, X., Shen, H., Su, C., Zhang, B., Zhang, Q., et al. (2021). Pullulanase modification of granular sweet potato starch: Assistant effect of dielectric barrier discharge plasma on multi-scale structure, physicochemical properties. *Carbohydrate Polymers, 272*, Article 118481. https://doi.org/10.1016/j.carbpol.2021.118481.
33. Goldman, M., Goldman, A., & Sigmond, R. S. (1985). The corona discharge, its properties and specific uses. *Pure and Applied Chemistry, 57*(9), 1353–1362.
34. Grgić, I., Ačkar, Đ., Barišić, V., Vlainić, M., Knežević, N., & Medverec Knežević, Z. (2019). Nonthermal methods for starch modification—A review. *Journal of Food Processing and Preservation, 43*(12), Article 14242. https://doi.org/10.1111/jfpp.14242.
35. Guo, C., Tang, F., Chen, J., Wang, X., Zhang, S., & Zhang, X. (2015). Development of dielectric-barrier-discharge ionization. *Analytical and Bioanalytical Chemistry, 407*(9), 2345–2364.
36. Guo, Z., Gou, Q., Yang, L., Yu, Q. L., & Han, L. (2022). Dielectric barrier discharge plasma: A green method to change structure of potato starch and improve physicochemical properties of potato starch films. *Food Chemistry, 370*, Article 130992. https://doi.org/10.1016/j.foodchem.2021.130992.
37. Han, Y., Cheng, J. -H., & Sun, D. -W. (2019). Activities and conformation changes of food enzymes induced by cold plasma: A review. *Critical Reviews in Food Science and Nutrition, 59*(5), 794–811.
38. Han, Z., Shi, R., & Sun, D. W. (2020). Effects of novel physical processing techniques on the multi-structures of starch. *Trends in Food Science & Technology, 97*, 126–135.
39. Huang, S., Chao, C., Yu, J., Copeland, L., & Wang, S. (2021). New insight into starch retrogradation: The effect of short-range molecular order in gelatinized starch. *Food Hydrocolloids, 120*, Article 106921. https://doi.org/10.1016/j.foodhyd.2021.106921.
40. Jane, J. (2006). Current understanding on starch granule structures. *Journal of Applied Glycoscience, 53*(3), 205–213.
41. Jane, J. L., Kasemsuwan, T., Leas, S., Zobel, H., & Robyt, J. F. (1994). Anthology of starch granule morphology by scanning electron microscopy. *Starch – Stärke, 46*(4), 121–129.

42. Townsend, J. S. E., & Tizard, H. T. (1913). The motion of electrons in gases. *Proceedings of the Royal Society of London. Series A, Containing Papers of a Mathematical and Physical Character, 88*(604), 336–347.
43. Kalaivendan, R. G. T., Mishra, A., Eazhumalai, G., & Annapure, U. S. (2022). Effect of atmospheric pressure non-thermal pin to plate plasma on the functional, rheological, thermal, and morphological properties of mango seed kernel starch. *International Journal of Biological Macromolecules, 196*, 63–71.
44. Kamgang-Youbi, G., Herry, J. M., Bellon-Fontaine, M. N., Brisset, J. L., Doubla, A., & Naïtali, M. (2007). Evidence of temporal post discharge decontamination of bacteria by gliding electric discharges: Application to *Hafnia alvei*. *Applied and Environmental Microbiology, 73*(15), 4791–4796.
45. Khalili, F., Shokri, B., Khani, M. R., Hasani, M., Zandi, F., & Aliahmadi, A. (2018). A study of the effect of gliding arc non-thermal plasma on almonds decontamination. *AIP Advances, 8*(10), Article 5044476. https://doi.org/10.1063/1.5044476.
46. Khan, M. S. I., Lee, E. J., & Kim, Y. J. (2016). A submerged dielectric barrier discharge plasma inactivation mechanism of biofilms produced by *Escherichia coli* O157, *Cronobacter sakazakii*, and *Staphylococcus aureus*. *Scientific Reports, 6*(1), Article 37072. https://doi.org/10.1038/srep37072.
47. Kim, H. S., & Min, S. C. (2017). Effects of microwave-discharged cold plasma on synthesis and characteristics of citrate derivatives of corn starch granules. *Food Science and Biotechnology, 26*(3), 697–706.
48. Kong, X., Kasapis, S., Bertoft, E., & Corke, H. (2010). Rheological properties of starches from grain amaranth and their relationship to starch structure. *Starch – Stärke, 62*(6), 302–308.
49. Lan, X., Liu, X., Yang, Y., Wu, J., & Wang, Z. (2017). The effect of lamellar structure ordering on the retrogradation properties of canna starch subjected to thermal and enzymatic degradation. *Food Hydrocolloids, 69*, 185–192.
50. Laroque, D. A., Seó, S. T., Valencia, G. A., Laurindo, J. B., & Carciofi, B. A. M. (2022). Cold plasma in food processing: Design, mechanisms, and application. *Journal of Food Engineering, 312*, Article 110748. https://doi.org/10.1016/j.jfoodeng.2021.110748.
51. Laroussi, M., & Akan, T. (2007). Arc-free atmospheric pressure cold plasma jets: A review. *Plasma Processes and Polymers, 4*(9), 777–788.
52. Lee, K. H., Kim, H. J., Woo, K. S., Jo, C., & Kim, J. K. (2016). Evaluation of cold plasma treatments for improved microbial and physicochemical qualities of brown rice. *LWT – Food Science and Technology, 73*, 442–447.
53. Leins, M., Kopecki, J., Gaiser, S., Schulz, A., & Walker, M. (2014). Microwave plasmas at atmospheric pressure. *Contributions to Plasma Physics, 54*(1), 14–26.
54. Li, C., & Gong, B. (2021). Relations between rice starch fine molecular and lamellar/crystalline structures. *Food Chemistry, 353*, Article 129467. https://doi.org/10.1016/j.foodchem.2021.129467.
55. Liu, J., Wang, R., Chen, Z., & Li, X. (2021). Effect of cold plasma treatment on cooking, thermomechanical and surface structural properties of Chinese milled rice. *Food and Bioprocess Technology, 14*(5), 866–886.
56. Lokeswari, R., Sharanyakanth, P. S., Jaspin, S., & Mahendran, R. (2021). Cold plasma effects on changes in physical, nutritional, hydration, and pasting properties of pearl millet (*Pennisetum glaucum*). *IEEE Transactions on Plasma Science, 49*(5), 1745–1751.

57. Lu, P., Cullen, P. J., & Ostrikov, K. (2016). Atmospheric pressure nonthermal plasma sources. In *Cold Plasma in Food and Agriculture: Fundamentals and Applications* (Vol. 2, pp. 83–116). Elsevier.
58. Lu, X., Laroussi, M., & Puech, V. (2012). On atmospheric-pressure non-equilibrium plasma jets and plasma bullets. *Plasma Sources Science and Technology, 21*(3), Article 034005. https://doi.org/10.1088/0963-0252/21/3/034005.
59. Ma, S., Zhu, P., & Wang, M. (2019). Effects of konjac glucomannan on pasting and rheological properties of corn starch. *Food Hydrocolloids, 89*, 234–240.
60. Mandal, R., Singh, A., & Pratap Singh, A. (2018). Recent developments in cold plasma decontamination technology in the food industry. *Trends in Food Science & Technology, 80*, 93–103.
61. Martin, D., Sullivan, S., Bose, I. R., Landesberger, C., & Wieland, R. (2020). Dicing of MEMS devices. In *Handbook of Silicon Based MEMS Materials and Technologies*, pp. 677–689, Edition: 3. Elsevier. https://doi.org/10.1016/B978-0-12-817786-0.00031-1.
62. Milella, A., & Palumbo, F. (2014). Cold plasma. In *Encyclopedia of Membranes* (pp. 1–2). Springer.
63. Milella, A., & Palumbo, F. (2016). Cold plasma. In *Encyclopedia of Membranes* (Vol. 70, pp. 429–431). Springer.
64. Misra, N. N., Kaur, S., Tiwari, B. K., Kaur, A., Singh, N., & Cullen, P. J. (2015). Atmospheric pressure cold plasma (ACP) treatment of wheat flour. *Food Hydrocolloids, 44*, 115–121.
65. Misra, N. N., Pankaj, S. K., Segat, A., & Ishikawa, K. (2016). Cold plasma interactions with enzymes in foods and model systems. *Trends in Food Science & Technology, 55*, 39–47.
66. Misra, N. N., Schlüter, O., & Cullen, P. J. (2016). Plasma in food and agriculture. In *Cold Plasma in Food and Agriculture*, pp. 1–16, ISBN: 978-0-12-801365-6. Elsevier. https://doi.org/10.1016/B978-0-12-801365-6.00001-9.
67. Misra, N. N., Tiwari, B. K., Raghavarao, K. S. M. S., & Cullen, P. J. (2011). Nonthermal plasma inactivation of food-borne pathogens. *Food Engineering Reviews, 3*(3–4), 159–170.
68. Misra, N. N., Yadav, B., Roopesh, M. S., & Jo, C. (2019). Cold plasma for effective fungal and mycotoxin control in foods: mechanisms, inactivation effects, and applications. *Comprehensive Reviews in Food Science and Food Safety, 18*(1), 106–201.
69. Mitsugi, F., Wago, M., Sakamoto, R., Nishida, K., & Kawasaki, T. (2021). Influence of retrogradation on ki-starch visualization of reactive oxygen species emitted by plasma jets. *IEEE Transactions on Plasma Science, 49*(7), 2141–2147.
70. Momeni, M., Tabibiazar, M., Khorram, S., Zakerhamidi, M., & Mohammadifar, M. (2018). Pectin modification assisted by nitrogen glow discharge plasma. *International Journal of Biological Macromolecules, 120*, 2572–2578.
71. Moreau, M., Feuilloley, M. G. J., Veron, W., Meylheuc, T., Chevalier, S., et al. (2007). Gliding arc discharge in the potato pathogen *Erwinia carotovora* subsp. *atroseptica*: Mechanism of lethal action and effect on membrane-associated molecules. *Applied and Environmental Microbiology, 73*(18), 5904–5910.
72. Nasiru, M. M., Frimpong, E. B., Muhammad, U., Qian, J., & Mustapha, A. T. (2021). Dielectric barrier discharges cold atmospheric plasma: Influence of processing parameters on microbial inactivation in meat and meat products. *Comprehensive Reviews in Food Science and Food Safety, 20*(3), 2626–2659.

73. Nawaz, H., Waheed, R., Nawaz, M., & Shahwar, D. (2020). Physical and chemical modifications in starch structure and reactivity. In *Chemical Properties of Starch* (Vol. 1, pp. 1–21).
74. Nemtanu, M. R., & Minea, R. (2006). Functional properties of corn starch treated with corona electrical discharges. *Macromolecular Symposia, 245*(1), 525–528.
75. Niemira, B. A. (2012). Cold plasma decontamination of foods. *Annual Review of Food Science and Technology, 3*(1), 125–142.
76. Noora, B., Sudheesh, C., Sangeetha, N., & Sunooj, K. V. (2019). Effect of isolation methods on the crystalline, pasting, thermal properties and antioxidant activity of starch from queen sago (*Cycas circinalis*) seed. *Journal of Food Measurement and Characterization, 13*(3), 2147–2156.
77. Okyere, A. Y., Bertoft, E., & Annor, G. A. (2019). Modification of cereal and tuber waxy starches with radio frequency cold plasma and its effects on waxy starch properties. *Carbohydrate Polymers, 223,* Article 115075. https://doi.org/10.1016/j.carbpol.2019.115075.
78. Okyere, A. Y., Rajendran, S., & Annor, G. A. (2022). Cold plasma technologies: Their effect on starch properties and industrial scale-up for starch modification. *Current Research in Food Science, 5*, 451–463.
79. Ottenhof, M. A., & Farhat, I. A. (2004). Starch retrogradation. *Biotechnology and Genetic Engineering Reviews, 21*(1), 215–228.
80. Pan, Y., Cheng, J. H., & Sun, D. W. (2019). Cold plasma-mediated treatments for shelf-life extension of fresh produce: A review of recent research developments. *Comprehensive Reviews in Food Science and Food Safety, 18*(5), 1312–1326.
81. Pankaj, S. K., & Keener, K. M. (2018). Cold plasma processing of fruit juices. In *Fruit Juices*, pp. 529–537. Elsevier. https://doi.org/10.1016/B978-0-12-802230-6.00026-6.
82. Pedrow, P., Hua, Z., Xie, S., & Zhu, M. J. (2019). Engineering principles of cold plasma. In *Advances in Cold Plasma Applications for Food Safety and Preservation* (pp. 3–48). Elsevier. https://doi.org/10.1016/B978-0-12-814921-8.00001-3.
83. Perinban, S., Orsat, V., & Raghavan, V. (2019). Nonthermal plasma–liquid interactions in food processing: A review. *Comprehensive Reviews in Food Science and Food Safety, 18*(6), 1985–2008.
84. Potluri, S., Sangeetha, K., Santhosh, R., Nivas, G., & Mahendran, R. (2018). Effect of low-pressure plasma on bamboo rice and its flour. *Journal of Food Processing and Preservation, 42*(12), Article e13846. https://doi.org/10.1111/jfpp.13846.
85. Raghunathan, R., Pandiselvam, R., Kothakota, A., & Mousavi Khaneghah, A. (2021). The application of emerging non-thermal technologies for the modification of cereal starches. *LWT, 138*, Article 110795. https://doi.org/10.1016/j.lwt.2020.110795.
86. Rajan, A., Boopathy, B., Radhakrishnan, M., Rao, L., Schlüter, O. K., & Tiwari, B. K. (2023). Plasma processing: A sustainable technology in agri-food processing. *Sustainable Food Technology, 1*(1), 9–49.
87. Ratnayake, W. S., & Jackson, D. S. (2008). Starch gelatinization. In *Advances in Food and Nutrition Research* (Vol. 55, pp. 221–268).
88. Richard, F., Cormier, J. M., Pellerin, S., & Chapelle, J. (1996). Physical study of a gliding arc discharge. *Journal of Applied Physics, 79*(5), 2245–2250.
89. Roy Choudhury, A. K. (2017). Various ecofriendly finishes. In *Principles of Textile Finishing*, pp. 467–525. Elsevier. https://doi.org/10.1016/B978-0-08-100646-7.00014-X.

90. Roy Goswami, S., Dumont, M. J., & Raghavan, V. (2016). Starch to value added biochemicals. *Starch/Staerke, 68*(3–4), 274–286.
91. Sakudo, A., Misawa, T., & Yagyu, Y. (2019). Equipment design for cold plasma disinfection of food products. In *Advances in Cold Plasma Applications for Food Safety and Preservation,* pp. 289–307. Elsevier. https://doi.org/10.1016/B978-0-12-814921-8.00010-4.
92. Sarangan, A. (2016). Nanofabrication. In *Fundamentals and Applications of Nanophotonics*, pp. 149–184, DOI: https://doi.org/10.1016/B978-1-78242-464-2.00005-1. Elsevier. ISBN 9781782424871.
93. Sarangapani, C., Devi, Y., Thirundas, R., Annapure, U. S., & Deshmukh, R. R. (2015). Effect of low-pressure plasma on physicochemical properties of parboiled rice. *LWT – Food Science and Technology, 63*(1), 452–460.
94. Saremnezhad, S., Soltani, M., Faraji, A., & Hayaloglu, A. A. (2021). Chemical changes of food constituents during cold plasma processing: A review. *Food Research International, 147*, Article 110552. https://doi.org/10.1016/j.foodres.2021.110552.
95. Schirmer, M., Jekle, M., & Becker, T. (2015). Starch gelatinization and its complexity for analysis. *Starch/Staerke, 67*(1–2), 30–41.
96. Shah, N., Mewada, R. K., & Mehta, T. (2016). Crosslinking of starch and its effect on viscosity behaviour. *Reviews in Chemical Engineering, 32*(2), 265–270. https://doi.org/10.1515/revce-2015-0047.
97. Shen, H., Ge, X., Zhang, B., Su, C., & Zhang, Q. (2021). Understanding the multi-scale structure, physicochemical properties and in vitro digestibility of citrate naked barley starch induced by non-thermal plasma. *Food & Function, 12*(17), 8169–8180.
98. Shen, H., Guo, Y., Zhao, J., Zhao, J., & Ge, X. (2021). The multi-scale structure and physicochemical properties of mung bean starch modified by ultrasound combined with plasma treatment. *International Journal of Biological Macromolecules, 191*, 821–831.
99. Sifuentes-Nieves, I., Hernández-Hernández, E., Neira-Velázquez, G., Morales-Sánchez, E., Méndez-Montealvo, G., & Velazquez, G. (2019). Hexamethyldisiloxane cold plasma treatment and amylose content determine the structural, barrier and mechanical properties of starch-based films. *International Journal of Biological Macromolecules, 124*, 651–658.
100. Sifuentes-Nieves, I., Méndez-Montealvo, G., Flores-Silva, P. C., Nieto-Pérez, M., & Neira-Velazquez, G. (2021). Dielectric barrier discharge and radio-frequency plasma effect on structural properties of starches with different amylose content. *Innovative Food Science & Emerging Technologies, 68*, Article 102630. https://doi.org/10.1016/j.ifset.2021.102630.
101. Sifuentes-Nieves, I., Velazquez, G., Flores-Silva, P. C., Hernández-Hernández, E., Neira-Velázquez, G., et al. (2020). HMDSO plasma treatment as alternative to modify structural properties of granular starch. *International Journal of Biological Macromolecules, 144*, 682–689.
102. Sruthi, N. U., Josna, K., Pandiselvam, R., Kothakota, A., Gavahian, M., & Mousavi Khaneghah, A. (2022). Impacts of cold plasma treatment on physicochemical, functional, bioactive, textural, and sensory attributes of food: A comprehensive review. *Food Chemistry, 368*, Article 130809. https://doi.org/10.1016/j.foodchem.2021.130809.
103. Sudheesh, C., Sunooj, K. V., Sinha, S. K., George, J., & Kumar, S. (2019). Impact of energetic neutral nitrogen atoms created by glow discharge air plasma on the

physicochemical and rheological properties of kithul starch. *Food Chemistry, 294*, 194–202.
104. Sun, X., Sun, Z., Guo, Y., Zhao, J., Zhao, J., et al. (2021). Effect of twin-screw extrusion combined with cold plasma on multi-scale structure, physicochemical properties, and digestibility of potato starches. *Innovative Food Science & Emerging Technologies, 74*, Article 102855. https://doi.org/10.1016/j.ifset.2021.102855.
105. Tabibian, S. A., Labbafi, M., Askari, G. H., Rezaeinezhad, A. R., & Ghomi, H. (2020). Effect of gliding arc discharge plasma pretreatment on drying kinetic, energy consumption and physico-chemical properties of saffron (*Crocus sativus* L.). *Journal of Food Engineering, 270*, Article 109766. https://doi.org/10.1016/j.jfoodeng.2019.109766.
106. Taslikh, M., Abbasi, H., Mortazavian, A. M., Ghasemi, J. B., Naeimabadi, A., & Nayebzadeh, K. (2022). Effect of cold plasma treatment, cross-linking, and dual modification on corn starch. *Starch – Stärke*, pp. 2200008. https://doi.org/10.1002/star.202200008.
107. Thirumdas, R., Deshmukh, R. R., & Annapure, U. S. (2016). Effect of low temperature plasma on the functional properties of basmati rice flour. *Journal of Food Science and Technology, 53*(6), 2742–2751.
108. Thirumdas, R., Kadam, D., & Annapure, U. S. (2017). Cold plasma: An alternative technology for the starch modification. *Food Biophysics, 12*(1), 129–139.
109. Thirumdas, R., Trimukhe, A., Deshmukh, R. R., & Annapure, U. S. (2017). Functional and rheological properties of cold plasma treated rice starch. *Carbohydrate Polymers, 157*, 1723–1731.
110. Townsend, J. S. (1915). *Electricity in gases*. Oxford: The Clarendon Press.
111. Waghmare, R. (2021). Cold plasma technology for fruit-based beverages: A review. *Trends in Food Science & Technology, 114*, 60–69.
112. Wang, J. H. (2012). Surface preparation techniques for biomedical applications. In *Coatings for Biomedical Applications* (pp. 143–175).
113. Wang, J. M., Zhuang, H., Lawrence, K., & Zhang, J. H. (2018). Disinfection of chicken fillets in packages with atmospheric cold plasma: Effects of treatment voltage and time. *Journal of Applied Microbiology, 124*(5), 1212–1219.
114. Wu, T. -Y., Chang, C. -R., Chang, T. -J., Chang, Y. -J., Liew, Y., & Chau, C. -F. (2019). Changes in physicochemical properties of corn starch upon modifications by atmospheric pressure plasma jet. *Food Chemistry, 283*, 46–51.
115. Wu, T. -Y., Sun, N. -N., & Chau, C. -F. (2018). Application of corona electrical discharge plasma on modifying the physicochemical properties of banana starch indigenous to Taiwan. *Journal of Food and Drug Analysis, 26*(1), 244–251.
116. Xie, S. X., Liu, Q., & Cui, S. W. (2005). Starch modification and applications. In *Food Carbohydrates: Chemistry, Physical Properties, and Applications*, pp. 131–149. Springer. https://doi.org/10.1007/978-981-15-0622-2_8.
117. Yan, S., Chen, G., Hou, Y., & Chen, Y. (2020). Improved solubility of banana starch by dielectric barrier discharge plasma treatment. *International Journal of Food Science & Technology, 55*(2), 641–648.
118. Yan, Y., Zhou, Y., Shi, M., Liu, H., & Liu, Y. (2019). Influence of atmospheric pressure plasma jet on the structure of microcrystalline starch with different relative crystallinity. *International Journal of Food Science & Technology, 54*(2), 567–575.

119. Zare, L., Mollakhalili-Meybodi, N., Fallahzadeh, H., & Arab, M. (2022). Effect of atmospheric pressure cold plasma (ACP) treatment on the technological characteristics of quinoa flour. *LWT, 155*, Article 112898. https://doi.org/10.1016/j.lwt.2021.112898.
120. Zhang, B., Chen, L., Li, X., Li, L., & Zhang, H. (2015). Understanding the multi-scale structure and functional properties of starch modulated by glow-plasma: A structure-functionality relationship. *Food Hydrocolloids, 50*, 228–236.
121. Zhou, Y., Yan, Y., Shi, M., & Liu, Y. (2019). Effect of an atmospheric pressure plasma jet on the structure and physicochemical properties of waxy and normal maize starch. *Polymers, 11*(1), 1–8.
122. Zhu, F. (2017). Plasma modification of starch. *Food Chemistry, 232*, 476–486.
123. Zou, J. -J., Liu, C. -J., & Eliasson, B. (2004). Modification of starch by glow discharge plasma. *Carbohydrate Polymers, 55*(1), 23–26.

CHAPTER 6

APPLICATIONS OF COLD PLASMA TECHNOLOGY IN SOLID FOOD PRODUCTS

SANGEETHA KARUNANITHI, GNANA MOORTHY ESWARAN, PREM PRAKASH SRIVASTAV, and PROSHANTA GUHA

ABSTRACT

In this chapter, the application of cold plasma (CP) technology in solid food products focuses on the basics of cold plasma technology in food processing, types of CP, principles, and methods of CP generation, equipment, factors affecting CP efficiency (intrinsic, extrinsic, and processing parameters), application of CP in the food industry such as cereals, fruits, vegetables, meat, seafood, milk, and spices. This chapter also covers the principles of the mechanism of CP such as sterilization, germination, extraction, pesticide reduction, surface modification, and starch modification. Cold plasma also influences the food quality parameters after treatments. Therefore, this chapter discusses the effects of CP on physical, biochemical, and sensory characteristics; food safety aspects; and consumer acceptability.

6.1 INTRODUCTION

Solid, liquid, gas, and plasma are the four primary states of matter that can be formed by increasing the energy level. As the amount of energy exerted on atoms rises, the thermal motion of atoms in solids becomes more pronounced, eventually overcoming the restricted relationship between atoms in solids such as the ionic bond and forming a liquid. Similarly, when liquid atoms get enough energy to overcome the restricting Van der Waals forces from nearby atoms, they will transform into a gas, which is obviously considerably higher

Nonthermal Plasma Technology for Food and Food Products.
R. Mahendran, Preeti Birwal, and Megh R. Goyal (Eds.)

than in liquid and solid. The electron will be taken away when the energy is high enough for it to overcome the electrostatic potential barrier, leaving positively charged ions and free electrons, which is called an ionization process [37].

Plasma is a neutral ionized gas comprised of positive ions, electrons, and neutral particles that exist in abundance. Plasma exhibits unique properties consisting of various excited atoms, ions, molecules, and radical species along with numerous reactive radical species, including wide electromagnetic radiation (visible light and UV photons), free radicals, ground/excited state molecules, gas atoms, positive and negative ions, and electrons [32]. These chemical species are responsible for various unique functions of plasma, such as being internally interactive, electrically conductive, having a strong electromagnetic field response, and efficient antimicrobial action [78].

At the nano, micro, millisecond timescale, the air plasma contains about 75 unique species and over 500 chemical species. The chemical reactive species are produced by the process of electrical excitation, which allows the production of energetic electrons and passing their energy to neutral gas atoms. Negative charges are normally carried by electrons with one unit of negative charge. Atoms or molecules without those electrons give off a positive charge. Electrons detached from one molecule/atom and attached to nearby molecules cause the production of both positive ions and negative ions. Heating gas to high temperatures under controlled pressure causes collisions of atoms/molecules, which produce the outcomes of ripping free electrons, creating plasma discharge. A comparable process occurs within stars. As photons from sunlight as well as stars are captured by an excited gas, the reaction becomes photoionization, which emits electrons, the most common plasma creation process in space. Because of this, the sun and starlight continuously ionize nearby matter. So the plasma is considered to be totally ionized in such instances.

When the temperature rises, matter transitions from a solid state to a liquid state to a gaseous phase to an ionized plasma state. The thermal motions of electrons as well as heavier particles (ions and atoms) regulate the temperature of the plasma. The electron volt (eV), defined as the energy acquired by an electron under vacuum conditions when accelerated by more than 1 V of electric potential, is a useful metric for measuring temperature in plasma research. When T is stated in Kelvin, $W = T/12{,}000$ equals the temperature, W, measured in electron volts. These values are used to calculate the amount of energy essential for eliminating one electron from a molecule/atom [89]. Self-ionization occurs at temperatures ranging from 2.5 to 8 electron volts.

The proportion of charged elements against the total amount of elements in plasma, including neutrals and ions, is defined as the degree of ionization: $\alpha = n^*/ (n + n^*)$, where n = quantity of neutral particles; and n^* = quantity of charged particles. The degree of ionization indicates when a gas generates plasma; and it can be as low as 10^{-6} [68].

This chapter focuses on the application of plasma on solid foods, its working mechanism within solid food matrices, the factors influencing the process, and its various applications.

6.2 HISTORY OF PLASMA STATE

The plasma state, as we know it now, is a relatively new notion, originating from the early 1950s. Many disciplines were intertwined throughout its history. Electric discharges, magnetohydrodynamics (the study of a conducting fluid such as mercury), and kinetic theory are three core disciplines of study that contributed significantly to the early development of plasma physics as a subject. Electric discharge events have been studied since the early 1800s, with three English physicists – Michael Faraday in the 1830s, Joseph John Thomson, and John Sealy Edward Townsend laying the groundwork for today's knowledge of the phenomenon. Irving Langmuir discovered plasma in 1923 while exploring electric discharges. He coined the name in 1929 with another American physicist, Lewi Tonks, to define the areas of a discharge where periodic oscillations of negatively charged electrons can occur. The wide applicability and concept of plasma were first investigated by two other American physicists, David Pines and David Bohm, who explained the combined behavior of electrons in metals as opposed to ionized gases [235].

6.3 TYPES OF PLASMA TECHNOLOGY

Plasma is classified as "thermal" when it is in thermodynamic equilibrium due to electrons and the presence of other gaseous species. When it is not in equilibrium, it is classified as "non-thermal." Cold or low-temperature plasmas can be created at low pressures or at atmospheric pressure [240]. For decades, low-pressure cold plasma (LPCP) has been used for fabricating semiconductor devices. Generally, atmospheric pressure cold plasma (APCP) has been used to modify the hydrophilic or hydrophobic nature of materials. Cold plasma used to be created under low-pressure settings, limiting its applicability. Recent developments in plasma engineering have enabled the creation of cold plasma at atmospheric pressure, leading to increased research at the interfaces of many life sciences. Thermal equilibrium is achieved for all particles when there are intense collisions among electrons and heavy particles at high particle densities. This type of plasma reaches very high temperatures, often in the thousands of degrees [163].

When a plasma discharge occurs rapidly at atmospheric pressure, however, another class of plasma emerges in which heavy particles and electrons are in thermal non-equilibrium. The heavy particles have a lower temperature compared to electrons, resulting in the development of a cold region. Therefore, "Cold atmospheric plasmas" will be the name given to these plasmas (CAP). CAP's heavy particles have a temperature of 25 to 45°C. CAP generates an extensive collection of active radical species, including nitrogen-based reactive radicals, oxygen-based radicals, and other substances [160, 166]. As we know, there are two types of plasma and their differences are described in Table 6.1 [253].

TABLE 6.1 Difference Between Hot and Cold Plasma

Hot Plasma	Cold Plasma
Thermodynamic equilibrium	Thermodynamic non-equilibrium
Thermal behavior of ions is inclusive.	Thermal behavior of ions is excluded.
The electric, pressure and the magnetic force acts on the particles.	Electric force is the only force that acts on the particles whereas pressure and magnetic force is nil.
$Te \simeq Tg$	$Te >> Ti >> Tg$
High degree of ionization as close to one (i.e., 100%). The stellar interiors, solar wind, and plasma fusion are all examples.	Low degree of ionization. weakly ionized between 10^{-6} and 10^{-1}. Degree of ionization is below 10^{-4}. The degree of ionization in partially ionized plasma is less than one. The ionosphere as well as the gas discharge tube are examples.
High temperature (10^5 K)	Low temperature (300 K to 600 K)

6.4 IMPORTANCE OF COLD PLASMA IN FOOD INDUSTRY

CAP interacts with a range of biological systems, including cells and tissues, due to its complicated chemistry [145, 244]. The energy of CP components is significantly greater than the energy of conventional chemical reactions. Chemically and physically, these particles can be used to modify polymer surfaces. There is a demand for safe foods with improved qualities such as convenience, nutritional profile, freshness, sensory attributes, extended shelf-life, additive-free, eco-friendly green processing techniques, and reduced cost compared to food processed through conventional methods [76]. Heat treatments have already been extensively used to prolong shelf-life and improve food safety for many years, and they are universally approved among consumers. However, these methods have certain disadvantages, including nutritional degradation and a significant decrease in the organoleptic aspect of certain foods [110].

The growth and focus on non-thermal processing of food products have received a high level of awareness among the food business sector in the past two decades. As a result, substantial thought has recently been given to the deployment of novel preservation systems based on novel technology, popularly known as "Emerging Food Preservation Technologies," with the mission of inactivating microbes while retaining their nutritional, sensory, and functional characteristics.

High-pressure processing, UV radiation, pulsed electric field, ultrasound, high-intensity pulsed light, and, most recently cold plasma (CP) are some of the most successful techniques [90, 116]. These alternative technologies have the capability to enhance efficacy as well as storability while decreasing detrimental effects on food nutrients and natural flavor. The prospect of utilizing plasma in surface decontamination was originally addressed in a patent filed in the United States in the late 1960s [153, 157]. The process of producing cold plasma under vacuum conditions and its expensiveness define its implacability for commercial applications; hence NTAP has yet to be accepted in the food industry. Nonetheless, in the late 1990s, technological advances in plasma generation sources enabled the development of equipment that generated plasma under atmospheric conditions, removing the demand for vacuum conditions and allowing for continuous treatment of materials with comparatively simple and low-cost equipment. The food business benefits greatly from NTAP, making it a viable new food preservation method. It allows for rapid processing, for example. In addition, very brief treatment durations (from a few seconds to minutes) have been demonstrated to yield more than a 5 log reduction in bacteria. Its effectiveness at room temperature makes it ideal for products of heat-sensitive nature, and it could be used to disinfect pre-packaged meals to minimize cross-contamination during processing [255].

In recent decades, researchers have been looking into utilizing NTAP for preserving food, and a few studies are underway to quantify its antimicrobial efficacy and pinpoint the inactivation processes involved. Cold Plasma (CP) is a cutting-edge technique that has received extensive importance among food sectors for processing and protecting products from spoilage, as well as ensuring safety with slight variation in certain quality parameters. Its effectiveness in inactivating microbial spores is attributed to its simple design, ease of use, low operating costs, and rapid treatment timeframes [147].

6.5 ADVANTAGES OF CP APPLICATIONS IN SOLID FOODS OVER LIQUID FOODS

The important differences between plasma uses on solid or dry medium and liquefied products are diffusion depths and dispersion of plasma species, which alter plasma chemistry. The CP is generally used for processing food at the surface level in the case of solid-state products. A number of factors influence the depth of reactive species penetration. Their components, such as water content and porosity, are the most essential ones. According to a few studies, the distance of plasma diffusion and the effectiveness of radical active substances (such as O_3, RNS, ROS, and UV) are restricted to a few millimeters deep within biofilms. The diffusion level of radicals developed by the PJ system inside bacterial biofilms has been explored using a microscope and concludes that efficient disruption of biofilm is achieved to a maximum level of 15 μm [275]. Another study using a portable air-based PJ reactor disrupted the biofilm up to 25.5 μm [190]. Additional limiting criteria in terms of penetration depth include the plasma's composition and its flow rate [38]. Another essential consideration is the half-life of reactive species, which has a direct impact on the depth of penetration into solid meals. For example, the half-life of H_2O_2 is 1 ms with a penetration depth of 1 μm and is stable, but other ROS have a short half-life (1 μs for O ~ 30 nm, 1 ns for OH ~ 1 nm, 1 μs for O* ~ 30 nm) causing comparatively less diffusion [168]. The shallow penetration depth is desirable in several instances, like mild destruction of microbes and preserving nutrients. A short penetration distance is suitable for retaining natural food nutrients during the process of targeting surface decontamination. If microbial inactivation is the primary goal for liquid food treatment, the depth of penetration is unimportant because all parts of the liquid interact with plasma species. However, the plasma components have a deleterious effect on the nutrients in liquid foods.

6.6 PRINCIPLE OF COLD PLASMA ON SOLID FOOD PRODUCTS

Several types of actions occur concurrently or successively when CP interacts with a material placed within it (substrate). Plasma species when comes in contact with solid surface it produces various reactions such as surface etching, depolymerization, functionalization, deposition, structural modification etc., for performing enormous unique properties in solid food products.

6.6.1 ETCHING

Etching is the process of creating active sites over the substrate's surface (ablation of surface) and volatile species that diffuses into gas phase by combining reactive gas phase species within atoms or groups of atoms on the substrate. The main principle involved in plasma process is surface etching also called plasma etching. This plasma etching process utilized for roughening of surface at microscopic scale. The CP reactive substance bombard continuously to specimen surface cause etching of exposed surface atoms then remove them into the gas phase which cracks the surface of the materials [274]. Plasma etching works by two mechanisms 1) Ion bombardment at the surface hence removing the materials required to form cracks/ trenches. 2) The gas monomers in CP system interact with sample molecules, forming volatile byproducts which will be removed [77, 184].

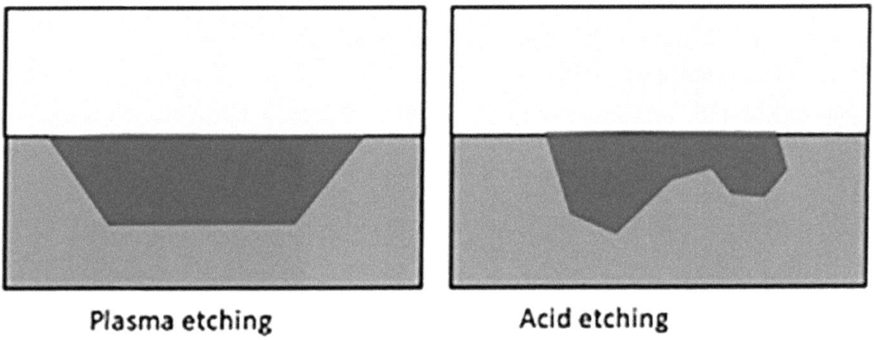

FIGURE 6.1 Etching phenomena on surface after plasma and acid treatment.

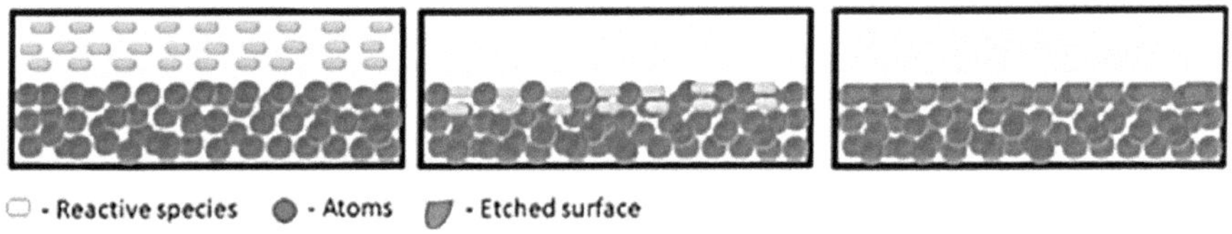

FIGURE 6.2 Mechanisms of etching by plasma species.

This method etching is more ecofriendly than the acid etching methods. This plasma etching produces deep straight trenches for increased roughness thus makes its way in industrial revolution (Figure 6.1).

Etching cause by Plasma is also called dry etching. In this process, the reactive species generated and it is diffused to the surface of the materials. The surface material absorbs the reactive species which leads to the chemical reactions. The byproducts generate on surface is desorbed and dispersed towards gas space. This how the plasma creates etching in the materials [77]. The surface etching behavior of plasma can be visualized in Figure 6.2.

Through the stimulation of chemical reactions ions bombardment plays a crucial role in etching. The ion assisted gas surface chemistry mechanisms is perhaps the appropriate word for plasma surface etching which can harm or change the surface properties during etching or create opening/cracks on the surface. Etching is beneficial to surface cleaning or surface structuring process by removing the organic contaminants and surface sterilization. The surface etching removes top layers and restructures the surface, resulting in increased corrugation and surface area.

6.6.2 PLASMA DEPOLYMERIZATION

Ionization has been the most significant factor in the depolymerization mechanism, which results in starch modification [68]. Plasma ion bombardment promotes depolymerization of the side chains of amylose as well as amylopectin of starch into smaller fragments has been depicted in Figure 6.3.

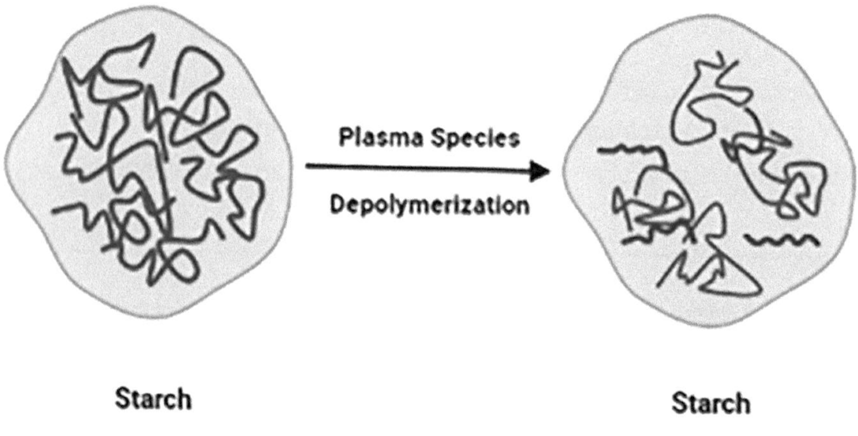

FIGURE 6.3 Depolymerization of starch by cold plasma species.

Also at low energy levels, a tiny amount of depolymerization can occurs. The cold plasma induced scission of a biopolymer by free radicals, which break down the glyosidic bond at the C1 point of a molecule of glucose in a starch molecule. Plasma's energetic reactive species destroy the starch crystalline structure's inter helical water molecules, producing oxygen, hydroxyl, and hydrogen free radicals that impair the starch's crystallinity [130]. The mechanism of starch depolymerization following plasma is similar to that of X-ray and gamma ray irradiation. Depolymerization is influenced by plasma reactive species, free radicals produced, ion concentration, starch type, and shape. Starch derived from potato highly responsive towards depolymerization reaction than starch of rice, corn and waxy corn. End products of starch polymerization are mostly found to be Maltose, malto triose, and malto tetrose [250].

6.6.3 CROSS LINKING OR POLYMERIZATION

During plasma development, cross-linking occurs between polymeric chains found in starch molecules, resulting in the synthesis of free radicals as well as energetic electrons. Cross-linking occurs through collisions among the gas reactive substances with the molecules of starch polymers [169]. The development of free radicals by plasma generation results in cross-linking and the creation of active sites for subsequent reactions, which substantially rely on the type of gas and power level. The relative strength of OH groups decreased in plasma-treated samples, demonstrating that OH groups were depleted. The cleavage of two polymeric chains at their reducing ends (C-OH) created a C-O-C ring among those ends of chains as a function of cross-linking through the elimination of an H_2O molecule [290].

Hydroxyl radicals (*OH) produced by the breakdown of water molecules and the O_2 present during plasma formation cleave the glyosidic bonds. Cross-linking (in the case of a polymeric surface): the bonds between atoms on the substrate's surface are opened (under the influence of reactive plasma species), allowing them to rearrange (by bridging between the macromolecular chains). The cross-linking behavior of the polymeric chain is shown in Figure 6.4.

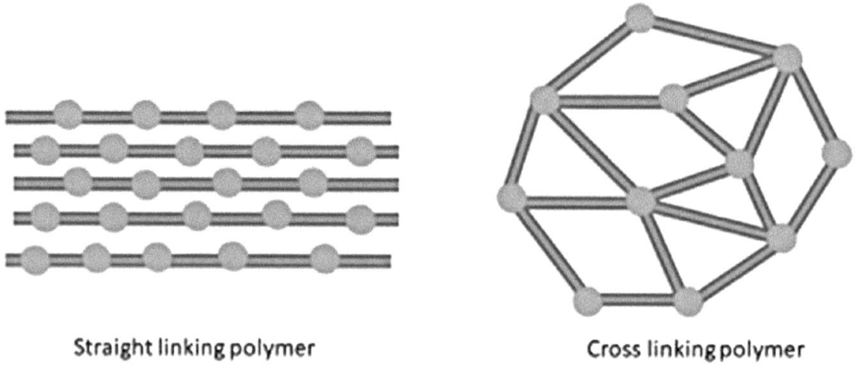

FIGURE 6.4 Molecules of straight and cross-linking polymer chain.

Chemical crosslinking is the process of joining two or more molecules through the influence of any cross-linking agents, forming covalent bonds. Crosslinking of polymer surfaces is beneficial for further processing, such as improving wettability and printability. The adhesion and wear properties of polymers used in various industrial and biomedical applications can be significantly influenced by surface crosslinking. The excited energy emitted during plasma development can impact this cross-linking behavior. Energetic plasma particles, such as ions, vacuum ultraviolet, uncharged particles including neutrons, radicals, metastable species, and UV photons, have enough energy to break down C-H bonds and form radicals [270]. Cross-linking behavior among molecular rings inside polymeric surfaces results from the recombination of these radicals.

6.7 MECHANISM OF COLD PLASMA GENERATION

Gas breakdown and electron avalanche are two key approaches for transforming a quasi-gaseous state into conductive gases intended for carrying electrons. Streamer theory and Townsend theory explain the process of gas breakdown under a variety of conditions, including temperature, electrode field configuration, air pressure, properties of electrode surfaces, and availability of initial conducting particles [273]. According to the Townsend theory (Figure 6.5), discharging all types of plasma requires the generation and amplification of so-called electron avalanches during gas breakdown.

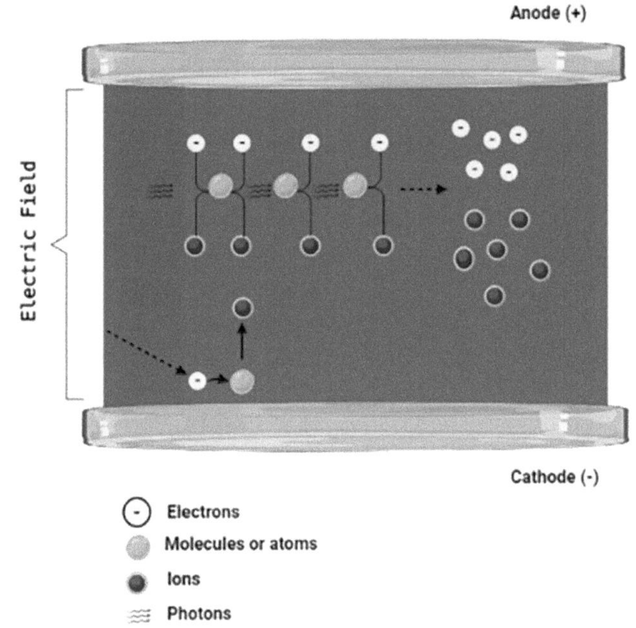

FIGURE 6.5 Electron avalanche by reactive species.

Assume the cathode emits a number of electrons, one of which collides with a neutral particle, resulting in the formation of a positive atom and an electron. When this happens, an ionization collision occurs. When energy is applied between two electrodes, it causes growth of kinetic energy of the molecules and discharge of electrons from the surface of the cathode contrary to the electric current region. The electrical current rises by providing voltage until it achieves a saturation state and is said to be steady when it is consistent. As they reach the anode, the electrons accelerate and under this specific condition, collisions are elastic (no change in internal energy), and inadequate electron energy restricts the ionization of nearby substances (molecules).

The collision of electrons continues until they accumulate enough energy for ionizing atoms in inelastic collisions, sufficient for efficient transfer of energy. The molecules and atoms can be dissociated and converted into ions and electrons if the collisions are intense enough. The movement of ions and electrons creates the current which is driven into the electric field, impacting and ionizing other molecules and atoms causing an explosion of intensive electrons, positive ions, and an electron avalanche. Electrons migrate to the avalanche head respective of their lower mass and higher speed while positive ions move to the tail.

Avalanches occur when the ions pull fresh electrons from the surface of the cathode. The gas becomes entirely disturbed and conductive when the ionization is powerful enough. Following the breakdown, a low-pressure glow discharge (GD) is formed in the electrode gap when electrons are captured by the anode and generate a significant amount of positively charged ions through the electrodes compared to micro-discharges. The ions produced combine with the free electrons, thus producing photons, causing photoionization of the surrounding gas and the generation of further electrons. As a result, multiple avalanches form (secondary avalanches). Recombining electrons with positively charged ions produces secondary avalanches, which again join the main avalanche, thus causing the emission of photons. The development of subsequent avalanches results in creating a highly conductive channel by atmospheric conditions known as streamer discharge [35, 54]. The overall process flow of plasma generation is shown in Figure 6.6.

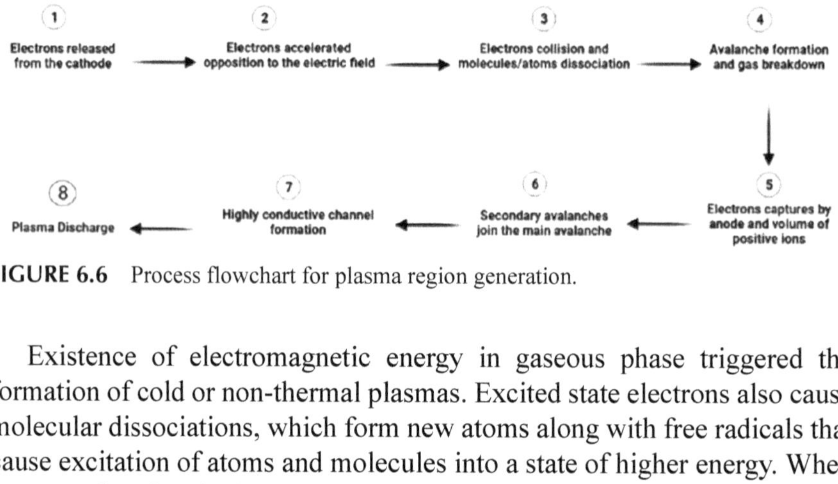

FIGURE 6.6 Process flowchart for plasma region generation.

Existence of electromagnetic energy in gaseous phase triggered the formation of cold or non-thermal plasmas. Excited state electrons also cause molecular dissociations, which form new atoms along with free radicals that cause excitation of atoms and molecules into a state of higher energy. When atoms and molecules in an excited state bounce back to a stable state, they release excess electromagnetic energy, particularly UV light. As a result, the plasma discharge is found to have a combination of excited molecules, atoms, and reactive species [35].

6.8 METHODS OF COLD PLASMA GENERATION

Cold plasma generating systems are used in a variety of sectors and come in a variety of shapes and sizes. Corona discharges, radio frequency plasmas, microwave plasmas, inductively coupled plasmas, electron cyclotron resonance plasmas, dielectric barrier discharge plasmas, and capacitively coupled plasmas are just a few examples of what can be done. The dielectric barrier discharge (DBD), corona discharge (CD), microwave (MW) and radiofrequency (RF), and plasma jet (PJ) are the most often employed plasma production technologies in food processing because of their simplicity, versatility, and adjustable design and operating conditions.

6.8.1 DIELECTRIC BARRIER DISCHARGE

A dielectric is an electrical insulator that can be polarized to conduct electricity when put in an electric field. Huge electric fields provide free charges (electrons) that can freely flow through the material and carry current. Dielectric breakdown or dielectric barrier refers to the transition of the dielectric from an insulator to a conductor. The DBD plasma generated by passing the electric current through electrodes contains insulating material which is depicted in Figure 6.7. The technique employs a high-voltage or voltage amplitude oscillating between 1 and 100 kV, low-frequency (10 to 50 MHz), pulsed direct current (DC) or alternating current and 1×10^4 to 1×10^6 Pa gas pressures [74].

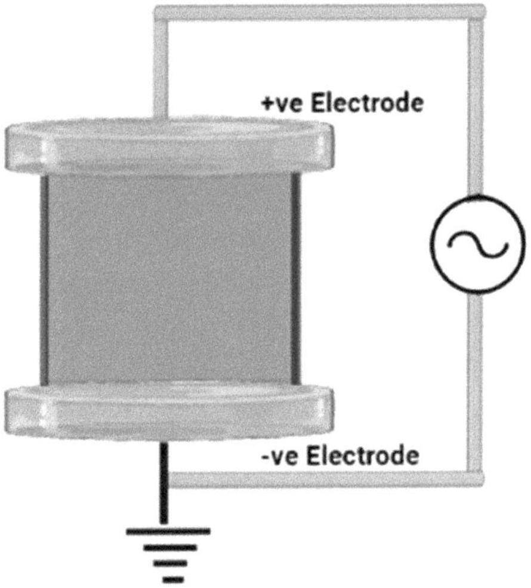

FIGURE 6.7 Configuration of DBD plasma system.

This technique employs high-voltage, low-frequency alternating current (Figure 6.7). DBD devices are typically constructed of plate electrodes separated by a 0.1 mm thick dielectric layer. In a DBD plasma system, the electrodes are attached with different dielectric substances such as polymer, glass, quartz, and ceramic [26]. Because barrier plasma operates at atmospheric pressure, it requires a lot of energy to function. The most important aspect of using DBD plasma is to ensure that the membrane is tightly attached to the electrode and that the gas flow around it is smooth. Plasma production using DBD is becoming more popular due to its low expenses on an industrial scale. The promising qualities of DBD plasma, such as configuration technique and flexibility in the shape of the electrode and the type of dielectric material utilized, make it an appropriate technique for plasma creation [167]. A wide range of gases are suitable for generating plasma (noble gases, special mixtures of precursors, air or water vapor, and almost all combinations of gases). The geometry of electrodes and the homogeneous discharge of gases over a large area all contribute to DBDs' adaptability. The minimal electrode distance and the need for maximum input power (10 kV voltage) are drawbacks that require the use of safety precautions or isolation. For large surfaces, DBD is a suitable plasma source [192].

The DBD plasma technique can affect the structural, mechanical, and physicochemical features of soy protein isolate. Reactive oxygen species (ROS), in addition to reactive nitrogen species (RNS) and other active chemical radicals, exist during DBD plasma and might react with soy protein isolate, altering its physicochemical and functional properties. The results showed that exposure of soy protein isolate to DBD plasma alters its structural, physicochemical, and functional properties with respect to the time and voltage of the generated plasma. Exposure of the protein to DBD plasma caused disruption of the protein structure. When the plasma voltage increased from 18 to 20 kV, it enhanced the development of intramolecular hydrogen bonds, leading to the aggregation of the protein. Consequently, exposure of dry protein to DBD plasma affects the characteristics of proteins through chemical reactions between the components of the protein, mechanical breakdown of the structure, and heat generated during the processes [227]. The study investigated the influence of a DBD-based plasma system for destroying *Escherichia coli* and *B. cereus,* which contaminate pepper powder stored at 25°C. During storage, *E. coli* was destroyed more quickly after a 15-minute DBD plasma treatment due to the disruption of bacterial cell membranes, resulting in cell death. A 15-minute DBD plasma treatment is sufficient for destroying microbes present in pepper powder without causing any quality changes [122].

In-package food processing technique has been done with the creation of CP within a sealed package, considered to be a promising application that paves the way for utilizing the DBD system in the area of food preservation. This approach permits the reactive species to have a longer action duration on microorganisms while also preventing post-process contamination. The DBD reactor is mostly preferred among the food industry sector for CP development [286].

6.8.2 PLASMA JET

Plasma jets are available in a variety of designs that enable gas discharge and plasma species projection in an open environment using non-sealed ("open") electrode configurations (Figure 6.8) [176].

Non-thermal plasma jets are capacitive linked devices where high-speed gas is supplied between coaxial electrodes, and they produce small "plasma fires" that often start in the radio frequency band at atmospheric pressure. The central electrodes are triggered by an RF wave at 13.56 MHz, while the outer electrodes are grounded. Free particle electrons are triggered by RF energy whenever electrons come into contact with components of the gas mixture. The interactive chemical substances (energetic atoms along with molecules) produced through the unbendable collisions might immediately exit from the jet [163]. Increased flow velocities of gas (He/Ar) greater than 10 s/m are regularly applied for the PJ operation. This is an expensive technique, but it is appropriate for certain biological uses apart from the food sector, where budget is a major influence. Plasma jets also exceed the allowable treatment rate, and the small size is a disadvantage. The high diffusivity and direct exposure become the benefits of this PJ plasma system [176]. At atmospheric pressure, the PJ provides a stable, homogeneous, and homogeneous discharge [146]. When targeting a large area, this plasma is limited; however, in this case, a series of jets must be used [30].

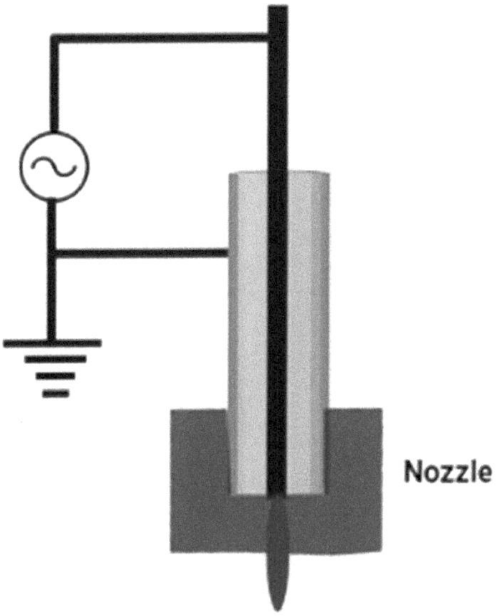

FIGURE 6.8 Configuration of plasma jets (PJ) system.

The pilot-scale ACP-PJ method for inactivating indigenous microbes, which contaminate solid foods, was created [30]. The plasma jets in the PJ treatment chamber had a 1 kV output voltage, a 30 Hz frequency, and a plasma discharge length of 18 mm. Natural bacteria were inactivated by 1.4 log in cereals (rice germ). The ACP-PJ technique could be utilized to microbially disinfect the particle food, according to these investigators. The PJ system, on the other hand, has been used more frequently in research work at the lab. The PJ method is intended to be employed for minimizing allergens present in seafood, developed at 7 kV DC output voltage with 10 A electric current, and a 50 kHz frequency [70].

6.8.3 CORONA DISCHARGES

The Corona discharge (CD) is a slightly glowing ejection that arises in the field of electric energy with non-uniformity, with a piercing electrode like thinner wires/edges/point configurations under atmospheric pressure. The creation of electrons occurs randomly while having a high electric field, resulting in the process of atmospheric gas ionization [55]. A common configuration used in CD is point-to-plane, like a strongly arc-shaped electrode positioned in the opposite direction of a flat electrode and cylindrical arrangements (Figure 6.9).

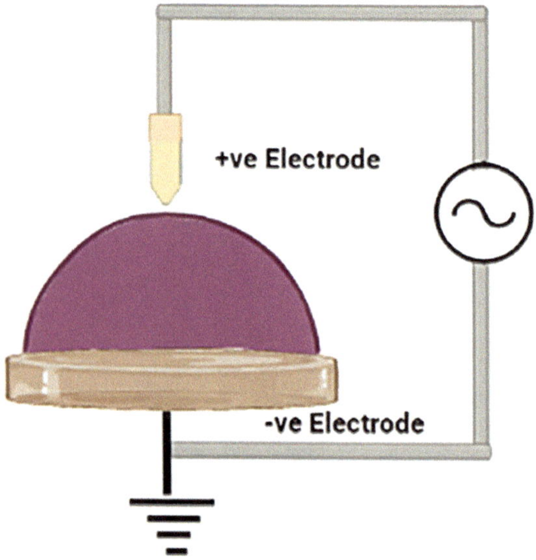

FIGURE 6.9 Configuration of corona discharges (CD) system.

The sharp electrode is primarily responsible for ionization and brightness with positive or negative potential, and these setups operate with either AC or DC. The CD plasma system has low complexity and works well with a normal system with a small setup and minimum operating expenses. The key disadvantage is that it only addresses a tiny area and does so in an inconsistent manner [221]. An asymmetric electrode pair (plate with point) forms plasma whenever the electrical current in a certain geographical area crosses a threshold failure. A weakly ionized plasma arises when the electrode's high electric field surpasses the gas breakdown strength. A diffusive release induced by CD, which focuses on delivering a much more powerful and intense plasma over DBD, encompasses a broader proportion of the specimen surface compared with needle-pointed electrodes [219]. The investigators evaluated the effectiveness of CD on prominent peanut allergenic species using a broad spacing (70 mm) multi-hop plate-to-point system. The plasma region generated with specifications of 32 kV voltage, 118s switching frequency, 1 kHz emission frequencies, 52 kHz operating frequency at atmospheric conditions [261]. The outcomes exhibited CD-based CP treatment as a viable solution for minimizing peanut allergy symptoms. The usual operating parameters for CD utilization are the distance between the electrode and sample (10–30 mm) as well as voltage (60–100 kV). The originally stated multipoint-plate reactor featured a minimum voltage of approximately 32 kV with a maximum distance of 70 mm. This innovation highlighted the fusion of the reactor with the generator, as well as a detailed evaluation of the impact of material, electrode configuration, and electrical parameters, among other things, on CP applications in food processing.

6.8.4 RADIO FREQUENCY

The three primary types of RF plasmas are capacitive coupled plasmas (CCP), helicon wave sources, and inductive coupled plasma (ICP). The manufacturing sector prefers CCP and ICP systems [30]. In a vacuum chamber, the RF is created between the electrodes while passing AC through the CCP system. There are a variety of ways to generate radiofrequency (RF) electromagnetic fields, distributing RF energy across coils or antennae which are either immersed in the plasma system or isolated through DBD material, or delivering the RF energy across parallel electrodes (Figure 6.10). The RF is generated at 13.36 MHz (1–500 MHz) with 1 kW power used for operating electrodes [52].

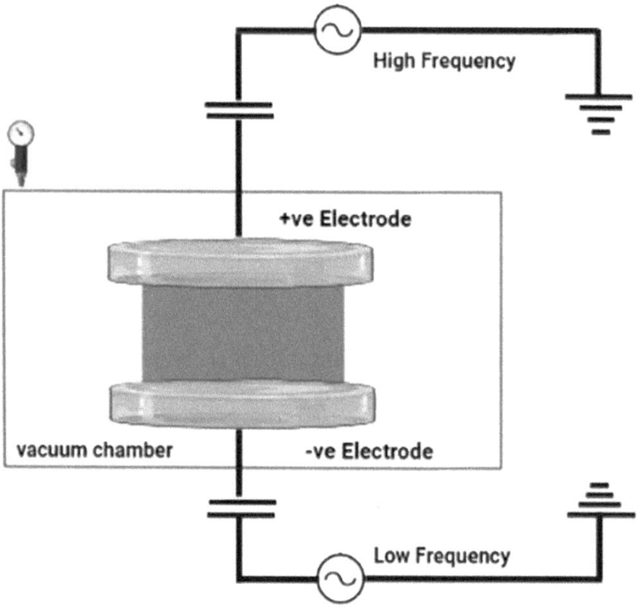

FIGURE 6.10 Configuration of radio frequency (RF) system.

Electrons are connected with electromagnetic fields in the plasma, allowing energy to be supplied to them and the plasma to continue to exist. The efficacy of power coupling into charged particles and plasma uniformity is determined by how RF stimulation is constructed [40]. When gases are ionized by electromagnetic pulse waves, the density of the particles improves in the range of 10^{15}–10^{16} m^3 in the plasma. In ICP systems, two RF power supplies are frequently used: one drives a coil that is normally in the outer region of the plasma, which is divided by a DBD sheet. The other one drives a coil that is normally inside the plasma and isolated from it by a dielectric; the other biases the substrate holder and controls the ion energy [40]. Both the cell-wall structure and nucleic acids appear to be affected when bacteria are inactivated by RF plasma. The RF-based plasma system operated at 60 MHz for generating ACP facilitates the destruction of *E. coli,* where UV radiation and other radicals have been demonstrated to play substantial and secondary roles [229]. Furthermore, these inactivating agents were found to work in *E. coli* cells, and B. subtilis endospores were discovered to be inactivated by oxygen radicals produced by an RF plasma (based on He gas mixed with different amounts of O_2) [102]. The maize starch modified by RF technique could be used to create high thermal stability of food.

6.8.5 MICROWAVE (MW)

In a microwave discharge plasma system, also called an electrode-free system, the magnetron assists in creating plasma, and cables with waveguide assist in conveying plasma discharge into the microwave operating area where collisions occur between gas and electrons (Figure 6.11) [256]. Ionization processes occur when electrons absorb microwave energy, intensifying kinematic energy in addition to inelastic collisions that release energy as UV and visible light photons [220].

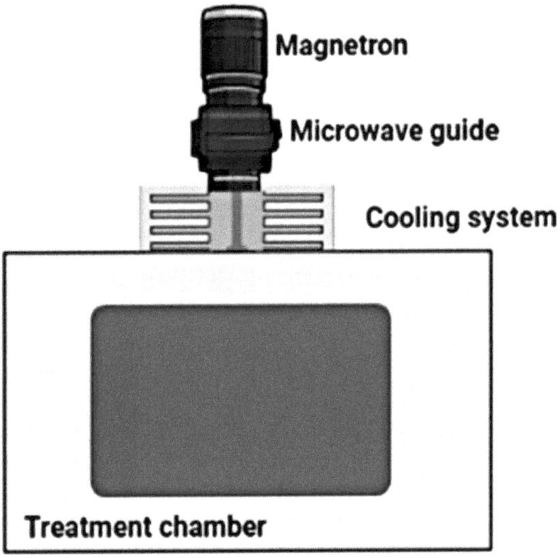

FIGURE 6.11 Configuration of microwave (MW) plasma system.

The electrode-free, easy-to-handle designs of MW-driven discharges, higher density of electrons, and greater productivity are all significant advantages of MW plasma systems. Even with particular admixtures, precursors, or water vapor, they can be ignited in an open atmosphere. However, due to its high cost and the care required for its functioning, it is currently underutilized. The drawback of MW plasma is the space restriction. The intensified discharge is necessary for achieving greater deactivation on big-sized samples.

Therefore, a wide range of plasma discharges is created for eliminating microbes present in vast regions [69]. The researchers [14] created a pilot-plant scale to generate CP with a 2.45 GHz frequency followed by 1.1 kW power for treating water samples. The resultant water was utilized for washing sliced lettuce, resulting in a 2 log reduction in the microbial count. Furthermore, 45 kg of shredded lettuce required 1,500 L of plasma-treated water for decontamination, which is prepared by the developed pilot plant, demonstrating that this technique has significant potential for industrial usage in the disinfection of crops.

6.9 FACTORS INFLUENCING COLD PLASMA EFFICIENCY

Several research studies have focused on enhancing CP treatment in recent years by developing new plasma equipment and evaluating various process factors in a variety of settings [14, 160, 286]. The ability of plasma species is determined by electron energy. The energy of electrons changes depending on numerous aspects such as gas chemical composition, voltage, scattering of charge, posing a barrier to plasma equipment. The quantity of active radicals and their half-life, size, density, as well as temperature are also related to plasma properties [35]. The method of CP discharge is determined based on the temperature of gas with electron density (Table 6.2) [35]. A streamer discharge's high electron density makes it simple for arcs to form, but it degrades electrodes due to high gas temperature, resulting in less usage of this discharge. A plasma reactor should be built to avoid this type of discharge by selecting the shape of electrodes [165].

TABLE 6.2 Density of Electron with Temperature of Gas at ACP System

Plasma System	Electron Density (m^{-3})	Gas Temperature (°C)
Arc discharge	$10^{21.5}$ to 10^{26}	12,000 to 40,000
DC Corona discharge	10^{14} to $10^{16.5}$	25 to 180
Glow discharge	$10^{14.5}$ to 10^{19}	25 to 900
Microwave discharge	10^{20} to 10^{24}	600 to 14,000
Spark discharge	10^{21} to 10^{24}	300 to 2,500
Streamer discharge	$10^{18.25}$ to $10^{20.8}$	25 to 300
Townsend discharge	10^{12} to 10^{14}	25 to 170

Noble gases have a lower start voltage and better thermal conductivity than most molecular gases. People will be able to use less gas to heat their homes as a result. The temperature of the gas can be regulated by adjusting the flow rate of gas, increasing the surface-to-volume ratio of plasma, and ensuring uniform discharge of plasma. The effectiveness of CP is based on the chemical composition produced during gas ionization. CP stability is greatly influenced by the purity of gas, which regulates the voltage breakdown, and accurately measuring the pure gas breakdown voltage is difficult [33]. When an electron collides with gas molecules at a low enough velocity to create ionization, energy loss is significant, and the collapse of the electric field is huge. Less electric field is essential, and the greater breakdown potential produces higher gas ionization potential [273].

6.9.1 INTRINSIC FACTORS

6.9.1.1 PRODUCT PARAMETERS

The nature and initial microbial count on food will determine cold plasma efficiency. Because of the higher resistive stability of microorganism species and initial concentration, cold plasma inactivates most bacteria when they are in exponential stage and spore form, compared to smaller reduction in stationary stage and vegetative microorganisms. Higher initial concentrations weaken CP destruction effectiveness. Gram-negative bacteria are more sensitive to CP compared to Gram-positive bacteria depending on cell-wall thickness. Therefore, the gram-negative bacteria have a higher inactivation efficiency [139].

The antibacterial behavior of CP varies due to different food shape and porosity presence. It is difficult to work with bulky and unevenly shaped foods. Some products' rough surfaces give several places for bacteria to adhere and perhaps will escape antimicrobial treatment. Microorganisms' ability to hide in the microstructures of foods and therefore avoid plasma action has also been observed [76].

When using a plasma jet, it is best to expose all sides of the meat surface to the plasma to provide homogeneous treatment and increased inactivation. It was discovered that the exposure time for achieving a higher destruction rate declined to 2.5 minutes when all parts of the food sample were exposed to CP compared to 5 minutes for one side exposure of the sample to CP [278]. The composition of food, including moisture, protein, fat, and other elements, influences plasma efficacy. The reactive oxygen species generated in high-fat dairy products can cause oxidation. Oxidation of other macromolecules like amino acids, proteins, and carbohydrates, and destruction of bonds can all be caused by cold plasma [217].

6.9.2 EXTRINSIC FACTORS

6.9.2.1 GAS COMPOSITION

Chemistry of the plasma functions based on gaseous properties when CP is applied in a sealed container. Any gas could theoretically be used to produce

ACP, and the rate of microbe inactivation would differ depending on the gas. Argon (Ar) [70, 75], Helium (He) [120, 236], and Air are the most frequently employed gases in CP food applications [66, 287]. Because of its low cost and ease of usage, air provides advantages. Changes in composition, on the other hand, may affect the speed of gas breakdown as well as reactive radical production. Understanding the composition and flow of the working gas is critical. During ACP therapy, reactive species are generated at a higher level. Among the reactive species, Ozone persists with a lengthy half-life, which is extremely cause oxidation followed by the creation of peroxide when it reacts with water. So, oxygen causes the formation of ROS (i.e., ozone, peroxides), which remains one of the main reasons for microbial inactivation [230].

Argon was also used in the creation of DBD ACP. After 20 seconds of ACP treatment with 70% argon and 30% oxygen, RTE beef contaminated with *L. innocua* concentration declined to log of 1.6 [276]. The researcher [170] investigated yeast inactivation effect by utilizing ACP created with Ar + O_2, and discovered that destruction rates were greater with oxygen than with argon. The Oxygen-based ACP system declined the D value significantly to 3.2 and 3.4 min compared to Ar-based ACP system to 7.3 and 4.7 min for *S. cerevisiae* and *M. frigida* correspondingly. The DNA and protein oxidation by ozone, etching caused by erosion with ROS are primary foundations proving that O_2 plasma deactivation effectiveness [170]. Pure argon was found to have a weaker antibacterial impact for in-package ACP treatment than 93% Ar + 7% CO_2 [51]. This highlights the prominence of oxygen utilized for antibacterial methods. The He-based ACP system employed to expand protection of ham as well as cut cheese [237]. It was found that the decline of *L. monocytogenes* beyond 8 logs in cheese and 1.73 logs in ham. In order to achieve the appropriate inactivation of target microorganisms, the quantity of gaseous employed developing plasma is crucial. High antimicrobial effect obtained with a more oxygen environment facilitates for generating a large number of ROS (O_3, O^+, etc.) [170].

6.9.2.2 RELATIVE HUMIDITY (RH)

Plasma physics and chemistry are influenced by the relative humidity of the environment. Microbial inactivation under ACP treatment has been observed to increase when RH levels rise. This could be owing to greater RH creating an increase in OH (ROS) radical concentrations generated by the breakdown of H_2O compounds in air which induced the higher level of oxidation [29]. Water vapor, on the other hand, reduces micro-discharges by lowering dielectric resistance of surface due to water molecule adsorption

[71]. Interaction among the molecules and electrons declines the energy of electrons, where high humidity exists in the discharge, these extinguishing reactions reduce the effectiveness of plasma [34].

Water vapors release OH radicals, which diminish the quantity of micro discharges, thus minimizing the overall quantity of reacting gases inside the reactor. Adsorption of water vapor by the dielectric barrier covers the dielectric surface, lowering resistance and increasing effective dielectric capacity. The discharge homogeneity decreases as the gas humidity rises [34]. A study [172] showed that an increase in the relative humidity of air considerably decreases the efficiency of decontamination. Electrons and water molecules combine to generate negative ions in air with a high concentration of water vapor. Negative ions move slowly in the electric field, which reduces the likelihood of ionization and necessitates a greater breakdown voltage due to fewer electrons causing avalanches. The low water vapor concentration in the system results in a minimum voltage breakdown. Hydrogen (H_2) and oxygen (O_2) gases are formed from water molecules. As a result, when H_2 is mixed with nitrogen, the breakdown voltage is reduced due to the low concentration of H_2 (N_2) [273]. A study [204] discovered that increasing the RH level from 35 to 65% (at 25°C) enhanced the concentration of OH radicals, intensified the inactivation of *S. enteritidis* on eggshell (4.5 from 2.5 log). At 70% of RH, the inactivation of *Aspergillus Niger* conidiophores is above 80%. These outcomes are being ascribed to the oxidation of water to form OH to facilitate the production of H_2O_2. Another study [172] discovered that an RH of 70% was optimal for the inactivation of *A. niger* conidiospores. Surprisingly, when the RH was higher than 70%, the inactivation effectiveness was reduced. In addition, the higher endospores inactivation level of *Bacillus subtilis* attained with dry atmospheric conditions than in wet air, illustrating the need for defining the gas composition for each application for enhanced accomplishment. This demonstrates that the effectiveness of inactivation is governed by water vapor quantity along with the characteristics of microbes. When employed under ambient circumstances, the CP quality changes because of the instability around the atmospheric RH. As a result, exclusively meant for biotic uses, relative humidity needs to be monitored as well as maintained.

6.9.3 PROCESSING FACTORS

6.9.3.1 AIRFLOW RATE AND TEMPERATURE

Since dense material within plasma commonly exists around lower temperatures, the CP operations are quite often carried out at ambient temperature levels (within 70°C), hence the term "Cold Plasma" arises. As a result, temperature inactivation is habitually overlooked [60]. The surface temperature of ham increased slightly from 21°C to 26.1°C during treatment and 25.1°C after treatment [276]. Increasing the speed of airflow to a high level helps control the temperature rise; consequently, the gas temperature decreases until it reaches room temperature (23°C) [131]. As the flow velocity increases, the decrease in oxygen intensity has been observed.

6.9.3.2 ELECTRODES

The decontamination efficiency of ACP treatment depends on the type of materials used for electrode manufacturing process. The highly energy-efficient materials are aluminum and stainless steel at 123%, whereas the least energy-efficient materials are copper followed by iron. However, the temperature of the aluminum outer electrode was higher at 129°C than stainless steel at 77°C. High-temperature electrodes can promote more particle collisions, and the degradation of generated particles can be faster at high temperatures [109]. At high applied voltages, the current intensity of steel electrodes was larger than that of aluminum electrodes, implying that the collisions between oxygen molecules and O_3 generation increase as the current intensity increases. However, due to the greater temperature, the O_3 breakdown was quicker for steel electrodes. Though the electrodes provide the greatest energy, other factors such as the goal temperature will influence the material selection, which could affect the food product's quality [197]. A researcher [203] studied the efficiency of decontamination through the use of air for generating ACP by changing electrode materials like brass/glass, stainless steel, silver to achieve *L. monocytogenes* and *E. coli* inactivation with CP exposure of 20 to 60 minutes. Significant differences in *L. monocytogenes* reduction were reported after 40 and 60 minutes of treatment, with the electrodes made of Brass followed by silver achieving higher reductions approximately 8 log CFU/ml) than other electrodes made from brass/glass followed by stainless steel.

The higher antibacterial properties may be due to molecule migration from silver and brass as the TT prolonged irrespective of pH, nitrate, ozone, and nitrite concentration, electric features of discharge. It's worth mentioning that the authors reported no significant changes in *E. coli* decontamination efficacy when implementing various electrodes. The DBD electrode's dielectric material might also have an impact on the effectiveness of decontamination. Variation among the properties of dielectric material and interaction between the surface with plasma might be responsible for the influence of dielectric material [126].

A study was conducted for ACP generation with different materials like Pyrex, quartz, alumina, mullite. Among those, alumina produced the strongest dielectric material because of the quantity and duration of micro-discharges, followed by quartz [183]. Plasma efficiency is also influenced by dielectric material aging caused by UV degradation or plasma etching. Furthermore, as the dielectric barrier thickness increases, the decline of dielectric capacitance with no changes in electrical charge has been noted, resulting in an increase in voltage among DBD systems followed by the production of highly uniform micro-discharges which may improve antimicrobial efficiency [183]. The CP effectiveness declined due to the discharge of plasma non-uniformly and rejoining of reactive species before reaching the sample surface caused by the higher level of distance between the electrodes as well as the high distance from the sample source to the electrode point [104, 140].

The literature has a variety of electrode structures including plate-to-plate electrode arrangement. Typically, one electrode is a plate while the other has a different configuration, such as multipoint or mesh [246]. The multipoint and mesh electrode designs reduce the breakdown voltage, improve discharge dispersion, and increase the potential in industrial applications as compared to the conventional plane-to-plane electrode. Bacillus subtilis spores were deactivated while using mesh-to-plate structured electrodes with less energy and a brief time based on the power density of plasma [194]. The author investigated O_3 production using DBD units and drew the conclusion that multipoint-to-plane configured electrodes produced a greater method to generate a high amount of ozone (80–135 g/kWh) compared to electrodes made in plane-to-plane configurations. According to the authors, the multipoint electrode generated a number of advantages in terms of less energy and voltage consumption (7.3 kV), minimum dielectric loss in the barrier, and higher micro-discharges (2,700 compared to 50 for plane-to-plane) [246].

6.9.3.3 DURATION OF PLASMA TREATMENT

The inactivation rate enhanced after DBD ACP exposure through extended treatment duration. The decline level of *S. aureus* was enhanced up to 4.95 from 0.09 log CFU/mL by extending TT up to 45 from 5s [140]. Other researchers [87, 104] have shown a similar association. Because high-concentration bacteria reside in multiple levels, ACP treatment times must be extended for deactivation to reach up to the bottom layer in foods. Lengthy TT results in larger reactive species concentrations, lower pH, and higher pathogen death, which explains why inactivation rates are higher [6, 87,

140]. The prolonged CP exposure time, on the other hand, causes qualitative degradation in treated items [29, 138].

The antibacterial effectiveness of ACP radiation against the development of microorganisms on fresh chicken slices sealed with air then stored at refrigerated conditions was unaffected by prolonging the duration of exposure up to 9 min produced on 80 kV voltage [272]. Ozone is a highly reactive gas that, once reaching a certain concentration, interacts with the nearby gaseous active species created in a sealed pack, resulting in the development of new radical species. The TT greater than 3 min led to considerable decreases in a and b values, which resulted in alterations in meat appearance [266]. This concludes that 3 minutes of exposure time is sufficient for bacterial inactivation. The treatment duration beyond 3 minutes leads to quality deterioration of treated products. Reactive species produced by plasma species cause cellular membrane destruction, ensuring cell demise [139]. After plasma treatment, the products kept at 4°C retard the recovery of sub-lethally wounded cells, resulting in reduced *L. innocua* concentrations. The time of storage after plasma treatment decides the efficacy of cold plasma inactivation, as longer storage allows reactive species to penetrate into food. By preserving engendered active chemical species inside a sealed package for 24 hours and boosting the distribution of long-lived reactive species into food substances, results in a higher effect on bacterial cells. Long-term reactions among bacteria with high half-life reactive species (such as ozone) within the container may result in higher *E. coli* deactivation efficiency achieved on 24 hours storage than 0.5 hours storage, preventing microbial cell recovery from damage and stress [289].

6.9.3.4 DIRECT AND INDIRECT CONTACT OF FOOD WITH PLASMA

The performance of plasma treatment is influenced by the methods of exposure, whether the food particles directly or indirectly contact with the plasma species. Exposure of plasma on all three sides of foods is difficult for large-sized irregularly shaped foods due to less space among electrodes and the plasma source, which limits its application. This can be rectified by implementing remote plasma exposure. However, this remote/indirect method has certain limits like incomplete inactivation of microbes because the sample experiences minimal temperature and the charged particles recombine before reaching the surface of the sample due to the short half-life of reactive chemical species. Therefore, for inactivating microbes, the direct mode of plasma is effective [166].

A researcher [86] used direct and indirect contact ACP techniques to study the deactivation of *S. typhimurium* on the egg surface. After 10 minutes and 25 minutes, the *Salmonella* count has been decreased to 102 cells/egg, which is beyond the level of recognition in both direct and indirect approaches. Direct techniques use less time than indirect contact methods to accomplish the same microbial inactivation [97, 282]. However, the indirect mode plasma exposure reduced the microorganism count higher than direct exposure due to the spore coat protein damage [144].

6.10 ROLE OF COLD PLASMA IN SOLID FOOD PRODUCTS

6.10.1 STERILIZATION/DECONTAMINATION

Plethora of pathogens and food-spoiling microbes have been successfully inactivated by employing non-thermal cold plasma processing methods. A number of research studies have been reported on the applicability of CP techniques in the food processing sector, and they provide a picture of progress. Recognizing the plasma inactivation mechanisms of microbial pathogens has garnered a great deal of interest, despite the fact that the majority of existing research remains focused on exploring and continuing to improve plasma disinfecting activity for different foods targeting different pathogenic organisms. The involvement of reactive oxygen species in microbial inactivation has already been recognized, resulting in cellular destruction followed by high oxidative stress due to deactivating enzymes and LPOD processes. For many decades, the medical field has already used plasma techniques in sterilization procedures, but recent studies have proven its decontamination capability over food matrices. Traditional techniques of controlling microorganisms are challenging since there is a risk of leaving trace components in food matrices, which might have a detrimental effect on the food chain. The following are two basic concepts for gas-induced cell death: First, the overall electric force created by surface charge buildup may be larger than the entire tensile tension upon the cell membrane, causing electrostatic disturbance. Another concept is that gas discharge generates reactive radical species, including highly energized ions, all of which oxidize and destroy the cell membrane and other cellular components.

Various generated substances were depicted to influence the CP-mediated bacterial inactivation; however, the specific pathways are still being investigated. Charged particles, ultraviolet radiation, ROS, as well as RNS are all abundant inside the plasma gaseous state. Singlet oxygen, O_3, superoxide, atomic oxygen, hydroxyl radicals, and peroxide are only a few of the ROS that are believed to be involved in the deactivation of microbes [7, 111].

The biological membrane chemically reacts with charged particles (ions and electrons), energized active substances (NO, O_3, O_2), and radicals (OH, O), which are generated by the breakdown of air molecules like superoxide, hydroxyl radicals, peroxides, atomic oxygen, and O_3 responsible for microbial deactivation. Meanwhile, NO and nitrogen dioxide NO_2 disintegrate the chemical components of microorganisms like lipids, proteins, and nucleic acid, leading to microbial inactivation [3]. According to [148], the principal disinfection agents in He/O_2 mixed gases are O_3 and the metastable state oxygen O_2*. Through the use of CP technique, ultraviolet radiation could contribute to the destruction of microorganisms. UV photons destroy genetic components directly, preventing DNA replication. The study identified multiple mechanisms by which plasma damages DNA, including photon-induced creation of thymine dimers, nucleobase alterations, and reactive species-induced nucleotide oxidation. Accelerated bacteria DNA damage revealed that the principal cause of plasma's antibacterial effectiveness upon surfaces enveloped with microbes (*Staphylococcus aureus*, *E. coli*, *P. aeruginosa*, and *B. cereus*). Furthermore, decomposition of protein and lipid in CP-treated meals has been suggested to influence the complete microbial elimination. The hydroxyl radicals formed by plasma contribute to the deterioration of cell membranes through a sequence of oxidation processes [83]. Reactive oxygen (O_3) and nitrogen radicals, according to [165] the study, can speed up the lipid peroxidation process. The author [63] discovered that O_3 as well as NO were the predominant reactive compounds developed using Helium mixed with oxygen gas at atmospheric plasma, contributing to the destruction of cellular proteins.

The author [243] discovered alteration of microbial protein structure after plasma exposure due to the interaction of plasma species with amino acid composition, which leads to microbial spore death. The plasma species cause the cell membrane destruction on the surface of yeast like Candida and Saccharomyces after exposure revealed destruction of yeast. The oxidation process alters the functioning of biotic tissues membrane and macromolecules compounds present in cells through the interaction of DNA, protein, and lipid with plasma species. ACP's strong electric field breaks the cell tissue, resulting in cell lysis followed by deactivating cells [160].

Charged particles, electrons, and ions bombarding the cell wall can disrupt chemical structure bonds, causing corrosion, forming scratches, and creating pores in the cell membrane. This allows hazardous chemical species from plasma to penetrate further inside the bacterial cell. Gram-negative bacteria are believed to be easier to inactivate through erosion due to their thinner cell wall compared to the thicker cell wall of Gram-positive bacteria. However, Gram-positive bacteria continue to experience high intracellular destruction due to the presence of higher intracellular ROS levels. The mechanism of microbial decontamination is illustrated in Figure 6.12.

CAP is considered a good solution for eliminating microorganisms in products without affecting the product's quality. The efficacy of CAP is influenced by two aspects: the food's topography and surface, as seen in lettuce, and the multidimensional character of the strawberry surface [238]. These studies have shown that certain characteristics of the food surface, such as the waxy cuticle, may allow plasma to function as a mild decontaminant. In most cases, microbial breakdown and inactivation depend on the duration of exposure and the disinfectant gas used. While a high exposure dose can lead to sterilization, there is a possibility of alteration in bioactive molecules in some instances. The electrostatic force causes charged particles to disrupt the membrane [207].

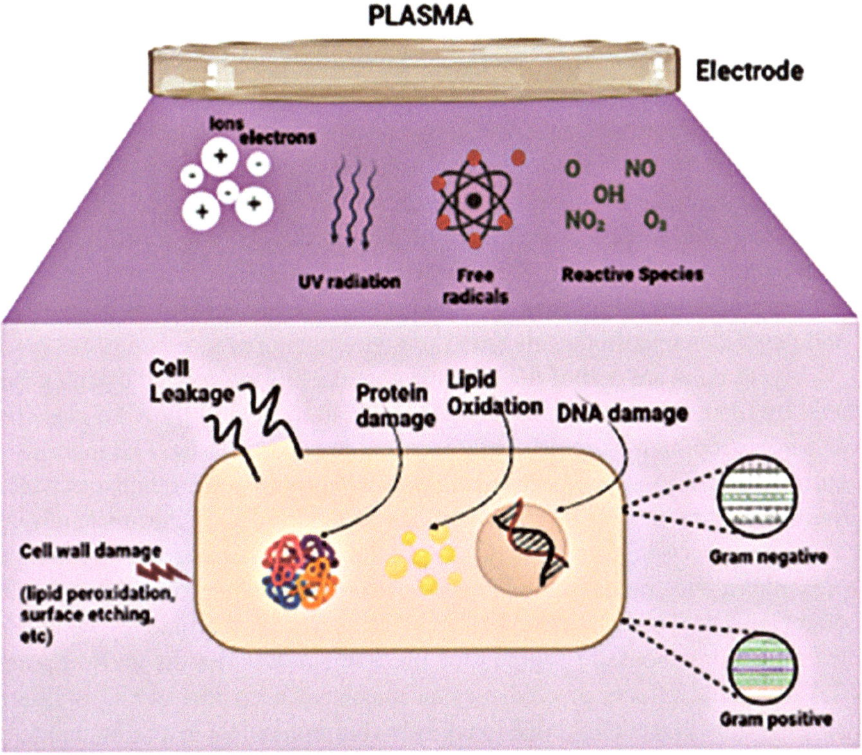

FIGURE 6.12 Mechanism of microbial decontamination.

Numerous latest investigations demonstrate the potential of cold plasma for the destruction of food pathogens. For example, strawberries, lettuce, tomatoes, and carrots tainted with *L. innocua* and *E. coli*, fresh vegetables infected with *Salmonella* species, and eggshells tainted with *Salmonella enteritidis* were all successfully treated with CAP [84, 127, 288]. The bacterial population and ecology, plasma gas composition, different types of exposure and treatment period, all have a role in decontamination. Reduction in cell culturability and metabolic functions, cellular membrane alteration, EPS breakdown, cell damage, and decrease in thickness of biofilm are some of the reasons for decontamination [288].

6.10.2 PESTICIDE REDUCTION

Pesticides are a popular agricultural technique that has considerably reduced crop losses due to insects, resulting in increased yields on a worldwide scale. Pesticides, on the other hand, are hazardous to people and must be strictly regulated in all food items below the residual toxicity level. Toxins, both natural and manmade, are abundant in our environment. Consumers are concerned about the health and safety of food products, mainly about pesticides, mycotoxins, endocrine disruptors, and other chemicals. There are currently just a few non-thermal solutions available for controlling and degrading hazardous chemicals found in food. For some time, cold plasma has been found to be a possible application in the decline of different foodstuff contaminants, piquing the interest of food scientists in recent years. The majority of current food research focuses on the breakdown of mycotoxins and pesticides in cold plasma.

Mycotoxins are one of the major outbreaks in food-related toxins generated by filamentous fungi as secondary metabolites, and they are classified as carcinogenic, mutagenic, and genotoxic in several cases. The inability to entirely inactivate mycotoxins using standard processing methods, as well as their resistance to heat degradation, has prompted significant research into alternate processing methods. Cold plasma is used to degrade mycotoxins present in a wide range of foodstuffs. Researchers are currently concentrating on refining process parameters to improve degradation efficiency while also learning more about the degradation mechanism. Besides the various potential beneficial effects of cold plasma that have been identified, additional investigation is recommended to entirely comprehend the reaction manner of various reactive species, as well as a toxicity study. The study demonstrated that a diversity of pesticides on food products can be decomposed using cold plasma [13, 17, 164, 215].

The formation of different active gaseous molecular species, particularly molecular species (H_2O_2, O_3), ions (OH, H_3O^+, O, O^+, N_2^+), as well as active radicals, has been ascribed to the proficiency of CP on pesticide breakdown (O, OH, NO radical). The factors that determine its consequences on pesticide residues have been interconnected towards the plasma gas, the emergence of reactive intermediates, the intensity of newly formed radical intermediates, as well as the electron's mean energy. These factors have been related to the success of CP in disintegrating pesticides. In most of the cold plasma gases, the average energy of electrons ranges from 0 to 10 eV. All secondary metabolites possessing breakdown of bond and ionizing energy corresponding to plasma gas were anticipated to disintegrate as a consequence of plasma exposure [17].

The pesticide molecules are first dissociated by the cold plasma treatment. Simultaneously, plasma treatment produces many free radicals with high oxidation potentials (E_0), such as: 2.8, 2.1, and 1.8 for OH•, O_3•, and H_2O_2•, respectively, which may result in pesticide breakdown [17, 18]. These reactions eventually disrupt a few pesticide bonds, resulting in the formation of new chemicals that are usually less dangerous or innocuous. The state of plasma, kind of pesticides, food composition influences the reactions to determine the complexities of pesticide decomposition and the exact pathway. Creating free radicals and energizing electrons in a plasma treatment are thought to be critical elements of the breakdown of pesticides like dichlorvos and organophosphorus [105].

Researchers have examined the possibility of using cold plasma for disintegrating pesticides including omethoate, paraoxon, parathion, fludioxonil, azoxystrobin, dichlorvos, pyriproxyfen, as well as cyprodinil. In a recent study, the author successfully used CP for disintegrating pesticides called organophosphorus in vegetables like cucumbers as well as apples [171]. It is believed that the primary plasma producing reactive chemical species (molecular nitrogen, hydroxide radicals, atomic oxygen) are accountable for parathion as well as paraoxon oxidation. Corresponding to the above, it was suggested that both dichlorvos and omethoate sprayed onto maize kernels were entirely eliminated after 2 min of plasma exposure at 120 W power [17].

The author [164] examined the possibility of pesticide breakdown on the surface of strawberries by applying in-pack gas plasma. Decomposition of pesticides is a function which depends on both voltage and exposure duration. According to the investigation, improving pesticide decomposition demanded extending the treatment duration (1–5 min) or expanding voltage (60–80 kV). The best results were obtained when the plasma was generated at 80 kV and treated for 5 min, which decreased the pesticide concentration by 45% (cyprodinil), 46% (azoxystrobin), 69% (pyriproxyfen), and 71% (fludioxonil).

6.10.3 EXTRACTION OF BIOACTIVE COMPOUNDS

Cold plasma is a cost-effective technology used to modify a variety of material surfaces, such as improving the surface roughness and surface area through etching phenomena [196], altering the wettability of polymer surfaces [64], etc. Cold plasma has been utilized for commercial sterilization of food products, in which plasma reactive species attack bacteria's cell membrane, disrupting it and causing cell leakage [95]. Due to the deterioration of the cuticle layer, which comprises hydrophobic cuticle and epicuticle waxes, the surfaces of fresh lettuce and white grapes become more hydrophilic after plasma treatments [92, 106].

Plasma's ability to produce cellular damage and surface alterations suggests that it has the capacity to reduce internal molecule diffusion resistance and increase the extractability of hydrophilic substances, facilitating phenolic extraction from biomass. DBD plasma microdischarges first cause damage to the peel's surface layer, which covers the oil glands. The pretreatment accelerated the amount of essential oil extraction by improving the diffusion rate of essential oil to the exterior of the oil glands. These features of cold plasma show its potential for treating agricultural byproducts for enhancing bioactive compounds recovery rate because demonstrated capabilities of CP for changing biomass by forming ruptures on its cell-wall structure and raising its surface hydrophilicity. This newly designed extraction approach may more successfully valorize food processing byproducts while using less solvent, which enhances its sustainability at an industrial scale. It also offers a lot of possibilities for recovering other bioactives from various agricultural wastes, which may be used as functional food ingredients and nutraceuticals with significant nutritional and economic value without creating more adverse effects on the environment. The He-based PJ-CP system destructs cell wall configuration directly in plant epidermal glands [225]. Plasma has been discovered to cause different forms of surface morphological changes in plant tissue, such as roughening lettuce surfaces, cracking white grape skin, and scorching lemon peel surfaces, which enhances the extraction efficiency

[124]. Furthermore, the discharge of liberated polyphenols from the cellular cytoplasmic compartment towards the extracting medium is influenced by fissures on cells of grape pomace when exposed to CP [195]. The covalent bonds are cleaved by plasma reactive species, tending to release bioactive chemicals which are covalently linked to cell wall polysaccharides, contributing to the greater extraction yields observed. CP technology enhances TPC (10.9 to 22.8%) and antioxidant activity (16.7 to 34.7%) due to the improved extraction rate of anthocyanin concentrations. Recent findings demonstrate the efficiency of CP techniques, which effectively promote bioactive component extraction yield with upgraded nutritional values [21].

The N_2-DBD-CP system with 15 W power for 15 min TT applied to green leaves for enhancing antioxidant capacity (41.06%) and improving TPC value (41.14%) [114]. The surface morphology of galactomannan was also altered by HVACP treatment due to plasma etching without affecting its molecular or crystalline structures, which shows that HVACP can be a green approach to effectively improve the extraction efficiency of fenugreek galactomannan and modify its functional characteristics while avoiding toxic chemical materials [209].

6.10.4 SURFACE MODIFICATION WITH CP METHOD

Surface modification with CP technique can increase functionalization while cleaning packaging materials. Functionalization of the surface is the process of imbuing a packaging material's surface with various functional compounds to improve antibacterial properties and mechanical capabilities. Surface functionalization of polymers is commonly used to improve wettability, printability, sealability, adhesion, glazing resistance, dye uptake without sacrificing appropriate polymer characteristics. Surface functionalization is desirable for the production of antimicrobial packaging materials with improved barrier quality used in the food industry [184].

Surface modifications also eliminate impurities and clean polymeric materials by etching. Cold plasma treatment improves moisture barrier qualities while also increasing tensile strength and elongation. When stored at 4°C under refrigerated circumstances, cold plasma therapy can also help minimize lipid oxidation. The plasma species create etching on the polymer surface which facilitates the deposition of desirable coating material to improve its physical and mechanical strength, as well as to improve antimicrobial properties, reduce gas permeability rate, and clean by removing undesirable material with no residues. However, the less penetration depth might be a drawback on an industrial scale that should be researched more. The cold plasma treatment of LDPE film causes a positive alteration on polymer surfaces by reducing roughness and surface energy, decreasing contact angle, and increasing crystallinity, which eventually causes functionalization of the surface. The CO_2 plasma species with broad energy levels induce surface functionalization along with carboxylic acid.

The cold plasma technology on polymer surfaces induces less improvement in crystallinity and propagates defects in trans conformation chains, which promote the development of extremely ordered crystalline structures. The plasma species from the surface impart cross-linking pathways and produce active surface radicals which help to modify the structure of polymers [155].

For decades, cold plasma has been employed in the food packaging industry. Sterilization of surface etching, packing materials, surface functionalization, surface deposition, and surface activation have all been done using it. In addition to ordinary polymers, it has lately been employed to modify the biofilm's surface [178]. Current research has revealed the potential benefits of CP which can be utilized for production of active as well as smart packaging with improved quality parameters [185, 186].

6.10.5 STARCH MODIFICATION

Several foods contain carbohydrates in starch form, comprised of molecules like amylose and amylopectin, which have wide usage in various industries. To enhance their functionality and broaden their applicability, natural starch is treated with many technologies to address drawbacks such as high retrogradation rate, increased viscosity, and low solubility at low temperatures. According to various experts, plasma processing can bring about favorable changes in native components [250, 271]. Improvements in starch, in terms of pasting qualities, enzyme susceptibility, swelling power, composition, thermal characteristics, structural characteristics, water absorption, and solubility, are all influenced by reactive nitrogen species (RNS), reactive oxygen species (ROS), and other reactive species generated by cold plasma (CP) [254]. Depolymerization, increased surface energy, cross-linking, addition of new functional compounds, and modification of hydrophilic properties are important mechanisms for altering starch [271].

CP can also cause modification through granular corrosion, etching, and molecular disintegration. Within a three-minute timescale, CD-CP reactive species interact with banana starch, resulting in surface etching that modifies starch functions, such as a declining trend in peak viscosity and an increasing trend in pasting temperature (88.8–92.1°C), swelling resistance, and crystallinity improvement [271]. However, outcomes may vary depending on plasma supply and generation method, treatment time, composition, and starch type [271].

Apart from surface modification, the CP can also alter grain surfaces, which results in reducing cooking period (30.25 to 20.45 min) and improving cooking quality by reducing hardness (22.50 to 12.36 N) [218]. Furthermore, after 4 hours of soaking, the hydrophilicity improved due to the synthesis of simple sugars like dextrin, glucose, maltose though reduction of contact angle decreased (85.4 to 59.7°). In brown rice, α-amylase content and rate of water absorption are enhanced by depolymerization and etching phenomena [136].

The CP usage in grains is beneficial for the production of products like flour, pellets, and other items. The water retention capacity of basmati rice flour, for example, improves as plasma power and exposure time rise. Additionally, the researchers discovered a significant improvement in swelling power and starch solubility, lower peak temperatures, and reduced pasting viscosity (692–591 BU). In rice flour, there is a decreasing trend of amylose (30.34–27.89%) and crystallinity (28.25–24.36) after plasma treatment, owing to improved flour hydration characteristics and liberation of phenolic content [249]. After being exposed to cold plasma, rice starch depolymerized and changed in functional properties such as gel hydration, synergies, and turbidity. The pasting temperature and the peak gelatinization temperature both dropped.

6.10.6 ENZYME INACTIVATION

Some food products like bananas, apples, and mushrooms are susceptible to discoloration due to enzymes in the post-harvest process like drying followed by storage [72]. Peeling and cutting activate a cascade of events that result in an instantaneous increase in respiration rate and endogenous metabolic functions by causing tissue rupture and enzyme-substrate interaction [107]. The enzymatic oxidation of phenolics into quinones causes browning in bananas; brown products are then polymerized [174]. Polyphenol oxidase (PPO) and peroxidase (POD) enzymes are involved in the enzymatic browning of bananas. Free phenols can be converted to quinone by PPO, resulting in colorful products. POD increases a number of processes, including lipid and phenol oxidation, and has an undesirable influence on food value [248].

Several strategies have been developed to prevent enzymatic browning in fruits and vegetables, including enzyme inactivation, incorporation of antioxidants, and keeping the enzyme away from its substrate. In order to inactivate enzymes, thermal treatment as well as chemical usage may induce degradation of food quality. However, the majority of customers demand products that are free of artificial preservatives. As a result, new approaches for reducing enzymatic browning are recommended. CP technology has shown tremendous promise in inactivating food enzymes, particularly in thermolabile fruits and vegetables, as a unique non-thermal strategy. The existence of active sites in an enzyme determines its activity. The formation of alpha helical structures is aided by some amino acids found in the active sites. The decrease in α-helical structure is beneficial in inactivating enzyme activity [39, 163].

The beta sheet (β-sheet) structures containing pleated peptide chains were more resistant to heat and chemical agents [243]. However, after treatment with CA-PJ, the reduction in beta-strands and a falling trend in beta structure, which accounted for the loss of enzyme activity [16]. The changes to random regions of both α-sheets & β-sheets, which are components of secondary structures, inhibit the enzyme activity. The interaction of plasma species with secondary structures of protein induces oxidation of side-chain amino acids, which results in the reduction of browning reaction. Inactivating enzymes like peroxidase (POD), pectin methylesterase (PME), and polyphenol oxidase (PPO), which cause quality degradation in fresh vegetable products, may be made easier by changing protein structure. For example, bananas are processed with CP techniques for up to 2 minutes to decrease browning reactions. As the ROS concentration increased in cells due to plasma exposure, the quinones produced by enzymatic browning decreased with time.

Due to enzyme inactivation, CP treatment improved the color of bananas on the surface. So, adaptation of CP usage prior to dielectric drying will eliminate the enzymes without affecting food quality. Similarly, MW-CP for 10 minutes on potato and apple had a significant reduction of POD and PPO [39]. The secondary enzyme structure loss is caused by chemical bond cleavage based on voltage, permeability of cells, temperature, design of enzyme structure, and enzyme stability [163].

The reduction of enzyme activity also depends on the susceptibility of cell tissue, concentration of RNS & ROS found in RF-CP exposed carrot slices [16]. Stimulating certain enzyme systems can help avoid oxidative stress caused by plasma active species. The enzymes are completely inhibited whenever the concentration of RNS & ROS exceeds the strength of cells.

6.11 APPLICATIONS OF COLD PLASMA IN INDUSTRY FOR SOLID FOOD

Cold plasma has shown extended applications in various industries like treatment of textiles, biomedical equipment, automotive, medical devices, aerospace, packaging, and electronics. However, CP has recently been used in the food industry to reduce microbes, degrade mycotoxins, inactivate enzymes, increase the concentration of bioactive compounds, enhance antioxidant activity, and reduce pesticides and allergens in food products (Tables 6.3–6.8).

6.11.1 APPLICATION OF COLD PLASMA IN CEREALS

Cereal grains such as wheat, rice, corn, and other grains are staple foods derived from the seeds of gramineous plants. These cereal grains play a crucial role in human diet, nutrition, and health regulation. However, these cereal grains often come into contact with natural environmental conditions like water, air, soil, insects, and rodents, which may affect the cereal grain coat by adhering pesticide residues, mycotoxins, and fungi, negatively influencing food safety.

Apart from this, the cereals contain a hard outer layer, seed coat, and germ in the cereal grains that hinder the utility of cereal grains due to the short shelf-life, difficulty in maturity, and post-processing operation. This problem can be rectified by adopting suitable scientific and reasonable technology. There are several suitable methods for cereal grain processing like thermal sterilization (radiofrequency and infrared sterilization) and chemical processing (triazolinthione fungicide) which may result in sensory loss, nutritional loss, and chemical residues on the surface. Irradiation sterilization typically requires the creation of related workshops and infrastructure, as well as the hiring of skilled practitioners and a high level of consumer approval [2]. High concentrations of ozone are toxic to the human body, while insufficient ozone revives bacteria after ozone disinfection. As a result, scientists must focus on breakthrough methods to protect cereal grains in order to meet consumer demand.

The modern techniques should be chemical-free, residue-free, and operate at room temperature. Because they can work at low temperatures for short periods of time without hurting plants, commodities, seeds, consumers, or the ecosystem. The CP offers wide applications in agriculture. The discharge of RNS & ROS is possible at low temperature and pressure, which is a key component of cold plasma invention. Cold plasma reduces energy consumption, is safe, and simple to handle, making it a superior technique to conventional grain processing techniques [242]. This section includes research on how cold plasma technology can boost cereal grain foods.

6.11.1.1 DECONTAMINATION OF CEREAL AND ITS PRODUCTS

Cereals and grains are comprised of three components: endosperm, bran, and germ along with the cortex, which plays a role as a defensive layer of grain, thus restricting the entry of microbes in grains. However, the cortex is easily infected by a vast variety of microorganisms, particularly mycotoxins, as well as other impurities that cause detrimental effects on human health, resulting in a severe negative impact on the storage of cereals and grains. The higher production of cereals leads to storage needs for future consumption. During storage, several microorganisms cause undesirable changes in grain quality, such as color reduction and loss of nutritional values. Microorganisms like *Aspergillus spp., Penicillium spp., Tribolium castaneum, Bacillus cereus, Pseudomonadaceae, Micrococcaceae, Lactobacillaceae*, bacterial pathogens, and spore-forming species are responsible for affecting the quality of grains and processed products [128]. There is difficulty in completely decontaminating individual grains since the grain size is small. However, processing conditions can overcome the insufficient decontamination efficiency. The optimal preservation technology is one that maintains good quality without damaging the nutritive characteristics of the product. The existence of microbes should be suppressed or eradicated as much as possible on the basis of processed food quality and nutrition. Various methods like mechanical, chemical, and biological techniques are employed to inhibit microorganisms in cereals, causing chemical residues on the surface, degradation of the inner part of the nutrient, and environmental dependence with high cost, respectively. As a result, a fast, minimal energy consumption, and expensive approach for grain preservation techniques is demanded without sacrificing the organoleptic and nutritional properties.

Cold plasma decontaminates grains by inactivating microbes. Generally, grain products take a longer time to achieve the desired inactivation. The author [224] suggested that a 1 kHz frequency and 20 kV voltage under cold plasma conditions are needed for the inactivation of *Penicillium spp.* In grains, the author [202] found that grain flour required even less processing conditions like a 50 Hz frequency and 2 kV voltage for 5 to 10 minutes of processing, conditions sufficient for high decontamination efficacy.

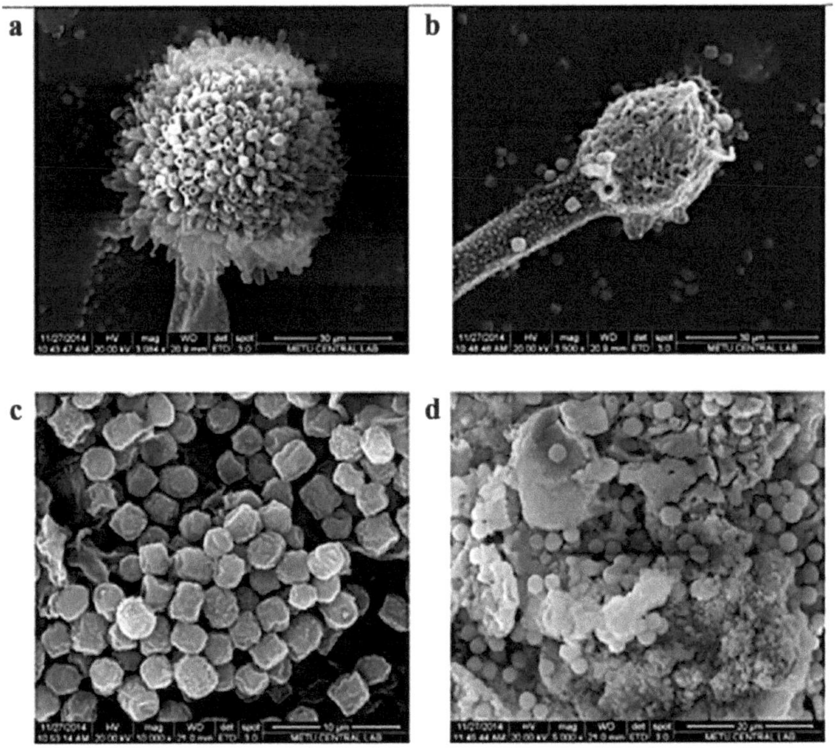

FIGURE 6.13 SEM image of *A. flavus*: a) Control c) plasma treated. *A. parasiticus*: c) control d) plasma treated.

Source: Reprinted with permission from Ref. [59]. Copyright, 2017. Elsevier.

Bacillus cereus is another microorganism responsible for the spoilage of brown rice, which can further affect the quality characteristics like composition, color, eating, and cooking parameters. This can be inactivated by exposure to plasma species generated by a 13.56 MHz frequency [136]. The author [173] reported that the plasma reactive species migrate into the spores and cause DNA destruction, which ultimately affects the bacterial endospores' genome after 5 seconds of exposure to plasma conditions. The tempered wheat grains utilized for the processing of wheat flour are affected by Shiga toxin-generating natural microflora, *E. coli,* and *Salmonella enterica*. These spoilage microorganisms and enteric pathogens on tempered wheat are effectively inactivated by the application of cold plasma. Similarly, during post-harvest operations, transportation, and storage, the grains are susceptible to highly toxigenic fungal infections which produce mycotoxins at the end of the exponential phase. This mycotoxin production has to be restricted by inactivating the fungus with cold plasma treatment.

Generally, cold plasma inactivates microorganisms through various mechanisms such as direct chemical reactions between cells and plasma reactive species, destruction of cellular components and membranes, and disintegration of DNA strands by UV radiation. The effective inactivation efficiency can be achieved through this multiple antimicrobial mechanism. *A. flavus* and *A. parasiticus* are two common species responsible for destroying grain quality. The author applied atmospheric cold plasma for the destruction of microorganisms. Scanning electron microscopy (SEM) shows the effect of plasma reactive species on the structure of Aspergillus species (Figure 6.13) [59].

The destruction caused on the conidial head structure of *A. flavus* could be clearly depicted; the destruction of phialide and metulae structures of *A. flavus* were entirely destroyed and disintegrated due to the interaction of reactive plasma species. *Aspergillus parasiticus* was detected as separate spread spheres, but the integrity of spore cells was disrupted due to the plasma process (65 W, 30 sec), and the cell contents were scattered as clusters. The disruption of cell integrity followed by cell death occurred due to plasma species, which clearly exhibits evidence of powerful erosion leading to significant perforation and disintegration as depicted by SEM images. The erosion phenomena are most likely caused by chemical sputtering, a process in which defects are formed by the collision of oxygen species and ions reacting with these defects to generate volatile molecules such as CO_2, H_2O, and CO. This shows that plasma treatment was discovered to have a good inhibitory effect on yeast and mold on maize kernels [59] and wheat grains [280], keeping grain products safe from microbial spoilage.

The author [151] investigated the mortality of *Tribolium castaneum* with Maida flour by using cold plasma with different processing parameters like voltage, exposure time, electrode distance. It was concluded that the highest mortality of insects was achieved by 30 kV voltage, 3 minutes of exposure time, and a 4 cm gap of electrode distance without changing the quality of treated flour. This shows that cold plasma is an effective method for controlling insects in stored grain products. Bacterial foodborne illness linked to raw wheat products, for example, has heightened interest in decontaminating wheat grains used to manufacture flour and other processed foods.

The author [255] evaluated the ability of ACP to eliminate Shiga-toxin-producing natural microflora, *Salmonella enterica*, and *E. coli* on tempered wheat grains. To achieve an initial count of 7 $\log_{10}$ CFU/g, portions (10 g) of grains were infected with a 5-strain combination of *S. enterica* & E. coli.

Wheat grains treated with 44 kV ACP-DBD for different treatment times and found that initial viable counts declined up to 3.09 log CFU/g after 20 minutes of treatment. Therefore, ACP has a high potential for eliminating gastrointestinal pathogens as well as spoiling bacteria on tempered grains intended for wheat flour production.

6.11.1.2 PESTICIDE REDUCTION IN CEREALS

There is a huge production of cereal grains in the world to meet consumer demands. These grains are left with significant pesticide residue after harvesting, which harms biological systems. Therefore, there is a demand to reduce the pesticide residue on cereal grains to ensure food safety. Generally, malathion, dichlorvos, endosulfan, and chlorpyrifos are often used in the production of agricultural commodities. Much research has been conducted on cold plasma use in reducing pesticides and has found that temperature and voltage significantly influence the decline of pesticides [214]. The cold plasma reactive species are responsible for degrading pesticide compounds through oxidation and further chemical reactions, resulting in the production of less toxic molecules. For example, Malathion has been converted into malaoxan through the process of oxidative desulfuration, which produces simpler chemical groups [285]. In other words, the chemical bond of malathion (P=5) is converted into (P=0) dimethoxyphosphorylthio maleate (malaoxan) through the oxidation process caused by the energetic oxygen species of plasma. Further disintegration then occurs at P-S and S-C bonding.

The author [17] used cold plasma at 120 W power for 120 s to reduce 95% of pesticide residue like Dichlorvos (DDVP) and omethoate, which are applied in maize production. Similarly, chlorpyrifos pesticide present in soybean was reduced by applying cold plasma at 2mbar pressure, 2 kV power for 30 min and compared with ozone treatment. Also, ozone took 30 min to achieve 50% pesticide reduction, whereas cold plasma only took 6 min to achieve the same pesticide reduction [13]. This proves that cold plasma is effectively utilized for reducing pesticides through UV radiation impact and etching caused by the plasma species.

6.11.1.3 INACTIVATION AND MODIFICATION OF ENZYMES CEREALS AND ITS PRODUCTS

Generally, the cereal grain contains few lipid contents which may oxidize in further processes and reduce the shelf-life of cereal grains. The endogenous enzymes in food like peroxidase, lipase, polyphenol oxidase, α-amylase, and lipoxygenase present in cereal grains induce oxidation and cause a huge impact on the quality parameters like flavor, texture, and color of the grain foods. There are many technologies developed for the inactivation of those enzymes like microwave treatment. However, this heat treatment method develops negative impacts on the product such as loss of nutrients and heat restricts the broad application in grain products. So, the cold plasma has been applied to inactivate those endogenous enzymes in grain products as described in this part. Generally, the protein is constructed by primary, secondary, tertiary, and quaternary structure. The application of cold plasma species on the grains significantly alters the conformational structure of enzymes or proteins. Particularly, the secondary structure results in the inactivation of enzymes.

The wheat germ contains enzymes (like lipase and lipoxygenase), which may oxidize and decrease the shelf-life of wheat germ. To prolong wheat germ shelf-life, it is important to inactivate or reduce the activities of those enzymes. So, cold plasma has been applied to wheat germ at 24 kV power for 25 minutes [257]. The author concluded that the cold plasma treatment resulted in a decline of lipoxygenase and lipase activity of wheat germ to 49.98% and 25.03% of the initial activity, respectively, which can be further increased by extending the contact period. However, cold plasma does not permanently inactivate the enzymes. Although enzyme activity has been recovered after storage (>30 days), the enzyme activity is significantly lower than in untreated wheat germ. The reduction in lipase activity was higher than the reduction in lipoxygenase activity due to differences in the structure of the enzymes. α-amylase is widely used in the food industry and is present in cereal grains like rice, where it is utilized to convert or hydrolyze starch, glycogen, and polysaccharides into oligosaccharides, maltose, and maltotriose.

The application of cold plasma species on brown rice enhanced the α-amylase to 162% higher than untreated brown rice which influences the germination and speeds up the seed growth. This improvement in α-amylase activity is mainly due to the development of fissures on the brown rice bran layer [45, 134].

6.11.1.4 ENHANCEMENT OF GERMINATION IN CEREALS

Germination is the crucial step in cereal processing which enhances the nutritional compounds of the particular grain. The cereal grains are degraded during germination by respiration. Also, the grains lead to the development of new cell components for synthesizing embryos, resulting in a significant effect on the physicochemical, nutritional, and sensory attributes of grains [45]. During germination, GABA and other important amino acids are produced, providing a variety of health benefits such as blood pressure stabilization, heart rate regulation, anxiety and pain relief, and improved pancreas insulin output [2]. GABA is a chemical found in germinated cereal grains that contributes significant benefits inside the circulatory system and key inhibitory neurotransmitters in the sympathetic nervous system [265].

When cold plasma is applied to the cereal grains, the rate of fermentation increases, which further increases the accumulation or deposition of GABA compounds on germinated grains. The author [45] reported that cold plasma-treated grains have a higher GABA content than untreated grains. The generation of GABA content is based on some compounds like GAD (glutamate decarboxylase), glutamate, and GABA transaminase. This component (GAD) present in the seed has been deactivated by the application of cold plasma reactive species, which ultimately influences the high GABA content in cold plasma-treated germinated grains. The plasma species like electrons, ions, and radicals UV radiation interact with the seed coat, resulting in the reduction of germination periods.

The Chen et al. [45] applied cold plasma on brown rice and found that cold plasma treated brown rice took a very short period to sprout, achieved a greater germination ratio, and longer sprout length. Similarly, Será et al. [226] found that microwave-discharged cold plasma exposure on seeds resulted in a 50% increase in sprout length. In the same way, the cold plasma treated maize roots increased in length by 21% and in fresh weight by 10% [98]. However, germination depends on a number of conditions, including seed type, discharge modes, gas, exposure time, and voltage.

6.11.1.5 *STARCH MODIFICATION OF CEREALS AND ITS PRODUCTS*

Starch is the biopolymer which is extensively utilized in industrial scale. The major sources of starch are wheat, corn, rice, and vegetable crops. The extracted starch is dried into a white powder that is insoluble in organic solvents, ethanol, and cold water. Starch is a polysaccharide molecule comprised of linear structured amylose (20–25%) and branched structured amylopectin (75–80%). Physical, chemical, and enzymatic techniques are applied to enhance the characteristics of the original starch. However, due to the drawbacks of toxicity and high cost, there is a demand to find alternative methods. In this way, the utilization of cold plasma (CP) has emerged as a new technique for enhancing the functional characteristics of starch.

The author [169] reported that starch can be modified through three mechanisms: cross-linking, depolymerization, plasma etching, and introduction of new functional groups, where each method induces different modifications in starch. Cold plasma active species interact with the starch biopolymers, resulting in the induction of polar hydrophilic groups. Different gas plasmas form different functional groups; for example, oxygen-based plasma treatment introduces carboxylic acid, hydroxyl groups, and peroxide groups, while CO_2 gas-based cold plasma forms aldehydes, ketones, hydroxyls, and esters, and N_2 and ammonia-based plasma incorporates primary, secondary, and tertiary amines. The crosslinking of starch molecules is the main reason for starch modification, which occurs by cleavage between two polymer chains (C-OH) and the introduction of a new C-O-C bond among these two chains through the removal of water compounds. Simultaneously, by disintegration of H_2O molecules, OH– is generated due to oxygen plasma species, which induce glyosidic bond cleavages. The author [115] found that beneficial modification occurs in starch molecules when treated by cold plasma, such as changes in the polarity of starch, improving solubility, increasing crystallinity, declining swelling capability of quinoa, wheat, sorghum starch, and Tartary buckwheat. The cold plasma species cause hydrolysis of starch, which increases starch digestibility and reduces viscosity [81].

Enhancement of gelatinization in starch, maize, waxy rice is due to the crosslinking of amylopectin side chains. Simultaneously, the cold plasma species also degrades amylose content due to crosslinking behavior [250], and gelatinization temperature depends on the degree of crystallinity. Starch with a low gelatinization temperature indicates a lower degree of crystallinity. However, starch modification depends on factors like voltage, gas, and exposure time. Therefore, plasma-assisted modified starch grains will receive more attention in the future due to their eco-friendly, non-toxic, chemical-free characteristics with unique properties.

6.11.1.6 TEXTURE MODIFICATION OF CEREALS

Cereal grains are unpleasant for food processing due to the presence of seed coat and germ. Due to their rough texture and chewiness in sensory qualities, cereal grain consumption has dropped in numerous regions. The structure and construction of amylose and amylopectin ratio determine the grain's hardness and stickiness. Hardness is reduced and stickiness is improved when the starch structure is disrupted. The inherent morphology of grains, such as black gram, can be influenced by cold plasma [218], brown rice [46], bean seed [262], and so on.

Generally, the grain surface or rice bran surface is compact with shallow destruction of the morphological structure containing protein, fat, fiber, and ash which resist hydration properties. However, the plasma species react with grain surface, resulting in the etching process which creates fissures with narrow and deep destruction. This etching effect on the hard texture of the grain improved the hydration properties (water absorption rate) of the kernel that can lead to the reduction of hardness of the grain. The cold plasma species like ROS, RNS, and so on affect the structure of starch and promote leakage of amylopectin followed by amylase, which is responsible for the reduction of hardness. Also, the CP treatment may disrupt the structure of the protein which causes fluctuation of viscoelastic behavior of dough responsible for the reduction of hardness and chewiness after mastication [161].

The decrease in hardness is related to the reduction of chewiness, which obviously improves consumer acceptability and utility. The plasma also induces depolymerization of starch molecules and leaching of amylose and short amylopectin, which improves the gelatinization and starch fluidity of the cooked rice after plasma treatment. This results in a reduction of hardness and an increase in the stickiness or adhesiveness of the starch products. This result also coincides with the plasma-treated black gram. The plasma reactive species induce changes in the viscoelastic characteristics of dough, which reduce mixing time and improve dough strength due to modification of secondary protein structure, such as an increase in α-helix and β-turns while reducing β-sheet structure after plasma treatment of wheat flour.

6.11.1.7 IMPROVEMENT OF WATER ADSORPTION RATE AND REDUCTION IN COOKING TIME

The cold plasma species induce surface fissures or etching on the surface of kernels like rice. Because of this etching, the cold plasma-treated brown rice tends to absorb more moisture or water from the surroundings. Apart from this, cold plasma tends to form simple sugars such as maltose, dextrin, and glucose, which contribute to the higher water absorption rate leading to their high affinity towards water [80]. This water absorption rate also depends on the permeability of the seed coat, the structure of the seed, and the availability of H_2O [154]. This confirms that cold plasma improves the higher water absorption rate. Similar results have been found by [198, 218, 281].

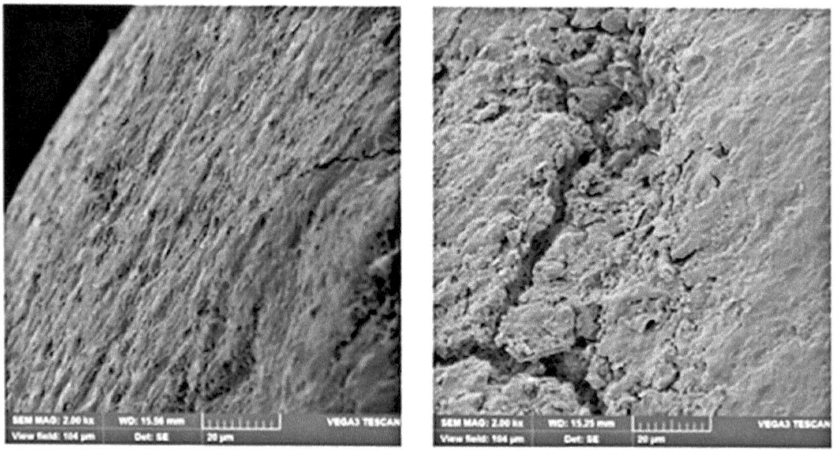

FIGURE 6.14 SEM image of bamboo rice before and after plasma treatment.

Source: Reprinted with permission from Ref. [199]. Copyright, 2018. John Wiley and Sons.

Generally, certain rice varieties like brown rice and bamboo rice have hard outer bran layers and take longer time to cook. Cold plasma has unique properties of surface etching which can create fissures on the outer hard layer due to the reaction between the plasma species and the surface of the kernel, as determined by SEM (Figure 6.14).

This reaction leads to physical damage and splitting of the surface's chemical bonds. The plasma also induces oxidation and starch depolymerization, forming carboxylic starch. These fissures on the rice surface allow water to penetrate into the endosperm easily, significantly reducing the cooking time. The author [198] applied cold plasma to bamboo rice to reduce the cooking time and found that the bamboo rice after CP reduced the cooking time by 12 minutes compared to the control rice. Similar results were found in many rice varieties like parboiled rice [249] and black gram [218]. Cold plasma can be effectively used to reduce cooking time.

In conclusion, it has been proven that cold plasma is effectively used in grain processing to modify surfaces, enhance the functional properties of cereals, extend shelf life, and more.

TABLE 6.3 Application of Cold Plasma on Cereals and its Products

Raw Material	Purpose	Conditions	Results	Quality Changes	Inference	Ref.
Brown rice	To improve germination	LPCP – 1 to 3 kV for 10 min	Optimum condition is 3-kV for 10 min. Germination percentage (↑) 40 → 62% after 24 h germination time.	GABA level (↑) 19 →28 mg/100 g. α-amylase activity ↑ with a high value 162% than control. TPC (↑) 1.84 → 2.30 mg GAE/g. Higher antioxidant capacity: 6.71 → 10.54 (ABTS) and 11.9 → 20.6 (DPPH) mg Trolox eq/g.	Surface etching. LPCP is excellent at stimulating the efficiency synthesis of GABA in brown rice, provide nutrition to consumers.	[45]
Brown rice	To Surface modification, to reduce cooking time, to improve texture properties	LPCP – 1.2 Ma, voltage: 1–3 kV, TT: 30 min.	Better efficiency @ 1 kV. Surface etching – wide and shallow destruction, Hardness (↓) and adhesiveness (↑), gumminess (↓) and chewiness (↓), Cooking time (↓) 24.8→17.2 min.	Moisture content (↓), No changes in Proximate composition, water absorption rate (↑), Soft texture, cooking loss (↓), brightness (L- ↓) and yellow value (b -↓).	As a result, plasma treatment could be utilized to enhance the cooking characteristics and quality of brown rice.	[46]
Chick pea wheat, soy bean, corn, oat, rye, barley, bean & lentil.	Decontamination (*Penicillum* & *Aspergillus* species)	LPCP- 1 kHz, 20 kV, 300 W at ~30 s to 30 min exposure	Significant (↓) by 3-log with 15 min of treatment.	Preserving germination quality. No significant changes in quality of grains except wheat	Suitable treatment for surface disinfecting of seeds	[224]
Maize	To degrade pesticides	AP O_2 plasma – power: 30 to 120 W, TT: 30 to 120 s.	Optimum Condition 120 W, 120 seconds 140 cm^3/min of O_2 flux, over 90% of DDVP and 95% of omethoate has been reduced from the initial concentration.	Pesticides eventually degrade into a sequence of less dangerous chemicals than the original pesticides compounds. Changes in quality still in investigation	Cold plasma generated by O_2 has potential to minimize pesticides in agricultural commodities.	[17]

TABLE 6.3 *(Continued)*

Raw Material	Purpose	Conditions	Results	Quality Changes	Inference	Ref.
Refined wheat flour.	To control the insects (*Red flour beetles, Tribolium castaneum)*	LPCP- voltage (500 to 3,500 V), exposure time (2 to 20 min), distance between electrodes (2 to 6 cm).	Optimized condition – 2,500 V, 15 min, gap: 3.7 cm	Deformation changes and insect egg cell membrane damage, No significant color change of flour	Cold plasma can be applicable as novel non-thermal and nonchemical method for integrated pest management for stored products	[210]
Rice starch	To modify of starch	Atmospheric air pressure plasma – 40 W and 60 W power for 5 and 10 min	Best result at 40 W–10 min. Creation of fissures on the surface of starch. Increased final viscosity.	Gel hydration (↑), syneresis (↑), and turbidity (↑). Peak gelatinization (↓) & pasting temperature (↓). Formation of new compounds like carboxyl and hydroxyl groups. No starch's crystalline pattern changes.	Cold plasma utilized to change structural, rheological and functional characteristics of native rice by Plasma etching, depolymerization and cross linking phenomena.	[254]
Tartary buckwheat, sorghum, wheat and quinoa starches	To alter viscosity and digestibility of starch.	APCP – 20 kV, 1 kHz, 30 s, 2 mm discharge distance.	More fissures and holes on starch. starch digestibility (↑) by increased enzyme accessibility.	Solubility (↑) and swelling capacity (↓), crystallinity of starch (↑) but no changes in crystalline type, Gelatinization temperature (↑) and viscosity (↓).	The DBD plasma treatment on starch modification used for several industrial processes that require starch hydrolysis acceleration, including bioethanol production, brewing, and food fermentation.	[81]

TABLE 6.3 *(Continued)*

Raw Material	Purpose	Conditions	Results	Quality Changes	Inference	Ref.
Waxy rice, maize and potato	To prevent bread staling, to withstand retrogradation by starch modification.	CO_2-Ar gas RF (13.56 MHz) cold plasma: 120 W, 30 min TT.	Starch stabilization improved via cross-linking for better retrogradation.	Resistant starch (↑), decreases setback and final viscosities (↓), Crystallinity: (↓) only for potato.	CO_2-Ar RF cold plasma was an effective method for changing the characteristics of the three waxy starches.	[180]
Wheat germ	To inactivates enzymes & increase stability.	AP DBD Plasma – 20 and 24 kV of 50 HZ, TT (min): 5 to 25.	↓ residual activity: lipoxygenase (49.98%) & lipase (39.97%) at 24 kV at 25 min.	No significant changes in DPPH radical scavenging activity and TPC value	The CP provide opportunity to stabilize the wheat germ while preserve up to 30 days by inactivating endogenous enzymes especially lipase and lipoxygenase.	[257]
Wheat, barley	Decontamination of grains	APCP – 80 kV, Electrode gap: 10 mm TT: 5 to 20 min.	↓ Microbial population at 20 min TT: Barley: 2.4 (bacteria) and 2.1 (yeast), wheat: 1.5 (bacteria) and 2.5 (yeast) log_{10} CFU/g.	Less changes: rate of germination, Surface hydrophobicity (↓).	Effective treatment for cereal grain decontamination.	[143]

TABLE 6.4 Applications of Cold Plasma on Fruits and Vegetables

Raw Material	Purpose	Conditions	Results	Quality Changes	Inference	Refs.
Apple	To inactivate microbes and remove pesticide	AP remote plasma – voltage – 4.0 kV, TT: 5 to 120 s.	95% reduction of microbial population (*S. Typhimurium, E. Coli, B. Subtilis, & S. Aureus)* with 120 s treatments. Concentrations of paraoxon on Apple ↓ to 83% after 60 s exposure of remote plasma.	Pesticides eventually degrade into a sequence of less dangerous chemicals than the original pesticides compounds. Changes in quality still in investigation	Remote plasma treatment in a closed space exhibits tremendous opportunity for application of CP by the food industry in prolonging the product shelf-life through reducing food spoiling by microorganisms and improving food safety in terms of agricultural chemicals.	[99]
Fresh fruit and vegetable slices (pears, cucumbers and carrots)	To disinfect *Salmonella*	PJ- V: 500, TT: 1–8 min.	Inactivation efficiency of *Salmonella*: pear slice: 40%, cucumber: 60%, carrot: 90%.	No significant difference: color, moisture & vitamin C.	Reactive oxygen species are major cause of bactericidal agents to disinfect fruits and vegetables.	[267]
Radish paocai	Decontamination of spoilage organism	LPCP DBD – 60 kV, 50 Hz, 60 s for 3 time with 60s gap interval.	Eliminated gas-producing yeast. Retained lactic acid bacteria 4.54–5.61 log CFU/g from paocai sample.	Total acid content (↑) to 72.34%, pH (↓) to 3.98 after 7 days. Less nitrite content (48.54%), firmness (↑) to 12.53%, b* value (↓) to 26.26% than pasteurization, softening (↓) and browning (↓).	CP has the potential to increase the storage stability of packaged paocai while maintaining the same level of safety as pasteurization, highlighting its utility in maintaining packaged fermented vegetables.	[286]

TABLE 6.4 *(Continued)*

Raw Material	Purpose	Conditions	Results	Quality Changes	Inference	Refs.
Spinach, lettuce, tomato and potato	To sanitize vegetables against *S. Typhimurium*	13.56 MHz RF-APCP -100–300 W power, 180 to 800 sec, gab between electrodes: 2.5 cm	The complete sanitization at 800 s better than that of 3% H_2O_2 (<1 log CFU)	CP modify the cuticle layer wax in tomato by chemical oxidation as well as breakdown of carbon chain, which has been easily and fully eliminated by washing.	Non-thermal oxygen plasma has the potential to be a new and innovative method of sanitizing fresh vegetables.	[282]
Strawberries	To degrade pesticides	APCP – 60, 70 and 80 kV, 1 to 5 min	After 5 min of exposure at 80 kV, the maximum reductions reached 69%, 45%, 71%, and 46% for azoxystrobin, cyprodinil, fludioxonil, and pyriproxyfen, respectively.	Pesticides eventually degrade into a sequence of less dangerous chemicals than the original pesticides compounds. Changes in quality still in investigation	In addition to its established microbicide effects, non-thermal plasma treatments has the potential to ensure food safety by the breakdown the pesticide residues on fresh vegetables.	[164]
Tissue of potato & apple cubes	To reduce browning reaction by inactivating PPO & POD	AP MW plasma torch – 2.45 g MHz, 1.2 kw for 2.5, 5, 7 & 10 min.	PPO: potato (77%) & apple (62%) POD: potato (89%) & apple (65%) at 10 min TT.	↓: pH (1.5). & Discoloration. No changes: dry matter content & cell integrity.	Cold plasma can be effectively utilized for reducing browning reaction and extending quality without significantly affecting quality of fruits and vegetables.	[39]

TABLE 6.4 *(Continued)*

Raw Material	Purpose	Conditions	Results	Quality Changes	Inference	Refs.
Tomato	To reduce pesticides (chlorpyrifos)	LP DBD- powers -2 to 5 W, Exposure time of 4–6 min, initial pesticide concentrations- 0.6 to 0.8 ppm	Maximum pesticide reduction of 89.18% with 5 W at 6 min	(↑): Tomato color index, (↓): TPC, bio yield point, carotenoids, firmness,	Cold plasma proves to become a potential non-thermal method for degrading and reducing the pesticide chlorpyrifos in tomato.	[208]

TABLE 6.5 Application of Cold Plasma on Meat and its Products

Raw Material	Purpose	Conditions	Results	Quality Changes	Inference	Refs.
Beef loins (longissimus muscle)	Inactivation of microorganisms	AP DBD- 9 kHz, 2 cm, 29.9 W, 5 to 15 min	*Staphylococcus aureus, Listeria monocytogenes* and two *E. coli* – ↓ >2 log after 10 min of treatment	No adverse effect major physical-chemical properties. Beef quality characteristics,	Potential techniques to sanitize meet surfaces.	[25]
Chicken breast	To improve microbial safety	AP DBD-15 khz, 100 W, 1 to 10 min.	Reduction of *L. monocytogenes*:2.14 *E. coli:* 2.73, & *S. Typhimurium*: 2.71 Total aerobic bacteria: 3.36 Log CFU/g on 10 min CP.	L* & b* (↑) whereas a* (↓), No Lipid oxidation, flavor (↓) and off-flavor (↑), No changes in acceptability, cohesiveness (↑),	Improve microbiological safety with minimal changes in sensory.	[134]
Chicken eggs	Inactivation of *Salmonella Enteritidis*	AP DBD – 60 Hz, 85 kV for 5, 10 and 15 min	Direct treatment – ↓ up to 1.84 & Indirect treatment – ↓ up to 5.65 log CFU/egg.	Egg quality remains same.	CP is a non-thermal method effective for minimizing *Salmonella* from packed chicken eggs	[263]

TABLE 6.5 *(Continued)*

Raw Material	Purpose	Conditions	Results	Quality Changes	Inference	Refs.
Emulsion-type sausages	To use PAW as an alternative nitrite source for curing	AP DBD- 15 kHz, 200 W, 30 to 120 min.	The PTW improved nitrite (782 ppm) after 120 min treatment	↓ concentrations of residual nitrite than chemical curing agents, no changes in sensory properties. No changes – peroxide values, color and microbes.	Emulsion-type sausage can be prepared by addition of PTW instead of normal water to reduce the use of chemicals.	[112]
Ground pork	To improve curing process for meat.	AP DBD – 25 kHz, 7 kV, TT: 10–60 min.	nitrite content – ↑ (0.64 to 60.50 mg/Kg) at 60 min CP.	No changes: pH, color, nitrosyl hemochrome, residual nitrite, texture, LPOD and PNOD. ↓: bacteria count, ↑ taste, overall acceptability than those cured with sodium nitrite.	The APP procedure can cure canned ground ham without the use of nitrite chemicals. The industrial sector may cure meat instead of using nitrite additives	[135]
Pork ham	Reduction of *L. Innocua*	APCP-DBD-0–28 kV, ~2 mm gap between electrode and ham, 30 to 180 s,	High log reduction value: 1.43 CFU/cm^2 at 180 s.	No significant changes in L* & b* values. a* values (↓). Concentration of (TBARS) ↑ after 180 s. water content (↓), Lipid oxidation (↑).	Used for surface decontamination of RTE ham.	[276]
Pork longis-simus dorsi muscle	Inactivation of microorganisms	LPCP – 0.8 MPa, 5 to 10 min time,	Psychrotroph bacteria counts: ↓ 3 log & total plate count: ↓ 2 log CFU/cm^2 at 10 min CP.	No changes: color, & Ph.	Cold plasma had antimicrobial activity and it could be a useful bio decontamination technique.	[258]

TABLE 6.6 Application of Cold Plasma on Sea Food and its Products

Raw Material	Purpose	Conditions	Results	Quality Changes	Inference	Refs.
Chub Mackerel	To preserve seafood	AP -DBD -50 Hz, 0 to 75 kV, 0 to 75 sec,	After 45 seconds of exposition at 60 kV, TVC ↓ – 5.02 to 2.64 log CFU/g.	No significant changes in sensory, proximate value, total viable count (TVC), and total volatile basic nitrogen (TVB-N). Slow the process of myofibrillar protein breakdown and improve tissue structural stability	Shelf-life prolonged to 14 days than control 6 days.	[48]
Fish nuggets	To Decontaminate RTE fish products against *S. Aureus* and *L. Monocytogenes*	AP -DBD-Helium and argon, 2 kV & 18 kHz for helium, 22 kV & 32 kHz for argon, 4, 6 and 10 min exposure	↓ *S. Aureus* survival. Ar – (1.04–0.19 Log cycles), He- (0.55 to 0.03 Log cycles)	Quality analysis ion future studies	ACP is effective in suppressing the growth of several microorganisms on RTE fish samples.	[94]
Herring (Clupea harengus)	TO increase Shelf-life	AP-DBD: 50 Hz, 70 to 80 kV, TT of 5 min.	Optimized condition – 70 kV, microbial load ↓ (total aerobic psychrotrophics, & mesophilic, *Pseudomonas, Enterobacteriaceae* & lactic acid bacteria).	Lower oxidation and less color modification.	CP exposure preserve fish for 11 days.	[5]

TABLE 6.6 *(Continued)*

Raw Material	Purpose	Conditions	Results	Quality Changes	Inference	Refs.
Pacific white shrimp *Litopenaeus vannamei*)	To preserve of shrimp quality	AP -DBD-10 min, 500 Hz, 40 kV	Reduced initial microbial load of 16.15%.	Stabilizing the pH, WHC (↑), weight losses (↓) after cooking, and ↓ variation of color during storage. Melanosis ↓	Shelf-life increased to 14 days than of 9 days in control. Cold plasma aided in the improvement of physicochemical qualities throughout storage days.	[62]

TABLE 6.7 Application of Cold Plasma on Milk Products

Raw Materials	Purpose	Conditions	Results	Quality Changes	Inference	Ref.
Non-fat dry milk powder	To inactivate microbial pathogens.	APCP, 4.4 kV for 0–120 s	C. Sakazakii ↓ by 1.17–3.27 $\log_{10}$ after 20–120 s of CAP treatment.	Slight change – color, No change – TPC, crystallinity, amino acid composition.	Antimicrobial activity of CP is efficient while having low impact on a few physicochemical characteristics	[43]
Queso fresco cheese	To decontaminate cheese	APCP-60, 80, or 100 kV, 60 Hz, 5 min	100 kV (60 Hz) for 5 minutes. ↓ to 1.4 (L. Innocua) and 3.5 (E. Coli) $\log_{10}$CFU/g.	↓ changes – moisture, pH, color and LPOD	CP as a promising quasi technique for fresh cheese processing, with negligible effects on product quality.	[264]
Sliced Adobera cheese	Microorganism inactivation	RF-DBD CP-13.56 MHz, 850 V, 1 to 15 min.	The populations of *E. coli, Salmonella* and *S. aureus* was reduced 5 log cycles after 15 minutes of CP treatment.	Alteration of odor & in taste by fatty acid generation & PNOD.	CP is an excellent treatment for bacterial inactivation.	[4]

TABLE 6.7 *(Continued)*

Raw Materials	Purpose	Conditions	Results	Quality Changes	Inference	Ref.
Sliced cheese	To improve microbial safety	AP DBD-3.5 kV, 50 kHz, 11 to 15 min.	Significant decreases in *S. aureus* – He (0.05 to 0.45 log cycles) & He/O_2 (0.08 to 0.91 log cycles)	L* (↓) and b* (↑). Significant decreases in sensory quality, such as flavor, odor, and acceptability.	Quality and shelf-life increased with He/O_2 than He alone.	[133]
Sliced cheese	To inactivate pathogens.	AP DBD- 250 W, 15 kHz	60 s TT for 2.67, 45 s TT for 3.10, 7 min TT for 1.65 decimal reduction.	After 10 minutes of treatment, there was no visible damage to the cheese slices.	Extending the post-treatment time improved inactivation effectiveness.	[277]

TABLE 6.8 Application of Cold Plasma on Spices and Herbs

Raw Material	Purpose	Conditions	Results	Quality Changes	Inference	Ref.
Black peppercorns	To inactivate indigenous mesophilic aerobic bacteria	UVCP -9.7–10.6 kV, 7.9 to 22.1 min, 15 kHz, 4 cm gap.	Inactivation rate: ↑ (1.7 log CFU/g), aw: ↑ (0.1–0.7) @ 10.3 kV and 22.1 min.	No color changes.	CP – ↓ microbial contamination. Used to preserve spices and its color.	[20]
Fenugreek	To increase extraction rate	AP DBD- 80 kV, 30 min, 60 Hz.	Improved yield of Galactomannan: Soaked seeds – 122% Dry seeds: 67%	↑ – swelling index, water-binding capacity, and viscosity. plasma etching	HVACP can become a green technique in reduction in chemical, use to successfully improve fenugreek galactomannan extraction efficiency and adjust its functional characteristics.	[209]

TABLE 6.8 *(Continued)*

Raw Material	Purpose	Conditions	Results	Quality Changes	Inference	Ref.
Green tea leaves	To increase extraction rate	Nitrogen DBD CP- 5 to 15 W, 5 to 15 min	TPC as well as antioxidant activity improved by 41.14% and 41.06%, respectively at 15 W for 15 minutes.	The catechin ↑ by 103.12%. Noticeable changes: L* & ΔE. Etching of surface observed.	N_2-DBD-CP is Promising approach for the ecologically friendly and beneficial extraction techniques for bioactive compounds extraction from the food industry.	[114]
Hazelnuts	To detoxify mycotoxin	O_2 Gases: 0.1–21%, power: 400 –1,150 W, TT: 1–12 min.	At 1000 W for 12 min, aflatoxins: >70% (↓).	Maintain the organoleptic characteristics	A potential technique for aflatoxin removal from food,	[232]
Saffron (*Crocus sativus* L.)	To disinfect sensitive foods	RF LPCP, 13.6 mhz, 70 to 110 W for 5 to 30 min.	TVC (log CFU/g): 3.5, yeasts: 4.12 (↓) @ CP with 110 W for 30 min.	↑ – color (L*, b*, a*, chroma, ΔE), ↓ – picrocrocin, safranal, & crocin.	LPCP is a potential approach for decontaminating saffron, improving color intensity, and preserving saffron quality attributes.	[58]

6.12 APPLICATIONS OF CP ON FRUITS AND VEGETABLES

Fruits and vegetables are the principal foundation of fiber, minerals, and vitamins, becoming an essential component of the human nutritional composition. It has been shown that consuming fresh fruits and vegetables on a regular basis can decrease the incidence of cancer and heart disease [182]. Fruits and vegetables are most vulnerable to microbial infection and deterioration. The percentage of vegetables decomposing due to microbial infection was almost as high as 12%, resulting in major financial losses and food safety risks [149]. Furthermore, inadequate cleanliness might lead to cross-contaminations. For the microbiological disinfection of fruits and vegetables, chemical sanitizers, particularly chlorine-based solutions, were frequently employed. However, the possible problems associated with the creation of hazardous chlorinated organic substances raise doubts about public healthiness and the ecosystem. Thermal processing is quite successful in decontamination; however, it may degrade food components and quality. To address these issues, innovative types of technology such as ionizing radiation, ultraviolet light, high hydrostatic pressure, ultrasound, and more recently, cold plasma, have been implemented to control the microbiological load without negatively impacting phytochemical constituents and sensory characteristics [159, 216].

6.12.1 DECONTAMINATION OF FRUITS AND VEGETABLES

Cold plasma is currently being used to efficiently disinfect fresh vegetables. On a variety of food matrices, such as apples, mangoes, blueberries, pears, strawberries, grapes, mandarins, fresh-cut cantaloupe, and kiwi, cold plasma treatment has effectively inactivated commonly occurring pathogenic organisms such as *Listeria spp., E. coli, Salmonella* spp., molds, and yeast [56]. Since lettuce and tomatoes are in high demand in packaged RTE salads, the studies focus on them, followed by salad leaves, carrots, cabbage, cucumbers, and radicchio. The RTE vegetables are preserved successfully with CP by inactivating certain microbes causing spoilage such as *Salmonella* species, *E. coli, Listeria* species, microflora, mesophilic bacteria, yeasts, molds, and fungi [152].

The PJ system for 120 seconds shows significant reduction efficiency of microbes like *L. monocytogenes, E. coli, Salmonella, and Cladosporium fulvum* present in tomatoes [147]. When compared to untreated samples, the

PJ process reduced 40% of organisms *Pectobacterium carotovorum subsp. carotovorum*.

The author [191] discovered that the deactivation capability of plasma jet varied among bacterial species. *S. cerevisiae* & *E. coli* were reduced to an undetermined point with 5 s PJ exposure. However, *G. liquefaciens* and *P. agglomerans* reached the undetermined level within 2.5 s PJ exposure. Furthermore, PJ efficiency for decontamination also depends on the exposed surface properties of the sample [31]. PJ reduced the *S. Typhimurium* by 2.72 (lettuce), 0.94 (potato), and 1.76 (strawberry) after 15 min TT [76].

6.12.2 ENZYMATIC INACTIVATION OF FRUITS AND VEGETABLES

Fruits and vegetables are more susceptible to browning reactions due to endogenous enzymes. Several research studies have proven the potential of CP used to deactivate enzymes and minimize browning [39, 248]. The PJ was utilized to address HRP and it was discovered that along with structure disintegration and microstructure deformation, the residual productivity of HRP was reduced to approximately 17% [97]. The microwave plasma torch decreased the activity of PPO (potato – 77% and apple – 62%) as well as POD (potato – 89% and apple – 65%) after 10 minutes of CP exposure [39].

When DBD was applied to apple, the activity of PPO declined by 42% proportionally as the treatment time increased (from 10 to 30 minutes) [247]. The same trend was noticed for watermelon [248] in reducing the activity of POD & PME. All of these findings confirm the possibility of CP for successful inactivation of enzymes in fruits and vegetables, hence extending their freshness.

6.13 APPLICATIONS OF CP ON MEAT AND MEAT PRODUCTS

The meat business industry proved as the least recognized of the food industries, which has been connected to growing food scarcity [91]. Government authorities, businesses, and consumers are all concerned about food safety. Meat's nature not only makes it perishable, but it also makes it extremely sensitive to loss of sensory characteristics when exposed to standard sterilization technologies, posing a unique difficulty in decontamination of meat foods.

Salmonella spp, *E. coli, L. monocytogenes,* and *Campylobacter jejuni* on meat surface cause significant food poisoning among consumers. The meat industry frequently employs food processing technology such as dehydration, freezing, cooling, and other thermal processing. Another typical strategy is to utilize hurdle technology, which involves active packaging, pH stabilizer, antimicrobials, a_w depressors (sugars and salts), and spices [23]. Changes in food structure and texture are frequent during thermal treatment of meat due to LPOD causing off-flavors and rancidity as storage increases. This is especially true for meats with high levels of unsaturated fatty acids, such as chicken. Furthermore, thermally processed meat food produces undesired chemicals (e.g., N-nitroso compounds, heterocyclic amines, and polycyclic aromatic hydrocarbons) [28]. Nowadays technologies are rapidly spreading, and people are more worried about possible risks due to poisonous substances.

6.13.1 DECONTAMINATION OF MEAT AND MEAT PRODUCTS

Recent research has aimed for the utilization of CP as a potential decontamination technology in various food sectors, including cereals, milk products, seafood, spices, fruits, vegetables, as well as meat [212, 260]. Many studies have shown that CP technologies improve meat's shelf-life by inactivating spoilage microorganisms. Several research studies imply that exposure of CP to meat products kills pathogens, such as *C. jejuni* (chicken thigh) & *S. enterica* (skinless chicken breast) [65], *S. aureus* (beef jerky) [121], *S. typhimurium* (bacon), *E. coli,* and *L. monocytogenes* (raw pork loin) [110]. Moreover, air CAP treatment on the surface of the meat, which is regarded as a possible disinfectant, results in a considerable reduction in the overall population of *Listeria innocua* (0.85 log). The microbial reduction of CP treatment is further enhanced with a combination of other techniques like the addition of peracetic acid (100 to 200 ppm) (5.3 log reduction) compared to CP treatment alone (2.3 log reduction of *S. typhimurium*) [41]. Additional manipulations to the gas mixture affect meat's water activity. When a 25% O_2 and 75% Ar gas combination was utilized in the CAP, water activity (a_w) was reduced to 0.71 [88]. Reduced water activity has been found to be crucial in avoiding bacterial growth in meat products. Adding O_2-CAP to the meat, on the other hand, increases L* while Ar-CAP lowers it because of moisture evaporating from the surface after the application of CAP [88].

In the decontamination treatment technology, UV photons, OH, O_3, NO, O, H_2O_2, charged particles, NO_2, and electric fields all play a significant role [134]. The strategies adopted by CP to protect the safety of the products in foods include PNOD, erosion of bacterial cell, damage of nucleic acid, morphological alteration, cell viability impairment, LPOD, and etching. The design of in-package CP treatment avoids post-process infection, improves microbiological safety, and keeps RTE chicken foods toxicologically safe [132].

After CP processing with a power of 24 kV at TT of 3 min, the microbes on the chicken sample significantly declined for *Salmonella* (1.45 to 0.05), indigenous mesophilic bacteria (0.70 to 0.12), and Tulane virus (1.08 to 0.15 log CFU/cube) [132]. After 10 min of CP exposure, the levels of microbes on chicken breast declined like *E. coli* (5.84 to 3.11), *S. typhimurium* (5.48 to 2.77), & *L. monocytogenes* (5.88 to 3.74 log CFU/g) [134].

Several current investigations on pork, chicken, and beef have proved the possibilities of CP innovation approach for assuring meat security [65, 121]. This is attributable to cold plasma's potential to successfully inactivate a diverse variety of microorganisms in a number of foods, especially spores, biofilms, as well as certain viruses [117]. Some recent studies have demonstrated the pathogens inactivation and decontamination by CP techniques [166]. ROS generated from plasma have a negative impact on important cellular biological molecules, proteins, DNA, and enzymes. The potentiality of ROS can alter the biological cell membrane functions by interacting with fats, which produce peroxidation of unsaturated fatty acid, protein (amino acid) oxidation, destruction of bacterial cell outer membranes due to electrostatic tension, structure bond breakage (C-C, C-N, and C-O), modification of morphological characteristics of bacterial cells or lysis, and loss of functionality are the important mechanisms for decontamination of meat products [279].

The H_2O_2 interacts with myoglobin to produce choleglobin, which gives a unique color to poultry meat. The alteration in color value is also based on the applied voltage. When chicken meat is exposed to 110 kV, the L* and b* values rise followed by a decreasing trend as the storage days extend to 9 days due to slime formation [132]. However, chicken breast treated with 70 kV, the color value changes insignificantly [287].

6.13.2 CURING OF MEAT AND MEAT PRODUCTS

The nitrite (NO_2) is utilized as a curing agent in the meat industry, which contributes to producing a unique meat color, good organoleptic quality, inhibiting LPOD, reducing pathogenic organisms, and spores of *Clostridium botulinum* present in the meat curing process [259].

Currently, meat products are cured by traditional chemical methods (sodium nitrite) and natural curing processes (vegetable concentrates) for the production of nitrite. However, these methods have drawbacks such as chemical residue, lower microbial decontamination efficiency, long incubation time, high cost, etc. Therefore, the development of a unique curing technique or a potential source of nitrite that is acceptable to both customers and the industry is required. It is known that there is no direct interaction of CP with liquid microbes. Conversely, the plasma interacts with the liquid and causes acidification of the liquid as well as the synthesis of RNS & ROS (NO_3, NO_2), which eliminates microbes [177]. Researchers expect that plasma-treated water containing nitrite might be employed as a source of nitrite for curing beef based on previous studies. The PAW could be employed for the beef batter curing process, in addition to microbial reduction.

The author [113] looked into the effect of CP on beef batter during mixing. The interaction of plasma and aqueous medium causes many nitrogen species, especially nitrite (NO_2). After 30 minutes of plasma treatment, the nitrite level in meat composition gradually increased to approximately 65.96 ppm. After heating, this amount of nitrites produced a unique cured pink color, indicating that cold plasma therapy can be used instead of nitrite addition to beef products. CAP is made up of a variety of RNS that dissolve to nitrogen oxide as given in Eqn. 1 followed by breakdown into nitric acid with H_2O as given in Eqn. 2. The plasma-activated brine (PAB) was used to cure beef jerky, and it was revealed that the development of nitrites in PAB is greater in air-plasma than in N_2 CP, with 26–180 ppm of residual nitrites. The direct interaction of nitrite with the meat component is demonstrated by this nitrite depletion [108].

$$NO_2 + H_2O + HNO_2 \rightarrow NO_2^- + NO_3^- + 2H^+ \tag{1}$$

$$2HNO_2 \rightarrow NO + NO_2 + H_2O \tag{2}$$

Recent research investigated the effectiveness of the curing process with PTW in emulsion-based pork sausage batter [112]. The researchers used a DBD-CP system to treat water for up to 120 minutes containing 1% sodium pyrophosphate, yielding 358 ppm of nitrate and 782 ppm of nitrite. The sodium pyrophosphate was used to retain the nitrous acid produced during the cold plasma process from decomposing. Both subjective and objective color measurements demonstrated that PTW successfully imparted the required red color to the beef product. The pork batter cured with PTW had a different color than that cured with a chemical curing agent (nitrite). However, the PTW-cured sample had less toxic chemical residue, a lower peroxide value, and higher sensory acceptability than the conventionally cured sample. In the meat sector, color and texture are significant quality characteristics. Increased plasma treatment exposure time on pigs and cattle resulted in a lower a* value (redness) [110]. The plasma treatments reduced the lightness value of pork loin, most likely due to moisture loss [117].

Color changes in meat are caused by the interaction of H_2O_2 with myoglobin while releasing choleglobin [79]. Unlike color, no significant variations in texture have been observed after the CP process of beef loin, pig [110], and beef jerky [121].

The oxidation of polyunsaturated fatty acids in meat is evaluated by quantifying the TBARS value. Research has found that when plasma treatment periods are prolonged (about 10 minutes), the TBARS value of beef rises. Similar ratings for color, appearance, odor, and overall acceptability except taste have been observed for pork and beef being pretreated using DBD plasma. However, certain off-flavors have been detected in chicken breasts treated with flexible DBD [134]. Some recent investigations have revealed that O_2 CP promotes LPOD in fish oil depending on the fatty acid profile. After 10 minutes of plasma treatment, DBD plasma has been shown to have a deleterious effect on consumer taste preferences. However, numerous recent studies have shown that CP is successfully applied for inactivation in meat foods.

6.14 APPLICATIONS OF CP IN SEA FOOD PRODUCTS

The safety of seafood is increasingly becoming an important concern, especially when raw seafood is recommended. Fish, crustaceans, mollusks, and echinoderms all contain significant amounts of protein, fat, vitamins, and minerals. Because of its high nutritious content, neutral pH, and high moisture content, seafood has a limited shelf-life [93]. The fast microbiological and metabolic processes that occur in shellfish quickly after death degrade nutritional and organoleptic characteristics, reducing shelf-life. The breakdown or decomposition of nitrogenous substances produces unpleasant odors or flavors such as trimethylamine, ammonia, and other compounds [181].

Seafood has a lot of polyunsaturated fatty acids, which induce LPOD. *Campylobacter spp., Norovirus, Salmonella, Staphylococcus aureus, Clostridium perfringens,* and *Listeria monocytogenes* are the predominant bacteria contributing the most sickness and mortality every year worldwide as per the Centers for Disease Control and Prevention. These bacteria have unfortunately been identified in seafood and fisheries processing regions. COVID-19 was recently discovered in Ecuadorian frozen white shrimp packaging and in a container of exported frozen shrimp in China. As a result, it is critical to create technologies and enhancers that ensure seafood's safety while also increasing its shelf life without affecting its organoleptic characteristics. Fish, beef, shrimp, and fresh agricultural products are successfully preserved using CP technique. CP produces a variety of species, comprising of RNS and ROS (reactive nitrogen and oxygen species), like singlet oxygen, ozone, peroxide, and nitrogen oxides, that are crucial for their superior microorganism inhibitory properties [171].

Chemical reactions and indigenous enzymes are accountable for the quick freshness loss in fish, while bacterial metabolic action is accountable for total spoiling. *Photobacterium phosphoreum, Listeria monocytogenes,* and *Lactobacillus sakei* become inactivated by Dielectric Barrier Discharge produced plasma using Ar for specific intervals (60 or 120 s) [51]. A rapid reduction in microbial population of *Penicillium citrinum* and *Cladosporium cladosporioides* on CP-exposed (0–20 min) dried filefish fillets has been observed [188].

Based on the type of food, DBD and JP can be utilized as an alternate approach for seafood preservation. ROS and RNS increase the breakdown and alteration of unsaturated fatty acids into lipid peroxides, which enhance lipid oxidation (LPOD) [181]. As a result of CP, a rise in LPOD has been seen in a variety of seafood.

The author demonstrated that mackerel (*Scombruss combrus*) processed with in-bag based DBD-CP system produced using ambient at 70 to 80 kV and treatment time (TT) of 1–5 min had a greater conjugated dienes and peroxide value. In mackerel slices exposed to CP, reduced levels of eicosapentaenoic acid and oleic acid were reported when compared with untreated equivalent. In CP Pacific white shrimp, the peroxide value, free fatty acids, & TBARS increased as TT rises [291].

The ROS and RNS of CP induce LPOD in CP exposed seafood without considering TT. As a result, optimizing the process parameter for CP is crucial which maximizes the effectiveness of CP in destroying both disease-causing germs and spoilage pathogens while maintaining quality qualities, which is especially important for raw seafood. Antimicrobial active species, on the other hand, have a negative impact on PNOD & LPOD, therefore processing seafood with natural antioxidants may assist in decreasing this occurrence. The PNOD (alteration of covalent bond) caused by direct or indirect contact protein molecules with oxidized products produced by CP reactive species is one of the technique's drawbacks within the seafood sector. Protein backbones and amino acid side chains are oxidized in the presence of reactive agents. As a result of these conditions, protein–protein crosslinking and/or protein fragmentation may occur [283].

When compared to commonly used thermal procedures such as canning and pasteurization, the appearance of seafood changes dramatically when cooked, and water loss is linked to protein denaturation. When seafood is dried, it loses its juiciness and develops a hard texture. As a result, heat has a considerable impact on consumer satisfaction and the quality of cooked seafood. The effect of CP on enzyme inactivation has been demonstrated, with structural changes caused by protein oxidation in enzymes accounting for the majority of their deactivation. Melanosis is a process of post-mortem changes in biochemical reactions inside the cephalothorax or carapace caused by PPO inhibition, specifically during storage above 5°C after CP exposure for 10 minutes at 500 Hz [62]. Apart from PPO, CP has been shown to disrupt a number of other enzymes, but not all of them. DBD-CP quickly inactivated alkaline phosphatase when used at different treatment times of 15 seconds and 5 minutes with 40–60 kV [223]. Although the color variations in Alaska Pollock shreds caused by CP usage may not be noticeable, an increase in the ΔE value with treatment time has been observed [53]. It is possible that the color variations in seafood are caused by the O_2 CP system. Greater water holding capacity in seafood is one of the desirable qualities related to juiciness and soft texture in seafood. High surface area exposed to CP produces extensive water retention ability in CP-treated samples. CP has a significant impact on the sensory properties of seafood, particularly under extreme conditions or with a substantial reactive species concentration, which is attributable to high oxidation of lipids in seafood. Development of various non-thermal techniques is being considered to preserve and increase the shelf life of seafood. Due to its superior inhibitory efficiency against a varied spectrum of bacteria, CP has recently been acknowledged as a potential non-thermal preservation technique for seafood. However, the applicability of this technique has been limited due to a number of processes that occur in seafood, including protein and lipid oxidation triggered by CP ROS. As a consequence, the CP treatment in seafood can be improved by choosing the correct combination of gas, independent process variables that will eliminate the negative impact on seafood quality and nutritional changes.

6.15 APPLICATIONS OF CP IN MILK PRODUCTS

Infectious diseases associated with cheese have been observed in multiple nations as dairy intake has increased [123]. Heat has traditionally been used to inactivate or destroy germs in foods. However, heat has unfavorable impacts on the functional, sensory, and nutritional characteristics of foods. Food quality protection is critical, and the method used must be successful in killing germs without causing undesirable changes in food quality.

Each method of food processing has its own set of benefits and drawbacks. As a novel processing technology, cold plasma has both beneficial and negative impacts. The results of this investigation show that CP has a strong potential for decontaminating and sterilizing dairy products. CP appears to be effective against microbial species present in different dairy foods. CP ensures sterilization at room temperatures without leaving residues in the food system. After CP treatment, the nutritional and sensory properties are largely retained. In certain cases, there were also negative impacts in terms of flavor and scent. The goal is to create NTCP discharges at atmospheric pressure to render decontamination as practical, with minimal quality losses and cost-effectiveness.

Ready-to-eat foods such as sliced cheese and ham were associated with foodborne *listeriosis* [67]. Listeriosis is consistently related to food items that promote the development of *Listeria monocytogenes* to concentrations of more than 100 CFU/g [49]. The goal of the research conducted by the author [237] was to assess the APP for effective decontamination of *L. monocytogenes*-inoculated ham and sliced cheese with three power levels of 75 to 150 W and treatment times of 60 to 120 seconds. More than an 8 log reduction can be achieved in cheese when treated for 120 seconds at 150 W. In sliced ham treated with CP for 120 seconds, the microbial count reduced to 1.73 from 0.25 log. Similarly, *E. coli* present in cheese has been reduced to 1.47 from 0.09 log and to 1.98 from 0.05 log with He & He/O_2 combination used for cold plasma generation, respectively.

Cronobacter sakazakii potentially leads to life-threatening infectious diseases in newborns. Consumption of infected powdered foods, particularly powdered milk, makes individuals susceptible to infection caused by *C. sakazakii*. Using a fluidized CP reaction system for uniform exposure, the research examines the efficiency of CAP against *C. sakazakii* in nonfat dry milk (NFDM) powder. *C. sakazakii* was reduced by 1.17–3.27 log_{10} after 20–120 s of CAP treatment. The deactivation of *C. sakazakii* increased as the flow rate varied from 8 to 20 L/min. There were no significant color changes (E 1.5) in the quality characteristics of NFDM following the CAP treatments. Furthermore, after a 120s CAP treatment, there were alterations in the composition of amino acids, crystallinity, and TPC value. These data imply that the combination of CAP with a fluidized system delivers good antibacterial activity while having little impact on NFDM powder physicochemical properties [43]. Further optimization and controlled experimental circumstances can be used to mitigate the unfavorable consequences. When compared to thermal processing, cold plasma technology allows for more precise control.

In general, the exposure of plasma lowers the pH value because reactive species solubility causes the generation of H_2O_2 and nitric acid [251]. The production of hydrogen ions in plasma is possibly responsible for decreasing trends in pH compared to the control sample. The minor difference is related to the formation of phosphate compounds and casein proteins, which resist changes in pH when compared to water. The enhancement of acidity is due to factors like the rate of gas usage, flow rate of gas, and generation of heat. A substantial reduction of pH value in cheese occurs after 5 minutes of treatment with DBD plasma [278]. The research showed that the pH lowering was caused by the development of acid (HNO_3) during CP exposure. Likewise, CP treatment reduced the pH value of whey protein isolate (WPI) [222]. The shorter treatment time with N2 contributed to fewer changes in the color of milk powder exposed to ACP [43]. The WPI exposed to ACP shows that the yellowness increased as the treatment time went on, and pigment development caused the yellow color [222]. The sensory characteristics and overall cheese quality exposed to (He + O_2) gas were much superior to samples exposed to He alone [133]. However, no obvious changes in appearance were seen at 3.5 kV of DBD-CP exposure.

The DBD-CP reduced the flavor of Cheddar cheese significantly while having no effect on appearance, color, or overall acceptability [278]. Even after being subjected to cold atmospheric plasma for 120 seconds, showed no noticeable alteration in the composition of amino acids present in nonfat dry milk [43]. These changes are due to the low humid environment (35%) developed during the CP process resulting in less production of atomic oxygen, as well as the shorter treatment time and lower temperature surroundings. For cheese subjected to CP exposure up to 10 minutes, the level of lipid oxidation rose dramatically compared to the control sample [277]. It was recognized that DBD-CP induces oxidation by plasma species that produce oxidized products depending on gas, treatment time, fat content, and plasma type. Because thermal treatment has been a component of the dairy industry for decades, upgrading this with non-thermal procedures may provide some ways to prevent issues. For large dairy product processors to pursue CP treatment, it must be proved to be effective and comparable to thermal treatments. Because it is a modern technology, the initial cost is still high, but it does have an advantage over thermal approaches due to the good benefits on sensory characteristics. Due to a variety of designs and applications, large-scale commercial machinery is not readily accessible. The regulation of CP-treated foods and their consumption is still being worked out. Non-thermal technologies in the dairy industry have a bright future, but they must be implemented on a broad scale to be successful in eco-friendly environments.

6.16 APPLICATIONS OF CP IN SPICES AND HERBS

A big epidemic of food pathogens has emerged among herbs, spices, nuts, and their products, prompting a significant impact on products. Extended shelf-life and safety of the products can be achieved by CP food-processing techniques. A wide range of active substances produced through cold plasma may interact with microorganisms at room temperature, as well as complicated geometric forms and heat-sensitive materials, with chemical residue free. This section describes the applications of CP techniques for an effective decontamination process on the surface of spices, nuts, and herbs.

6.16.1 INACTIVATION OF MICROORGANISMS IN SPICES

Microorganisms present on the surface of almonds and peanuts can be inactivated with cold plasma (CP) treatment. Almonds soaked in water and contaminated with *E. coli* were exposed to CP for 30 seconds, resulting in a 5-log reduction in microbial count [63]. Another study showed similar results with *Salmonella* (NTP) [175]. Researchers demonstrated that exposing pistachios to plasma created by pure oxygen and argon for 1 minute reduced the quantity of *E. coli* on the pistachio surface by about 4 logs [193].

Toxic fungi can contaminate nuts and other foods, leading to the presence of unavoidable aflatoxins [15]. Cold plasma technology has potential applications in detoxifying mycotoxins. Aflatoxins (AFs) can be found in various foods, such as nuts, cereals, and spices [200]. The antifungal activity of low-pressure cold plasma (LPCP) against *Aspergillus parasiticus* in pistachios, peanuts, and hazelnuts has been extensively studied. Research indicates a 5-$\log_{10}$ reduction in fungal population per gram after 10 minutes of exposure [24].

The shelf-life of walnuts was significantly extended through cold plasma processing. Cold plasma demonstrates bactericidal properties that can be an alternative method for food decontamination. Studies have shown that increasing voltage and frequency enhances the bactericidal effect of plasma. Cold plasma effectively eliminates contaminants from the target surface.

The study reports that no noticeable alteration in qualities of (texture, odor, & appearance) between CP exposed and unexposed nuts (hazelnuts, peanuts, pistachios) [24]. However, the plasma generating gas can significantly alter the color of the products. The L* and b* values were reduced while using $N_2{}^{+}O_2$ as gas in plasma system, resulting in darker unpeeled almonds. When the nuts were merely handled with N_2 or air, the color changed significantly. The hue of the almond remained consistent for O_2, CO_2, and 90% CO_2 + 10% Ar [100]. Furthermore, other specialists observed the TPC & antioxidant capacity of walnuts were unaffected by plasma jet system for 11 min of exposure, however both improved during storage period [11]. CP processing has minimal effect on cashew nut composition and nearly little effect on cashew nut sensitization. However, the oil and anacardic acid levels of cashew nuts treated with plasma were enhanced [10].

Black pepper is a spice that is widely used across the globe and has a significant flavor-enhancing impact. Black pepper is usually collected by hand and then organically dried, a procedure that can quickly get infected with germs. As a result, spices sold in retail stores may include a variety of microorganisms [82]. The author [118] examined the influence of MW-CP exposure for decontamination of powdered onion against *A. brasiliensis*. When compared with hot air drying, CP exposure followed by vacuum drying had significant inhibitory activity against *A. brasiliensis*. Another author [241] discovered that inoculating *Salmonella* on black pepper and treating it for 60–80 seconds with atmospheric plasma (APP) dramatically decreased the *Salmonella* population in the range of 4.5 to 5.0 $\log_{10}$ CFU/g.

The mechanism of CP species reacting on microbes is complex and varied depending on gas composition, plasma discharge method, and microbes including bacteria as well as fungi. Initial microbial concentration and treatment duration are crucial parameters to achieve an effective disinfection rate in mold/yeast [12]. The complete inactivation of mold and yeast is achieved after 7 minutes of CP exposure when the initial microbial concentration is 3.30 (mold) and 3.0 (yeast) CFU/g. No noticeable color alteration is observed in black pepper while using the microwave plasma jet system [101]. The b* value of CP-treated oregano leaves decreased due to chlorophyll degradation. The significant color changes in red pepper powder after plasma exposure result in an increase in L* and b* value with a reduction in a* value, indicating that the red color became a light yellowish color

[101]. The author [119] found that the antioxidant activity of CP-treated onion powder improved without altering quality parameters and sensory characteristics. Furthermore, the antioxidant capacity of red pepper powder was found to be positively related to the length of CP exposure [284]. This might be due to the fact that CP processing can degrade the surface structure by etching, which enables the bioactive compounds to be released or leached out into the solvent [269].

6.16.2 EXTRACTION BIOACTIVE COMPOUND FROM SPICES AND HERBS

Spices are a vital commodity that have a substantial influence on the economics of many nations. Spices and their products are commonly utilized as effective flavoring agents in the food industry and act as potential antimicrobial agents in preserving food products [201].

Cuminum cyminum L., often known as cumin, is a famous seed spice noted for its unique fragrance, flavor, and therapeutic characteristics with organic acids, phytochemicals, and nutritional content [239]. Cuminaldehyde (4-isopropyl benzaldehyde) is the major element of cumin essential oil, and the hydrodistillation method of extraction results in higher flavor constituent retention [27]. This essential oil composition and yield can be further improved by the application of suitable pretreatment. Cold plasma is very effective in surface etching, creating the way for releasing more internal compounds like essential oil and other constituents. The morphological changes in lemon peel surface due to CP etching were investigated by SEM analysis, which demonstrated the emergence of fissures and fractures due to plasma reactive species that resulted in a greater extraction yield. Plasma treatment at 2.5 kV resulted in a greater extraction yield than traditional hydrodistillation (149.34% increase) [199]. Similarly, when compared to their untreated control cumin seed sample, the enhancement of extraction rate in essential oil recovery (19.5%) is achieved by subjecting the sample to a plasma bubbling system for up to 90 minutes. The production of hydroxyl radicals and their improved dispersion as a result of uniform plasma bubbling contributed to an increase in cuminaldehyde content in essential oil compared to the control [228]. Therefore, DBD plasma might be a potential approach for improving essential oil extraction.

Consumer desires for nutritious, quality-rich food necessitate scientists and food processors to adopt modern technology integrated with global commerce. Overall, the application of CP in heat-sensitive herbs and spices has succeeded in microbial eradication without compromising on quality characteristics. This significant element raises awareness among industries to adopt novel techniques in food processing.

6.17 EFFECTS OF CP ON FOOD COMPOSITION

Food is essential for living because it supplies both energy and nutrients to all creatures. Food generally consists of many components, such as vitamins, phenols, and antioxidants, that promote various metabolic functions accountable for the body's growth and repair. The application of food processing techniques frequently results in either a rise or fall in the nutritional content of foods, although it mostly contributes to the preservation of nutrients that might otherwise be lost during storage. The most popular technique in food preservation is utilized with or without heat. However, NTP treatment is a revolutionary method with several applications in diverse sectors, including microbial disinfection, enzyme inactivation, and functional customization in food.

CP treatment is a well-studied technology used in food sectors as a replacement for established thermal processing approaches for food quality preservation. The CP species provide considerable contributions to progressive alterations in the food matrix, which eventually impact food quality. Furthermore, the interactions are prone to be impacted by a variety of elements such as input power, plasma source, processing time, gas flow rate, gas components, discharge gap, and the sample [158]. Though certain quality characteristics of cold plasma-treated foods are discussed in the above section, it is necessary to discuss the overall changes caused by cold plasma in different sectors where food is essential. Therefore, major quality parameters like acidity, pH, texture, antioxidant, antimicrobial, etc., have been described here.

6.17.1 ACIDITY AND PH

Food scientists are interested in the influence of CP on food parameters in addition to microbial inactivation. pH and acidity are important factors in determining the processed quality of food because they influence the processing parameters that are necessary for producing safe products. When nitrogen was used as a working gas, pretreatment with CP resulted in a higher pH value than the untreated sample [234].

The creation of nitrogen oxides and other compounds arises from the collision of high-energy electrons with nitrogen molecules inside the gaseous state, as well as a series of subsequent interactions, which could have led to an increase in pH values in plasma-treated samples [55]. The noticeable increase in pH had also been described on the basis of a modification in metabolism, moisture loss, and intrinsic buffering during plasma exposure [162]. Following plasma treatment, equivalent pH values were shown in fig fruits and cherry tomatoes [1].

High plasma discharge rates result in the generation of more acidogenic molecules such as nitric acids, singlet oxygen, and nitrous oxide, as well as its products (NO_2 and NO_3), which react with water to form nitric acid, lowering the pH of food products [278]. The breakdown of hydroxyl radicals created in CP discharge can also affect the acidity of fruits [187].

Several other investigations have found that plasma treatment increased acidity and decreased pH in bayberries, fresh-cut potatoes, and apples [39, 150, 272]. Low pH, on the other hand, permits reactive species to get through cell walls, lowering bacteria's acid tolerance. Furthermore, considerable variances in farming operations based on variety variation and physical conditions of fruits and vegetables. As a result, it is feasible to deduce that processing time, food type, treatment intensity, buffering capacity, and physiological functions of biological cells all influence the effect of CP on the pH value of complicated food matrices.

6.17.2 TSS

TSS primarily consists of sugars, but it also contains organic acids, soluble amino acids, and many miscellaneous components (such as lipids, alcohol, minerals), and flavonoids [268]. °Brix is often used to calculate the amount of TSS in a substance. TSS value was lower in Chinese bayberries pretreated with PAW than the untreated sample due to the suppression of fruit respiration rate, which may have resulted in a decrease in sugar and acid utilization. In pumpkin puree, CP's impact on decontaminating *Escherichia coli* was attempted and physicochemical modifications were examined [213]. In contrast to the preceding findings, a substantial improvement in TSS values was seen in plasma-treated samples of blueberries [215] and fresh-cut pears [42], where the breakdown of starch into sugars was linked to an increase in TSS. Similar findings were made in strawberry and apple samples, with the increase related to water loss and associated metabolic changes after the CP treatment [207, 247].

6.17.3 COLOR

Color influences food choice, preference, acceptability, taste thresholds, sweetness perception, and pleasantness. The presence of natural and synthetic pigments, as well as apparent chemical reactions during processing, are important factors in the distinctive color appearance of foods [22]. Although color would be an indicator of nutritional and functional modifications in food throughout processing, yet it is primarily seen as a quality inspection variable by academia and industry because any change in color offers data on the impact of process parameters.

There was an alteration in the color characteristics of strawberries after CP exposure. However, the changes are insignificant in terms of specific color characteristics, such as brightness, redness, and greenness. As the storage time increases, noticeable changes have been observed. The plasma species (O_3) cause disintegration of the conjugated double bond present in the carotenoid structure, leading to an increase in the L* value, which is responsible for the light yellow color in plasma-treated flour [156]. The modification had been connected to pigment isomerization along with oxidation reactions. When Chinese bayberries are exposed to PAW, the color changes from red/violet to dark violet due to an increase in anthocyanin content compared to the control sample [150].

Fruit discoloration was attributed to moisture loss and cutaneous wax melting on the surface of fruits [215]. The greatest whiteness index (WI) value was found in plasma-treated flour (88.31±0.055). For all flour samples, the WI was enhanced with the duration of plasma exposure relative to the control (86.70 ± 0.003). Several researchers have shown a significant increase in WI in cereals when treated with CP due to etching of the bran layer [218, 254].

6.17.4 TEXTURE

The structure of the food is directly influenced by texture to assess the food quality [47]. Food texture is used for sensory experience, while structure is used for material property that impacts textural characteristics. Many research findings have also shown that CP processing retains food texture after treatment. The shiitake mushrooms showed insignificant differences between treated and control samples after plasma exposure for up to one week of storage period [85].

Insignificant texture changes have been observed for fruits and vegetable cut surfaces due to no changes in the cell-wall polymer. Blueberries subjected to cold plasma for an extended time showed a decline in firmness, bruising, and skin ruptures [129]. However, such modifications seem to be insignificant on a macroscopic scale. Then, this CP provides several benefits over traditional thermal treatments in applications involving cell disintegration, concentration, and extraction. The hardness of almond slices increased with increasing time of plasma exposure due to the availability of moisture and its interaction with reactive species, which convert to other substances [231].

The authors [50] achieved consistent findings when they used CP to dehydrate wet noodles, as reactive species cause oxidation of the free sulfhydryl groups of glutenin during dehydration, forming disulfide bonds, resulting in a tight structure and increased hardness. However, the production of denser protein in hairtail muscle due to cross-linking and oxidation phenomena, the texture has been improved after plasma treatment [125]. If cereal products like rice and wheat are exposed to plasma, the hardness values have been reduced due to surface etching causing fissures and pores, which enhance the mass transfer rate in drying and reduce energy consumption [218]. It is concluded that the changes in texture properties depend on the product categories.

6.17.5 PROXIMATE COMPOSITION

Protein, carbohydrate, fiber, moisture, ash, and fat are significant factors that can affect food quality. The application of any technology for processing leads to modification of the food components in terms of nutrients. The exposure to CP influences the reduction of water content in starch due to the interaction of plasma species, causing the biodegradation of water molecules (H_2O) into oxygen radicals (such as OH^-, O_3, and O^-) [290]. Moreover, when treatment power and duration are increased, it causes water loss. However, complete removal of moisture is not possible with cold plasma techniques [252].

Reduction in moisture is achieved for different cereals and starches like corn, wheat, black gram, basmati, and brown rice [141, 218, 245, 249]. Alteration of protein and fat value is caused by the reduction of moisture or oxidation of free radicals and the development of intermediate compounds from lipid oxidation. The reduction in carbohydrate has been caused by the interaction of plasma species leading to depolymerizing starch. These changes are insignificant based on the voltage, time, and other factors. There is no noticeable difference in fat and ash content in plasma-treated samples. The results demonstrate that the application of plasma does not affect the nutritional value of cereals [44, 46, 249].

6.17.6 ANTIOXIDANTS

Fruits contain phenolic compounds and ascorbic acid, which can influence the antioxidant activities of the food. Such bio-substances have the potential to absorb harmful toxins and minimize the chances of illnesses caused by oxidation. CP exposure does not influence the antioxidant activity of certain foods like kiwifruits, onion powder, radish sprouts, and red chicory [119, 179, 189, 205].

The germinated rice improved its antioxidant activity due to plasma species interaction resulting in etching phenomena, thus causing it to absorb water and improve its bioactive compounds. However, the antioxidant activity of plasma-treated samples after 7 days decreased due to the reduction of antioxidant bio-compounds by scavenging free radicals; and the loss of phenolic molecules is the primary reason for the decrease in antioxidant capacity.

On long-term exposure, researchers have also found a decrease in antioxidant capacity in apples, white grapes, and cashew apples after CP treatments due to degradation of antioxidant compounds [187, 206, 211]. The author [8] found that direct plasma treatment reduced the antioxidant capacity, whereas indirect plasma treatment had no significant effects. Several findings reported that CP caused insignificant changes in the capacity of antioxidants in various foods based on products and processing parameters [12, 103, 206]. According to the studies, the plasma production source, food types, method of exposure, and independent variables influence the efficiency of antioxidant activity in CP-treated foods.

6.17.7 ANTIMICROBIAL ACTIVITIES

The CP can be effectively applied to inactivate a wide range of microbes in contaminated food products. Simultaneously, the CP is utilized to extract valuable bioactive compounds from plants as well as fruits and vegetable byproducts. Enhancement of total phenolic (41.13%) and antioxidant content (41.05%) has been observed when green tea leaves were exposed to CP for 15 minutes at 15 W due to the extraction of high catechin (103.12%) that is responsible for high antimicrobial and antioxidant capacity [114]. The researcher applied cold plasma-treated extract for preserving food products like salami and found that the *Salicornia neei* extract extracted after CP treatment has greater antimicrobial activity towards *S. aureus* and *E. coli* compared to the control extract depicted by the diameter of the inhibitory zone [73]. The presence of phenolic compounds and their synergistic impact with fatty acids and other osmotic chemicals such as betaine is responsible for the higher antibacterial activity [9].

6.17.8 PHENOLIC CONTENT

Phenolic compounds are crucial bioactive compounds in plant foods having the capacity to fight against oxidative stress and prevent metabolic disorders. Plants are self-protective in nature against the stress caused by environments due to the deposition and accumulation of major phenolic compounds in plant tissues.

Exposure of cold plasma to plants creates an impact on abiotic elicitors, allowing the formation of secondary metabolites such as phenolic compounds. Cold plasma can stimulate the biosynthesis of phenolics in fruits by increasing the adenosine triphosphate content and speeding up the usage of sugars [142]. Cold plasma treatment of pitaya fruit increased total phenolic content by accumulating and forming phenolic substances such as p-coumaric acid, caffeic acid, p-hydroxybenzoic acid, and gallic acid [137].

The increase in phenolic content is also due to the degradation of cell-wall membranes by cold plasma etching to create a way for the easy release of phenolic molecules. The total phenolic content was increased after dielectric barrier discharge plasma treatment due to the accumulation of phenolic molecules in germinated rice [19]. However, there is a significant decrease in phenolic concentration in rice paddies due to the inactivation of phenylalanine ammonia lyase, which is used for the synthesis of phenolic compounds and the decomposition of already present compounds [61]. Therefore, various studies confirm that the total phenolic content depends on the type and variety of plant, and cold plasma conditions.

6.17.9 SENSORY ATTRIBUTES

Food odors, aroma, flavor, color, texture, mouthfeel, and overall appearance have been demonstrated to impact food choices, caloric intake, and sometimes even develop a unique appetite for certain foods. Odor compounds contain chemical characteristics that activate chemical receptors including aroma and flavor [36]. The concentration of the volatile substances depending on the processing conditions contributes to aroma and flavor changes that are either increased or decreased to a certain level. As a result, CP is employed to alter its flavor and aroma according to customer preference. C5–C11 aldehydes, the most prevalent linoleic and oleic acid oxidation products, were found in plasma-treated almond samples [231]. Octane, nonanal, hexanal can produce an odor, which interferes with the aroma and quality of almonds that continue to rise throughout storage and oxidation.

Development of off-flavor has been observed while applying cold plasma on sliced cheese for more than 10 minutes to inactivate microbes [133]. Because cheese includes a lot of fat, the off-flavor is generated by lipid oxidation byproducts produced by oxygen radicals. When lipid oxidation induced by free radicals occurs, products called hydroperoxides responsible for a similar effect on the flavor of cheddar cheese were seen [278] due to the production of secondary oxidative compounds, such as alcohols, acids, alkanes, ketones, and aldehydes. Exposure to CP may induce significant changes in flavor, which are likely to be influenced by essential constituents in the sample, such as proteins and fatty acids. As a result, determining the chemical species is crucial to prevent any negative treatment effects and to produce more appealing food by focusing on the creation of desirable flavor qualities.

The color and texture have already been discussed in this chapter. No noticeable changes in odor and appearance were observed in CP-exposed radish sprouts, showing that CPT had no negative effect on the sensory features of the radish sprouts [179]. Likewise, the lettuce treated with N_2-based CP had no effect on the organoleptic characteristics like texture, flavor, and appearance [147]. However, using a single parameter to determine freshness is insufficient, and it should be combined with other quality attributes because both plasma-treated and untreated products remained organoleptically acceptable even when changes were induced within the limit.

6.18 FOOD SAFETY AND CONSUMER ACCEPTABILITY OF COLD PLASMA TREATED FOOD PRODUCTS

Only a few studies have looked into the production of harmful compounds in CP-treated foods. The oral acute and subacute toxicity in a soy-based edible film treated with CP has been studied on rats [96]. The study confirms the film had extremely low toxicity and insignificant harmful risks. CP enhanced interconnections among starch granules without generating new harmful compounds, which implies that residue-free and safe techniques are applied for starch modification [233]. The research statement in ACP is the production of ROS, including ozone. Ozone is an FDA-approved antimicrobial compound for use in foods. However, remaining reactive species do not have any regulations [57].

In general, CP is considered non-toxic under optimal processing conditions. However, during future industrial adaption, each individual operation and final product quality should be checked to identify hazardous chemicals. Many companies across the world are now developing CP equipment with potential culinary applications. Low operating costs, prolonged shelf-life, and food quality and safety are advantages of this technology due to the use of air as a working gas and no residue. However, because regulatory permission has not been sought and toxicity testing is still ongoing, this equipment is exclusively for research and development purposes.

Consumers are more concerned with the techniques and processes employed across the production and processing chain, not just the nutrition, source, and quality of the food they buy. Consumers, on the other hand, are preferring minimally processed foods, which reduce cooking time without sacrificing the taste and nutrient content. Consumer survey reports repeatedly reveal that consumers remain unaware of an extensive variety of breakthrough food-processing technologies, resulting in potential innovations being lost or adoption being constrained. It is vital to discuss technology, application, potential benefits, and constraints as early as possible to secure customer acceptance. For the effective acceptance and development of innovative processing technologies in the food sector, it is crucial to address customer expectations and create awareness. Consumers are more inclined to adopt revolutionary processing systems if something enables the launch of new products with tremendous benefits or delivers a product or process that is safer than present technology [57]. However, there is insufficient data on the organoleptic attributes of foods processed by CP, with a large number of research focusing on objective analytical techniques for texture, color, and aroma. To secure approvals, such data imbalance must be addressed in subsequent research.

6.19 SUMMARY

Cold plasma (CP) has recently gained popularity in the food processing sector. It is a novel and breakthrough approach for safety and preservation of food products with minimal influence on food quality. The primary objective of using CP on the surface is to achieve microbiological disinfection. The shallow penetration depth is desirable in various cases. A shallow penetration depth protects the majority of the nutrients within the food while allowing the surface to be completely decontaminated in food structure, which has more surface-to-volume ratio.

Plasma species cause a variety of reactions, such as surface etching (creation of pores/tissue damage), depolymerization (formation of new compounds through starch breakdown), cross-linking (cleavage of C-OH polymeric chains for adding new C-O-C bond by eliminating water). The plasma zone is created by breaking gas molecules and forming an electron avalanche which converts gas (non-conductive mode) into electrons (conductive mode). The DBD plasma is generated between two electrodes containing insulating material by electrical discharge. Plasma jet systems are capacitive-coupled devices where gas is passed between two coaxial electrodes with high velocity, producing miniature "plasma fires" that typically originate in the radio frequency band. CD system has point-to-plate geometries (a strongly curved electrode positioned in the opposite direction of a flat electrode) and cylindrical arrangements. RF-CP systems create electromagnetic fields by providing RF currents flowing in coils or antennas which are either submerged in plasma or isolated from it by a dielectric barrier. In MWPS, the electrons absorb the microwave energy, resulting in an increase in kinetic energy and inelastic collisions releasing energy as photons of UV and visible light which leads to ionization processes.

The great density of electrons in streamer discharge creates arcs to form, producing gas heating and electrode damage. Based on the cell wall thickness, Gram-positive bacteria are more resistant to plasma compared to gram-negative species. The D-values of *S. cerevisiae* and *M. frigida* declined as oxygen concentration increased. This could be due to increased ion and reactive species generation, such as ozone and O^+, when higher oxygen concentrations are used. At 70% RH, the inactivation of *Aspergillus niger* conidiophores is greater than at 80% RH. This result was attributed to water degradation into OH and (H_2O_2) generation from OH. When RH was higher than 70%, the inactivation effectiveness was reduced.

At high applied voltages, the current intensity of steel electrodes was higher than that of aluminum electrodes, suggesting that the collisions between oxygen molecules and O_3 generation increase as the current intensity rises. However, due to the higher temperature, the breakdown of O_3 was faster for steel electrodes. Larger gaps between the powered electrode and the substrate, as well as between electrodes, can result in decreased treatment efficacy due to non-uniform plasma formation and active species recombination before reaching the sample. In addition, the dielectric loss in the barrier was reduced with the multipoint electrode, resulting in a larger number of micro-discharges from 50 (plate-plate) to 2,700 (multipoint-plate) with a lower energy amount.

In contrast, persistent CP exposure promotes qualitative degradation in plasma-treated goods. Long-lived reactive species (e.g., ozone) inside the package may result in increased *E. coli* inactivation after 24 hours compared to 0.5 hours, inhibiting microbial cell recovery from damage and stress.

The ACP's strong electric field causes cell membrane destruction, resulting in leakages from bacterial cells and loss of cell functioning. Reactive species created in plasma interact with amino acids in proteins, causing structural changes in proteins that lead to microbial spore death. The pesticide molecules are first dissociated by the cold plasma treatment. Simultaneously, plasma treatment produces many free radicals with high oxidation potentials, such as O_3•, OH•, and H_2O_2•, which may result in pesticide breakdown. The above-mentioned reactions eventually disrupt some of the chemical bonds of pesticides, resulting in the formation of new chemicals that are usually less dangerous.

As plasma can cause cellular damage and surface changes, it may be capable of minimizing internal molecule diffusion resistance and increasing the extractability of hydrophilic compounds, thus increasing phenolic extraction from biomass. By increasing the diffusion rate of essential oil to the exterior of the oil glands, the pretreatment accelerated the amount of essential oil extracted.

CP-generated active species modify the starch structure in grains, resulting in changes in swelling power, pasting qualities, water absorption, enzyme susceptibility, solubility, and compositional, structural, and thermal properties. Granular corrosion and etching, molecular disintegration, cross-linking, and depolymerization can also be caused by CP. Reduction in hardness and cooking time of solid food with improved hydration qualities is also achieved by plasma treatment.

Microorganisms found in cereals, fruits and vegetables, seafood, cheese, spices, herbal leaves, and nuts are effectively inactivated by cold plasma. It also deactivates enzymes found in cereals, fruits, and vegetables. Cold plasma is effective for increasing the germination rate, converting hazardous pesticide molecules into less toxic ones, and increasing the extraction rate by destroying hard tissues that limit mass flow rate. Cold plasma produces slight changes in food quality such as color and pH, as well as a large increase in antioxidant, antibacterial, and phenolic content. Cold plasma has numerous applications in the solid food business. However, the consumer acceptability of cold plasma-treated food must be extensively investigated in the future.

KEYWORDS

- **cold plasma**
- **food composition**
- **food safety**
- **fresh produce**
- **fruits and vegetable products**
- **inactivation**
- **processing factors**

REFERENCES

1. Abbaszadeh, R., Alimohammad, K., & Zarrabi Ekbatani, R. (2018). Application of cold plasma technology in quality preservation of fresh fig fruit (Siyah): A feasibility study. *International Journal of Horticultural Science and Technology, 5*(2), 165–173.
2. Adeghate, E., & Ponery, A. S. (2002). GABA in the endocrine pancreas: Cellular localization and function in normal and diabetic rats. *Tissue and Cell, 34*(1), 1–6.
3. Afshari, R., & Hosseini, H. (2014). Non-thermal plasma as a new food preservation method, its present and future prospect. *Journal of Paramedical Sciences Winter, 5*(1), 2005–2015.
4. Aguilar Uscanga, B. R., Calderón Santoyo, M., Ragazzo Sánchez, J. A., Alemán Duarte, M. I., & Pérez Montaño, J. A. (2022). Effect of the application of cold plasma energy on the inactivation of microorganisms, proteins, and lipids deterioration in adobera cheese. *Journal of Food Quality, 2022*, E-article ID 8230955. https://doi.org/10.1155/2022/8230955.
5. Albertos, I., Martin-Diana, A. B., Cullen, P. J., Tiwari, B. K., & Ojha, K. S. (2019). Shelf-life extension of herring (*Clupea harengus*) using in-package atmospheric plasma technology. *Innovative Food Science and Emerging Technologies, 53*, 85–91.
6. Albertos, I., Martín-Diana, A. B., Cullen, P. J., Tiwari, B. K., & Ojha, S. K. (2017). Effects of dielectric barrier discharge (DBD) generated plasma on microbial reduction and quality parameters of fresh mackerel (*Scomber scombrus*) fillets. *Innovative Food Science and Emerging Technologies, 44*, 117–122.
7. Alkawareek, M. Y., Gorman, S. P., Graham, W. G., & Gilmore, B. F. (2014). Potential cellular targets and antibacterial efficacy of atmospheric pressure non-thermal plasma. *International Journal of Antimicrobial Agents, 43*(2), 154–160.

8. Almeida, F. D. L., Cavalcante, R. S., Cullen, P. J., Frias, J. M., & Bourke, P. (2015). Effects of atmospheric cold plasma and ozone on prebiotic orange juice. *Innovative Food Science and Emerging Technologies, 32*, 127–135.
9. Alonso-Montemayor, F. J., Reyna-Martínez, R., Neira-Velázquez, M. G., Sáenz-Galindo, A., Aguilar, C. N., & Narro-Céspedes, R. I. (2021). A review on antibacterial and therapeutic plasma-enhanced activities of natural extracts. *Journal of King Saud University – Science, 33*(6), E-article ID 101513. https://doi.org/10.1016/j.jksus.2021.101513.
10. Alves Filho, E. G., Silva, L. M. A., Oiram Filho, F., Rodrigues, S., & Fernandes, F. A. N. (2019). Cold plasma processing effect on cashew nuts composition and allergenicity. *Food Research International, 125*, E-article ID 108621. https://doi.org/10.1016/j.foodres.2019.108621.
11. Amini, M., & Ghoranneviss, M. (2016). Effects of cold plasma treatment on antioxidants activity, phenolic contents and shelf life of fresh and dried walnut (*Juglans regia* L.) cultivars during storage. *LWT – Food Science and Technology, 73*, 178–184.
12. Amini, M., & Ghoranneviss, M. (2016). Black and green tea decontamination by cold plasma. *Research Journal of Microbiology, 11*(1), 42–46.
13. Anbarasan, R., Jaspin, S., Bhavadharini, B., Pare, A., Pandiselvam, R., & Mahendran, R. (2022). Chlorpyrifos pesticide reduction in soybean using cold plasma and ozone treatments. *LWT – Food Science and Technology, 159*, E-article ID 113193. https://doi.org/10.1016/j.lwt.2022.113193.
14. Andrasch, M., Stachowiak, J., Schlüter, O., Schnabel, U., & Ehlbeck, J. (2017). Scale-up to pilot plant dimensions of plasma processed water generation for fresh-cut lettuce treatment. *Food Packaging and Shelf Life, 14*, 40–45.
15. Arrus, K., Blank, G., Abramson, D., Clear, R., & Holley, R. A. (2005). Aflatoxin production by Aspergillus flavus in Brazil nuts. *Journal of Stored Products Research, 41*(5), 513–527.
16. Attri, P., Venkatesu, P., Kaushik, N., Han, Y. G., & Nam, C. J. (2012). Effects of atmospheric-pressure non-thermal plasma jets on enzyme solutions. *Journal of the Korean Physical Society, 60*(6), 959–964.
17. Bai, Y., Chen, J., Mu, H., Zhang, C., & Li, B. (2009). Reduction of dichlorvos and omethoate residues by O_2 plasma treatment. *Journal of Agricultural and Food Chemistry, 57*(14), 6238–6245.
18. Bai, Y., Chen, J., Yang, Y., Guo, L., & Zhang, C. (2010). Degradation of organophosphorus pesticide induced by oxygen plasma: Effects of operating parameters and reaction mechanisms. *Chemosphere, 81*(3), 408–414.
19. Banchuen, J., Thammarutwasik, P., Ooraikul, B., Wuttijumnong, P., & Sivongpaisal, P. (2010). Increasing the bio-active compounds contents by optimizing the germination conditions of southern Thai brown rice. *Songklanakarin Journal of Science and Technology, 32*(3), 219–230.
20. Bang, I. H., Kim, Y. E., Lee, S. Y., & Min, S. C. (2020). Microbial decontamination of black peppercorns by simultaneous treatment with cold plasma and ultraviolet C. *Innovative Food Science and Emerging Technologies, 63*, E-article ID 102392. https://doi.org/10.1016/j.ifset.2020.102392.
21. Bao, Y. (2020). Innovative cold plasma-assisted extraction for bioactive compounds from agricultural byproducts (PhD Dissertation, Purdue University Graduate School). Retrieved from: https://hammer.purdue.edu/articles/thesis/

INNOVATIVE_COLD_PLASMA-ASSISTED_EXTRACTION_FOR_BIOACTIVE_COMPOUNDS_FROM_AGRICULTURAL_BYPRODUCTS/12250031/files/22537967.pdf (accessed on 16 June 2024).
22. Barba, F. J., Esteve, M. J., & Frígola, A. (2012). High pressure treatment effect on physicochemical and nutritional properties of fluid foods during storage: A review. *Comprehensive Reviews in Food Science and Food Safety, 11*(3), 307–322.
23. Barbosa-Cánovas, G. V., Medina-Meza, I., Candoğan, K., & Bermúdez-Aguirre, D. (2014). Advanced retorting, microwave assisted thermal sterilization (MATS), and pressure assisted thermal sterilization (PATS) to process meat products. *Meat Science, 98*(3), 420–434.
24. Basaran, P., Basaran-Akgul, N., & Oksuz, L. (2008). Elimination of Aspergillus parasiticus from nut surface with low pressure cold plasma (LPCP) treatment. *Food Microbiology, 25*(4), 626–632.
25. Bauer, A., Ni, Y., Bauer, S., Paulsen, P., & Modic, M. (2017). The effects of atmospheric pressure cold plasma treatment on microbiological, physical-chemical and sensory characteristics of vacuum packaged beef loin. *Meat Science, 128*, 77–87.
26. Becker, K. H., Kogelschatz, U., Schoenbach, K. H., & Barker, R. J. (2004). *Non-Equilibrium Air Plasmas at Atmospheric Pressure*, pp. 1–700, https://doi.org/10.1201/9781482269123. CRC Press. ISBN 9780429175084.
27. Behera, S., Nagarajan, S., & Jagan Mohan Rao, L. (2004). Microwave heating and conventional roasting of cumin seeds (*Cuminum cyminum* L.) and effect on chemical composition of volatiles. *Food Chemistry, 87*(1), 25–29.
28. Behsnilian, D., Butz, P., Greiner, R., & Lautenschlaeger, R. (2014). Process-induced undesirable compounds: Chances of non-thermal approaches. *Meat Science, 98*(3), 392–403.
29. Berardinelli, A., Vannini, L., Ragni, L., & Guerzoni, M. E. (2012). Impact of atmospheric plasma generated by a DBD device on quality-related attributes of "abate fetel" pear fruit. In NATO Science for Peace and Security Series A: Chemistry and Biology; Springer. https://doi.org/10.1007/978-94-007-2852-3_35.
30. Bermudez-Aguirre, D. (2019). *Advances in Cold Plasma Applications for Food Safety and Preservation*, ISBN:9780128149225, pp. 1–410. Academic Press.
31. Bhide, S., Salvi, D., Schaffner, D. W., & Karwe, M. V. (2017). Effect of surface roughness in model and fresh fruit systems on microbial inactivation efficacy of cold atmospheric pressure plasma. *Journal of Food Protection, 80*(8), 1337–1346.
32. Bourke, P., Ziuzina, D., Boehm, D., Cullen, P. J., & Keener, K. (2018). The potential of cold plasma for safe and sustainable food production. *Trends in Biotechnology, 36*(6), 615–626.
33. Brandenburg, R., Navrátil, Z., Jánský, J., St'Ahel, P., Trunec, D., & Wagner, H. E. (2009). The transition between different modes of barrier discharges at atmospheric pressure. *Journal of Physics D: Applied Physics, 42*(8), 1–11. https://doi.org/10.1088/0022-3727/42/8/085208.
34. Bruggeman, P., Iza, F., Guns, P., Lauwers, D., & Kong, M. G. (2010). Electronic quenching of OH(A) by water in atmospheric pressure plasmas and its influence on the gas temperature determination by OH(A-X) emission. *Plasma Sources Science and Technology, 19*(1), 015016. https://doi.org/10.1088/0963-0252/19/1/015016.

35. Bruggeman, P. J., Iza, F., & Brandenburg, R. (2017). Foundations of atmospheric pressure non-equilibrium plasmas. *Plasma Sources Science and Technology, 26*(12), aa97af. https://doi.org/10.1088/1361-6595/aa97af.
36. Buettner, A., & Beauchamp, J. (2010). Chemical input – Sensory output: Diverse modes of physiology–flavor interaction. *Food Quality and Preference, 21*(8), 915–924.
37. Burm, K. T. A. L. (2012). Plasma: The fourth state of matter. *Plasma Chemistry and Plasma Processing, 32*(2), 401–407.
38. Bußler, S. (2017). Cold atmospheric pressure plasma treatment of food matrices (M.Sc. Thesis). pp. 1–120, Publisher: Technische Universitat Berlin, https://doi.org/10.14279/depositonce-5709
39. Bußler, S., Ehlbeck, J., & Schlüter, O. K. (2017). Pre-drying treatment of plant related tissues using plasma processed air: Impact on enzyme activity and quality attributes of cut apple and potato. *Innovative Food Science and Emerging Technologies, 40*, 78–86.
40. Chabert, P., & Braithwaite, N. (2011). *Physics of Radio-Frequency Plasmas*. Cambridge University Press. pp. 1–383, ISBN:9780511974342, DOI: https://doi.org/10.1017/CBO9780511974342.
41. Chaplot, S., Yadav, B., Jeon, B., & Roopesh, M. S. (2019). Atmospheric cold plasma and peracetic acid-based hurdle intervention to reduce Salmonella on raw poultry meat. *Journal of Food Protection, 82*(5), 878–888.
42. Chen, C., Liu, C., Jiang, A., Guan, Q., & Sun, X. (2019). The effects of cold plasma-activated water treatment on the microbial growth and antioxidant properties of fresh-cut pears. *Food and Bioprocess Technology, 12*(11), 1842–1851.
43. Chen, D., Peng, P., Zhou, N., Cheng, Y., & Min, M. (2019). Evaluation of *Cronobacter sakazakii* inactivation and physicochemical property changes of non-fat dry milk powder by cold atmospheric plasma. *Food Chemistry, 290*, 270–276.
44. Chen, H. H. (2014). Investigation of properties of long-grain brown rice treated by low-pressure plasma. *Food and Bioprocess Technology, 7*(9), 2484–2491.
45. Chen, H. H., Chang, H. C., Chen, Y. K., Hung, C. L., Lin, S. Y., & Chen, Y. S. (2016). An improved process for high nutrition of germinated brown rice production: Low-pressure plasma. *Food Chemistry, 191*, 120–127.
46. Chen, H. H., Chen, Y. K., & Chang, H. C. (2012). Evaluation of physicochemical properties of plasma treated brown rice. *Food Chemistry, 135*(1), 74–79.
47. Chen, J., & Rosenthal, A. (2015). Food texture and structure. In *Modifying Food Texture: Novel Ingredients and Processing Techniques*, pp. 3–24. Elsevier. https://doi.org/10.1016/B978-1-78242-333-1.00001-2.
48. Chen, J., Wang, S. Z., Chen, J. Y., Chen, D. Z., Deng, S. G., & Xu, B. (2019). Effect of cold plasma on maintaining the quality of chub mackerel (*Scomber japonicus*): Biochemical and sensory attributes. *Journal of the Science of Food and Agriculture, 99*(1), 39–46.
49. Chen, Y., Ross, W. H., Scott, V. N., & Gombas, D. E. (2003). Listeria monocytogenes: Low levels equal low risk. *Journal of Food Protection, 66*(4), 570–577.
50. Chen, Y., Zhang, Y., Jiang, L., Chen, G., & Yu, J. (2020). Moisture molecule migration and quality changes of fresh wet noodles dehydrated by cold plasma treatment. *Food Chemistry, 328*, 127053. https://doi.org/10.1016/j.foodchem.2020.127053.
51. Chiper, A. S., Chen, W., Mejlholm, O., Dalgaard, P., & Stamate, E. (2011). Atmospheric pressure plasma produced inside a closed package by a dielectric barrier discharge in

Ar/CO_2 for bacterial inactivation of biological samples. *Plasma Sources Science and Technology, 20*(2), 025008. https://doi.org/10.1088/0963-0252/20/2/025008.

52. Chizoba Ekezie, F. G., Sun, D. W., & Cheng, J. H. (2017). A review on recent advances in cold plasma technology for the food industry: Current applications and future trends. *Trends in Food Science & Technology, 69*, 46–58.
53. Choi, S., Puligundla, P., & Mok, C. (2016). Microbial decontamination of dried Alaska pollock shreds using corona discharge plasma jet: Effects on physicochemical and sensory characteristics. *Journal of Food Science, 81*(4), M952–M957.
54. Conrads, H., & Schmidt, M. (2000). Plasma generation and plasma sources. *Plasma Sources Science and Technology, 9*(4), 441–454.
55. Coutinho, N. M., Silveira, M. R., Rocha, R. S., Moraes, J., & Ferreira, M. V. S. (2018). Cold plasma processing of milk and dairy products. *Trends in Food Science & Technology, 74*, 56–68.
56. Critzer, F. J., Kelly-Wintenberg, K., South, S. L., & Golden, D. A. (2007). Atmospheric plasma inactivation of foodborne pathogens on fresh produce surfaces. *Journal of Food Protection, 70*(10), 2290–2296.
57. Cullen, P. J., Lalor, J., Scally, L., Boehm, D., & Milosavljević, V. (2018). Translation of plasma technology from the lab to the food industry. *Plasma Processes and Polymers, 15*(2), 00085. https://doi.org/10.1002/ppap.201700085.
58. Darvish, H., Ramezan, Y., Khani, M. R., & Kamkari, A. (2022). Effect of low-pressure cold plasma processing on decontamination and quality attributes of saffron (*Crocus sativus* L.). *Food Science & Nutrition, 10*(6), 2082–2090.
59. Dasan, B. G., Boyaci, I. H., & Mutlu, M. (2017). Nonthermal plasma treatment of Aspergillus spp. spores on hazelnuts in an atmospheric pressure fluidized bed plasma system: Impact of process parameters and surveillance of the residual viability of spores. *Journal of Food Engineering, 196*, 139–149.
60. De Geyter, N., & Morent, R. (2012). Nonthermal plasma sterilization of living and nonliving surfaces. *Annual Review of Biomedical Engineering, 14*, 255–274.
61. De La Hera, E., Gomez, M., & Rosell, C. M. (2013). Particle size distribution of rice flour affecting the starch enzymatic hydrolysis and hydration properties. *Carbohydrate Polymers, 98*(1), 421–427.
62. De Souza Silva, D. A., Da Silva Campelo, M. C., De Oliveira Soares Rebouças, L., De Oliveira Vitoriano, J., & Junior, C. A. (2019). Use of cold atmospheric plasma to preserve the quality of white shrimp (*Litopenaeus vannamei*). *Journal of Food Protection, 82*(7), 1217–1223.
63. Deng, S., Ruan, R., Mok, C. K., Huang, G., Lin, X., & Chen, P. (2007). Inactivation of *Escherichia coli* on almonds using nonthermal plasma. *Journal of Food Science, 72*(2), M62–M66.
64. Desmet, B. G. (2011). Surface modification of polymers using atmospheric pressure microplasmas. In *Polymer Science: Research Advances, Practical Applications and Educational Aspects*, pp. 1–170, Publisher: Ghent University Library. ISBN 978-84-942134-8-9.
65. Dirks, B. P., Dobrynin, D., Fridman, G., Mukhin, Y., Fridman, A., & Quinlan, J. J. (2012). Treatment of raw poultry with nonthermal dielectric barrier discharge plasma to reduce *Campylobacter jejuni* and *Salmonella enterica*. *Journal of Food Protection, 75*(1), 22–28.

66. Dong, X. Y., & Yang, Y. L. (2019). A novel approach to enhance blueberry quality during storage using cold plasma at atmospheric air pressure. *Food and Bioprocess Technology, 12*(8), 1409–1421.
67. Donnelly, C. W. (1998). Factors associated with the microbiological safety of cheeses prepared from raw milk: A review. *Bulletin of the International Dairy Federation, 16*-27.
68. Ehiasarian, A. P., New, R., Münz, W. D., Hultman, L., Helmersson, U., & Kouznetsov, V. (2002). Influence of high-power densities on the composition of pulsed magnetron plasmas. *Vacuum, 65*(2), 147–154.
69. Ehlbeck, J., Schnabel, U., Polak, M., Winter, J., & Von Woedtke, T. (2011). Low temperature atmospheric pressure plasma sources for microbial decontamination. *Journal of Physics D: Applied Physics, 44*(1), 013002. https://doi.org/10.1088/0022-3727/44/1/013002.
70. Ekezie, F. G. C., Sun, D. W., & Cheng, J. H. (2019). Altering the IgE binding capacity of king prawn (*Litopenaeus vannamei*) tropomyosin through conformational changes induced by cold argon-plasma jet. *Food Chemistry, 300*, 125143. https://doi.org/10.1016/j.foodchem.2019.125143.
71. Falkenstein, Z. (1997). The influence of ultraviolet illumination on OH formation in dielectric barrier discharges of Ar/O_2/H_2O: The Joshi effect. *Journal of Applied Physics, 81*(11), 7158–7162.
72. Farhaninejad, Z., Fathi, M., Shahedi, M., & Sadeghi, M. (2017). Osmotic dehydration of banana slices using direct and indirect sonication: Optimization and microstructure analysis. *Journal of Food Process Engineering, 40*(1), e12336. https://doi.org/10.1111/jfpe.12336.
73. Faria, G. Y. Y., Souza, M. M., Oliveira, J. R. M., Costa, C. S. B., Collares, M. P., & Prentice, C. (2020). Effect of ultrasound-assisted cold plasma pretreatment to obtain sea asparagus extract and its application in Italian salami. *Food Research International, 137*, 109435. https://doi.org/10.1016/j.foodres.2020.109435.
74. Feizollahi, E., Misra, N. N., & Roopesh, M. S. (2020). Factors influencing the antimicrobial efficacy of dielectric barrier discharge (DBD) atmospheric cold plasma (ACP) in food processing applications. *Critical Reviews in Food Science and Nutrition, 61*(4), 666–689.
75. Feng, X., Ma, X., Liu, H., Xie, J., He, C., & Fan, R. (2019). Argon plasma effects on maize: Pesticide degradation and quality changes. *Journal of the Science of Food and Agriculture, 99*(12), 5491–5498.
76. Fernández, A., Noriega, E., & Thompson, A. (2013). Inactivation of Salmonella enterica serovar Typhimurium on fresh produce by cold atmospheric gas plasma technology. *Food Microbiology, 33*(1), 24–29.
77. Flamm, D. L., Donnelly, V. M., & Ibbotson, D. E. (1983). Basic chemistry and mechanisms of plasma etching. *Semiconductor International, 1*(4), 136–143.
78. Fridman, A., Chirokov, A., & Gutsol, A. (2005). Non-thermal atmospheric pressure discharges. *Journal of Physics D: Applied Physics, 38*(2), R1–R24. https://doi.org/10.1088/0022-3727/38/2/R01.
79. Fröhling, A., Durek, J., Schnabel, U., Ehlbeck, J., Bolling, J., & Schlüter, O. (2012). Indirect plasma treatment of fresh pork: Decontamination efficiency and effects on quality attributes. *Innovative Food Science & Emerging Technologies, 16*, 381–390.
80. Gani, A., Gazanfar, T., Jan, R., Wani, S. M., & Masoodi, F. A. (2013). Effect of gamma irradiation on the physicochemical and morphological properties of starch extracted

from lotus stem harvested from Dal Lake of Jammu and Kashmir, India. *Journal of the Saudi Society of Agricultural Sciences, 12*(2), 109–115.

81. Gao, S., Liu, H., Sun, L., Liu, N., & Wang, J. (2019). The effects of dielectric barrier discharge plasma on physicochemical and digestion properties of starch. *International Journal of Biological Macromolecules, 138*, 819–830.
82. García, S., Iracheta, F., Galván, F., & Heredia, N. (2001). Microbiological survey of retail herbs and spices from Mexican markets. *Journal of Food Protection, 64*(1), 99–103.
83. Gaunt, L. F., Beggs, C. B., & Georghiou, G. E. (2006). Bactericidal action of the reactive species produced by gas-discharge nonthermal plasma at atmospheric pressure: A review. *IEEE Transactions on Plasma Science, 34*(4), 1257–1269.
84. Gavahian, M., Peng, H. J., & Chu, Y. H. (2019). Efficacy of cold plasma in producing Salmonella-free duck eggs: Effects on physical characteristics, lipid oxidation, and fatty acid profile. *Journal of Food Science and Technology, 56*(12), 5271–5281.
85. Gavahian, M., Sheu, F. H., Tsai, M. J., & Chu, Y. H. (2020). The effects of dielectric barrier discharge plasma gas and plasma-activated water on texture, color, and bacterial characteristics of shiitake mushroom. *Journal of Food Processing and Preservation, 44*(1), e14316. https://doi.org/10.1111/jfpp.14316.
86. Georgescu, N., Apostol, L., & Gherendi, F. (2017). Inactivation of Salmonella enterica serovar Typhimurium on egg surface, by direct and indirect treatments with cold atmospheric plasma. *Food Control, 76*, 52–61.
87. Ghomi, H., Safa, N. N., Ramezani, K., & Karimi, M. (2009). Bacterial inactivation by DBD plasma in atmospheric pressure. In *19th International Symposium on Plasma Chemistry* (ISPC19), 81(1), 3–5.
88. Gök, V., Aktop, S., Özkan, M., & Tomar, O. (2019). The effects of atmospheric cold plasma on inactivation of *Listeria monocytogenes* and *Staphylococcus aureus* and some quality characteristics of pastırma—A dry-cured beef product. *Innovative Food Science & Emerging Technologies, 56*, 102188. https://doi.org/10.1016/j.ifset.2019.102188.
89. Gordillo-Vázquez, F. J. (2008). Air plasma kinetics under the influence of sprites. *Journal of Physics D: Applied Physics, 41*(23), 234016. https://doi.org/10.1088/0022-3727/41/23/234016.
90. Govaert, M., Smet, C., Graeffe, A., Walsh, J. L., & Van Impe, J. F. M. (2020). Inactivation of *L. monocytogenes* and *S. Typhimurium* biofilms by means of an air-based cold atmospheric plasma (CAP) system. *Foods, 9*(2), 157–165.
91. Grasso, S., Brunton, N. P., Lyng, J. G., Lalor, F., & Monahan, F. J. (2014). Healthy processed meat products—Regulatory, reformulation and consumer challenges. *Trends in Food Science & Technology, 39*(1), 4–17.
92. Grzegorzewski, F., Rohn, S., Kroh, L. W., Geyer, M., & Schlüter, O. (2010). Surface morphology and chemical composition of lamb's lettuce (*Valerianella locusta*) after exposure to a low-pressure oxygen plasma. *Food Chemistry, 122*(4), 1145–1152.
93. Gudipati, V. (2017). Role of plant extracts as natural additives in fish and fish products: A review. *Fishery Technology, 54*, 145–154.
94. Hajhoseini, A., Sharifan, A., & Yousefi, H. R. (2020). Effects of atmospheric cold plasma on microbial growth of *Listeria innocua* and *Staphylococcus aureus* in ready-to-eat fish products. *Iranian Journal of Fisheries Sciences, 19*(1), 262–271.
95. Han, L., Patil, S., Boehm, D., Milosavljević, V., Cullen, P. J., & Bourke, P. (2016). Mechanisms of inactivation by high-voltage atmospheric cold plasma differ for

Escherichia coli and *Staphylococcus aureus*. *Applied and Environmental Microbiology, 82*(2), 450–458.

96. Han, S. H., Suh, H. J., Hong, K. B., Kim, S. Y., & Min, S. C. (2016). Oral toxicity of cold plasma-treated edible films for food coating. *Journal of Food Science, 81*(12), T3052–T3057.
97. Han, Y. X., Cheng, J. H., & Sun, D. W. (2019). Changes in activity, structure, and morphology of horseradish peroxidase induced by cold plasma. *Food Chemistry, 301*, 125240. https://doi.org/10.1016/j.foodchem.2019.125240.
98. Henselová, M., Slováková, Ľ., Martinka, M., & Zahoranová, A. (2012). Growth, anatomy, and enzyme activity changes in maize roots induced by treatment of seeds with low-temperature plasma. *Biologia, 67*(3), 490–497.
99. Heo, N. S., Lee, M. K., Kim, G. W., Lee, S. J., Park, J. Y., & Park, T. J. (2014). Microbial inactivation and pesticide removal by remote exposure of atmospheric air plasma in confined environments. *Journal of Bioscience and Bioengineering, 117*(1), 81–85.
100. Hertwig, C., Leslie, A., Meneses, N., Reineke, K., Rauh, C., & Schlüter, O. (2017). Inactivation of *Salmonella enteritidis* PT30 on the surface of unpeeled almonds by cold plasma. *Innovative Food Science & Emerging Technologies, 44*, 242–248.
101. Hertwig, C., Reineke, K., Ehlbeck, J., Knorr, D., & Schlüter, O. (2015). Decontamination of whole black pepper using different cold atmospheric pressure plasma applications. *Food Control, 55*, 221–229.
102. Hong, Y. F., Kang, J. G., Lee, H. Y., Uhm, H. S., Moon, E., & Park, Y. H. (2009). Sterilization effect of atmospheric plasma on *Escherichia coli* and *Bacillus subtilis* endospores. *Letters in Applied Microbiology, 48*(1), 33–37.
103. Hosseini, S. I., Mohsenimehr, S., Hadian, J., Ghorbanpour, M., & Shokri, B. (2018). Physico-chemical induced modification of seed germination and early development in artichoke (*Cynara scolymus* L.) using low energy plasma technology. *Physics of Plasmas, 25*(1), 135–145.
104. Hu, M., & Guo, Y. (2012). The effect of air plasma on sterilization of *Escherichia coli* in dielectric barrier discharge. *Plasma Science and Technology, 14*(8), 735–740.
105. Hu, Y., Bai, Y., Li, X., & Chen, J. (2013). Application of dielectric barrier discharge plasma for degradation and pathways of dimethoate in aqueous solution. *Separation and Purification Technology, 120*, 191–197.
106. Huang, C. C., Wu, J. S. B., Wu, J. S., & Ting, Y. (2019). Effect of novel atmospheric-pressure jet pretreatment on the drying kinetics and quality of white grapes. *Journal of the Science of Food and Agriculture, 99*(11), 5102–5111.
107. Hui, Y. H. (2006). *Handbook of Fruits and Fruit Processing* (Vol. 129). Blackwell Publishing. pp. 1–697, ISBN:9780813819815 DOI:10.1002/9780470277737.
108. Inguglia, E. S., Oliveira, M., Burgess, C. M., Kerry, J. P., & Tiwari, B. K. (2020). Plasma-activated water as an alternative nitrite source for the curing of beef jerky: Influence on quality and inactivation of *Listeria innocua*. *Innovative Food Science & Emerging Technologies, 59*, 102276. https://doi.org/10.1016/j.ifset.2019.102276.
109. Jahanmiri, A., Rahimpour, M. R., Mohamadzadeh Shirazi, M., Hooshmand, N., & Taghvaei, H. (2012). Naphtha cracking through a pulsed DBD plasma reactor: Effect of applied voltage, pulse repetition frequency and electrode material. *Chemical Engineering Journal, 191*, 416–425.

110. Jayasena, D. D., Kim, H. J., Yong, H. I., Park, S., & Kim, K. (2015). Flexible thin-layer dielectric barrier discharge plasma treatment of pork butt and beef loin: Effects on pathogen inactivation and meat-quality attributes. *Food Microbiology, 46*, 51–57.
111. Joshi, S. G., Cooper, M., Yost, A., Paff, M., & Ercan, U. K. (2011). Nonthermal dielectric-barrier discharge plasma-induced inactivation involves oxidative DNA damage and membrane lipid peroxidation in *Escherichia coli*. *Antimicrobial Agents and Chemotherapy, 55*(3), 1053–1062.
112. Jung, S., Kim, H. J., Park, S., In Yong, H., Choe, J. H. (2015). The use of atmospheric pressure plasma-treated water as a source of nitrite for emulsion-type sausage. *Meat Science, 108*, 132–137.
113. Jung, S., Lee, J., Lim, Y., Choe, W., Yong, H. I., & Jo, C. (2017). Direct infusion of nitrite into meat batter by atmospheric pressure plasma treatment. *Innovative Food Science & Emerging Technologies, 39*, 113–118.
114. Keshavarzi, M., Najafi, G., Ahmadi Gavlighi, H., Seyfi, P., & Ghomi, H. (2020). Enhancement of polyphenolic content extraction rate with maximal antioxidant activity from green tea leaves by cold plasma. *Journal of Food Science, 85*(10), 3415–3422.
115. Khorram, S., Zakerhamidi, M. S., & Karimzadeh, Z. (2015). Polarity functions' characterization and the mechanism of starch modification by DC glow discharge plasma. *Carbohydrate Polymers, 127*, 72–78.
116. Kim, C., Lee, T., Puligundla, P., & Mok, C. (2020). Effect of relative humidity on the inactivation of foodborne pathogens by corona discharge plasma jet (CDPJ). *LWT, 127*, 109379. https://doi.org/10.1016/j.lwt.2020.109379.
117. Kim, H. J., Yong, H. I., Park, S., Choe, W., & Jo, C. (2013). Effects of dielectric barrier discharge plasma on pathogen inactivation and the physicochemical and sensory characteristics of pork loin. *Current Applied Physics, 13*(7), 1420–1425.
118. Kim, J. E., Choi, H. S., Lee, D. U., & Min, S. C. (2017). Effects of processing parameters on the inactivation of Bacillus cereus spores on red pepper (*Capsicum annum* L.) flakes by microwave-combined cold plasma treatment. *International Journal of Food Microbiology, 263*, 61–66.
119. Kim, J. E., Oh, Y. J., Won, M. Y., Lee, K. S., & Min, S. C. (2017). Microbial decontamination of onion powder using microwave-powered cold plasma treatments. *Food Microbiology, 62*, 112–123.
120. Kim, J. H., & Min, S. C. (2018). Moisture vaporization-combined helium dielectric barrier discharge-cold plasma treatment for microbial decontamination of onion flakes. *Food Control, 84*, 321–329.
121. Kim, J. S., Lee, E. J., Choi, E. H., & Kim, Y. J. (2014). Inactivation of *Staphylococcus aureus* on the beef jerky by radio-frequency atmospheric pressure plasma discharge treatment. *Innovative Food Science & Emerging Technologies, 22*, 124–130.
122. Kim, Y. H., Lee, C., Lee, S. J., & Yoon, K. S. (2022). Reduction of E. coli O157: H7 and Bacillus cereus levels in red pepper powder using dielectric barrier discharge (DBD) plasma for enhanced quality. *Innovative Food Science & Emerging Technologies, 76*, 102916. https://doi.org/10.1016/j.ifset.2022.102916.
123. Koch, J., Dworak, R., Prager, R., Becker, B., & Brockmann, S. (2010). Large listeriosis outbreak linked to cheese made from pasteurized milk, Germany, 2006-2007. *Foodborne Pathogens and Disease, 7*(12), 1581–1584.

124. Kodama, S., Thawatchaipracha, B., & Sekiguchi, H. (2014). Enhancement of essential oil extraction for steam distillation by DBD surface treatment. *Plasma Processes and Polymers, 11*(2), 126–132.
125. Koddy, J. K., Miao, W., Hatab, S., Tang, L., & Xu, H. (2021). Understanding the role of atmospheric cold plasma (ACP) in maintaining the quality of hairtail (*Trichiurus lepturus*). *Food Chemistry, 343*, 128418. https://doi.org/10.1016/j.foodchem.2020.128418.
126. Kogelschatz, U. (2003). Dielectric-barrier discharges: Their history, discharge physics, and industrial applications. *Plasma Chemistry and Plasma Processing, 23*(1), 1–46.
127. Kumar Mahnot, N., Siyu, L. P., Wan, Z., Keener, K. M., & Misra, N. N. (2020). In-package cold plasma decontamination of fresh-cut carrots: Microbial and quality aspects. *Journal of Physics D: Applied Physics, 53*(15), 155301. https://doi.org/10.1088/1361-6463/ab6cd3.
128. Laca, A., Mousia, Z., Díaz, M., Webb, C., & Pandiella, S. S. (2006). Distribution of microbial contamination within cereal grains. *Journal of Food Engineering, 72*(4), 332–338.
129. Lacombe, A., Niemira, B. A., Gurtler, J. B., Fan, X., & Sites, J. (2015). Atmospheric cold plasma inactivation of aerobic microorganisms on blueberries and effects on quality attributes. *Food Microbiology, 46*, 479–484.
130. Laovachirasuwan, P., Peerapattana, J., Srijesdaruk, V., Chitropas, P., & Otsuka, M. (2010). The physicochemical properties of a spray dried glutinous rice starch biopolymer. *Colloids and Surfaces B: Biointerfaces, 78*(1), 30–35.
131. Laroussi, M., & Leipold, F. (2004). Evaluation of the roles of reactive species, heat, and UV radiation in the inactivation of bacterial cells by air plasmas at atmospheric pressure. *International Journal of Mass Spectrometry, 233*(1–3), 81–86.
132. Lee, E. S., Cheigh, C. I., Kang, J. H., Lee, S. Y., & Min, S. C. (2020). Evaluation of in-package atmospheric dielectric barrier discharge cold plasma treatment as an intervention technology for decontaminating bulk ready-to-eat chicken breast cubes in plastic containers. *Applied Sciences, 10*(18), 6301. https://doi.org/10.3390/app10186301.
133. Lee, H. J., Jung, S., Jung, H. S., Park, S. H., & Choe, W. H. (2012). Evaluation of a dielectric barrier discharge plasma system for inactivating pathogens on cheese slices. *Journal of Animal Science and Technology, 54*(3), 191–198.
134. Lee, H., Yong, H. I., Kim, H. J., Choe, W., & Yoo, S. J. (2016). Evaluation of the microbiological safety, quality changes, and genotoxicity of chicken breast treated with flexible thin-layer dielectric barrier discharge plasma. *Food Science and Biotechnology, 25*(4), 1189–1195.
135. Lee, J., Jo, K., Lim, Y., Jeon, H. J., & Choe, J. H. (2018). The use of atmospheric pressure plasma as a curing process for canned ground ham. *Food Chemistry, 240*, 430–436.
136. Lee, K. H., Kim, H. J., Woo, K. S., Jo, C., & Kim, J. K. (2016). Evaluation of cold plasma treatments for improved microbial and physicochemical qualities of brown rice. *LWT, 73*, 442–447.
137. Li, X., Li, M., Ji, N., Jin, P., & Zhang, J. (2019). Cold plasma treatment induces phenolic accumulation and enhances antioxidant activity in fresh-cut pitaya (*Hylocereus undatus*) fruit. *LWT, 115*, 108447. https://doi.org/10.1016/j.lwt.2019.108447.
138. Liao, X., Li, J., Muhammad, A. I., Suo, Y., & Chen, S. (2018). Application of a dielectric barrier discharge atmospheric cold plasma (DBD-ACP) for *Escherichia coli* inactivation in Apple Juice. *Journal of Food Science, 83*(2), 401–408.

139. Liao, X., Liu, D., Xiang, Q., Ahn, J., & Chen, S. (2017). Inactivation mechanisms of non-thermal plasma on microbes: A review. *Food Control, 75*, 83–91.
140. Liao, X., Xiang, Q., Liu, D., Chen, S., Ye, X., & Ding, T. (2017). Lethal and sublethal effect of a dielectric barrier discharge atmospheric cold plasma on Staphylococcus aureus. *Journal of Food Protection, 80*(6), 928–932.
141. Lii, C. Y., Liao, C. D., Stobinski, L., & Tomasik, P. (2002). Effects of hydrogen, oxygen, and ammonia low-pressure glow plasma on granular starches. *Carbohydrate Polymers, 49*(4), 449–456.
142. López-Gámez, G., Elez-Martínez, P., Martín-Belloso, O., & Soliva-Fortuny, R. (2020). Enhancing phenolic content in carrots by pulsed electric fields during post-treatment time: Effects on cell viability and quality attributes. *Innovative Food Science & Emerging Technologies, 59*, 102252. https://doi.org/10.1016/j.ifset.2019.102252.
143. Los, A., Ziuzina, D., Akkermans, S., Boehm, D., & Cullen, P. J. (2018). Improving microbiological safety and quality characteristics of wheat and barley by high voltage atmospheric cold plasma closed processing. *Food Research International, 106*, 509–521.
144. Los, A., Ziuzina, D., Boehm, D., Cullen, P. J., & Bourke, P. (2017). The potential of atmospheric air cold plasma for control of bacterial contaminants relevant to cereal grain production. *Innovative Food Science & Emerging Technologies, 44*, 36–45.
145. Lu, H., Patil, S., Keener, K. M., Cullen, P. J., & Bourke, P. (2014). Bacterial inactivation by high-voltage atmospheric cold plasma: Influence of process parameters and effects on cell leakage and DNA. *Journal of Applied Microbiology, 116*(4), 784–794.
146. Lu, P., Cullen, P. J., & Ostrikov, K. (2016). Atmospheric pressure nonthermal plasma sources. In *Cold Plasma in Food and Agriculture* (pp. 83–116). Elsevier.
147. Lu, Q., Liu, D., Song, Y., Zhou, R., & Niu, J. (2014). Inactivation of the tomato pathogen *Cladosporium fulvum* by an atmospheric-pressure cold plasma jet. *Plasma Processes and Polymers, 11*(11), 1028–1036.
148. Lu, X., Ye, T., Cao, Y., Sun, Z., & Xiong, Q. (2008). The roles of the various plasma agents in the inactivation of bacteria. *Journal of Applied Physics, 104*(5), 053309. https://doi.org/10.1063/1.2977674.
149. Lung, H. M., Cheng, Y. C., Chang, Y. H., Huang, H. W., Yang, B. B., & Wang, C. Y. (2015). Microbial decontamination of food by electron beam irradiation. *Trends in Food Science & Technology, 44*(1), 66–78.
150. Ma, R., Yu, S., Tian, Y., Wang, K., & Sun, C. (2016). Effect of non-thermal plasma-activated water on fruit decay and quality in postharvest Chinese bayberries. *Food and Bioprocess Technology, 9*(11), 1825–1834.
151. Mahendran, R., Ratish Ramanan, K., Sargunam, R., & Sarumathi, R. (2016). Effect of cold plasma on mortality of *Tribolium castaneum* on Maida flour. *Journal of Stored Products Research, 77*, 126–134.
152. Mao, L., Mhaske, P., Zing, X., Kasapis, S., Majzoobi, M., & Farahnaky, A. (2021). Cold plasma: Microbial inactivation and effects on quality attributes of fresh and minimally processed fruits and ready-to-eat vegetables. *Trends in Food Science & Technology, 116*, 146–175.
153. Mastanaiah, N., Johnson, J. A., & Roy, S. (2013). Effect of dielectric and liquid on plasma sterilization using dielectric barrier discharge plasma. *PLoS ONE, 8*(8), e70840. https://doi.org/10.1371/journal.pone.0070840.

154. McDonald, M. B. (2015). Seed germination and seedling establishment. In *Physiology and Determination of Crop Yield* (pp. 53–84). https://doi.org/10.2134/1994.physiologyanddetermination.c3.
155. Médard, N., Soutif, J. C., & Poncin-Epaillard, F. (2002). Characterization of CO_2 plasma-treated polyethylene surface bearing carboxylic groups. *Surface and Coatings Technology, 160*(2–3), 197–205.
156. Mei, J., Liu, G., Huang, X., & Ding, W. (2016). Effects of ozone treatment on medium hard wheat (*Triticum aestivum* L.) flour quality and performance in steamed bread making. *CYTA – Journal of Food, 14*(3), 449–456.
157. Menashi, W. (1968). Treatment of surfaces. pp. 1–6, Patent no as: US3383163A, URL: https://patentimages.storage.googleapis.com/7a/44/9e/59f851ab3001ad/US3383163.pdf.
158. Mir, S. A., Shah, M. A., & Mir, M. M. (2016). Understanding the role of plasma technology in the food industry. *Food and Bioprocess Technology, 9*(5), 734–750.
159. Mir, S. A., Siddiqui, M. W., Dar, B. N., Shah, M. A., & Wani, M. H. (2020). Promising applications of cold plasma for microbial safety, chemical decontamination, and quality enhancement in fruits. *Journal of Applied Microbiology, 129*(3), 474–485.
160. Misra, N. N., & Jo, C. (2017). Applications of cold plasma technology for microbiological safety in the meat industry. *Trends in Food Science & Technology, 64*, 74–86.
161. Misra, N. N., Kaur, S., Tiwari, B. K., Kaur, A., Singh, N., & Cullen, P. J. (2015). Atmospheric pressure cold plasma (ACP) treatment of wheat flour. *Food Hydrocolloids, 44*, 115–121.
162. Misra, N. N., Keener, K. M., Bourke, P., Mosnier, J. P., & Cullen, P. J. (2014). In-package atmospheric pressure cold plasma treatment of cherry tomatoes. *Journal of Bioscience and Bioengineering, 118*(2), 177–182.
163. Misra, N. N., Pankaj, S. K., Segat, A., & Ishikawa, K. (2016). Cold plasma interactions with enzymes in foods and model systems. *Trends in Food Science & Technology, 55*, 39–47.
164. Misra, N. N., Pankaj, S. K., Walsh, T., O'Regan, F., Bourke, P., & Cullen, P. J. (2014). In-package nonthermal plasma degradation of pesticides on fresh produce. *Journal of Hazardous Materials, 271*, 33–40.
165. Misra, N. N., Schlüter, O., & Cullen, P. J. (2016). *Cold Plasma in Food and Agriculture: Fundamentals and Applications*. Academic Press. pp. 1–368, ISBN: 978-0-12-801365-6, https://doi.org/10.1016/C2014-0-00009-3.
166. Misra, N. N., Tiwari, B. K., Raghavarao, K. S. M. S., & Cullen, P. J. (2011). Nonthermal plasma inactivation of food-borne pathogens. *Food Engineering Reviews, 3*(3–4), 159–170.
167. Misra, N. N., Yepez, X., Xu, L., & Keener, K. (2019). In-package cold plasma technologies. *Journal of Food Engineering, 244*, 21–31.
168. Møller, I. M., Jensen, P. E., & Hansson, A. (2007). Oxidative modifications to cellular components in plants. *Annual Review of Plant Biology, 58*, 459–481.
169. Morent, R., De Geyter, N., Desmet, T., Dubruel, P., & Leys, C. (2011). Plasma surface modification of biodegradable polymers: A review. *Plasma Processes and Polymers, 8*(3), 171–190.
170. Morgan, N., & Morgan, N. N. (2009). Atmospheric pressure dielectric barrier discharge chemical and biological applications modified polymer membrane using plasma technique for fuel cells applications view project atmospheric pressure dielectric barrier

discharge chemical and biological applications. *International Journal of Physical Sciences, 4*(13), 885–892.
171. Moutiq, R., Misra, N. N., Mendonça, A., & Keener, K. (2020). In-package decontamination of chicken breast using cold plasma technology: Microbial, quality and storage studies. *Meat Science, 159*, 107942. https://doi.org/10.1016/j.meatsci.2019.107942.
172. Muranyi, P., Wunderlich, J., & Heise, M. (2008). Influence of relative gas humidity on the inactivation efficiency of a low temperature gas plasma. *Journal of Applied Microbiology, 104*(6), 1659–1666.
173. Muranyi, P., Wunderlich, J., & Langowski, H. C. (2010). Modification of bacterial structures by a low-temperature gas plasma and influence on packaging material. *Journal of Applied Microbiology, 109*(6), 1875–1885.
174. Nguyen, T. B. T., Ketsa, S., & Van Doorn, W. G. (2004). Effect of modified atmosphere packaging on chilling-induced peel browning in banana. *Postharvest Biology and Technology, 31*(3), 313–317.
175. Niemira, B. A. (2012). Cold plasma reduction of Salmonella and *Escherichia coli* O157 on almonds using ambient pressure gases. *Journal of Food Science, 77*(3), M171–M175.
176. Nishime, T. M. C., Borges, A. C., Koga-Ito, C. Y., Machida, M., Hein, L. R. O., & Kostov, K. G. (2017). Non-thermal atmospheric pressure plasma jet applied to inactivation of different microorganisms. *Surface and Coatings Technology, 312*, 19–24.
177. Oehmigen, K., Hähnel, M., Brandenburg, R., Wilke, C., Weltmann, K. D., & Von Woedtke, T. (2010). The role of acidification for antimicrobial activity of atmospheric pressure plasma in liquids. *Plasma Processes and Polymers, 7*(3–4), 250–257.
178. Oh, Y. A., Roh, S. H., & Min, S. C. (2016). Cold plasma treatments for improvement of the applicability of defatted soybean meal-based edible film in food packaging. *Food Hydrocolloids, 58*, 150–159.
179. Oh, Y. J., Song, A. Y., & Min, S. C. (2017). Inhibition of Salmonella typhimurium on radish sprouts using nitrogen-cold plasma. *International Journal of Food Microbiology, 249*, 66–71.
180. Okyere, A. Y., Bertoft, E., & Annor, G. A. (2019). Modification of cereal and tuber waxy starches with radio frequency cold plasma and its effects on waxy starch properties. *Carbohydrate Polymers, 223*, 115075. https://doi.org/10.1016/j.carbpol.2019.115075.
181. Olatunde, O. O., & Benjakul, S. (2018). Nonthermal processes for shelf-life extension of seafoods: A revisit. *Comprehensive Reviews in Food Science and Food Safety, 17*(4), 892–904.
182. Oyebode, O., Gordon-Dseagu, V., Walker, A., & Mindell, J. S. (2014). Fruit and vegetable consumption and all-cause, cancer and CVD mortality: Analysis of health survey for England data. *Journal of Epidemiology and Community Health, 68*(9), 856–862.
183. Ozkan, A., Dufour, T., Bogaerts, A., & Reniers, F. (2016). How do the barrier thickness and dielectric material influence the filamentary mode and CO2 conversion in a flowing DBD? *Plasma Sources Science and Technology, 25*(4), 045016. https://doi.org/10.1088/0963-0252/25/4/045016.
184. Pankaj, S. K., Bueno-Ferrer, C., Misra, N. N., Milosavljević, V., & O'Donnell, C. P. (2014). Applications of cold plasma technology in food packaging. *Trends in Food Science & Technology, 35*(1), 5–17.
185. Pankaj, S. K., Bueno-Ferrer, C., Misra, N. N., O'Neill, L., Bourke, P., & Cullen, P. J. (2017). Effects of cold plasma on surface, thermal and antimicrobial release properties of chitosan film. *Journal of Renewable Materials, 5*(1), 14–20.

186. Pankaj, S. K., Kelly, C. A., Bueno-Ferrer, C., Kerry, J. P., & Papkovsky, D. B. (2016). Application of phosphorescent oxygen sensors in in-package dielectric barrier discharge plasma environment. *Innovative Food Science & Emerging Technologies, 33*, 234–239.
187. Pankaj, S. K., Wan, Z., Colonna, W., & Keener, K. M. (2017). Effect of high voltage atmospheric cold plasma on white grape juice quality. *Journal of the Science of Food and Agriculture, 97*(12), 4016–4021.
188. Park, S. Y., & Ha, S. D. (2015). Application of cold oxygen plasma for the reduction of *Cladosporium cladosporioides* and *Penicillium citrinum* on the surface of dried filefish (*Stephanolepis cirrhifer*) fillets. *International Journal of Food Science & Technology, 50*(4), 966–973.
189. Pasquali, F., Stratakos, A. C., Koidis, A., Berardinelli, A., & Cevoli, C. (2016). Atmospheric cold plasma process for vegetable leaf decontamination: A feasibility study on radicchio (*red chicory, Cichorium intybus L.*). *Food Control, 60*, 552–559.
190. Pei, X., Lu, X., Liu, J., Liu, D., & Yang, Y. (2012). Inactivation of a 25.5µm *Enterococcus faecalis* biofilm by a room-temperature, battery-operated, handheld air plasma jet. *Journal of Physics D: Applied Physics, 45*(16), 165205. https://doi.org/10.1088/0022-3727/45/16/165205.
191. Perni, S., Shama, G., & Kong, M. G. (2008). Cold atmospheric plasma disinfection of cut fruit surfaces contaminated with migrating microorganisms. *Journal of Food Protection, 71*(8), 1619–1625.
192. Phan, K. T. K., Phan, H. T., Boonyawan, D., Intipunya, P., & Brennan, C. S. (2018). Non-thermal plasma for elimination of pesticide residues in mango. *Innovative Food Science & Emerging Technologies, 48*, 164–171.
193. Pignata, C., D'Angelo, D., Basso, D., Cavallero, M. C., & Beneventi, S. (2014). Low-temperature, low-pressure gas plasma application on *Aspergillus brasiliensis, Escherichia coli* and pistachios. *Journal of Applied Microbiology, 116*(5), 1137–1148.
194. Pina-Perez, M. C., Martinet, D., Palacios-Gorba, C., Ellert, C., & Beyrer, M. (2020). Low-energy short-term cold atmospheric plasma: Controlling the inactivation efficacy of bacterial spores in powders. *Food Research International, 130*, 108921. https://doi.org/10.1016/j.foodres.2019.108921.
195. Pinelo, M., Arnous, A., & Meyer, A. S. (2006). Upgrading of grape skins: Significance of plant cell-wall structural components and extraction techniques for phenol release. *Trends in Food Science & Technology, 17*(11), 579–590.
196. Poletti, G., Orsini, F., Raffaele-Addamo, A., Riccardi, C., & Selli, E. (2003). Cold plasma treatment of PET fabrics: AFM surface morphology characterisation. *Applied Surface Science, 219*(3–4), 311–316.
197. Pontiga, F., Soria, C., & Castellanos, A. (2004). Ozone generation in coaxial corona discharge using different material electrodes. In *Annual Report – Conference on Electrical Insulation and Dielectric Phenomena, CEIDP* (pp. 1364313). https://doi.org/10.1109/ceidp.2004.1364313.
198. Potluri, S., Sangeetha, K., Santhosh, R., Nivas, G., & Mahendran, R. (2018). Effect of low-pressure plasma on bamboo rice and its flour. *Journal of Food Process Preservation, 42*(12), 13846. https://doi.org/10.1111/jfpp.13846.
199. Pragna, C. H., Ranjitha Gracy, T. K., Mahendran, R., & Anandharamakrishnan, C. (2019). Effects of microwave and cold plasma assisted hydro distillation on lemon peel oil extraction. *International Journal of Food Engineering, 15*(10), 1–10. https://doi.org/10.1515/ijfe-2019-0093.

200. Prelle, A., Spadaro, D., Garibaldi, A., & Gullino, M. L. (2014). Co-occurrence of aflatoxins and ochratoxin A in spices commercialized in Italy. *Food Control, 39*(1), 192–197.
201. Pruthi, J. S. (1980). *Spices and Condiments: Chemistry, Microbiology, Technology*. Academic Press. Series: Advances in Food Research, Supplement 4, Vol. 1, pp. 1–449, ISBN 13: 9780120164646.
202. Radhakrishnan, M., Ramanan, K. R., Sargunam, R., & Sarumathi, R. (2016). Effect of cold plasma on mortality of *Tribolium castaneum* on refined wheat flour. In *Proceedings of the 10th International Conference on Controlled Atmosphere and Fumigation in Stored Products (CAF 2016)* (pp. 7–11).
203. Ragni, L., Berardinelli, A., Iaccheri, E., Gozzi, G., Cevoli, C., & Vannini, L. (2016). Influence of the electrode material on the decontamination efficacy of dielectric barrier discharge gas plasma treatments towards *Listeria monocytogenes* and *Escherichia coli*. *Innovative Food Science & Emerging Technologies, 37*, 170–176.
204. Ragni, L., Berardinelli, A., Vannini, L., Montanari, C., & Sirri, F. (2010). Non-thermal atmospheric gas plasma device for surface decontamination of shell eggs. *Journal of Food Engineering, 100*(1), 125–132.
205. Ramazzina, I., Berardinelli, A., Rizzi, F., Tappi, S., & Ragni, L. (2015). Effect of cold plasma treatment on physico-chemical parameters and antioxidant activity of minimally processed kiwifruit. *Postharvest Biology and Technology, 107*, 55–65.
206. Ramazzina, I., Tappi, S., Rocculi, P., Sacchetti, G., & Berardinelli, A. (2016). Effect of cold plasma treatment on the functional properties of fresh-cut apples. *Journal of Agricultural and Food Chemistry, 64*(42), 8010–8018.
207. Rana, S., Mehta, D., Bansal, V., Shivhare, U. S., & Yadav, S. K. (2020). Atmospheric cold plasma (ACP) treatment improved in-package shelf-life of strawberry fruit. *Journal of Food Science and Technology, 57*(1), 102–112.
208. Ranjitha Gracy, T. K., Gupta, V., & Mahendran, R. (2019). Influence of low-pressure nonthermal dielectric barrier discharge plasma on chlorpyrifos reduction in tomatoes. *Journal of Food Process Engineering, 42*(6), e13242. https://doi.org/10.1111/jfpe.13242
209. Rashid, F., Bao, Y., Ahmed, Z., & Huang, J. Y. (2020). Effect of high voltage atmospheric cold plasma on extraction of fenugreek galactomannan and its physicochemical properties. *Food Research International, 138*, 109776. https://doi.org/10.1016/j.foodres.2020.109776.
210. Ratish Ramanan, K., Sarumathi, R., & Mahendran, R. (2018). Influence of cold plasma on mortality rate of different life stages of *Tribolium castaneum* on refined wheat flour. *Journal of Stored Products Research, 77*, 126–134.
211. Rodríguez, Ó., Gomes, W. F., Rodrigues, S., & Fernandes, F. A. N. (2017). Effect of indirect cold plasma treatment on cashew apple juice (*Anacardium occidentale* L.). *LWT – Food Science and Technology, 84*, 457–463.
212. Rudy, M., Kucharyk, S., Duma-Kocan, P., Stanislawczyk, R., & Gil, M. (2020). Unconventional methods of preserving meat products and their impact on health and the environment. *Sustainability, 12*(15), 5948. https://doi.org/10.3390/su12155948.
213. Santos, L. C. O., Cubas, A. L. V., Moecke, E. H. S., Ribeiro, D. H. B., & Amante, E. R. (2018). Use of cold plasma to inactivate *Escherichia coli* and physicochemical evaluation in pumpkin puree. *Journal of Food Protection, 81*(11), 1897–1905.

214. Sarangapani, C., Misra, N. N., Milosavljevic, V., Bourke, P., O'Regan, F., & Cullen, P. J. (2016). Pesticide degradation in water using atmospheric air cold plasma. *Journal of Water Process Engineering, 9*, 225–232.
215. Sarangapani, C., O'Toole, G., Cullen, P. J., & Bourke, P. (2017). Atmospheric cold plasma dissipation efficiency of agrochemicals on blueberries. *Innovative Food Science & Emerging Technologies, 44*, 235–241.
216. Sarangapani, C., Patange, A., Bourke, P., Keener, K., & Cullen, P. J. (2018). Recent advances in the application of cold plasma technology in foods. *Annual Review of Food Science and Technology, 9*, 609–629.
217. Sarangapani, C., Ryan Keogh, D., Dunne, J., Bourke, P., & Cullen, P. J. (2017). Characterization of cold plasma treated beef and dairy lipids using spectroscopic and chromatographic methods. *Food Chemistry, 235*, 324–333.
218. Sarangapani, C., Yamuna Devi, R., Thirumdas, R., Trimukhe, A. M., Deshmukh, R. R., & Annapure, U. S. (2017). Physico-chemical properties of low-pressure plasma treated black gram. *LWT – Food Science and Technology, 79*, 102–110.
219. Scally, L., Behan, S., Aguiar de Carvalho, A. M., Sarangapani, C., & Tiwari, B. (2021). Diagnostics of a large volume pin-to-plate atmospheric plasma source for the study of plasma species interactions with cancer cell cultures. *Plasma Processes and Polymers, 18*(6), 1–12. https://doi.org/10.1002/ppap.202000250.
220. Schlüter, O., & Fröhling, A. (2014). Non-thermal processing: Cold plasma for bioefficient food processing. In *Encyclopedia of Food Microbiology, Second Edition*, pp. 948–953, Publisher: Elsevier. https://doi.org/10.1016/B978-0-12-384730-0.00402-X.
221. Scholtz, V., Pazlarova, J., Souskova, H., Khun, J., & Julak, J. (2015). Nonthermal plasma – A tool for decontamination and disinfection. *Biotechnology Advances, 33*(6), 1108–1119.
222. Segat, A., Misra, N. N., Cullen, P. J., & Innocente, N. (2015). Atmospheric pressure cold plasma (ACP) treatment of whey protein isolate model solution. *Innovative Food Science & Emerging Technologies, 29*, 247–254.
223. Segat, A., Misra, N. N., Cullen, P. J., & Innocente, N. (2016). Effect of atmospheric pressure cold plasma (ACP) on activity and structure of alkaline phosphatase. *Food and Bioproducts Processing, 98*, 181–188.
224. Selcuk, M., Oksuz, L., & Basaran, P. (2008). Decontamination of grains and legumes infected with *Aspergillus spp.* and *Penicillum spp.* by cold plasma treatment. *Bioresource Technology, 99*(11), 5104–5109.
225. Seol, Y. B., Kim, J., Park, S. H., & Chang, H. Y. (2017). Atmospheric pressure pulsed plasma induces cell death in photosynthetic organs via intracellularly generated ROS. *Scientific Reports, 7*(1), 1–11.
226. Será, B., Stranák, V., Serý, M., Tichý, M., & Spatenka, P. (2008). Germination of *Chenopodium album* in response to microwave plasma treatment. *Plasma Science and Technology, 10*(4), 506–511.
227. Sharafodin, H., & Soltanizadeh, N. (2022). Potential application of DBD Plasma Technique for modifying structural and physicochemical properties of Soy Protein Isolate. *Food Hydrocolloids, 122*, 107077.
228. Sharanyakanth, P. S., Lokeswari, R., & Mahendran, R. (2021). Plasma bubbling effect on essential oil yield, extraction efficiency, and flavor compound of *Cuminum cyminum* L. seeds. *Journal of Food Process Engineering, 44*(7), 1–8.

229. Sharma, A., Pruden, A., Stan, O., & Collins, G. J. (2006). Bacterial inactivation using an RF-powered atmospheric pressure plasma. *IEEE Transactions on Plasma Science, 34*(4 II), 1290–1296.
230. Shi, H., Ileleji, K., Stroshine, R. L., Keener, K., & Jensen, J. L. (2017). Reduction of aflatoxin in corn by high voltage atmospheric cold plasma. *Food and Bioprocess Technology, 10*(6), 1042–1052.
231. Shirani, K., Shahidi, F., & Mortazavi, S. A. (2020). Investigation of decontamination effect of argon cold plasma on physicochemical and sensory properties of almond slices. *International Journal of Food Microbiology, 335*, 1–7. https://doi.org/10.1016/j.ijfoodmicro.2020.108892.
232. Siciliano, I., Spadaro, D., Prelle, A., Vallauri, D., & Cavallero, M. (2016). Use of cold atmospheric plasma to detoxify hazelnuts from aflatoxins. *Toxins, 8*(5), 1–10. https://doi.org/10.3390/toxins8050125.
233. Sifuentes-Nieves, I., Mendez-Montealvo, G., Flores-Silva, P. C., Nieto-Pérez, M., & Neira-Velazquez, G. (2021). Dielectric barrier discharge and radio-frequency plasma effect on structural properties of starches with different amylose content. *Innovative Food Science & Emerging Technologies, 68*, 102630. https://doi.org/10.1016/j.ifset.2021.102630.
234. Silveira, M. R., Coutinho, N. M., Rocha, R. S., Moraes, J., & Esmerino, E. A. (2019). Guava flavored whey-beverage processed by cold plasma: Physical characteristics, thermal behavior and microstructure. *Food Research International, 119*, 564–570.
235. Šimek, M., & Homola, T. (2021). Plasma-assisted agriculture: History, presence, and prospects—A review. *European Physical Journal D, 75*(7), 210.
236. Smet, C., Baka, M., Steen, L., Fraeye, I., & Walsh, J. L. (2019). Combined effect of cold atmospheric plasma, intrinsic and extrinsic factors on the microbial behavior in/on (food) model systems during storage. *Innovative Food Science & Emerging Technologies, 53*, 3–17.
237. Song, H. P., Kim, B., Choe, J. H., Jung, S., & Moon, S. Y. (2009). Evaluation of atmospheric pressure plasma to improve the safety of sliced cheese and ham inoculated by 3-strain cocktail *Listeria monocytogenes*. *Food Microbiology, 26*(4), 432–436.
238. Song, Y., & Fan, X. (2020). Cold plasma enhances the efficacy of aerosolized hydrogen peroxide in reducing populations of *Salmonella typhimurium* and *Listeria innocua* on grape tomatoes, apples, cantaloupe and romaine lettuce. *Food Microbiology, 87*, 1–7. https://doi.org/10.1016/j.fm.2019.103391.
239. Sowbhagya, H. B. (2014). Chemistry, technology, and nutraceutical functions of celery (*Apium graveolens* L.): An overview. *Critical Reviews in Food Science and Nutrition, 54*(3), 389–398.
240. Stoica, M., Alexe, P., & Mihalcea, L. (2014). Atmospheric cold plasma as new strategy for foods processing – An Overview. *Innovative Romanian Food Biotechnology, 15*, 112–117.
241. Sun, S., Anderson, N. M., & Keller, S. (2014). Atmospheric pressure plasma treatment of black peppercorns inoculated with *Salmonella* and held under controlled storage. *Journal of Food Science, 79*(12), E2441–E2446.
242. Sureshkumar, A., Sankar, R., Mandal, M., & Neogi, S. (2010). Effective bacterial inactivation using low temperature radio frequency plasma. *International Journal of Pharmaceutics, 396*(1–2), 17–22.

243. Surowsky, B., Fischer, A., Schlüter, O., & Knorr, D. (2013). Cold plasma effects on enzyme activity in a model food system. *Innovative Food Science & Emerging Technologies, 19*, 146–152.
244. Surowsky, B., Schlüter, O., & Knorr, D. (2015). Interactions of non-thermal atmospheric pressure plasma with solid and liquid food systems: A Review. *Food Engineering Reviews, 7*(2), 82–108.
245. Sutar, S. A., Thirumdas, R., Chaudhari, B. B., Deshmukh, R. R., & Annapure, U. S. (2021). Effect of cold plasma on insect infestation and keeping quality of stored wheat flour. *Journal of Stored Products Research, 92*, 101774. https://doi.org/10.1016/j.jspr.2021.101774.
246. Takaki, K., Hatanaka, Y., Arima, K., Mukaigawa, S., & Fujiwara, T. (2008). Influence of electrode configuration on ozone synthesis and microdischarge property in dielectric barrier discharge reactor. *Vacuum, 83*(1), 128–132.
247. Tappi, S., Berardinelli, A., Ragni, L., Dalla Rosa, M., Guarnieri, A., & Rocculi, P. (2014). Atmospheric gas plasma treatment of fresh-cut apples. *Innovative Food Science & Emerging Technologies, 21*, 114–122.
248. Tappi, S., Gozzi, G., Vannini, L., Berardinelli, A., & Romani, S. (2016). Cold plasma treatment for fresh-cut melon stabilization. *Innovative Food Science & Emerging Technologies, 33*, 225–233.
249. Thirumdas, R., Deshmukh, R. R., & Annapure, U. S. (2016). Effect of low temperature plasma on the functional properties of basmati rice flour. *Journal of Food Science and Technology, 53*(6), 2742–2751.
250. Thirumdas, R., Kadam, D., & Annapure, U. S. (2017). Cold plasma: An alternative technology for starch modification. *Food Biophysics, 12*(1), 129–139.
251. Thirumdas, R., Kothakota, A., Annapure, U., Siliveru, K., & Blundell, R. (2018). Plasma activated water (PAW): Chemistry, physico-chemical properties, applications in food and agriculture. *Trends in Food Science & Technology, 77*, 21–31.
252. Thirumdas, R., Sarangapani, C., Ajinkya, M. T., Deshmukh, R. R., & Annapure, U. S. (2016). Influence of low-pressure cold plasma on cooking and textural properties of brown rice. *Innovative Food Science & Emerging Technologies, 37*, 53–60.
253. Thirumdas, R., Sarangapani, C., & Annapure, U. S. (2015). Cold plasma: A novel Non-Thermal Technology for Food Processing. *Food Biophysics, 10*(1), 1–11.
254. Thirumdas, R., Trimukhe, A., Deshmukh, R. R., & Annapure, U. S. (2017). Functional and rheological properties of cold plasma treated rice starch. *Carbohydrate Polymers, 157*, 1723–1731.
255. Thomas-Popo, E., Mendonça, A., Misra, N. N., Little, A., & Wan, Z. (2019). Inactivation of *Shiga-toxin-producing Escherichia coli, Salmonella enterica* and natural microflora on tempered wheat grains by atmospheric cold plasma. *Food Control, 104*, 231–239.
256. Tolouie, H., Mohammadifar, M. A., Ghomi, H., & Hashemi, M. (2018). Cold atmospheric plasma manipulation of proteins in food systems. *Critical Reviews in Food Science and Nutrition, 58*(15), 2583–2597.
257. Tolouie, H., Mohammadifar, M. A., Ghomi, H., Yaghoubi, A. S., & Hashemi, M. (2018). The impact of atmospheric cold plasma treatment on inactivation of lipase and lipoxygenase of wheat germs. *Innovative Food Science & Emerging Technologies, 47*, 346–352.

258. Ulbin-Figlewicz, N., Brychcy, E., & Jarmoluk, A. (2015). Effect of low-pressure cold plasma on surface microflora of meat and quality attributes. *Journal of Food Science and Technology, 52*(2), 1228–1232.
259. Vanderstoep, J. (2002). Nitrite curing of meat, the n-nitrosamine problem and nitrite alternatives. Elsevier. *Food Research International, 4*(35), 1–411. https://doi.org/10.1016/s0963-9969(01)00128-4.
260. Varilla, C., Marcone, M., & Annor, G. A. (2020). Potential of cold plasma technology in ensuring the safety of foods and agricultural produce: A review. *Foods, 9*(10), 1435. https://doi.org/10.3390/foods9101435.
261. Venkataratnam, H., Sarangapani, C., Cahill, O., & Ryan, C. B. (2019). Effect of cold plasma treatment on the antigenicity of peanut allergen Ara h 1. *Innovative Food Science & Emerging Technologies, 52*, 368–375.
262. Volin, J. C., Denes, F. S., Young, R. A., & Park, S. M. T. (2000). Modification of seed germination performance through cold plasma chemistry technology. *Crop Science, 40*(6), 1706–1718.
263. Wan, Z., Chen, Y., Pankaj, S. K., & Keener, K. M. (2017). High voltage atmospheric cold plasma treatment of refrigerated chicken eggs for control of *Salmonella Enteritidis* contamination on egg shell. *LWT – Food Science and Technology, 76*, 124–130.
264. Wan, Z., Misra, N. N., Li, G., & Keener, K. M. (2021). High voltage atmospheric cold plasma treatment of *Listeria innocua* and *Escherichia coli K-12* on Queso Fresco (fresh cheese). *LWT, 146*, 111406. https://doi.org/10.1016/j.lwt.2021.111406.
265. Wang, H. F., Tsai, Y. S., Lin, M. L., & Ou, A. S. M. (2006). Comparison of bioactive components in GABA tea and green tea produced in Taiwan. *Food Chemistry, 96*(4), 648–653.
266. Wang, J. M., Zhuang, H., Lawrence, K., & Zhang, J. H. (2018). Disinfection of chicken fillets in packages with atmospheric cold plasma: Effects of treatment voltage and time. *Journal of Applied Microbiology, 124*(5), 1212–1219.
267. Wang, R. X., Nian, W. F., Wu, H. Y., Feng, H. Q., & Zhang, K. (2012). Atmospheric-pressure cold plasma treatment of contaminated fresh fruit and vegetable slices: Inactivation and physiochemical properties evaluation. *European Physical Journal D, 66*(10), 1–7.
268. Wei, K., Ma, C., Sun, K., Liu, Q., Zhao, N., Sun, Y., Tu, K., & Pan, L. (2020). Relationship between optical properties and soluble sugar contents of apple flesh during storage. *Postharvest Biology and Technology, 159*, 1–9. https://doi.org/10.1016/j.postharvbio.2019.111021.
269. Won, M. Y., Lee, S. J., & Min, S. C. (2017). Mandarin preservation by microwave-powered cold plasma treatment. *Innovative Food Science & Emerging Technologies, 39*, 25–32.
270. Wongsagonsup, R., Deeyai, P., Chaiwat, W., Horrungsiwat, S., & Leejariensuk, K. (2014). Modification of tapioca starch by non-chemical route using jet atmospheric argon plasma. *Carbohydrate Polymers, 102*(1), 790–798.
271. Wu, T. Y., Sun, N. N., & Chau, C. F. (2018). Application of corona electrical discharge plasma on modifying the physicochemical properties of banana starch indigenous to Taiwan. *Journal of Food and Drug Analysis, 26*(1), 244–251.
272. Xiang, Q., Liu, X., Li, J., Liu, S., Zhang, H., & Bai, Y. (2018). Effects of dielectric barrier discharge plasma on the inactivation of *Zygosaccharomyces rouxii* and quality of apple juice. *Food Chemistry, 254*, 201–207.

273. Xiao, D. (2016). *Gas Discharge and Gas Insulation, Vol. 6*. Publisher: Springer Berlin, Heidelberg, pp. 1–362, DOI: https://doi.org/10.1007/978-3-662-48041-0. ISBN 978-3-662-48040-3.
274. Xin, Q., Li, N., Wang, J., Wang, B., & Li, G. (2015). Surface roughening of ground fused silica processed by atmospheric inductively coupled plasma. *Applied Surface Science, 341*, 142–148.
275. Xiong, Z., Du, T., Lu, X., Cao, Y., & Pan, Y. (2011). How deep can plasma penetrate into a biofilm? *Applied Physics Letters, 98*(22), 3597622. https://doi.org/10.1063/1.3597622.
276. Yadav, B., Spinelli, A. C., Govindan, B. N., Tsui, Y. Y., McMullen, L. M., & Roopesh, M. S. (2019). Cold plasma treatment of ready-to-eat ham: Influence of process conditions and storage on inactivation of *Listeria innocua*. *Food Research International, 123*, 276–285.
277. Yong, H. I., Kim, H. J., Park, S., Alahakoon, A. U., & Kim, K. (2015). Evaluation of pathogen inactivation on sliced cheese induced by encapsulated atmospheric pressure dielectric barrier discharge plasma. *Food Microbiology, 46*, 46–50.
278. Yong, H. I., Kim, H. J., Park, S., Kim, K., & Choe, W. (2015). Pathogen inactivation and quality changes in sliced cheddar cheese treated using flexible thin-layer dielectric barrier discharge plasma. *Food Research International, 69*, 57–63.
279. Yusupov, M., Neyts, E. C., Khalilov, U., Snoeckx, R., Van Duin, A. C. T., & Bogaerts, A. (2012). Atomic-scale simulations of reactive oxygen plasma species interacting with bacterial cell walls. *New Journal of Physics, 14*(9), 093043. https://doi.org/10.1088/1367-2630/14/9/093043.
280. Zahoranová, A., Henselová, M., Hudecová, D., Kaliňáková, B., & Kováčik, D. (2016). Effect of cold atmospheric pressure plasma on the wheat seedlings vigor and on the inactivation of microorganisms on the seeds surface. *Plasma Chemistry and Plasma Processing, 36*(2), 397–414.
281. Zargarchi, S., & Saremnezhad, S. (2019). Gamma-aminobutyric acid, phenolics and antioxidant capacity of germinated indica paddy rice as affected by low-pressure plasma treatment. *LWT – Food Science and Technology, 102*, 291–294.
282. Zhang, M., Oh, J. K., Cisneros-Zevallos, L., & Akbulut, M. (2013). Bactericidal effects of nonthermal low-pressure oxygen plasma on *Salmonella typhimurium* LT2 attached to fresh produce surfaces. *Journal of Food Engineering, 119*(3), 425–432.
283. Zhang, W., Xiao, S., & Ahn, D. U. (2013). Protein oxidation: Basic principles and implications for meat quality. *Critical Reviews in Food Science and Nutrition, 53*(11), 1191–1201.
284. Zhang, X. L., Zhong, C. S., Mujumdar, A. S., Yang, X. H., & Deng, L. Z. (2019). Cold plasma pretreatment enhances drying kinetics and quality attributes of chili pepper (*Capsicum annuum* L.). *Journal of Food Engineering, 241*, 51–57.
285. Zhang, Y., Xiao, Z., Chen, F., Ge, Y., Wu, J., & Hu, X. (2010). Degradation behavior and products of malathion and chlorpyrifos spiked in apple juice by ultrasonic treatment. *Ultrasonics Sonochemistry, 17*(1), 72–77.
286. Zhao, N., Ge, L., Huang, Y., Wang, Y., & Wang, Y. (2020). Impact of cold plasma processing on quality parameters of packaged fermented vegetable (*Radish paocai*) in comparison with pasteurization processing: Insight into safety and storage stability of products. *Innovative Food Science & Emerging Technologies, 60*, 102300. https://doi.org/10.1016/j.ifset.2020.102300.

287. Zhuang, H., Rothrock, M. J., Hiett, K. L., Lawrence, K. C., & Gamble, G. R. (2019). In-package air cold plasma treatment of chicken breast meat: Treatment time effect. *Journal of Food Quality, 2019*, 1837351. https://doi.org/10.1155/2019/1837351.
288. Ziuzina, D., Misra, N. N., Han, L., Cullen, P. J., & Moiseev, T. (2020). Investigation of a large gap cold plasma reactor for continuous in-package decontamination of fresh strawberries and spinach. *Innovative Food Science & Emerging Technologies, 59*, 102229. https://doi.org/10.1016/j.ifset.2019.102229.
289. Ziuzina, D., Patil, S., Cullen, P. J., Keener, K. M., & Bourke, P. (2014). Atmospheric cold plasma inactivation of *Escherichia coli, Salmonella enterica* serovar *typhimurium* and *Listeria monocytogenes* inoculated on fresh produce. *Food Microbiology, 42*, 109–116.
290. Zou, J. J., Liu, C. J., & Eliasson, B. (2004). Modification of starch by glow discharge plasma. *Carbohydrate Polymers, 55*(1), 23–26.
291. Zouelm, F., Abhari, K., Hosseini, H., & Khani, M. (2019). The effects of cold plasma application on quality and chemical spoilage of pacific white shrimp (*Litopenaeus vannamei*) during refrigerated storage. *Journal of Aquatic Food Product Technology, 28*(6), 624–636.

CHAPTER 7

EFFECTS OF COLD PLASMA ON LIQUID FOODS

ANJALY PAUL and APARAJHITHA SUDARSAN

ABSTRACT

Cold plasma is a novel and promising nonthermal technology with numerous applications in food processing and preservation. The antimicrobial effect of cold plasma in both solid and liquid foods has been widely investigated by researchers over the decades. Industrial applications of cold plasma in liquid foods require an in-depth understanding of the influence of equipment design, operating parameters, and liquid characteristics. This chapter reviews the effect of processing conditions like plasma sources, inducer gases, and other operating parameters in liquid foods such as fruit and vegetable juices, milk, and milk-based beverages. Microbial inactivation studies in liquid foods along with the changes in physicochemical properties, bioactive composition, and sensory attributes are also highlighted.

7.1 INTRODUCTION

Ensuring microbiological safety is an inevitable aspect of food processing, considering the prevalence of foodborne illnesses. This critical issue has been constantly investigated by the scientific community. High-temperature sterilization techniques are typically employed for microbial inactivation at the industrial level. However, food components are heat-sensitive and lose their authentic flavor, nutritional, and sensory attributes during thermal decontamination [22]. Consequently, numerous non-thermal food preservation techniques have gained increased attention among researchers to alleviate the undesirable effects of thermal processing on food and the environment [77].

Nonthermal Plasma Technology for Food and Food Products.
R. Mahendran, Preeti Birwal, and Megh R. Goyal (Eds.)

The recent development of cold plasma as a nonthermal intervention in foods has multiple applications across both solid and liquid foods, primarily in improved food safety. This non-equilibrium excited gas, whose bulk temperature is still maintained at room temperatures (~ 40 °C), is a mix of various reactive gas species (RGS) that can be either excited molecules, free radicals, free electrons, neutral species, ions, etc., depending on the gas, energy density, electrode shape, frequency, as well as a sample: gas volume ratio [62]. These RGS result in detrimental effects on microbial species through surface etching, electroporation, as well as membrane damage. These species are also capable of diffusing through the microbial membrane to induce mitochondrial DNA damage [21, 33, 76]. These interactions often result in irreparable damage or microbial death, leading to improved food safety. The RGS produced are highly reactive and unstable and have a half-life ranging usually between $10^{-9}$and 10^{-6} seconds before reverting to the neutral gas [44].

Food is composed of macro and micro molecules that together contribute to the texture, flavor, and color profile. It is easy to elucidate and is fairly straightforward to understand the role of an individual type of molecule, but foods are often a matrix of complex interactions between these molecules depending on their relative presence. The formation of cross-linkages or secondary by-products as a result of interactions with other molecules or ions is highly dependent on the matrix composition and is not predictable [64]. Cold plasma application to solid foods is primarily restricted to surface treatments since the penetration ability of RGS is highly influenced by properties such as porosity, water content, chemical composition, surface characteristics such as roughness, crevices, and presence of natural wax layers [62, 64]. If the end goal is to achieve a non-intensive microbial reduction without affecting any nutritional properties inside the food, a limited penetration depth application through cold plasma is a highly effective intervention [64].

In liquid foods, the RGS diffuse easily through the matrix and each volume component gets exposed to the cold plasma – including microbial cells as well as the food components [64]. Hence, the interactions of plasma species with the liquid food components are more intense, which triggers severe chemical changes and degradation reactions [57]. The concentration and stability of RGS often depend on the plasma processing conditions, necessitating careful consideration of treatment parameters to achieve optimal microbial decontamination while retaining food constituents. This chapter discusses the major factors affecting the efficacy of cold plasma processing of liquid foods including the plasma sources, voltage, inducer gases, treatment time, and liquid characteristics. The influence of these factors on microbial decontamination, physicochemical properties, bioactive components, and sensory profile is also discussed. The optimization of plasma sources and operating conditions on liquid foods to ensure microbial safety and maintain quality are highly relevant for industrial adaptation of the technique.

7.2 EFFECT OF PROCESSING CONDITIONS

As we all know, cold plasma has a complex chemical composition consisting of free electrons, ions, radicals, photons, and other reactive species. These reactive agents either act individually or in synergy for specific cold plasma applications [4]. The composition of cold plasma and the intensity of reactive gas species are often affected by equipment design and operating conditions such as electrode configuration, electrical parameters, treatment time, and environmental conditions. A thorough understanding of design considerations and optimum process parameters is required for the commercialization of cold plasma for liquid food processing. For example, Laroque et al. [32] suggested that plasma characteristics are influenced by the geometry, material, and condition (wear and tear) of the electrodes and recommended having focused studies on these electrode attributes to improve and optimize plasma devices for food processing. Environmental conditions such as temperature, pressure, and relative humidity also influence the efficiency of plasma treatments [16]. Most of the research in liquid food processing is dedicated to the influence of plasma sources, inducer gases, electrical parameters, treatment time, and liquid volume. Hence, the influence of these parameters is further elaborated in the following subsections. The factors affecting cold plasma processing of liquid foods are portrayed in Figure 7.1.

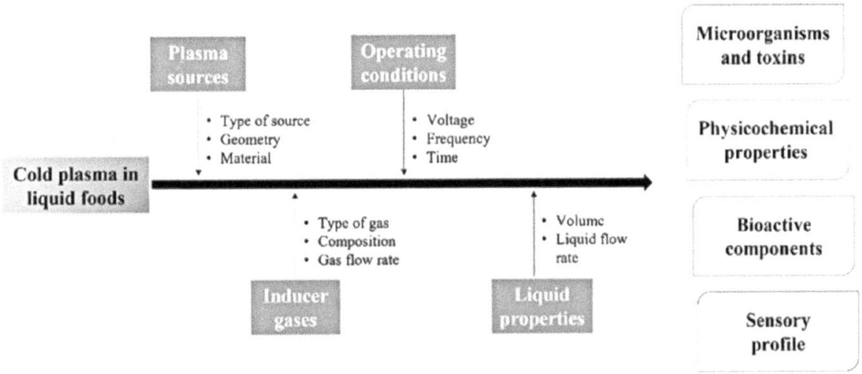

FIGURE 7.1 Influence of cold plasma on liquid foods.

7.3 PLASMA SOURCES

7.3.1 DIELECTRIC BARRIER DISCHARGE (DBD)

DBD discharges are the most stable cold plasma sources and are highly explored in food processing because of their reliability and convenience. In DBD, cold plasma is produced between two electrodes, aligned parallel to each other, with a dielectric barrier in between them. One of the electrodes will be powered and the other one will be grounded with a 0.1 mm to several centimeters gap in between [32]. Dielectric barriers are generally made of polymer, glass, quartz, or ceramic. DBD portrays numerous advantages over other plasma sources in terms of safety, flexibility in design, and cost-effectiveness and hence could be scaled up easily for commercial applications [16]. In liquid foods, DBD plasma studies reported better preservation of nutrient profile with reduced quality changes. Certain studies compared glow and DBD discharges in fruit juice processing and demonstrated the effectiveness of DBD over glow discharge sources in maintaining quality [1, 59]. An optimization study by Liu et al. [35] on kiwi juice suggested DBD plasma processing conditions to be 12 kV voltage, 1 min treatment time, and 18 mL juice volume for batch processing based on the sterilizing rate and juice quality.

7.3.2 PLASMA JET

Plasma jet has also found several applications in liquid food processing. It was one of the initial forms of plasma source for disinfection applications in solid foods. A plasma jet is highly suitable for applying various inducing gases. The main limitation of a plasma jet is the low area of exposure, which could be overcome by the application of multiarray jets. For instance, some studies Cao et al. [7], Furmanski et al. [19] found that the arrangement of quartz tubes in a honeycomb configuration resulted in an intensified plasma jet with twice the optical density of single plasma jets. However, optimization of jet configuration, gap, gas mixtures, and flow rates is necessary for the industrial application of these multi-jet arrays [12]. The actual food temperature in plasma jet applications should be carefully monitored, as the temperature of the plasma jet may rise to 60°C. Another common application of plasma jet is to generate plasma-activated liquids by direct treatment [66]. Hosseini et al. [26] found that the plasma jet was effective in reducing *E. coli* content without affecting pH, color, and phenol in sour cherry juice. The study also optimized the plasma jet operating conditions at 19.7 kV voltage, 18.27 kHz frequency, 0.18% oxygen in argon as the gas composition, and

5.32 L/min gas flow rate based on higher bacterial reduction and lower quality changes.

7.3.3 GLOW DISCHARGE

Another cold plasma source operated for liquid food processing is the glow discharge. The glow discharge is mostly affected by the pressure and gap between the electrodes. At higher pressures, glow discharges are often converted to streamer or spark [16, 68]. Gadri et al. [20] developed a glow discharge system operating at one-atmosphere pressure and it was found effective in surface microbial decontamination which can be applied in food industries. Micro discharges and hollow cathode discharges are different types of glow discharges, which are often applied for surface sterilization [16]. In a study on Camu Camu juice, the researchers found that adjusting the processing parameters of a glow discharge could regulate the aroma and flavor notes by interacting with the volatile compounds [6]. Rodriguez et al. [54] reported that glow discharge exposure has enhanced the antioxidant activity and phenolic content in fruit juices. Glow discharge treatment was also effective in retaining vitamin C and improving vitamin A and carotenoid content in acerola juice [18].

7.3.4 CORONA DISCHARGE

Corona discharges are produced by applying a high voltage between two asymmetrical electrodes of highly different radii of curvature, such as a thin wire or sharp needle to a plate electrode [55]. The corona discharge is unstable, with intense plasma streamers produced in the high electric field area close to the discharge electrode and low-intensity plasma produced by secondary photoionization processes in the remaining volume [68]. Even though corona discharge is not so popular in treating liquid foods, it has found various applications in wastewater treatment, destruction of toxic residues, and generation of ozone [3]. For instance, a pulsed corona discharge has exhibited a rapid reduction of methylene blue dye in water after 10 minutes of treatment [37]. Besides, Leon and Ramirez [25] reported complete deactivation of *E. coli* and *S. enterica* in 1,000 mL poultry wastewater at 20 kV and 360 seconds of low-temperature corona discharge treatment. Similar to corona discharges, gliding arc discharge plasma also finds applications mainly in effluent water treatment involving the degradation of microbes [31] and textile dyes.

7.4 INDUCER GASES

7.4.1 TYPE OF INDUCER GAS

The inducer/carrier gases affect the generation of charged species or plasma reactive components. Noble gases like argon and helium have been generally employed in cold plasma applications [36]. The RGS produced in the plasma are mainly responsible for its antimicrobial mechanisms and property modifications in foods [4, 67]. Consequently, the use of oxygen or air as an inducer gas or in combination with noble gases to form gas admixtures has gained a lot of attention in microbial decontamination and food safety [2]. The use of air has been actively researched because of its suitability in large-scale production, as air is a cheaper alternative to other gases [12]. On the other hand, the quality of fat-rich products like milk is in question because they are prone to oxidative reactions in the presence of oxygen. For example, in a study on the effect of nitrogen-oxygen gas mixtures on skim milk, browning and odor development were observed in N_2-O_2 gas admixtures, which were not found in the application of N_2 alone [58]. Furthermore, the lower dielectric strength of air (3 × 106 V/m) requires higher voltages or lower electrode gaps for breakdown and stable plasma discharges [12], limiting the area of plasma exposure.

The effect of inducer gases is further prominent in in-package cold plasma applications. Along with the air and other gases discussed before, modified packaging gas mixtures are also utilized in in-package treatments. There are several variations in the design consideration of cold plasma treatments within a sealed package, which include placing the package in between a volumetric DBD discharge, the surface DBD plasma discharge into the open side of the packaging, and the surface DBD electrodes placed inside the packaging [42]. The in-package studies available so far in liquid foods are by placing the surface DBD electrodes inside the sealed plastic or polymer bags/containers. Xu et al. [74] revealed that MA65 gas (65% O_2 + 30% N_2 + 5% CO_2 at <5% RH) was better than air in microbial inactivation (*Salmonella enterica* – 4.7 log reduction) of 50 ml orange juice placed in a sealed cryovac pouch, without a significant change in pH, Brix, or color. However, the vitamin content was decreased by the application of MA65. Kim et al. [30] also proved the effectiveness of in-package surface DBD in disinfection of milk without causing color changes or lipid oxidation, yet the pH was decreased due to the addition of plasma reactive species.

7.4.2 GAS COMPOSITION AND FLOW RATE

In addition to the type of inducer gases, the composition (gas admixtures), and gas flow rate also have a substantial impact on liquid food processing. The effect of oxygen in argon gas was studied in sour cherry juice [26] and found that increasing oxygen concentration reduces the plasma jet length. The study also found that the jet length increases when the gas flow rate was raised from 3 to 5 L/min but decreases with a further increase in flow rate. However, a higher bacterial reduction was obtained at higher oxygen concentrations. This is because of the ROS produced in the presence of oxygen, which is mainly responsible for the antimicrobial effect [45]. In a similar work on plasma-processed fruit juices, raising the nitrogen gas flow rate from 10 to 20 mL/min increased the phenolic content in sapota juice but reduced it in guava, Caja, and acai juices [54]. Gas flow rate has also affected the vitamins, phenolics, and antioxidant content in plasma-processed acerola juice [18]. Hence, inducer gases and gas flow rate should be selected depending on the application (microbial reduction or functional property modification), required plasma jet length, and final product quality.

7.5 OPERATING CONDITIONS

7.5.1 ELECTRICAL PARAMETERS

Voltage is one of the most studied electrical parameters. Higher voltage, power, or frequency levels are often beneficial for microbial inactivation [14, 26], but adversely affect the food quality [35]. Optimization studies are highly relevant considering that liquid foods are extremely susceptible to quality damages including a decrease in pH, color degradation, or vitamin loss along with lipid oxidation and off-odor development at higher voltages. The efficiency of plasma treatment also depends on the current, type of current (AC or DC), and the polarity of the electrodes.

7.5.2 TREATMENT TIME

There have been several investigations on the optimal plasma processing or treatment times for the application in liquid foods. Researchers have studied treatment times as low as 30 s to 30 min. As in the case of voltage, extended periods of plasma exposure would lead to undesirable quality changes in foods, which is again crucial for the disinfection. Rodriguez et al. [54] investigated the antioxidants and phenolics content of four fruit juices including acai, guava, Caja, and sapota. The study found that the antioxidants and phenolics content increased after 5–10 min of exposure but later decreased upon extended treatment times (15 min). The overexposure of plasma will produce more reactive species, accelerating plasma-liquid interactions, leading to the degradation and/or restructuring of food macro and micro molecules. A study on orange juice also reported reduction in vitamin C after raising the treatment time from 30 to 120 s [74].

7.6 LIQUID PROPERTIES

The product type, complexity, and composition of liquid foods affect the efficacy of plasma treatment. Fruit juices, milk, and other complex beverages react differently to plasma processing conditions. The fruit juices are rich in bioactive components. The fat and protein-rich cattle milk is highly susceptible to quality damage upon a slight change in treatment conditions. These changes will again influence the physical, chemical, and sensory attributes of liquid foods [49].

The volume of liquid food or liquid flow rate also has an impact on cold plasma applications. Liu et al. [35] analyzed the effect of juice volume on the bactericidal rate and the quality of kiwi juice and reported that the influence of the reactive species decreases at higher juice volume. This reduces the microbial inactivation rate and improves the chlorophyll content. Likewise, another study on orange juice observed higher microbial reduction in lower sample volumes [74]. Most of the studies on cold plasma treatment of liquid foods are performed in batches, but continuous processing lines are more effective from an industrial point of view. Liquid flow rate is a crucial parameter influencing the interaction time of reactive species with beverages and affecting plasma efficiency. Dharini et al. [39] investigated the milk flow rate at 3 and 6 mL/min in a continuous DBD plasma system. The study reported a similar microbial load irrespective of the flow rates; however, the viscosity and acidity of the milk were affected. Hence, further investigations must be performed in continuous systems at lower flow rates to reduce unfavorable quality changes for commercialization.

7.7 ANTIMICROBIAL EFFECT

The most common applications of nonthermal plasma in the food industry are to eradicate microorganisms in food products. The disinfection and sterilization effect of cold plasma has been keenly investigated even in liquid foods (Table 7.1). As discussed in previous sections, it is influenced

TABLE 7.1 Microbial Reduction in Liquid Foods by Cold Plasma Treatment

Microorganism	Product	Plasma Source – Gas	Maximum Log Reduction (log CFU/mL)	References
Total bacterial count	Sheep milk	DBD-air	2.98	[70]
Total bacterial count	Milk	DBD-air	1.2	[13]
Total bacterial count	Oat milk	Pins to plate – air	2.18	[14]
Total yeast and mold	Oat milk	Pins to plate – air	1.57	[14]
Bacillus sp.	Blueberry juice	Plasma jet – oxygen	7.2	[27]
Bacillus subtilis	A model liquid food system	Plasma jet-air	7.27	[8]
Escherichia coli	Chokeberry juice	Plasma jet with DBD-argon	2.83	[22]
Escherichia coli	Sour cherry juice	Plasma jet – argon	6	[26]
Escherichia coli	Tangerine juice	Plasma reactor – air	4.8	[75]
Escherichia coli	Milk	DBD – air	2.43	[30]
Escherichia coli	Milk	Corona-air	4.18	[24]
Listeria monocytogenes	Milk	DBD – air	2.46	[30]
Mesophilic bacteria	Guava-flavored whey beverage	Plasma etch PE50-nitrogen	1.82	[61]
Saccharomyces cerevisiae	Chokeberry juice	Plasma jet with DBD-argon	1.31	[22]
Saccharomyces cerevisiae	White grape juice	DBD-dry air	7.4	[51]
Salmonella typhimurium	Milk	Encapsulated DBD – air	2.4	[30]
Total aerobic bacteria	Strawberry juice	DBD-air	1	[40]
Total bacteria, molds, and yeast	Tiger nut milk	DBD-air	< DL	[46]
Zygosaccharomyces rouxii	Apple juice	Plasma bubbling – air	5.6	[72]
Zygosaccharomyces rouxii	Apple juice	DBD-air	5	[73]

*DL – detection limit.

by numerous factors including the complexity and volume of liquid, plasma source and operating conditions, the composition of inducer gas, and characteristics of microorganisms. The microbial reduction can be attributed to the individual and synergistic interactions of plasma components with structural and functional components of microorganisms leading to their destruction [45]. There are a variety of proposed antimicrobial mechanisms of cold plasma involving the influence of each plasma component. The interactions with RGS are the key factors contributing to the microbial inactivation [41]. The addition of oxygen or air to the inducer gas, leading to the production of higher RONS, has enhanced microbial reduction and toxin degradation in liquid foods [26, 48]. The reactive oxygen species like ozone, atomic oxygen, peroxide, superoxide, and hydroxyl radicals and reactive nitrogen species like nitrogen oxides, nitrous acids, and peroxynitrite ions are mainly involved in the decontamination [38]. The hydroxyl radical, produced in the presence of moisture, has also resulted in higher disinfection by initiating oxidative chain reactions. This can be highly beneficial in the case of liquid foods, as a higher amount of hydroxyl species is produced, as in the case of plasma-activated water [29]. Moreover, the stability of reactive species and reacted products like ozone and hydrogen peroxide is higher in liquids. The free electrons and ultraviolet photons are also involved in microbial deactivation by electroporation and disruption of the cell membrane [45].

7.8 EFFECT ON PHYSICOCHEMICAL PROPERTIES

Physicochemical properties like pH, acidity, soluble solids, and nutritional value are key parameters defining quality and are often evaluated in plasma-processed liquid foods. Most of the quality studies in liquid foods have reported a considerable decrease in pH upon plasma exposure. This effect was more evident in milk than in fruit juices (Table 7.2). The pH changes in juices are usually about 0.4–0.5 units, without affecting the taste and consumer acceptability [52]. Suwal et al. [65] reported a drastic decrease in the pH of plasma-activated water because of the acids produced by the solubilization of gaseous reactive oxygen and nitrogen species. Additionally, the degradation of amino acids to generate acidic by-products also decreases pH [5]. Hence, milk-based beverages are more susceptible to changes in acidity.

The total soluble solids (TSS) of fruit juices are mostly linked to sugars, carbohydrates, organic acids, and other macromolecules. The interaction of reactive species with these macromolecules is limited and thus causes only minute changes in the soluble solids [63]. Yet, an increase in TSS was observed for some fruit juices; this could be due to the reduction of moisture, dissociation of high molecular weight compounds, and release of cell-bound components upon disruption of cellular bonds and membranes during plasma treatment [34, 47]. Cold plasma treatment has also been reported to affect the sugar content and organic acids in fruit juices. Silva et al. [59] have reported a reduction in glucose, fructose, sucrose, and alanine after DBD plasma treatment with synthetic air in sapota juice. Similar results were obtained in cashew apple juice by [28]. However, Gan et al. [22] observed no changes in reducing sugar content in a plasma jet-DBD system with argon as an inducer gas. Malic acid content was increased in both apple juice [15] and cashew apple juice [28] upon plasma treatment.

In the case of plasma-treated milk and milk-based beverages, lipid oxidation reactions and subsequent quality changes are very common (Table 7.2). The oxidation of fat molecules is escalated in the presence of oxygen compared to using inert gases because of the chain reactions initiated by reactive oxygen species [58]. This will lead to secondary metabolite products and off-odor in plasma-treated products. Cold plasma treatment has also caused protein aggregation or denaturation in both plant-based and cattle milk upon longer processing times [14, 43, 70]. Hence, the process parameters should be carefully considered in milk processing. The use of lower voltages and discharge time with inert gases to induce plasma discharge could reduce these undesirable changes to a greater extent. The application of additives like coconut husk extract has been reported to mitigate plasma-induced lipid oxidation in fish [50]. Similar combination treatments could also be explored in milk processing to alleviate the unfavorable effects. In-package DBD treatment has also reduced lipid oxidation in milk [30].

7.9 EFFECT ON BIOACTIVE COMPONENTS

Bioactive constituents like polyphenols, antioxidants, and carotenoids have several functional and nutraceutical properties and are an essential part of the human diet. There have been varied effects of plasma treatment on the bioactive composition in beverages depending on the liquid type and treatment conditions (Table 7.2). In a study on whey beverage, plasma treatment preserved phenolics, vitamins, and carotenoids significantly better than the pasteurized sample, as high-temperature treatments negatively affect these thermally unstable compounds. Gan et al. [22] reported that the anthocyanin

TABLE 7.2 Recent Studies on the Effect of Cold Plasma on Liquid Food Quality

Product	Plasma Type	Factors	Key Findings	References
		Fruit and Vegetable Juices		
Acai, guava, Caja, sapota	Glow discharge-nitrogen	Plasma etch PE-50 Flow rate: 10–30 mL/min Time: 5 to 15 min	Increased TPC in sapota juice; little effect in others Improved antioxidants in fruit juices, except in Caja juice	[54]
Apple juice	DBD-air Glow discharge-synthetic air	DBD: Frequency: 50–900 Hz Gap: 1.5 cm Time: 15 min Juice volume: 20 mL Glow: Plasma etch PE-50 Voltage: 80 kV Gas flowrate: 10, 20, 30 mL/min Time: 10, 20, 30 min Juice volume: 40 mL	Sweetening power was decreased in DBD while increased in glow discharge Malic acid content was increased in both	[15]
Camu Camu juice	DBD-air Glow discharge (GP)-synthetic air	GP: Plasma Etch PE-100 Frequency: 80 kHz Gas flowrate: 10, 20, 30 mL/min Juice volume: 40 mL Time: 10, 20, 30 min DBD: Voltage: 24 kV Gap: 15 mm Frequency: 200–960 Hz Time: 15 min	DBD plasma (200 and 960 Hz) better retained volatile organic compounds and flavor profile than the glow discharge Vitamin C and ethanol were higher in GP In GP flowrate and treatment time affected the juice composition; 10 mL/min – 10 min and 20 mL/min – 20 min treatments better-preserved taste and aroma compounds	[1]
Cashew apple juice	DBD-air	Frequency: 200 and 700 Hz Voltage: 20 kV Gap: 15 mm gap Time: 15 min	Decreased glucose and fructose; increased malic acid and vitamin C No effect on TPC Improved bio-accessibility of vitamin C	[28]

TABLE 7.2 *(Continued)*

Product	Plasma Type	Factors	Key Findings	References
Chokeberry juice	Plasma jet with DBD-argon	Voltage: 9 kV Frequency: 1,500 Hz Gap: 1.5 cm Juice volume: 1–5 mL Time: 1–5 min	Optimized 2 mL juice volume and 4 min treatment time Improved anthocyanin, vitamin C, antioxidant capacity, and reduced color changes compared to thermal treatment (90°C for 15 s) Did not affect titratable acidity, soluble solids, and reducing sugar	[22]
Jujube juice	DBD-air	Voltage: 15–35 kV Treatment time: 30–180 s Brix: 8–20 at 20°C	Phenolics, flavonoids content, and antioxidant activity reduced; acidity increased; particle size reduced; color changed, especially lightness increased with voltage and time Alternaria mycotoxins degradation increased with voltage and decreased with Brix	[71]
Kiwi turbid juice	DBD-air	Voltage: 10, 25, 40 kV Juice volume: 10, 15, 20 mL Time: 1, 3, 5 min	Optimal conditions were 12 kV voltage, 18 mL juice volume and 1 min discharge time with 0.18 mg/g chlorophyll content, 3.47 OD280/g, and 83.97% TPC.	[35]
Sapota juice	DBD-air Vacuum-synthetic air	DBD Frequency: 50–960 Hz Time: 5–15 min Vacuum Gas flowrate: 10–30 mL/min Time: 10–30 min	Decrease in glucose, fructose, sucrose, and alanine Increased TPC and antioxidant activity in DBD; decreased in vacuum pH, TSS, and color were not significantly affected Better product quality in DBD	[59]
Sour cherry juice	Plasma jet- O_2 in argon	Voltage: 10, 15, 20 kV Frequency: 10, 15, 20 kHz Gas flowrate: 3, 5, 7 L/min Oxygen concentration: 0, 0.5, 1%	No significant change in color, pH, and phenol content	[26]
		Milk and Milk-Based Beverages		
Chocolate milk	Plasma etch-PE50-nitrogen	Gas flow: 5, 10, 15 mL/min Volume: 50 mL Time: 10, 20, 30 min	20 mL/min gas flow and 5 min treatment better-preserved phenolics, health indices, inhibited ACE activity, had higher fatty acid composition	[10]

TABLE 7.2 *(Continued)*

Product	Plasma Type	Factors	Key Findings	References
Chocolate milk	Plasma etch-PE50-nitrogen	Gas flow: 5, 10, 15 mL/min Volume: 40 mL Time: 10, 20, 30 min	20 and 30 mL/min and 10 min exposure has microstructure, consistency, flow behavior index, and particle size distribution similar to pasteurized No effect of cold plasma on melting temperature, enthalpy, and bound water	[11]
Guava-flavored whey beverage	Plasma etch-PE50-nitrogen	Gas flowrate: 10, 20, 30 mL/min Time: 5, 10, 15 min	Preserved phenolics, vitamin C, carotenoids, antioxidant activity, and volatile compounds compared to pasteurized sample Higher ACE inhibitory activity Higher flow rate and time adversely affected fatty acid profile	[60]
Guava-flavored whey beverage	Plasma etch-PE50-nitrogen	Gas flowrate: 10, 20, 30 mL/min Time: 5, 10, 15 min	10 mL/min and 5–10 min had microstructure and particle size distribution similar to pasteurized sample	[61]
Milk	Pin to plate-air	Voltage: 32 kV Gap: 4 cm Milk volume: 6 mL Time: 1, 5, 10 min	A lower degree of hydrolysis of Pseudomonas proteases upon plasma exposure (63.81% reduction for 10 min exposure after 8 days of storage). Caseins and whey protein degradation at 5- and 10-min exposure; increased aldehyde content	[43]
Milk	DBD-air	Voltage: 1, 1.5, 2 kV Gap: 1 cm Liquid flow rate: 3, 6, 9 mL/min Treatment time: 5, 2.5, 1 min	pH and viscosity decreased, color changed, conductivity increased for 5 min treatment	[13]
Oat milk	Pins to plate electrode with PET barrier-air	Voltage: 170, 200, 230 V Milk volume: 30 mL Time: 5, 10, 15 min	Total soluble solids, acidity, and sedimentation index increased pH, viscosity, and protein content decreased	[14]
Sheep milk	DBD-air	Power: 40 W Time: 30, 180, 300 s	Dispersion of protein structure from α-helix to β-sheet; reduced pH and a*;	[70]
Skim milk	Plasma jet-Nitrogen-Oxygen (0.9:0.1 mol)	Voltage: 20 kV Gas flowrate: 5 L/min Milk volume: 20 mL Time: 1–5 min	N_2–O_2: decreased pH, increased titrable acidity, protein aggregation, higher browning for 5 min; developed odor; N_2 alone was better	[58]

TABLE 7.2 *(Continued)*

Product	Plasma Type	Factors	Key Findings	References
Skim milk	DBD-Oxygen and air	Voltage: 60–80 kV Milk volume: 10–30 mL Gas composition: 65% oxygen in the air Time: 1–20 min	Aflatoxin (M1) was degraded by 65% in air and 78.9% in oxygen after 20 min. 83.9% reduction in aflatoxin toxicity after 5 min exposure. AFM1 reduction increased with voltage and decreased with juice volume No change in color and a slight reduction in pH	[48]
Tiger nut milk	DBD-air	Voltage: 30 V Gap: 6 mm Milk volume: 25 mL Time: 2–12 min	pH reduced for 8 and 12 min Acidity (titrable) increased Lipid oxidation intensified in all treatments Decreased protein content and peroxidase activity with time increment	[46]
Whey beverage (with Xylooligosaccharide)	Corona discharge-air	Voltage: 15 kV Volume: 200 mL Time: 0–15 min	Color intensity diminished and reduced consistency Similar viscosity, xylooligosaccharide stability, and sensory properties to that of untreated sample; Lower heat load indicators, whey protein nitrogen index Higher antioxidant activity and inhibition of enzymes compared to pasteurized sample 15 min exposure improved nutritional value	[53]
			Other Beverages	
Red wine	Plasma jet-argon	Batch Volume: 100 mL Power: 60 and 90 W Time: 1, 3, 5, 10 min Continuous Total volume: 750 mL Liquid flowrate: 1.2, 2.4 L/min Time: 25 min	Improved color intensity, phenolics, and anthocyanin content in batch treatment. Batch treatment yielded better product quality than continuous.	[56]
Tender coconut water	DBD-air	Voltage: 18, 23, 28 kV; Gap: 15 mm Volume: 15 mL	Polyphenol oxidase was better inactivated than peroxidase with half-maximal activity time of 1.35 min and 2.53 min, respectively	[9]

content, vitamin C, and antioxidant capacity were improved by DBD plasma treatment in the presence of argon as an inducer gas in chokeberry juice, whereas Wang et al. [71] observed a decline in phenolic and flavonoid contents after plasma exposure in a DBD source with air as the carrier gas. The changes in bioactive components during plasma treatment were reviewed in detail by several researchers [17, 52, 69].

7.10 EFFECT ON THE SENSORY ATTRIBUTES

The sensory profile of foods has been affected by plasma treatment, and this effect is more pronounced in liquid foods because of the interactions with plasma. Plasma exposure in liquids is known to increase acidity and initiate several chemical reactions [64]. For example, plasma discharge stimulated chemical transformations of terpenoids and sesquiterpene compounds through bond rupture, hydrolysis, and hydrogenation reactions in camu camu juice. The increased acidity in the juice due to plasma exposure resulted in the conversion of phytochemicals to alcohols. Consequently, secondary aroma compounds were produced, inducing changes in the aroma notes of the fruit juice [6]. Similarly, off-odor development and color changes were reported in milk treated with cold plasma [48, 58], due to the formation of lipid oxidation products such as hydrocarbons, carbonyls, and acids [49].

7.11 SUMMARY

Cold plasma as a food preservation tool has promising potential for several liquid foods such as milk, fruit juices, and drinking water. The commercial applications of cold plasma require an in-depth knowledge of the type of plasma generation sources, operating conditions, reactive gas chemistry, as well as interaction with the composition of the liquid food matrix. The reactive gas molecules diffuse through the bulk liquid and react with water and other molecular species to induce chain reactions that result in enzyme inactivation, and oxidation of macromolecules such as carbohydrates, sugars, proteins, and fats. More research is required to elucidate other simultaneous reactions that also occur between the reaction by-products that can further drive the reactive chemistry and determine the effectiveness of the process. The molecular-level interactions are quite complex to elucidate and require a careful design of the process parameters to achieve targeted action. This will be crucial in developing a scale-up of cold plasma systems for industrial-level processing of liquid foods.

KEYWORDS

- **cold plasma**
- **fruit juices**
- **liquids**
- **microbiological reduction**
- **milk**
- **operating conditions**
- **quality**

REFERENCES

1. Alves Filho, E. G., Silva, L. M. A., de Brito, E. S., Castro, D. R. G., Bezerra, J. A. (2021). Effect of glow and dielectric barrier discharges plasma on volatile and non-volatile chemical profiling of camu-camu juice. *Food and Bioprocess Technology, 14*(7), 1275–1286. https://doi.org/10.1007/s11947-021-02756-5.
2. Baier, M., Görgen, M., Ehlbeck, J., Knorr, D., Herppich, W. B., Schlüter, O. (2014). Non-thermal atmospheric pressure plasma: Screening for gentle process conditions and antibacterial efficiency on perishable fresh produce. *Innovative Food Science & Emerging Technologies, 22*, 147–157. https://doi.org/10.1016/j.ifset.2013.12.011.
3. Banaschik, R., Lukes, P., Jablonowski, H., Hammer, M. U., Weltmann, K. D., Kolb, J. F. (2015). Potential of pulsed corona discharges generated in water for the degradation of persistent pharmaceutical residues. *Water Research, 84*, 127–135. https://doi.org/10.1016/j.watres.2015.06.036.
4. Bourke, P., Ziuzina, D., Han, L., Cullen, P. J., Gilmore, B. F. (2017). Microbiological interactions with cold plasma. *Journal of Applied Microbiology, 123*(2), 308–324. https://doi.org/10.1111/jam.13469.
5. Bußler, S., Steins, V., Ehlbeck, J., Schlüter, O. (2015). Impact of thermal treatment versus cold atmospheric plasma processing on the techno-functional protein properties from *Pisum sativum* 'Salamanca.' *Journal of Food Engineering, 167*, 166–174. https://doi.org/10.1016/j.jfoodeng.2015.07.025.
6. Campelo, P. H., Alves Filho, E. G., Silva, L. M. A., de Brito, E. S., Rodrigues, S., Fernandes, F. A. N. (2020). Modulation of aroma and flavor using glow discharge plasma technology. *Innovative Food Science & Emerging Technologies, 62*, Article ID 102363. https://doi.org/10.1016/j.ifset.2020.102363.
7. Cao, Z., Nie, Q., Bayliss, D. L., Walsh, J. L., Ren, C. S. (2010). Spatially extended atmospheric plasma arrays. *Plasma Sources Science and Technology, 19*(2), Article ID 025003. https://doi.org/10.1088/0963-0252/19/2/025003.
8. Charoux, C. M. G., Free, L., Hinds, L. M., Vijayaraghavan, R. K., Daniels, S. (2020). Effect of non-thermal plasma technology on microbial inactivation and total phenolic

content of a model liquid food system and black pepper grains. *LWT, 118*, Article ID 108716. https://doi.org/10.1016/j.lwt.2019.108716.

9. Chutia, H., Kalita, D., Mahanta, C. L., Ojah, N., Choudhury, A. J. (2019). Kinetics of inactivation of peroxidase and polyphenol oxidase in tender coconut water by dielectric barrier discharge plasma. *LWT, 101*, 625–629. https://doi.org/10.1016/j.lwt.2018.10.084.
10. Coutinho, N. M., Silveira, M. R., Fernandes, L. M., Moraes, J., Pimentel, T. C. (2019). Processing chocolate milk drink by low-pressure cold plasma technology. *Food Chemistry, 278*, 276–283. https://doi.org/10.1016/j.foodchem.2018.10.123.
11. Coutinho, N. M., Silveira, M. R., Pimentel, T. C., Freitas, M. Q., Moraes, J. (2019). Chocolate milk drink processed by cold plasma technology: Physical characteristics, thermal behavior and microstructure. *LWT, 102*, 324–329. https://doi.org/10.1016/j.lwt.2018.12.027.
12. Cullen, P. J., Lalor, J., Scally, L., Boehm, D., Milosavljević, V. (2018). Translation of plasma technology from the lab to the food industry. *Plasma Processes and Polymers, 15*(2), Article ID 201700085. https://doi.org/10.1002/ppap.201700085.
13. Dharini, M., Jaspin, S., Mahendran, R. (2022). Optical emission spectra characterization of plasma discharge during the processing of milk and evaluation of its quality attributes. *IEEE Transactions on Plasma Science, 50*(4), 1128–1136. https://doi.org/10.1109/TPS.2022.3155645.
14. Eazhumalai, G., Ranjitha Gracy, T. K., Mishra, A., Annapure, U. S. (2021). Atmospheric pressure nonthermal pin to plate plasma system for the microbial decontamination of oat milk. *Journal of Food Processing and Preservation, 2021*(September), 1–12. https://doi.org/10.1111/jfpp.15722.
15. Farias, T. R. B., Alves Filho, E. G., Silva, L. M. A., De Brito, E. S., Rodrigues, S., Fernandes, F. A. N. (2021). NMR evaluation of apple cubes and apple juice composition subjected to two cold plasma technologies. *LWT, 150*, Article ID 112062. https://doi.org/10.1016/j.lwt.2021.112062.
16. Feizollahi, E., Misra, N. N., Roopesh, M. S. (2020). Factors influencing the antimicrobial efficacy of dielectric barrier discharge (DBD) atmospheric cold plasma (ACP) in food processing applications. *Critical Reviews in Food Science and Nutrition, 61*(4), 666–689. https://doi.org/10.1080/10408398.2019.1568261.
17. Fernandes, F. A. N., Rodrigues, S. (2021). Cold plasma processing on fruits and fruit juices: A review on the effects of plasma on nutritional quality. *Processes, 9*(12), Article ID 2098. https://doi.org/10.3390/pr9122098.
18. Fernandes, F. A. N., Santos, V. O., Rodrigues, S. (2019). Effects of glow plasma technology on some bioactive compounds of acerola juice. *Food Research International, 115*, 16–22. https://doi.org/10.1016/j.foodres.2018.07.005.
19. Furmanski, J., Kim, J. Y., Kim, S. O. (2011). Triple-coupled intense atmospheric pressure plasma jet from honeycomb structural plasma device. *IEEE Transactions on Plasma Science, 39*(11, Part 1), 2338–2239. https://doi.org/10.1109/TPS.2011.2161546.
20. Gadri, R. B., Roth, J. R., Montie, T. C., Kelly-Wintenberg, K., Tsai, P. P.-Y. (2000). Sterilization and plasma processing of room temperature surfaces with a one atmosphere uniform glow discharge plasma (OAUGDP). *Surface and Coatings Technology, 131*(1–3), 528–541. https://doi.org/10.1016/S0257-8972(00)00892-1.
21. Gallagher, M. J., Vaze, N., Gangoli, S., Vasilets, V. N., & Gutsol, A. F. (2007). Rapid inactivation of airborne bacteria using atmospheric pressure dielectric barrier grating

discharge. *IEEE Transactions on Plasma Science, 35*(5 II), 1501–1510. https://doi.org/10.1109/TPS.2007.902080.

22. Gan, Z., Feng, X., Hou, Y., Sun, A., & Wang, R. (2021). Cold plasma jet with dielectric barrier configuration: Investigating its effect on the cell membrane of *E. Coli* and *S. cerevisiae* and its impact on the quality of chokeberry juice. *LWT, 136*(1), Article ID 110223. https://doi.org/10.1016/j.lwt.2020.110223.
23. Ghezzar, M. R., Abdelmalek, F., Belhadj, M., Benderdouche, N., & Addou, A. (2009). Enhancement of the bleaching and degradation of textile wastewaters by gliding arc discharge plasma in the presence of TiO_2 catalyst. *Journal of Hazardous Materials, 164*(2–3), 1266–1274. https://doi.org/10.1016/j.jhazmat.2008.09.089.
24. Gurol, C., Ekinci, F. Y., Aslan, N., & Korachi, M. (2012). Low temperature plasma for decontamination of *E. coli* in milk. *International Journal of Food Microbiology, 157*(1), 1–5. https://doi.org/10.1016/j.ijfoodmicro.2012.04.017.
25. Gutierrez-Leon, D. G., & Serrano-Ramirez, T. (2019). Low temperature plasma by corona discharge in water: Lethal effect on *Escherichia coli* and Salmonella enterica serotype Typhimurium. In *2019 IEEE International Conference on Applied Science and Advanced Technology (iCASAT)*, Publisher: IEEE, pp. 1–5. https://doi.org/10.1109/iCASAT48251.2019.9069516.
26. Hosseini, S. M., Hosseinzadeh Samani, B., Rostami, S., & Lorigooini, Z. (2021). Design and characterization of jet cold atmospheric pressure plasma and its effect on *Escherichia coli*, color, pH, and bioactive compounds of sour cherry juice. *International Journal of Food Science & Technology, 56*(10), 4883–4892. https://doi.org/10.1111/ijfs.14933.
27. Hou, Y., Wang, R., Gan, Z., Shao, T., & Zhang, X. (2019). Effect of cold plasma on blueberry juice quality. *Food Chemistry, 290*, 79–86. https://doi.org/10.1016/j.foodchem.2019.03.035.
28. Leite, A. K. F., Fonteles, T. V., Miguel, T. B. A. R., Silvestre da Silva, G., & Sousa de Brito, E. (2021). Atmospheric cold plasma frequency imparts changes on cashew apple juice composition and improves vitamin C bio accessibility. *Food Research International, 147*, Article ID 110479. https://doi.org/10.1016/j.foodres.2021.110479.
29. Katsigiannis, A. S., Bayliss, D. L., & Walsh, J. L. (2022). Cold plasma for the disinfection of industrial food-contact surfaces: An overview of current status and opportunities. *Comprehensive Reviews in Food Science and Food Safety, 21*, 1086–1124. https://doi.org/10.1111/1541-4337.12885.
30. Kim, H. J., Yong, H. I., Park, S., Kim, K., Choe, W., & Jo, C. (2015). Microbial safety and quality attributes of milk following treatment with atmospheric pressure encapsulated dielectric barrier discharge plasma. *Food Control, 47*, 451–456. https://doi.org/10.1016/j.foodcont.2014.07.048.
31. Kim, H. S., Cho, Y. I., Hwang, I. H., Lee, D. H., & Cho, D. J. (2013). Use of plasma gliding arc discharges on the inactivation of *E. coli* in water. *Separation and Purification Technology, 120*, 423–428. https://doi.org/10.1016/j.seppur.2013.09.001.
32. Laroque, D. A., Seó, S. T., Valencia, G. A., Laurindo, J. B., & Carciofi, B. A. M. (2022). Cold plasma in food processing: Design, mechanisms, and application. *Journal of Food Engineering, 312*, Article ID 110748. https://doi.org/10.1016/j.jfoodeng.2021.110748.
33. Laroussi, M., Alexeff, I., & Kang, W. L. (2000). Biological decontamination by nonthermal plasmas. *IEEE Transactions on Plasma Science, 28*(1), 184–188. https://doi.org/10.1109/27.842821.

34. Liao, X., Li, J., Muhammad, A. I., Suo, Y., & Chen, S. (2018). Application of a dielectric barrier discharge atmospheric cold plasma (DBD-ACP) for *Escherichia coli* inactivation in apple Juice. *Journal of Food Science, 83*(2), 401–408. https://doi.org/10.1111/1750-3841.14005.
35. Liu, Z., Zhao, W., Zhang, Q., Gao, G., & Meng, Y. (2021). Effect of cold plasma treatment on sterilizing rate and quality of kiwi turbid juice. *Journal of Food Process Engineering, 44*(6), Article ID 13711. https://doi.org/10.1111/jfpe.13711.
36. Lu, X., Naidis, G. V., Laroussi, M., Reuter, S., Graves, D. B., & Ostrikov, K. (2016). Reactive species in non-equilibrium atmospheric-pressure plasmas: Generation, transport, and biological effects. *Physics Reports, 630*, 1–84. https://doi.org/10.1016/j.physrep.2016.03.001.
37. Magureanu, M., Bradu, C., Piroi, D., Mandache, N. B., & Parvulescu, V. I. (2013). Pulsed corona discharge for degradation of methylene blue in water. *Plasma Chemistry and Plasma Processing, 33*(1), 51–64. https://doi.org/10.1007/s11090-012-9382-8.
38. Mandal, R., Singh, A., & Singh, A. P. (2018). Recent developments in cold plasma decontamination technology in the food industry. *Trends in Food Science & Technology, 80*, 93–103. https://doi.org/10.1016/j.tifs.2018.08.003.
39. Manoharan, D., Stephen, J., & Radhakrishnan, M. (2021). Study on low-pressure plasma system for continuous decontamination of milk and its quality evaluation. *Journal of Food Process Engineering, 45*(2), Article ID 15138. https://doi.org/10.1111/jfpp.15138.
40. Mehta, D., & Yadav, S. K. (2020). Impact of atmospheric non-thermal plasma and hydrothermal treatment on bioactive compounds and microbial inactivation of strawberry juice: A hurdle technology approach. *Food Science and Technology International, 26*(1), 3–10. https://doi.org/10.1177/1082013220912555.
41. Misra, N. N., Tiwari, B. K., Raghavarao, K. S. M. S., & Cullen, P. J. (2011). Nonthermal plasma inactivation of food-borne pathogens. *Food Engineering Reviews, 3*(3–4), 159–170. https://doi.org/10.1007/s12393-011-9042-0.
42. Misra, N. N., Yepez, X., Xu, L., & Keener, K. (2019). In-package cold plasma technologies. *Journal of Food Engineering, 244*, 21–31. https://doi.org/10.1016/j.jfoodeng.2018.09.017.
43. Mohammadpour, H., Zarei, M., Cullen, P. J., Valtchev, P., Schindeler, A., & Dehghani, F. (2021). Potential application of non-thermal atmospheric plasma in reducing the activity of Pseudomonas-secreted proteases in milk. *International Dairy Journal, 120*, Article ID 105078. https://doi.org/10.1016/j.idairyj.2021.105078.
44. Møller, I. M., Jensen, P. E., & Hansson, A. (2007). Oxidative modifications to cellular components in plants. *Annual Review of Plant Biology, 58*, 459–481. https://doi.org/10.1146/annurev.arplant.58.032806.103946.
45. Muhammad, A. I. (2022). Application of cold plasma in liquid food products. In *Applications of Cold Plasma in Food Safety*, pp. 269–291. Springer. https://doi.org/10.1007/978-981-16-1827-7_11.
46. Muhammad, A. I., Li, Y., Liao, X., Liu, D., & Ye, X. (2019). Effect of dielectric barrier discharge plasma on background microflora and physicochemical properties of tiger nut milk. *Food Control, 96*, 119–127. https://doi.org/10.1016/j.foodcont.2018.08.023.
47. Munekata, P. E. S., Domínguez, R., Pateiro, M., & Lorenzo, J. M. (2020). Influence of plasma treatment on the polyphenols of food products—A review. *Foods, 9*(7), Article ID 976. https://doi.org/10.3390/foods9070976.

48. Nguyen, T., Palmer, J., Phan, N., Shi, H., Keener, K., & Flint, S. (2022). Control of aflatoxin M1 in skim milk by high voltage atmospheric cold plasma. *Food Chemistry, 386*, Article ID 132814. https://doi.org/10.1016/j.foodchem.2022.132814.
49. Nikmaram, N., & Keener, K. M. (2022). The effects of cold plasma technology on physical, nutritional, and sensory properties of milk and milk products. *LWT, 154*, Article ID 112729. https://doi.org/10.1016/j.lwt.2021.112729.
50. Olatunde, O. O., Benjakul, S., & Vongkamjan, K. (2020). Microbial diversity, shelf-life and sensory properties of Asian sea bass slices with combined treatment of liposomal encapsulated ethanolic coconut husk extract and high voltage cold plasma. *LWT, 134*, Article ID 110232. https://doi.org/10.1016/j.lwt.2020.110232.
51. Pankaj, S. K., Wan, Z., Colonna, W., & Keener, K. M. (2017). Effect of high voltage atmospheric cold plasma on white grape juice quality. *Journal of the Science of Food and Agriculture, 97*(12), 4016–4021. https://doi.org/10.1002/jsfa.8297.
52. Pohl, P., Dzimitrowicz, A., Cyganowski, P., & Jamroz, P. (2022). Do we need cold plasma treated fruit and vegetable juices? A case study of positive and negative changes occurred in these daily beverages. *Food Chemistry, 375*, Article ID 131831. https://doi.org/10.1016/j.foodchem.2021.131831.
53. Ribeiro, K. C. S., Coutinho, N. M., Silveira, M. R., Rocha, R. S., & Arruda, H. S. (2021). Impact of cold plasma on the techno-functional and sensory properties of whey dairy beverage added with Xylo oligosaccharide. *Food Research International, 142*, Article ID 110232. https://doi.org/10.1016/j.foodres.2021.110232.
54. Rodriguez, Ó., Rodrigues, S., & Fernandes, F. A. N. (2022). Effect of glow discharge plasma technology on the phenolic content and antioxidant capacity of four tropical juices with different phenolic composition. *Journal of Food Processing and Preservation, 46*(1), 1–10. https://doi.org/10.1111/jfpp.15326.
55. Rycroft, M. J. (1993). Gas discharge physics. *Journal of Atmospheric and Terrestrial Physics, 55*(10), 1469–1482. https://doi.org/10.1016/0021-9169(93)90114-E.
56. Sainz-García, E., López-Alfaro, I., Múgica-Vidal, R., López, R., & Escribano-Viana, R. (2019). Effect of the atmospheric pressure cold plasma treatment on tempranillo red wine quality in batch and flow systems. *Beverages, 5*(3), Article ID 70. https://doi.org/10.3390/beverages5030070.
57. Saremnezhad, S., Soltani, M., Faraji, A., & Hayaloglu, A. A. (2021). Chemical changes of food constituents during cold plasma processing: A review. *Food Research International, 147*, Article ID 110552. https://doi.org/10.1016/j.foodres.2021.110552.
58. Sharma, S., & Singh, R. K. (2022). Effect of atmospheric pressure cold plasma treatment time and composition of feed gas on properties of skim milk. *LWT, 154*, Article ID 112747. https://doi.org/10.1016/j.lwt.2021.112747.
59. Silva, R. M., Filho, E. G. A., Campelo, P. H., Silva, F. E. F., & Zampieri, D. S. (2022). NMR spectroscopy and chemometrics to evaluate the effect of different non-thermal plasma processing on sapota-do-solimões (*Quararibea cordata* Vischer) juice quality and composition. *Food and Bioprocess Technology, 15*(4), 875–890. https://doi.org/10.1007/s11947-022-02845-8.
60. Silveira, M. R., Coutinho, N. M., Esmerino, E. A., Moraes, J., & Fernandes, L. M. (2019). Guava-flavored whey beverage processed by cold plasma technology: Bioactive compounds, fatty acid profile and volatile compounds. *Food Chemistry, 279*, 120–127. https://doi.org/10.1016/j.foodchem.2018.12.038.

61. Silveira, M. R., Coutinho, N. M., Rocha, R. S., Moraes, J., & Esmerino, E. A. (2019). Guava flavored whey-beverage processed by cold plasma: Physical characteristics, thermal behavior and microstructure. *Food Research International, 119*, 564–570. https://doi.org/10.1016/j.foodres.2019.01.006.
62. Singh, A. K., Sarkar, P., Janaswamy, S., Yao, Y., & Bhunia, A. K. (2014). *Novel Food Preservation and Microbial Assessment Techniques* (pp. 218–236). CRC Press.
63. Sruthi, N. U., Josna, K., Pandiselvam, R., Kothakota, A., Gavahian, M., & Mousavi Khaneghah, A. (2022). Impacts of cold plasma treatment on physicochemical, functional, bioactive, textural, and sensory attributes of food: A comprehensive review. *Food Chemistry, 368*, Article ID 130809. https://doi.org/10.1016/j.foodchem.2021.130809.
64. Surowsky, B., Bußler, S., & Schlüter, O. K. (2016). Cold plasma interactions with food constituents in liquid and solid food matrices. In *Cold Plasma in Food and Agriculture*, pp. 179–203. Elsevier. https://doi.org/10.1016/B978-0-12-801365-6.00007-X.
65. Suwal, S., Coronel-Aguilera, C. P., Auer, J., Applegate, B., Garner, A. L., & Huang, J. Y. (2019). Mechanism characterization of bacterial inactivation of atmospheric air plasma gas and activated water using bioluminescence technology. *Innovative Food Science & Emerging Technologies, 53*(January), 18–25. https://doi.org/10.1016/j.ifset.2019.03.010.
66. Thirumdas, R., Kothakota, A., Annapure, U., Siliveru, K., & Blundell, R. (2018). Plasma activated water (PAW): Chemistry, physico-chemical properties, applications in food and agriculture. *Trends in Food Science & Technology, 77*, 21–31. https://doi.org/10.1016/j.tifs.2018.04.005.
67. Thirumdas, R., Sarangapani, C., & Annapure, U. S. (2015). Cold Plasma: A novel Non-Thermal Technology for Food Processing. *Food Biophysics, 10*(1), 1–11. https://doi.org/10.1007/s11483-014-9364-3.
68. Turner, M. (2016). Physics of Cold Plasma. In *Cold Plasma in Food and Agriculture: Fundamentals and Applications*, pp. 17–51. Elsevier. https://doi.org/10.1016/B978-0-12-801365-6.00002-0.
69. Waghmare, R. (2021). Cold plasma technology for fruit-based beverages: A review. *Trends in Food Science & Technology, 114*, 60–69. https://doi.org/10.1016/j.tifs.2021.01.049.
70. Wang, S., Liu, Y., Zhang, Y., Lü, X., & Zhao, L. (2022). Processing sheep milk by cold plasma technology: Impacts on the microbial inactivation, physicochemical characteristics, and protein structure. *LWT, 153*, Article ID 112573. https://doi.org/10.1016/j.lwt.2021.112573.
71. Wang, X., Han, Y., Niu, H., Zhang, L. H., Xiang, Q., & Zong, W. (2022). Alternaria mycotoxin degradation and quality evaluation of jujube juice by cold plasma treatment. *Food Control, 137*(February), Article ID 108926. https://doi.org/10.1016/j.foodcont.2022.108926.
72. Wang, Y., Wang, Z., Zhu, X., Yuan, Y., Gao, Z., & Yue, T. (2020). Application of electrical discharge plasma on the inactivation of *Zygosaccharomyces rouxii* in apple juice. *LWT, 121*(December 2019), Article ID 108974. https://doi.org/10.1016/j.lwt.2019.108974.
73. Xiang, Q., Liu, X., Li, J., Liu, S., Zhang, H., & Bai, Y. (2018). Effects of dielectric barrier discharge plasma on the inactivation of *Zygosaccharomyces rouxii* and quality of apple juice. *Food Chemistry, 254*, 201–207. https://doi.org/10.1016/j.foodchem.2018.02.048.
74. Xu, L., Garner, A. L., Tao, B., & Keener, K. M. (2017). Microbial inactivation and quality changes in orange juice treated by high voltage atmospheric cold plasma.

Food and Bioprocess Technology, 10(10), 1778–1791. https://doi.org/10.1007/s11947-017-1991-0.
75. Yannam, S. K., Estifaee, P., Rogers, S., & Thagard, S. M. (2018). Application of high voltage electrical discharge plasma for the inactivation of *Escherichia coli* ATCC 700891 in tangerine juice. *LWT, 90*, 180–185. https://doi.org/10.1016/j.lwt.2017.12.014.
76. Yusupov, M., Bogaerts, A., Huygh, S., Snoeckx, R., van Duin, A. C. T., & Neyts, E. C. (2013). Plasma-induced destruction of bacterial cell wall components: A reactive molecular dynamics simulation. *Journal of Physical Chemistry C, 117*(11), 5993–5998. https://doi.org/10.1021/jp3106953.
77. Zhang, Z. H., Wang, L. H., Zeng, X. A., Han, Z., & Brennan, C. S. (2019). Non-thermal technologies and its current and future application in the food industry: A review. *International Journal of Food Science & Technology, 54*(1), 1–13. https://doi.org/10.1111/ijfs.13911.

CHAPTER 8

EMERGING COLD PLASMA STRATEGIES TO PREVENT OR TREAT MYCOTOXINS IN FOODS

LOKESWARI RAMIREDDY and PAULA BOURKE

ABSTRACT

Mycotoxins are the toxic secondary metabolites produced by fungi, and their residues are primarily found in grains, nuts, and spices. Commonly observed mycotoxins are aflatoxin, ochratoxin, fumonisin, deoxynivalenol, and zearalenone. Ingestion of these mycotoxins has diverse toxic impacts on humans and animals. Moreover, resistance to various environmental and processing conditions makes it hard to eliminate the residues in contaminated foods. Therefore, several regulatory bodies have established stringent regulations on the limits of mycotoxins in various foods to reduce their distribution and toxic effects. Hence, it is vital to prevent or remove mycotoxins from the food chain. Several studies have shown cold plasma as an efficient technique to reduce mycotoxin content with the potential to decontaminate foods. The reactive species such as aldehyde radicals, ozone, and hydroxyl species are considered the primary contributors to the decontamination of fungal cells as well as the disintegration of mycotoxin structure. These reactive species aid in the structural degradation of mycotoxin residues and can form end products with fewer toxic effects. Besides the decontamination effects, plasma exposure can improve seed responses, seedling growth, and plant health. This chapter discusses the occurrence of various mycotoxins in foods and the potential of novel non-thermal plasma technology in preventing the formation of mycotoxins and eliminating the preformed mycotoxins from the food chain.

Nonthermal Plasma Technology for Food and Food Products.
R. Mahendran, Preeti Birwal, and Megh R. Goyal (Eds.)

8.1 INTRODUCTION

Maintaining the quality and safety properties of food and ingredients are two major challenges faced by the global food processing industries. Moreover, there are regulatory pressures on the use of plant protection products and additives, as well as an increased health awareness among consumers for chemical-free, safe, and nutrition-rich foods. Apart from the nutritional profile, both chemical and microbial contamination remain as major concerns. Contamination can occur at different stages of food processing either by natural or human causes. Some of the major food contaminants include mycotoxins, phytotoxins, heavy metals, phycotoxins, and chemical contaminants. These are categorized according to the nature of risk and influence on health. However, based on the risk assessments, mycotoxins are considered an acute hazard associated with higher risk than other chemical contaminants like pesticides or food additives. Additionally, mycotoxins represent a primary concern across feed and food processing industries as they are associated with both acute and chronic health diseases.

Mycotoxicosis is an acute disease caused by the ingestion of mycotoxins. Mycotoxins enter the human body not only through plant-based foods but also from animal-based foods. Therefore, it is important to decontaminate or avoid feed that contains mycotoxin residues. In consideration of public health and safety, legislative bodies and countries have established maximum limits for mycotoxin residues in foods and feed [53]. The occurrence of mycotoxins and their maximum limits are discussed below.

Various detection, decontamination, and mycotoxin degradation techniques have been introduced in food and feed processing to control or eliminate mycotoxins. Chromatographic and immunochemical-based detection methods, as well as conventional, thermal, chemical, and absorbent decontamination and degradation technologies, have been comprehensively described in previous studies [6, 15, 36, 59].

Cold plasma technology is a novel non-thermal technology with reported promising effects on the inactivation of fungal cells or on the degradation of mycotoxins. Plasma is known as the fourth state of matter. As the input energy levels increase beyond the threshold levels of the gaseous state, it changes into plasma, which is a mixture of ions, free radicals, photons, and atoms [65, 74].

Cold plasma can be produced by applying electrical energy to carrier gases like air, nitrogen, oxygen, helium, or argon with the help of electrodes. Depending on the system configuration, there is a wide range of plasma generation systems like Dielectric Barrier Discharge (DBD), Radiofrequency plasma, Microwave plasma, Corona discharge, etc. The most common plasma reactive species are Reactive Oxygen Species (ROS) like O_3, O^-, O_2^-, $^1O_2$, O, and O_2, Reactive Nitrogen Species (RNS) like N, N^+, N_2^-, NO, and NO^+ [77]. The concentration and type of reactive species vary with the system design and operating parameters. Cold plasma has a wide range of applications in the agriculture food sector such as decontamination, accelerated seed germination, quality improvement, disinfestation, pesticides, and toxin degradation [4, 5, 13, 49]. Numerous studies have reported the efficacy of plasma in decontaminating foods from spoilage microflora and degrading pre-formed mycotoxins. The mechanisms of microbial decontamination and mycotoxin degradation are discussed in this chapter.

Current chapter focuses on the potential of cold plasma technology for prevention or mitigation of mycotoxins in the food chain.

8.2 OCCURRENCE AND TOXICITY OF MYCOTOXINS IN FOODS

Mycotoxins are the toxic metabolites produced by fungal or mold cells when exposed to environmental conditions favorable to their production. Consumption of these residues in adequate amounts can result in a range of toxicological effects. Some of the major mycotoxin-producing fungal species are *Aspergillus, Penicillium, Fusarium, and Alternaria* [31]. Fungal species can produce several types of mycotoxins relying on various biological, physical, and environmental conditions that accelerate the contamination and growth of fungi. Mycotoxin residues are correlated with diseased or infected crops [89]. Apart from the contaminated crops, mycotoxins enter the human body by consuming animal products like milk, meat, or eggs as a consequence of contaminated feed consumption by the animal [89].

Fungal contamination occurs at various stages of food production from the field through food processing and storage. Unhygienic conditions, room temperature, high humidity during processing and storage of produce favor the growth of fungal cells and mycotoxin production [89]. The favorable range of temperature, pH, and water activity for fungal cell development are 10 to 40°C, 8.4, and >0.7, respectively [50]. Furthermore, high temperature, moisture, and pest infestation accelerate mycotoxin production and accumulation among stored foods. Physical damage, gas concentrations, nutrient availability, and fungal cell concentration impact the level of contamination. The most common mycotoxin-infested crops are peanuts, cereal grains, cotton, corn, and nuts [51]. Most of the mycotoxin residues present in foods are in foods prepared from contaminated millets, peanuts, and cereals like maize, wheat, sorghum, spices, or leguminous seeds like soybeans [22].

Mycotoxins are resistant to various environmental and processing conditions including pasteurization, cooking, roasting, boiling, frying, and baking. Due to this heat-stable behavior, the residues of mycotoxins can appear even after the complete elimination of the producer fungal cells [30]. This makes the removal of pre-formed mycotoxins a challenging task, and significant efforts have been dedicated to detect and remove mycotoxin residues from foods. According to the report of the United Nations Food and Agriculture Organization (FAO), published prior to 1985, 25% of the world's crops may possibly be infested with mycotoxins, including pre- and post-harvest contamination. However, recent data show an increased possibility of contamination, i.e., up to 60 to 80% [23]. The five major groups of mycotoxins often found in foods are aflatoxins, ochratoxin, fumonisin, zearalenone, and nivalenol.

8.2.1 AFLATOXIN (AF)

The term aflatoxin (*Aspergillus flavus* toxins) is named after the causative fungus *Aspergillus flavus*. The aflatoxin residues have been reported in a wide range of foods including cereal grains, spices, nuts, oil seeds, dried fruits, and figures [76]. Although the contamination is highly reported in crops grown and stored in tropical regions, global trading results in the spread within importing countries [2]. There are numerous aflatoxins, and the prominent aflatoxins observed in foods are Aflatoxins B (AFB1 and B2), Aflatoxins G (G1 and G2), and Aflatoxins M (M1 and M2). The International Agency for Research on Cancer (IARC) categorized aflatoxins as group 1 carcinogens; Hepatocellular carcinoma (HCC) is a result of aflatoxin toxicity. Aflatoxin ingestion can also induce hepatotoxicity, teratogenicity, genotoxicity, and cytotoxicity [35]. The Aflatoxins B and G can be found in food grains, nuts, and dried fruits, whereas Aflatoxins M are the metabolic products of aflatoxins observed in milk, deriving from animal consumption of AFB1 or B2 contaminated food [35]. The USFDA proposed the permissible limits of aflatoxins in various foods except milk as 20 μg/kg [27].

8.2.2 OCHRATOXIN (OT)

Ochratoxins are the mycotoxins produced by several fungi, namely *Aspergillus ochraceus, Aspergillus Niger*, and *Penicillium sps*. Ochratoxin A (OTA), B (OTB), and C (OTC) are different forms of ochratoxin. OTA is the chlorinated form of OTB, and OTC is the ethyl ester of OTA [33]. However, OTA is highly toxic compared to other forms of ochratoxins and is responsible for adverse health effects in both humans and animals. OTA is produced under tropical and temperate conditions by *A. ochraceus* and *P. verrucosum*, respectively [3]. They frequently appear in a wide range of plant and animal-based foods such as cereals, cereal-based products, tea, coffee, fruits, vegetables, dried foods, milk, eggs, poultry, pork, and fish [45]. Unlike aflatoxins, IRAC has categorized OTA as class 2b, a potential human carcinogen. OTA is considered highly toxic compared to B and C. Moreover, the intake of OTA is associated with multiple health issues including neurotoxicity, nephrotoxicity, hepatotoxicity, immunotoxicity, and genotoxicity. OTA has a great affinity to proteins, resulting in accumulation and transfer to humans through animal-derived products [47]. Coffee and wine are reported as major contributors to OTA. OTA has a tolerance to both high temperatures and acidic conditions, making it difficult to eliminate during normal processing [11].

8.2.3 FUMONISIN

Fumonisins are hydrophilic toxic residues produced by pathogenic fungi *Fusarium spp.*, namely *F. verticillioides* and *F. proliferatum* [39]. Fumonisins are most commonly observed in corn and corn-based products. *Fusarium* spp. infest the tip region and tissue of corn below the pericarp, leading to contamination of the inner parts of the grain. The four categories of fumonisins are A, B, C, and P. Fumonisin B is the predominant fumonisin in foods and is categorized as group 2b by IARC [3]. The structural similarities between fumonisin and sphinganine lead to the interruption of sphingolipid metabolism. They result in toxic effects on targeted organs like the kidney and liver in animals [3]. So far, no studies have reported the kinetics and metabolism of fumonisin in humans. However, epidemiological studies conducted in several countries disclosed that the occurrence of esophageal cancer in association with *F. verticillioides* seemed to be correlative [84].

8.2.4 TRICHOTHECENES

These are a large group of mycotoxins composed of approximately 150 metabolites produced by fungal species like *Fusarium, Myrothecium, Trichoderma, Verticimonosporium, Trichotecium, Phomopsis,* and *Stachybotrys*. Some of the important and common trichothecenes are deoxynivalenol, nivalenol, diacetoxyscirpenol, toxin T2, and HT2. Trichothecenes are capable of inhibiting protein synthesis in eukaryotes. They interfere at initiation, elongation, and termination of protein synthesis [19].

Deoxynivalenol (DON) is the most common trichothecene observed in foods. DON produced by Fusarium species is mainly produced from *F. graminearum* and *F. Culmorum*. DON is a group B trichothecene, majorly observed in cereal grains like wheat, barley, oats, maize, and other cereal grains and in cereal-based foods. They are hard to remove because of their resistance to various processing conditions like milling, cooking, and high temperature [16]. DON reduces cell proliferation by inhibiting protein and DNA synthesis. Their intake is also associated with gastrointestinal problems and reproductive toxicity in humans [24]. Surveys on human epidemiology demonstrated high exposure to DON-contaminated water and foods have a high prevalence of bone malformation and Kashin-beck disease (KBD). Because of its toxicity, there are imposed legal limits for the concentration of DON in foodstuffs. According to the European Commission regulation 2007, the legal limits of DON concentration range between 200 and 1750 μg/kg depending on the type of commodity [93].

8.2.5 ZEARALENONE

Zearalenone (ZEN) is a non-steroidal estrogenic mycotoxin produced by *Fusarium spp.*, especially *F. graminearum, F. culmorum, F. equiseti,* and *F. verticillioides*. It is mainly found in cereal crops like maize, wheat, sorghum, barley, and rye. ZEN residues have also been discovered in milk when animals consume contaminated feed [41]. *Fusarium* contamination is commonly observed in plants grown in temperate climatic conditions. ZEA is resistant to high temperatures, up to 150°C, and various food processing techniques. ZEA has a strong binding affinity towards estrogenic receptors and exhibits estrogenic effects. Additionally, ZEA binds with HSDS (hydroxysteroid dehydrogenase) to produce endogenous hormones [73]. Due to these health risks, EU legislation has established maximum limits for processed and unprocessed cereals at 100–200 μg/kg and 75 μg/kg, respectively [20].

8.3 IMPACT OF MYCOTOXINS ON FOOD SAFETY

Although the occurrence of mycotoxins is observed in certain geographical and climatic conditions, the global food supply chain encourages distribution throughout the world. Due to the adverse health impacts of mycotoxins, several laws have been implemented worldwide by prescribing the maximum limits in food or feedstuffs. Frequent monitoring is required to ensure the safety of the food. As it is not possible to completely remove these natural toxin residues in the food, several government agencies have adopted permissible limits in various foods [91]. The permissible limits of various mycotoxins range from non-detectable levels to certain parts per billion (ppb) depending on the country's food laws and regulations.[52] The detectable levels may change with the detection method used, and as detection methods improve, the limits of detection may be subject to change.

Various factors have to be considered in establishing the limits and regulations like the availability of toxicological data, the occurrence and distribution of mycotoxins, the availability of infrastructure, permissible limits in the importing or exporting countries, and the availability of sufficient food [21]. The Codex Alimentarius Commission (CAC) established the permissible limits of various mycotoxins in food and feed to assist food safety during international trading [7]. Proposed maximum permissible limits of mycotoxins in various foods by different regulatory bodies are presented in Table 8.1.

Mycotoxin contamination is one of the major challenges of food safety, particularly in developing countries. Most developed countries have well-established infrastructure facilities for monitoring food standards, whereas enforcing and monitoring food quality and safety can be challenging to achieve in developing countries. Due to strict monitoring and high food quality standards, good quality crops tend to be exported to developed countries, which may contribute to concentrating the food safety risk among populations in developing countries [85]. Therefore, it is vital to introduce efficient and affordable techniques in the food processing chain to mitigate mycotoxin formation in the first instance by inactivating the responsible fungi or degrading the formed mycotoxins in the foods.

8.4 COLD PLASMA TREATMENT FOR FUNGAL DECONTAMINATION

Mycotoxins can be formed at various stages in the food production cycle. One of the sustainable approaches is to prevent the formation of mycotoxins or to prevent fungal infection. Foods that are frequently contaminated by fungi are grains, nuts, fruits, vegetables, and animal products. Fungi are eukaryotic organisms with single or multicellular structures and can contaminate food by entering the food chain through infected seeds or soil, water, or spores transferred through the air. Several physical, biological, and chemical methods have been used to inactivate fungal cells on contaminated samples. Although they can be important in inactivating fungal cells to a certain extent, issues such as

TABLE 8.1 Maximum Acceptable Limits of Mycotoxins in Various Foods by Different Regulatory Bodies

Regulatory Body	Mycotoxin	Type of Food	Maximum Limits (µg/kg)	References
Codex Alimentarius standard (CAC)	Aflatoxin (B1+B2+G1+G2)	Ready-to-eat (RTE) nut: almonds pistachios, Brazil nuts, hazelnuts and dried figs	10	[26]
		Intended for further processing almonds, pistachios, Brazil nuts, dried figs, and hazelnuts	15	
	Aflatoxin M1	Milks	0.5	
	DON	Cereal based products intended for infant and young children	200	
		Wheat, maize or barley derived products like flour, meal, flakes, semolina,	1,000	
		Cereal grains intended for further processing	2,000	
	Fumonisin (B1+B2)	Maize grain	4,000	
		Flour and meal derived from maize	2,000	
	Ochratoxin A	Wheat, barley and rye	5	
	Patulin	Apple juice	50	
European Commission	Aflatoxins (B1+B2+G1+G2)	Groundnuts intended for processing	15	
		Nuts, dry fruits intended for processing	10	
		Groundnuts intended for human consumption	4	
		Dried fruits intended for human consumption, Cereals and cereal derived products	4	
		Maize intended for processing	10	
		Spices – chilies, paprika, cayenne, white and black pepper, ginger, turmeric and nutmeg	10	

TABLE 8.1 *(Continued)*

Regulatory Body	Mycotoxin	Type of Food	Maximum Limits (µg/kg)	References
	Aflatoxin B1	Groundnuts intended for processing	8	
		Nuts, dry fruits intended for processing	5	
		Dried fruits intended for human consumption, Cereals and cereal derived products	2	
		Spices – chilies, paprika, cayenne, white and black pepper, ginger, turmeric and nutmeg	5	
		Cereal based Infant and baby foods	0.10	
		Maize intended for processing	5	
		Foods intended for medical purposes especially infants	0.10	
	Aflatoxin M1	Milk	0.050	
		Infant milk or infant formulae	0.025	
		Foods intended for medical purposes especially infants	0.025	
	Ochratoxin	Unprocessed cereal grains	5	
		Cereal based products	3	
		Dried vine fruits like currants, raisins and sultans	10	
		Roasted and ground roasted coffee beans	5	
		Instant coffee	10	
		Wines and fruit juices	2	
		Cereal based foods for infants	0.5	
	DON	Unprocessed cereals except durum wheat, maize and oats	1,250	
		Unprocessed durum wheat, maize and oats	1,750	

TABLE 8.1 *(Continued)*

Regulatory Body	Mycotoxin	Type of Food	Maximum Limits (µg/kg)	References
		Cereals, cereal flour and bran intended for human consumption and dry pasta	750	
		Breakfast cereals, bread, pastries, cereal snacks and biscuits	500	
		Cereal based infant foods	200	
	Zearalenone	Unprocessed cereals except maize	100	
		Unprocessed maize	200	
		Maize and derived products intended human consumption	200	
		Cereals, cereal flour and bran intended for human consumption and dry pasta	75	
		Breakfast cereals, bread, pastries, cereal snacks and biscuits	50	
		Cereal based foods for infant and young children	20	
	Fumonisin (B1+B2)	Unprocessed maize	2,000	
		Flour, grits, meal, germ and oil derived from maize	1,000	
		Maize based products intended for direct human consumption	400	
		Processed maize products for infants and young children,	200	
U.S. FDA maximum limits	Aflatoxin (B1+B2+G1+G2)	All food crops	20	[90]
	Fumonisin (B1+B2+B3)	Maize	4,000	
	DON	Wheat products intended for human consumption	1,000	
		Grains and grain by products	10,000	

processing time, chemical residues, or resultant changes in quality parameters can occur, which drives the need for novel decontamination technologies [34].

Cold plasma has been reported as an alternative to conventional decontamination techniques by several studies [56, 58, 60, 72]. One of the major advantages of using plasma technology is that cells cannot develop any resistance traits, unlike with specific fungicide treatments [70]. The combined and synergistic effects of UV, electric fields, charged particles, and generated reactive species contribute to the antimicrobial effect. The decontamination efficiency of cold plasma depends on system and process parameters like the type of working gas used, the exposure time, direct or indirect mode of exposure, and system power density. The target microorganism can also play a vital role, where the diversity of fungal species and their phenotype can vary across foods and environments.

Plasma reactive species trigger a series of biological reactions in microorganisms leading to their inactivation. Recent studies explain that cold plasma inactivates fungal cells by oxidizing or destroying their structure, obstructing cell metabolism and cellular components leading to cell death [95]. The initial mechanism of fungal cell inactivation is assumed to start with the loss of membrane integrity. Membrane integrity is very important for fungi to maintain the structure and to adapt to various environmental conditions. Plasma reactive species could act on the fungal cell membrane by breaking down the membrane, possibly due to the surface etching mechanism [48]. The free radicals and reactive plasma species induce surface etching of the outer membrane by disrupting the bonds and allowing the reactive species into fungal cells. Electrostatic interference can be another possible mechanism of membrane damage where the electrostatic force on the membrane exceeds the tensile strength of the membrane due to the accumulation of charged particles [75].

UV light produced during the generation of cold plasma aids in structural damage of DNA. The photons in the UV alter the DNA bonds by inducing dimerization of thymine leading to changes in the DNA strands and aids in the breakdown of DNA strands. Moreover, the diffusion of plasma reactive species into the cell membrane results in the oxidation of proteins and peroxidation of the membrane lipids. The ROS oxidizes the proteins and impacts the fungal cell vitality by altering the structure of cytosolic enzymes and proteins [48]. The oxidation of lipids results in the formation of lipid peroxides. These reactions alter the composition of the cell, disturb the cell functionality, eventually leading to cell death [57]. Los et al. [50] described the mechanisms of action of cold plasma inactivation of fungal cells as similar for both biofilms and those on aqueous spore suspensions, which comprise a significant reduction in the number and viability of spores as well as the release of intracellular material such as MDA (a product of membrane lipid peroxidation), DNA, and proteins, due to the partial or total disruption of cell walls resulting in the loss of cell wall integrity and, subsequently, cell leakage. The SEM examinations of A. flavus biofilms and spores subjected to ACP revealed a significant change in shape and structural damage to fungal cells, particularly spores. The mechanisms of inactivation of fungal cells were illustrated by Los et al. [50] as shown in Figure 8.1.

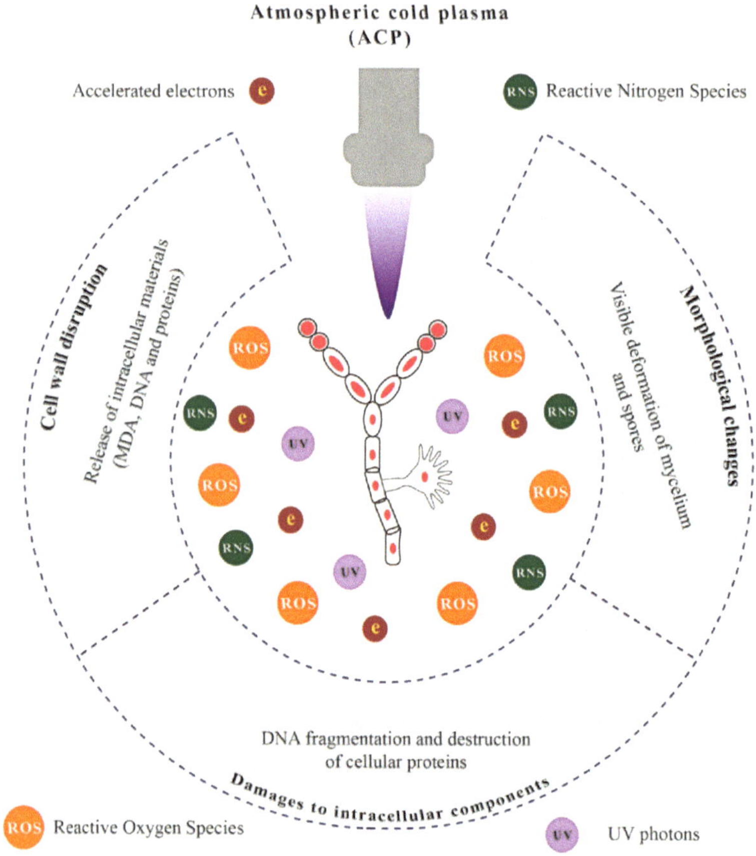

FIGURE 8.1 Inactivation mechanisms of fungal cells by atmospheric cold plasma treatment.
Source: Reprinted from Ref. [50]

A variety of legumes and grains contaminated with *Aspergillus* and *Penicillium spp.* and exposed to the low-pressure sulfur hexafluoride (SF6) plasma for 15 min has reported a 3-log reduction in both fungal cells. Moreover, the inactivation rates on the surface of chickpea and corn are higher when compared to barley and wheat [78].

The *A. parasiticus* contaminated nuts were exposed to air and SF6 plasma for up to 20 minutes; a 2 log and 5 log reduction, respectively, were observed [8]. *F. oxysporum* is one of the frequently observed soil-borne fungal pathogens and exerts great resistance to antifungal agents. Panngom et al. [66] reported that a 10-minute exposure of argon plasma to saline-suspended fungal spores resulted in a high degree of continuous inactivation of *F. oxysporum* spores during the incubation time. The inactivation might result from the high concentration of peroxynitrite and hydroxyl radical formation in the saline solution. Winter wheat was contaminated with nine different fungal species and exposed to cold plasma to analyze the efficiency of spore destruction and seed quality. The seeds were exposed for 3 to 30 seconds at 8 kV, and a significant reduction in CFU of fungi on treated grains was observed at 10 seconds [43].

8.4.1 INFLUENCE OF COLD PLASMA ON PLANT HEALTH

Apart from the nutritional and microbiological food safety perspective for humans, fungal-infected seeds can also impact seedling growth and lead to plant diseases and crop failures. Seeds contaminated with fungi result in seed-borne plant diseases and mycotoxin residues. These fungal species reduce the germination and development of seedlings, resulting in yield losses and food safety concerns. About 20 to 30% of the economic losses in agriculture occur due to fungal pathogen infection [64]. Fungi that contaminate plants at various stages are classified as plant pathogens; examples are *Fusarium moniliforme, A. flavus* (grows on stressed or senescent plants), and *A. flavus* (grows on plants before harvest and produces mycotoxins during processing) [89].

Although a wide range of pre-treatments were used to disinfect the seeds before sowing, the increasing demand for organic foods among consumers and various regulations enforced to reduce the utilization of chemical fungicides create the need for an alternative technology [83]. Cold plasma can also be used as a potential seed-priming technique as it can provide multiple benefits from decontaminating seeds to improving plant health and crop yield. Moreover, it has been noted to prevent the emergence of disease after germination, the spread of fungal pathogens, and post-harvest diseases [78].

Using plasma as a decontamination technique provides an added advantage by boosting plant health. Most of the studies conducted observed that plasma as a seed pre-treatment enhanced seed viability and seedling growth. The plasma exposure modifies the seed surface properties by creating micropores and improving the hydrophobicity of the seeds. A significant reduction in the contact angle, improved water absorption, and surface wettability observed in beans, wheat, and lentils with air plasma treatment due to the surface oxidation of the seeds [10]. The plasma reactive species like ROS and RNS play a prominent role in accelerating the germination of seedlings by influencing the physiological responses of seeds and aids in altering the metabolic and signaling pathways [70].

The influence of plasma exposure on seedling growth depends on various factors related to specific system and process parameters like the type of plasma, exposure time, and voltage as well as seed parameters like the type of seed, size, seed coat, location of the embryo, etc. This requires that the plasma treatment conditions may need to be optimized for each different seed. For instance, exposing soybeans to a lower dose of atmospheric pressure air and oxygen plasma increased the enzymatic activity of succinate dehydrogenase and increased the germination rate by 20%, whereas treating seeds at a higher dose of nitrogen plasma resulted in the inhibition of succinate dehydrogenase activity and damage to DNA structure but activated other enzyme activities like alcohol dehydrogenase and lactate [86].

Apart from the decontamination and germination acceleration, plasma exposure also exhibited a strong impact on the growth state of plants. Panngom et al. [66] examined the influence of plasma on the pathogenesis-related genes in an *F. oxysporum*-contaminated tomato plant cultivar. The study reported an up-regulation of pathogen-resistant gene expression in the roots of the treated plants with no damage to the plant. Lee et al. [46] exposed *Cylindrocarpon destructans* contaminated ginseng seeds to Ar and O_2 plasma and reported surface decontamination, increased germination, and a decrease in the disease severity of root rot. Though the plasma enhances the pathogen resistance in the treated seeds, the mechanisms behind this are not yet clearly elucidated and may vary across pathogens and plants.

8.5 DEGRADATION OF PRE-FORMED MYCOTOXINS IN STORED GRAINS AND FOOD PRODUCTS

Mycotoxins can be formed or enter into foods at different stages. The formation of mycotoxins in the field can be prevented at the pre-harvest stage by inculcating changes in agricultural practices like introducing crop rotation, tillage, and fungicides to reduce fungal attack on the crops. However, it is hard to achieve complete elimination of pre-formed mycotoxins because of their stability under various processing conditions [88]. Over recent decades, processing techniques including physical methods (such as dehulling, sieving, sorting, milling, steeping, germination, and thermal treatments), chemical methods (using oxidizing or reducing agents, acids like citric, acetic, and lactic), enzymatic methods (treating with proteases, amylases, and glucanases), and biological methods (fermentation and genetic engineering) have been used to remove preformed mycotoxins [40].

Some conventional techniques like washing, sorting, decortication, or dehulling can be helpful to remove preformed mycotoxins to a certain extent without forming any toxic by-products [1]. Studies have reported that soaking and washing can reduce the mycotoxin levels as they are concentrated more on the outer layer of the grains. The studies have reported a reduction in OTA and AFs in grains after washing [37, 50]. However, the efficiency of these methods relies on the penetration of fungal contamination and mycotoxins in the grains [36].

Thermal processing steps of dry heat, extrusion, and superheated steam have an impact on reducing mycotoxin residues. Though most mycotoxins are thermally resistant, temperature and moisture content have a significant impact on lowering the mycotoxin content. For example, extrusion has reported a 50 to 80% reduction in AF. Likewise, the combination of high temperatures (150 to 200°C) and humidity achieved a 79% reduction in AFB1 [76]. The degrading temperatures are different for various mycotoxins: AF's (268–269°C), OTA (169°C), DON (151–153°C), and Fumonisin (100–120°C) [37].

The Chemical treatments such as citric acid, lactic acid, hydrogen peroxide, and propionic acid can be used to degrade the mycotoxins into lower toxicity compounds. Various mechanisms of oxidation, hydrolysis, alkalization, reduction, and hydration are involved in degrading the mycotoxin structures [69]. For instance, lactic acid treatment led to the degradation of AFB1 to its degraded form AFD1 by hydrolyzing the lactone ring. By decarboxylation of the lactone ring in AFB1, this lowers its mutagenic toxicity by 18-fold [54].

Ozone, as an extensively used oxidizing agent, has been identified as an effective agent to degrade deoxynivalenol (DON), HT2, and T2 toxins [50, 69]. Meanwhile, several other factors like temperature, moisture, and the form of the chemical have an impact on the reduction of mycotoxins [36]. For example, the combination of high temperature with H_2O_2 and $NaHCO_3$ showed a 100% reduction in FB1 in corn [68]. However, chemical treatments are not recommended by various health organizations as the chances of penetrating the chemicals into the inner layers of grains and the residual chemicals may have negative health effects [55].

In contrast to the above-mentioned methods, an alternative inorganic or organic approach using binding agents like montmorillonite clays (MC), hydrated sodium calcium aluminosilicates, and yeast cell wall is sometimes used. They are designed to bind with the mycotoxins by reducing gastrointestinal absorption and inactivating the cytotoxicity of mycotoxins. The issue is that their efficacy is low and they need to be added in large quantities [69].

By considering the pros and cons of the above-mentioned techniques, non-thermal technologies can have significant advantages as eco-friendly, economical, and chemical-free alternatives. They are used in decontamination, preservation, and processing of foods. They have a positive impact on the quality and safety of the foods. Some of the non-thermal techniques proposed for use in degrading mycotoxins are irradiation, high-pressure processing, pulsed light, pulsed electric field, and cold plasma [1].

Irradiation implies subjecting food to ionizing radiation. It has been adopted by many countries in the world and successfully used as a phytosanitary technique to reduce fungal spores and mycotoxin residues [71]. The degradation percentage of mycotoxins varies depending on the dosage of irradiation. A dosage of up to 10 kGy is considered safe for food processing purposes [55]. Studies have reported that exposure to irradiation degraded 50% of OTA from food and feed. The complete elimination of FB1 in barley grains at 5 kGy, wheat and maize at 7 was achieved [8, 12]. Apart from detoxification, the overall effectiveness of irradiation is questionable due to its biological and chemical impacts on molecular reactions [50].

High-pressure processing subjects foods to high pressure ranging from 300 to 600 MPa. Recent studies reported inactivation of fungal spores and mycotoxin degradation by HPP. The use of 600 MPa for 5 minutes decreased 70% of patulin and 24% of AFB1 in various liquid foods [32, 63]. A similar study carried out at 550 MPa for 20 minutes reported almost complete elimination of DON and ZEA [38]. HPP can be applied in combination with other non-thermal technologies like US and PEF to achieve a higher percentage of deactivation [25, 62]. The challenge in HPP technology is that it is not efficient for treating low moisture content foods and also has limited packaging options [44]. Pulsed light or pulsed electric field involves subjecting food to short pulses of light or an electric field. However, the degradation rate of mycotoxins depends on the intensity of treatment, level of contamination, and moisture content of the sample [1].

Exposure of rice to pulsed light at 0.52 J/cm^2/pulse for various treatment times resulted in a 75% and 39% reduction in AFB1 and AFB2, respectively [92]. A 72% to 95% reduction in AFM1 was reported by exposing contaminated milk to PEF at 80 kV [42]. Low penetration depth is one of the major drawbacks of these technologies, which limits them to surface decontamination [17]. Considering the limitations of non-thermal techniques, cold plasma can be a possible alternative for degrading preformed mycotoxins in foods.

8.6 POSSIBLE DEGRADATION PATHWAYS OF MYCOTOXINS BY COLD PLASMA

The first study on mycotoxin degradation using plasma employed a microwave-argon plasma device and reported almost complete degradation of AFB1, DON, and NIV with 5 seconds of exposure [67]. Subsequently, Basaran et al. [9] examined the potential of cold plasma to degrade mycotoxins from nut surfaces. That study reported that 20 minutes of plasma exposure reduced almost 50% of aflatoxins. Following that, a study using double jet argon plasma reported complete elimination of FB2 and OTA on palm slices. Moreover, plasma exposure completely inactivated *A. Niger* spores [61]. Similar studies were conducted to eliminate various mycotoxins such as DON, ZEN, FB1, sterigmatocystin, T2 toxin, and AAL toxin. Key conclusions indicated that plasma interacts with the structure of mycotoxins, and their degradation rate depends on the molecular structure and type of mycotoxin. For example, the degradation rate of FB1 and AAL was observed to be higher than that of sterigmatocystin. It was also reported that the degradation efficiency of pure mycotoxins is higher compared to mycotoxin residues in foods [87].

The degradation of mycotoxins is certainly associated with plasma chemistry, the molecular structure of mycotoxins, and thus the rate of degradation inevitably depends on the interaction of plasma species with the structure of mycotoxins [57]. The degradation pathways of some mycotoxins by plasma exposure are described below.

8.6.1 AFLATOXINS

Numerous studies on aflatoxins and cold plasma technology are focused on the degradation of AFB1. Both direct and indirect plasma have a significant impact on the degradation of AFB1, indicating that long-lived plasma reactive species play a major role in its degradation [95]. Most studies carried out using air or nitrogen as a source of plasma have observed mixed results [80, 82]. However, the studies reported that aflatoxin B1 and G1 are comparably more sensitive to cold plasma than aflatoxin B2 and G2 because of their structural arrangements [80, 95]. The processing parameters of working gas, humidity, and exposure time play a key role in the percentage of degradation. Increasing the humidity of the input air increased aflatoxin degradation over dry air because of the higher concentrations of reactive species like NOx, ozone, and OH radicals that can be achieved using humidified air. The increase in RH of the gas from 5 to 40% significantly improved the degradation rates of aflatoxin in corn [81].

Controlling the degradation pathways of aflatoxin is challenging as the plasma chemistry involves a plethora of long-term and short-term lived species. In line with the degradation pathways, the AFB1 and AFG1 are highly sensitive to ROS, primarily ozone, in plasma directly compared to AFB2 and AFG2 due to the presence of a C_8–C_9 double bond. The C_8–C_9 double bond is renowned as the reactive site in the aflatoxin structure. The ozone reacts with the C_8–C_9 double bond and forms unstable intermediate compounds [18]. Moreover, the OH radicals in plasma act as a strong oxidizing agent and enhance the reactivity of ozone with the olefinic site in AFB1 [82].

To date, few studies have reported on the degradation pathway of AFB1 after plasma treatment. The primary study conducted by Wang et al. [94], where AFB1 was exposed to radiofrequency plasma at varied power levels from 100 to 300 W for 10 minutes, recorded an 88% degradation rate. The five reported degraded products of AFB1 are shown in Figure 8.3. The initiation of the degradation reaction begins with an addition reaction at C_8–C_9 bonds due to the presence of UV. Further, the reactive species like OH, H, O, OH aid the key role in the degradation process. Thus, the obtained degraded compounds lost the double bond in the terminal furan ring and have a lower toxic effect compared to AFB1 (Figure 8.2).

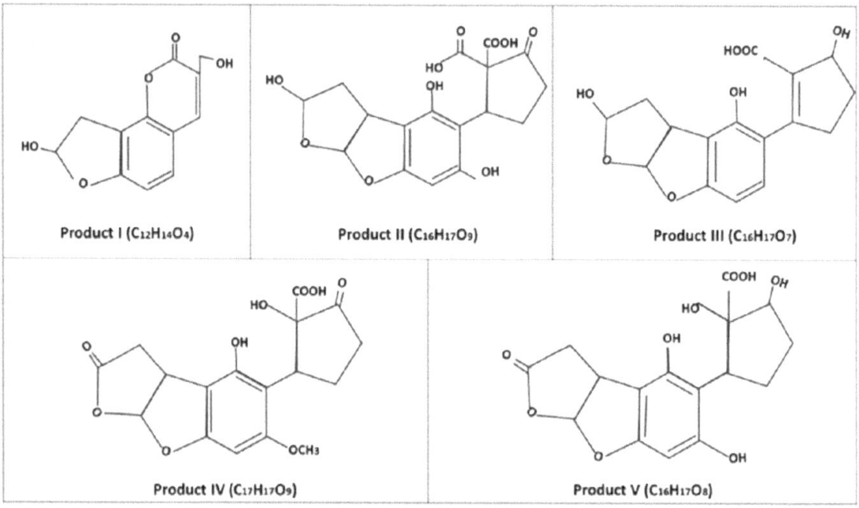

FIGURE 8.2 Proposed five degraded products of aflatoxin B (AFB) after plasma treatment [94].

Shi et al. [80] exposed AFB1 to high-voltage atmospheric cold plasma and observed a 76% reduction with a 5-minute exposure time. The study noticed six degraded products of AFB1 and proposed two degradation mechanisms. Among the six degraded products represented in Figure 8.3, two were formed by the action of O_3 whereas others were formed due to the presence of reactive species like H, OH, CHO, HO_2, H_2O_2. The initial pathway induces an addition reaction which includes H_2O, H, CHO, and the second pathway involves epoxidation and oxidation reactions due to HO_2, OH, H_2O_2, and ozone, respectively. Similar to the study by Wang et al. [94], Shi et al. [80] also reported the disappearance of C_8–C_9 double bonds at the terminal furofuran ring leading to reduced toxicity.

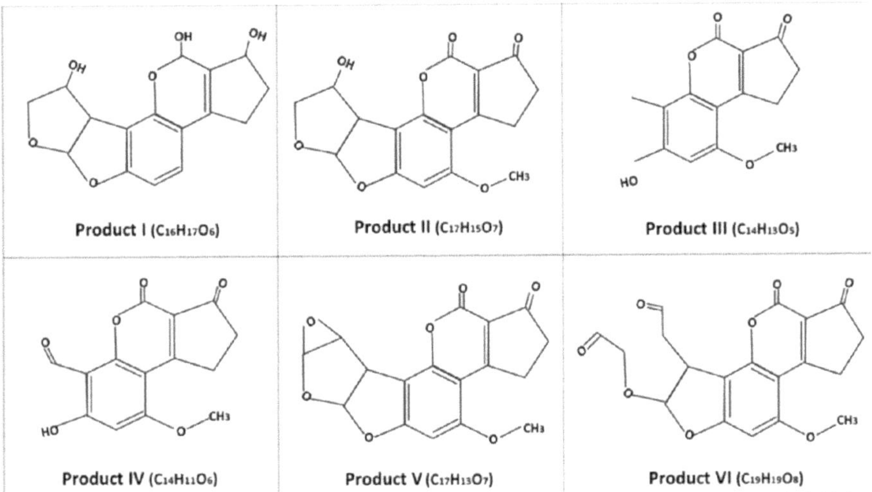

FIGURE 8.3 Proposed six degraded products of aflatoxin B (AFB) after plasma treatment [80].

8.6.2 *DEOXYNIVALENOL*

Exposing DON spiked barley to the DBD atmospheric cold plasma system operating at 34 kV for 6 and 10 minutes resulted in almost 49% and 54% degradation, respectively. The synergistic effect of UV, reactive oxygen, and nitrogen species contributes to the structural degradation of DON [28]. A 99% reduction in DON structure was observed when pure DON was exposed to HVACP at 85 kV for 20 minutes [60].

The possible degradation pathways of DON with plasma treatment were analyzed by Chen et al. [14]. Exposing standard DON to air plasma at 50 kV for 8 minutes resulted in almost 84% degradation of DON standard, and LC/TOF-MS was used to analyze the degraded products. Four degraded forms of DON were identified, namely $C_{15}H_{22}O_7$, $C_{15}H_{20}O_5$, $C_{14}H_{16}O_4$ and

$C_{15}H_{21}NO_9$. The possible reactions of degraded products tend to be due to addition, deoxygenation, and oxidation reactions. The addition reaction of the hydroxyl group of H_2O and C=C bonds in C_9–C_{10} positions of DON molecular structure by forming $C_{15}H_{22}O_7$. Deoxygenation reaction of epoxy groups of C_{12}–C_{13} results in less toxic compound $C_{15}H_{20}O_5$. Further degradation of $C_{15}H_{20}O_5$ results in the formation of $C_{14}H_{16}O_4$. Moreover, the addition of N, H and O_3 molecules result in the fourth degraded product $C_{15}H_{21}NO_9$. The possible molecular structures of degraded products were presented in Figure 8.4.

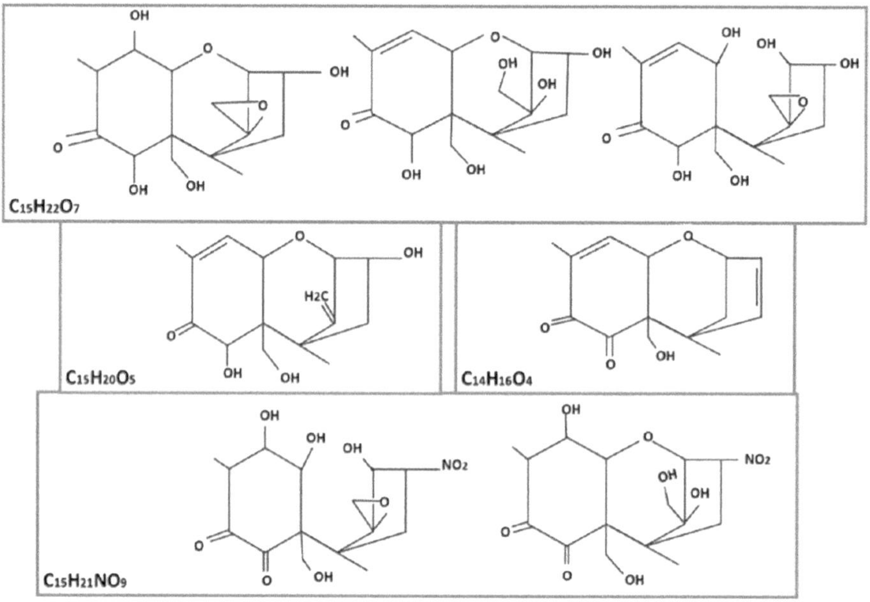

FIGURE 8.4 Proposed degraded don products and Possible molecular structures of degraded products [14].

8.6.3 *ZEARALENONE*

Cold plasma exposure has reported a great reduction in the ZEA concentrations. A 100% degradation of ZEA was observed with 8 minutes exposure to food contaminated by ZEA (2.5 mg/kg) [34]. A 3-minute exposure of contaminated barley grain, canola meal, and canola grain to atmospheric pressure DBD plasma resulted in 64%, 83%, and 91% reduction in the ZEA, respectively [29]. An increase in degradation percentage of zearalenone was observed with increasing voltage level and exposure time, and a 98%

degradation was observed in the sample exposed to 50 kV for 120 s. The cyclic olefin structure in the ZEA molecule decomposes in the presence of ROS, primarily O_3, OH, $^1O_2$, resulting in the degraded product $C_{18}H_{22}O_7$ as shown in Figure 8.5. The proposed degradation pathway was presented in Zheng et al. [96]. The ozonolysis of zearalenone molecule results in the breakdown of C=C to form trioxane cyclic compounds. Furthermore, the radical addition of plasma reactive species induces the formation of aldehydes at the end of the double bond, resulting in the degraded molecule $C_{18}H_{22}O_7$. The toxicity of the degraded form tends to be lower compared to the ZEA.

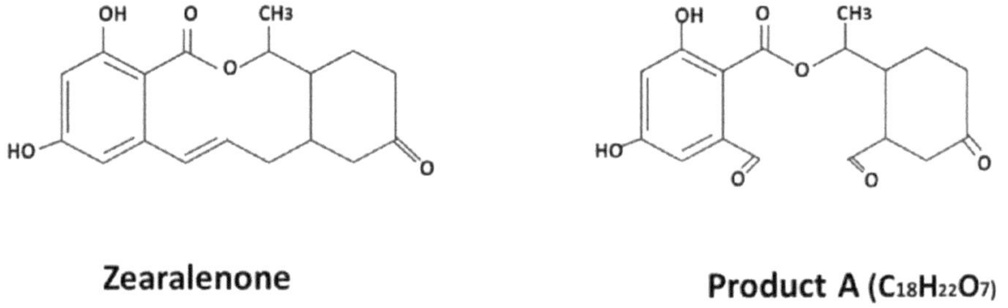

FIGURE 8.5 Proposed degraded product of zearalenone after plasma exposure [96].

8.7 SUMMARY

Mycotoxins are the secondary metabolites of fungi that occur widely in several foods and feeds, causing severe acute and chronic adverse health effects on humans and animals. Prevention or degradation of mycotoxins remains a challenge for food safety worldwide. Although there are several conventional and novel technologies that report significant reductions in mycotoxin levels and structural changes, they can have a negative impact on food quality and may have processing limitations. However, cold plasma has shown a positive impact on the prevention or degradation of mycotoxins by inactivating the mycotoxin-producing fungi or degrading the preformed mycotoxins into less toxic substances. Further studies on the degradation of pre-formed toxins across a wider range of foods may be warranted to evaluate the mycotoxin degradation patterns within contaminated foods and to expand on the findings for pure control compounds. Plasma reactive species play a major role in inactivating the fungi through mechanisms including surface etching and electrostatic interface mechanisms. The inactivation of fungal cells before storage or at various pre-processing stages may mitigate the formation of mycotoxins. Moreover, the plasma reactive species degrade the preformed mycotoxins by breaking down their structures and reducing them to derivative compounds. The toxicity of these derived compounds is comparably lower than that of the mycotoxins. Additionally, plasma has a positive impact on the quality and processing parameters, which helps overcome the limitations of other technologies. The interactions of plasma species with mycotoxins and the toxicity of the degraded products should be understood, along with structural profiling of degradants, in order to determine potential reduction, retention, or modified toxicity.

KEYWORDS

- **cold plasma**
- **decontamination**
- **dielectric barrier discharge**
- **Food and Agriculture Organization**
- **hepatocellular carcinoma**
- **reactive nitrogen species**
- **reactive oxygen species**

REFERENCES

1. Adebo, O. A., Molelekoa, T., Makhuvele, R., Adebiyi, J. A., & Oyedeji, A. B. (2021). A review on novel non-thermal food processing techniques for mycotoxin reduction. *International Journal of Food Science & Technology, 56*(1), 13–27. https://doi.org/10.1111/ijfs.14765.
2. Ahmad, A., Munir, B., & Abrar, M. (2012). Perspective of β-Glucan as functional ingredient for food industry. *Journal of Nutrition & Food Sciences, 2*(2), Article ID 1000133. https://doi.org/10.4172/2155-9600.1000133.
3. Alshannaq, A., & Yu, J. H. (2017). Occurrence, toxicity, and analysis of major mycotoxins in food. *International Journal of Environmental Research and Public Health, 14*(6), 632–640. https://doi.org/10.3390/ijerph14060632.
4. Anbarasan, R., Boopathy, B., Stephen, J., & Radhakrishnan, M. (2022). Cold plasma disinfestation of *Callosobruchus maculatus* infested soybeans: Its subsequent impact on soymilk extraction yield and quality. *Journal of Food Process Engineering, 46*(2), Article ID 14246. https://doi.org/10.1111/jfpe.14246.
5. Anbarasan, R., Jaspin, S., Bhavadharini, B., Pare, A., Pandiselvam, R., & Mahendran, R. (2022). Chlorpyrifos pesticide reduction in soybean using cold plasma and ozone treatments. *LWT, 159*, Article ID 113193. https://doi.org/10.1016/j.lwt.2022.113193.

6. Anfossi, L., Giovannoli, C., & Baggiani, C. (2016). Mycotoxin detection. *Current Opinion in Biotechnology, 37*, 120–126. https://doi.org/10.1016/j.copbio.2015.11.010.
7. Ashiq, S. (2015). Natural occurrence of mycotoxins in food and feed: Pakistan perspective. *Comprehensive Reviews in Food Science and Food Safety, 14*(2), 159–175. https://doi.org/10.1111/1541-4337.12134.
8. Aziz, N. H., El-Far, F. M., Shahin, A. A. M., & Roushy, S. M. (2007). Control of *Fusarium* moulds and fumonisin B1 in seeds by gamma-irradiation. *Food Control, 18*(11), 1337–1342. https://doi.org/10.1016/j.foodcont.2006.08.008.
9. Basaran, P., Basaran-Akgul, N., & Oksuz, L. (2008). Elimination of *Aspergillus parasiticus* from nut surface with low pressure cold plasma (LPCP) treatment. *Food Microbiology, 25*(4), 626–632. https://doi.org/10.1016/j.fm.2008.02.008.
10. Bormashenko, E., Grynyov, R., Bormashenko, Y., & Drori, E. (2012). Cold radiofrequency plasma treatment modifies wettability and germination speed of plant seeds. *Scientific Reports, 2*, Article 3–10. https://doi.org/10.1038/srep00412.
11. Bui-Klimke, T. R., & Wu, F. (2015). Ochratoxin A and human health risk: A review of the evidence. *Critical Reviews in Food Science and Nutrition, 55*(13), 1860–1869. https://doi.org/10.1080/10408398.2012.723475.
12. Calado, T., Venâncio, A., & Abrunhosa, L. (2014). Irradiation for mold and mycotoxin control: A Review. *Comprehensive Reviews in Food Science and Food Safety, 13*(5), 1049–1061. https://doi.org/10.1111/1541-4337.12096.
13. Chaple, S., Sarangapani, C., Jones, J., Carey, E., & Causeret, L. (2020). Effect of atmospheric cold plasma on the functional properties of whole wheat (*Triticum aestivum* L.) grain and wheat flour. *Innovative Food Science & Emerging Technologies, 66*, Article ID 102529. https://doi.org/10.1016/j.ifset.2020.102529.
14. Chen, X., Qiu, Y., Zhang, J., Guo, Y., Ding, Y., & Lyu, F. (2022). Degradation efficiency and products of deoxynivalenol treated by cold plasma and its application in wheat. *Food Control, 136*(February), Article ID 108874. https://doi.org/10.1016/j.foodcont.2022.108874.
15. Colović, R., Puvača, N., Cheli, F., Avantaggiato, G., & Greco, D. (2019). Decontamination of mycotoxin-contaminated feedstuffs and compound feed. *Toxins, 11*(11), Article 617–621. https://doi.org/10.3390/toxins11110617.
16. Darsanaki, R. K., Issazadeh, K., Aliabadi, M. A., & Chakoosari, M. M. D. (2015). Occurrence of deoxynivalenol (DON) in wheat flours in Guilan Province, northern Iran. *Annals of Agricultural and Environmental Medicine, 22*(1), 35–37. PMID: 25867418
17. Deng, L. Z., Tao, Y., Mujumdar, A. S., Pan, Z., & Chen, C. (2020). Recent advances in non-thermal decontamination technologies for microorganisms and mycotoxins in low-moisture foods. *Trends in Food Science & Technology, 106*(January), 104–112. https://doi.org/10.1016/j.tifs.2020.10.020.
18. Diao, E., Hou, H., & Dong, H. (2013). Ozonolysis mechanism and influencing factors of aflatoxin B1: A review. *Trends in Food Science & Technology, 33*(1), 21–26. https://doi.org/10.1016/j.tifs.2013.07.002.
19. Edite Bezerra da Rocha, M., Freire, F. da C. O., Erlan Feitosa Maia, F., Izabel Florindo Guedes, M., & Rondina, D. (2014). Mycotoxins and their effects on human and animal health. *Food Control, 36*(1), 159–165. https://doi.org/10.1016/j.foodcont.2013.08.020.
20. EFSA. (2011). Scientific Opinion on the risks for public health related to the presence of zearalenone in food. *EFSA Journal, 9*(6), 1–124. https://doi.org/10.2903/j.efsa.2011.2197.

21. Egmond, H. P. van, & Jonker, M. A. (2003). Current situation on regulations for mycotoxins. *Mycotoxins, 2003*(Suppl 3), 1–15.
22. El-Sayed, R. A., Jebur, A. B., Kang, W., & El-Demerdash, F. M. (2022). An overview on the major mycotoxins in food products: Characteristics, toxicity, and analysis. *Journal of Future Foods, 2*(2), 91–102.
23. Eskola, M., Kos, G., Elliott, C. T., Hajšlová, J., Mayar, S., & Krska, R. (2020). Worldwide contamination of food-crops with mycotoxins: Validity of the widely cited 'FAO estimate' of 25%. *Critical Reviews in Food Science and Nutrition, 60*(16), 2773–2789. https://doi.org/10.1080/10408398.2019.1617130.
24. European Food Safety Authority. (2013). Deoxynivalenol in food and feed: Occurrence and exposure. *EFSA Journal, 11*(10), Article ID 3379. https://doi.org/10.2903/j.efsa.2013.3379.
25. Evelyn, Kim, H. J., & Silva, F. V. M. (2016). Modeling the inactivation of *Neosartorya fischeri* ascospores in apple juice by high pressure, power ultrasound and thermal processing. *Food Control, 59*, 530–537.
26. FAO/WHO. (2005). Joint FAO/WHO Food Standards Programme by Codex Committee on Food Additives and Contaminants. *Codex Alimentarius Commission*, pp. 1–39, Vol. CODEX STAN 193–1995. Retrieved from: https://www.fao.org/fao-who-codexalimentarius/sh-proxy/en/?lnk=1&url=https%253A%252F%252Fworkspace.fao.org%252Fsites%252Fcodex%252FShared%2BDocuments%252FArchive%252FMeetings%252FCCFAC%252Fccfac37%252FFA37_33e.pdf (accessed on 16 June 2024).
27. FDA. (2011). FDA Mycotoxin Regulatory Guidance. pp. 1–15, Publisher: National Grain and Feed Association, Washington, D.C. Retrieved from: https://www.aflatoxinpartnership.org/wp-content/uploads/2021/05/NGFAComplianceGuide-FDARegulatoryGuidanceforMycotoxins8-2011.pdf (accessed on 16 June 2024).
28. Feizollahi, E., Iqdiam, B., Vasanthan, T., Thilakarathna, M. S., & Roopesh, M. S. (2020). Effects of atmospheric-pressure cold plasma treatment on deoxynivalenol degradation, quality parameters, and germination of barley grains. *Applied Sciences, 10*(10), Article ID 3530. https://doi.org/10.3390/app10103530.
29. Feizollahi, E., & Roopesh, M. S. (2021). Degradation of zearalenone by atmospheric cold plasma: Effect of selected process and product factors. *Food Bioprocess Technology, 14*(11), 2107–2119. https://doi.org/10.1007/s11947-021-02629-y.
30. González-Peñas, E. (2022). Mycotoxins: Classification, occurrence and determination. *Encyclopedia of Dairy Sciences, 4*(July), 586–592.
31. Groopman, J. D., Kensler, T. W., & Wu, F. (2012). Food safety: Mycotoxins – occurrence and toxic effects. *Encyclopedia of Human Nutrition, 2–4*, 337–341.
32. Hao, H., Zhou, T., Koutchma, T., Wu, F., & Warriner, K. (2016). High hydrostatic pressure assisted degradation of patulin in fruit and vegetable juice blends. *Food Control, 62*, 237–242.
33. Heussner, A. H., & Bingle, L. E. H. (2015). Comparative ochratoxin toxicity: A review of the available data. *Toxins, 7*(10), 4253–4282. https://doi.org/10.3390/toxins7104253.
34. Hojnik, N., Cvelbar, U., Tavčar-Kalcher, G., Walsh, J. L., & Križaj, I. (2017). Mycotoxin decontamination of food: Cold atmospheric pressure plasma versus "classic" decontamination. *Toxins, 9*(5), Article ID 151. https://doi.org/10.3390/toxins9050151.
35. Ismail, A., Gonçalves, B. L., de Neeff, D. V., Ponzilacqua, B., & Coppa, C. F. S. C. (2018). Aflatoxin in foodstuffs: Occurrence and recent advances in decontamination. *Food Research International, 113*(June), 74–85.

36. Jouany, J. P. (2007). Methods for preventing, decontaminating and minimizing the toxicity of mycotoxins in feeds. *Animal Feed Science and Technology, 137*(3–4), 342–362.
37. Kabak, B. (2009). The fate of mycotoxins during thermal food processing. *Journal of the Science of Food and Agriculture, 89*(4), 549–554.
38. Kalagatur, N. K., Kamasani, J. R., Mudili, V., Krishna, K., Chauhan, O. P., & Sreepathi, M. H. (2018). Effect of high-pressure processing on growth and mycotoxin production of *Fusarium graminearum* in maize. *Food Bioscience, 21*(2017), 53–59.
39. Kamle, M., Mahato, D. K., Devi, S., Lee, K. E., Kang, S. G., & Kumar, P. (2019). Fumonisins: Impact on agriculture, food, and human health and their management strategies. *Toxins, 11*(328), pp. 1–23. https://doi.org/10.3390/toxins11060328.
40. Karlovsky, P., Suman, M., Berthiller, F., De Meester, J., & Eisenbrand, G. (2016). Impact of food processing and detoxification treatments on mycotoxin contamination. *Mycotoxin Research, 32*(4), 179–205.
41. Karolina, R., & Magdalena, T. (2021). Zearalenone and its metabolites—General overview, occurrence, and toxicity. *Toxins, 13*(35), 1–12. https://doi.org/10.3390/toxins13010035.
42. Khoori, E., Hakimzadeh, V., Mohammadi Sani, A., & Rashidi, H. (2020). Effect of ozonation, UV light radiation, and pulsed electric field processes on the reduction of total aflatoxin and aflatoxin M1 in acidophilus milk. *Journal of Food Processing and Preservation, 44*(10), 1–8. https://doi.org/10.1111/jfpp.14935.
43. Kordas, L., Pusz, W., Czapka, T., & Kacprzyk, R. (2015). The effect of low-temperature plasma on fungus colonization of winter wheat grain and seed quality. *Polish Journal of Environmental Studies, 24*(1), 433–438.
44. Koutsoumanis, K., Alvarez-Ordóñez, A., Bolton, D., Bover-Cid, S., & Chemaly, M. (2022). The efficacy and safety of high-pressure processing of food. *EFSA Journal, 20*(3), Article ID 7128. https://doi.org/10.2903/j.efsa.2022.7128.
45. Kumar, P., Mahato, D. K., Sharma, B., Borah, R., & Haque, S. (2020). Ochratoxins in food and feed: Occurrence and its impact on human health and management strategies. *Toxicon, 187*, 151–162. https://doi.org/10.1016/j.toxicon.2020.10.023.
46. Lee, Y., Lee, Y. Y., Kim, Y. S., Balaraju, K., & Mok, Y. S. (2021). Enhancement of seed germination and microbial disinfection on ginseng by cold plasma treatment. *Journal of Ginseng Research, 45*(4), 519–526. https://doi.org/10.1016/j.jgr.2020.02.003.
47. Li, X., Ma, W., Ma, Z., Zhang, Q., & Li, H. (2021). The occurrence and contamination level of ochratoxin A in plant and animal-derived food commodities. *Molecules, 26*(22), 1–14. https://doi.org/10.3390/molecules26227012.
48. López, M., Calvo, T., Prieto, M., Múgica-Vidal, R., & Muro-Fraguas, I. (2019). A review on non-thermal atmospheric plasma for food preservation: Mode of action, determinants of effectiveness, and applications. *Frontiers in Microbiology, 10*, Article ID 622. https://doi.org/10.3389/fmicb.2019.00622.
49. Los, A., Ziuzina, D., Boehm, D., Cullen, P. J., & Bourke, P. (2019). Investigation of mechanisms involved in germination enhancement of wheat (*Triticum aestivum*) by cold plasma: Effects on seed surface chemistry and characteristics. *Plasma Processes and Polymers, 16*(4), 1–12. https://doi.org/10.1002/ppap.201800174.
50. Los, A., Ziuzina, D., Boehm, D., Cullen, P. J., & Bourke, P. (2020). Inactivation efficacies and mechanisms of gas plasma and plasma-activated water against *Aspergillus flavus*

spores and biofilms: A comparative study. *Applied and Environmental Microbiology, 86*(9), Article ID e02619. https://doi.org/10.1128/AEM.02619-19.
51. Marshall, H., Meneely, J. P., Quinn, B., Zhao, Y., & Bourke, P. (2020). Novel decontamination approaches and their potential application for post-harvest aflatoxin control. *Trends in Food Science & Technology, 106*, 489–496. https://doi.org/10.1016/j.tifs.2020.10.007.
52. Mazumder, P. M., & Sasmal, D. (2001). Mycotoxins – Limits and regulations. *Ancient Science of Life, 20*(3), 1–19.
53. McCormick, S. P. (2013). Microbial detoxification of mycotoxins. *Journal of Chemical Ecology, 39*(7), 907–918. https://doi.org/10.1007/s10886-013-0317-0.
54. Méndez-Albores, A., Arámbula-Villa, G., Loarca-Piña, M. G. F., Castaño-Tostado, E., & Moreno-Martínez, E. (2005). Safety and efficacy evaluation of aqueous citric acid to degrade B-aflatoxins in maize. *Food and Chemical Toxicology, 43*(2), 233–238. https://doi.org/10.1016/j.fct.2004.09.002.
55. Mir, S. A., Dar, B. N., Shah, M. A., Sofi, S. A., & Hamdani, A. M. (2021). Application of new technologies in decontamination of mycotoxins in cereal grains: Challenges, and perspectives. *Food and Chemical Toxicology, 148*, Article ID 111976. https://doi.org/10.1016/j.fct.2021.111976.
56. Misra, N. N. (2016). Quality of Cold Plasma Treated Plant Foods. In *Elsevier Inc.*, pp. 253–271.
57. Misra, N. N., Yadav, B., Roopesh, M. S., & Jo, C. (2019). Cold plasma for effective fungal and mycotoxin control in foods: Mechanisms, inactivation effects, and applications. *Comprehensive Reviews in Food Science and Food Safety, 18*(1), 106–120. https://doi.org/10.1111/1541-4337.12400.
58. Mravlje, J., Regvar, M., Starič, P., Mozetič, M., & Vogel-Mikuš, K. (2021). Cold plasma affects germination and fungal community structure of buckwheat seeds. *Plants, 10*(5), Article ID 851. https://doi.org/10.3390/plants10050851.
59. Nunes, V. M., Moosavi, M., Mousavi Khaneghah, A., & Oliveira, C. A. (2021). Innovative modifications in food processing to reduce the levels of mycotoxins. *Current Opinion in Food Science, 38*, 155–161. https://doi.org/10.1016/j.cofs.2021.07.007.
60. Ott, L. C., Appleton, H. J., Shi, H., Keener, K., & Mellata, M. (2021). High voltage atmospheric cold plasma treatment inactivates *Aspergillus flavus* spores and deoxynivalenol toxin. *Food Microbiology, 95*(November 2020), Article ID 103669. https://doi.org/10.1016/j.fm.2020.103669.
61. Ouf, S. A., Basher, A. H., & Mohamed, A. A. H. (2015). Inhibitory effect of double atmospheric pressure argon cold plasma on spores and mycotoxin production of *Aspergillus niger* contaminating date palm fruits. *Journal of the Science of Food and Agriculture, 95*(15), 3204–3210. https://doi.org/10.1002/jsfa.7231.
62. Pallarés, N., Berrada, H., Tolosa, J., & Ferrer, E. (2021). Effect of high hydrostatic pressure (HPP) and pulsed electric field (PEF) technologies on reduction of aflatoxins in fruit juices. *LWT, 142*(January), Article ID 111000. https://doi.org/10.1016/j.lwt.2021.111000.
63. Pallarés, N., Sebastià, A., Martínez-Lucas, V., González-Angulo, M., & Barba, F. J. (2021). High pressure processing impact on alternariol and aflatoxins of grape juice and fruit juice-milk based beverages. *Molecules, 26*(12), Article ID 26123769. https://doi.org/10.3390/molecules26123769.

64. Pańka, D., Jeske, M., Łukanowski, A., Baturo-Cieśniewska, A., & Prus, P. (2022). Can cold plasma be used for boosting plant growth and plant protection in sustainable plant production? *Agronomy, 12*(4), Article ID 12040841. https://doi.org/10.3390/agronomy12040841.
65. Pankaj, S. K., & Keener, K. M. (2017). Cold plasma: Background, applications and current trends. *Current Opinion in Food Science, 16*, 49–52. https://doi.org/10.1016/j.cofs.2017.08.008.
66. Panngom, K., Lee, S. H., Park, D. H., Sim, G. B., & Kim, Y. H. (2014). Non-thermal plasma treatment diminishes fungal viability and up-regulates resistance genes in a plant host. *PLoS One, 9*(6), Article ID e99300. https://doi.org/10.1371/journal.pone.0099300.
67. Park, B. J., Takatori, K., Sugita-Konishi, Y., Kim, I. H., & Lee, M. H. (2007). Degradation of mycotoxins using microwave-induced argon plasma at atmospheric pressure. *Surface and Coatings Technology, 201*(9-11 SPEC. ISS.), 5733–5737. https://doi.org/10.1016/j.surfcoat.2006.09.042.
68. Park, D. L., López-García, R., Trujillo-Preciado, S., & Price, R. L. (1996). Reduction of risks associated with fumonisin contamination in corn. *Advances in Experimental Medicine and Biology, 392*, 335–344.
69. Peng, W. X., Marchal, J. L. M., & van der Poel, A. F. B. (2018). Strategies to prevent and reduce mycotoxins for compound feed manufacturing. *Animal Feed Science and Technology, 237*(January), 129–153. https://doi.org/10.1016/j.anifeedsci.2017.12.009.
70. Pet'ková, M., Švubová, R., Kyzek, S., Medvecká, V., & Slováková, L. (2021). The effects of cold atmospheric pressure plasma on germination parameters, enzyme activities and induction of DNA damage in barley. *International Journal of Molecular Sciences, 22*(6), Article ID 1–16. https://doi.org/10.3390/ijms22063122.
71. Priyadarshini, A., Rajauria, G., O'Donnell, C. P., & Tiwari, B. K. (2019). Emerging food processing technologies and factors impacting their industrial adoption. *Critical Reviews in Food Science and Nutrition, 59*(19), 3082–3101. https://doi.org/10.1080/10408398.2018.1425970.
72. Puligundla, P., & Mok, C. (2017). Potential applications of nonthermal plasmas against biofilm-associated micro-organisms in vitro. *Journal of Applied Microbiology, 122*(5), 1134–1148. https://doi.org/10.1111/jam.13416.
73. Rai, A., Das, M., & Tripathi, A. (2020). Occurrence and toxicity of a *Fusarium* mycotoxin, zearalenone. *Critical Reviews in Food Science and Nutrition, 60*(16), 2710–2729. https://doi.org/10.1080/10408398.2019.1656270.
74. Rajan, A., Boopathy, B., Radhakrishnan, M., Rao, L., Schlüter, O. K., & Tiwari, B. K. (2023). Plasma processing: A sustainable technology in agri-food processing. *Sustainable Food Technology, 1*(1), 9–49. https://doi.org/10.1016/j.sift.2023.100002.
75. Ratish Ramanan, K., Sarumathi, R., & Mahendran, R. (2018). Influence of cold plasma on mortality rate of different life stages of *Tribolium castaneum* on refined wheat flour. *Journal of Stored Products Research, 77*, 126–134. https://doi.org/10.1016/j.jspr.2018.01.013.
76. Rushing, B. R., & Selim, M. I. (2019). Aflatoxin B1: A review on metabolism, toxicity, occurrence in food, occupational exposure, and detoxification methods. *Food and Chemical Toxicology, 124*(November 2018), 81–100. https://doi.org/10.1016/j.fct.2018.11.004.
77. Sarangapani, C., Patange, A., Bourke, P., Keener, K., & Cullen, P. J. (2018). Recent advances in the application of cold plasma technology in foods. *Annual*

Review of Food Science and Technology, 9, 609–629. https://doi.org/10.1146/annurev-food-030216-030244.
78. Selcuk, M., Oksuz, L., & Basaran, P. (2008). Decontamination of grains and legumes infected with *Aspergillus spp.* and *Penicillum spp.* by cold plasma treatment. *Bioresource Technology, 99*(11), 5104–5109. https://doi.org/10.1016/j.biortech.2007.09.061.
79. Shephard, G. S. (2006). Mycotoxins in the Context of Food Risks and Nutrition Issues. *The Mycotoxin Factbook: Food and Feed Topics*, Wageningen Academic Publishers, 1, 21–36. https://doi.org/10.3920/9789086865871_002, ISBN: 9789086860067.
80. Shi, H., Cooper, B., Stroshine, R. L., Ileleji, K. E., & Keener, K. M. (2017). Structures of degradation products and degradation pathways of aflatoxin B1 by high-voltage atmospheric cold plasma (HVACP) treatment. *Journal of Agricultural and Food Chemistry, 65*(30), 6222–6230. https://doi.org/10.1021/acs.jafc.7b02084.
81. Shi, H., Ileleji, K., Stroshine, R. L., Keener, K., & Jensen, J. L. (2017). Reduction of aflatoxin in corn by high voltage atmospheric cold plasma. *Food and Bioprocess Technology, 10*(6), 1042–1052. https://doi.org/10.1007/s11947-017-1913-6.
82. Siciliano, I., Spadaro, D., Prelle, A., Vallauri, D., & Cavallero, M. (2016). Use of cold atmospheric plasma to detoxify hazelnuts from aflatoxins. *Toxins, 8*(5), Article ID 125. https://doi.org/10.3390/toxins8050125.
83. Spadaro, D., & Gullino, M. L. (2005). Improving the efficacy of biocontrol agents against soilborne pathogens. *Crop Protection, 24*(7), 601–613. https://doi.org/10.1016/j.cropro.2005.01.004.
84. Stockmann-Juvala, H., & Savolainen, K. (2008). A review of the toxic effects and mechanisms of action of fumonisin B1. *Human and Experimental Toxicology, 27*(11), 799–809. https://doi.org/10.1177/0960327108096755.
85. Stoev, S. D. (2013). Food safety and increasing hazard of mycotoxin occurrence in foods and feeds. *Critical Reviews in Food Science and Nutrition, 53*(9), 887–901. https://doi.org/10.1080/10408398.2012.717652.
86. Švubová, R., Slováková, L., Holubová, L., Rovňanová, D., Gálová, E., & Tomeková, J. (2021). Evaluation of the impact of cold atmospheric pressure plasma on soybean seed germination. *Plants, 10*(1), Article ID 17. https://doi.org/10.3390/plants10010017.
87. Ten Bosch, L., Pfohl, K., Avramidis, G., Wieneke, S., Viöl, W., & Karlovsky, P. (2017). Plasma-based degradation of mycotoxins produced by Fusarium, Aspergillus and Alternaria species. *Toxins, 9*(3), Article ID 97. https://doi.org/10.3390/toxins9030097.
88. Thanushree, M. P., Sailendri, D., Yoha, K. S., Moses, J. A., & Anandharamakrishnan, C. (2019). Mycotoxin contamination in food: An exposition on spices. *Trends in Food Science & Technology, 93*(June), 69–80. https://doi.org/10.1016/j.tifs.2019.09.017.
89. Tola, M., & Kebede, B. (2016). Occurrence, importance and control of mycotoxins: A review. *Cogent Food & Agriculture, 2*(1), Article ID 1191203. https://doi.org/10.1080/23311932.2016.1191203.
90. US FDA. (2022). *Chemical Contaminants, Metals, Natural Toxins & Pesticides Guidance Documents & Regulations*. Retrieved from: https://www.fda.gov/food/guidance-documents-regulatory-information-topic-food-and-dietary-supplements/chemical-metals-natural-toxins-pesticides-guidance-documents-regulations (accessed on 16 June 2024).
91. Van Egmond, H. P., Schothorst, R. C., & Jonker, M. A. (2007). Regulations relating to mycotoxins in food: Perspectives in a global and European context. *Analytical and Bioanalytical Chemistry, 389*(1), 147–157. https://doi.org/10.1007/s00216-007-1317-9.

92. Wang, B., Mahoney, N. E., Pan, Z., Khir, R., & Wu, B. (2016). Effectiveness of pulsed light treatment for degradation and detoxification of aflatoxin B1 and B2 in rough rice and rice bran. *Food Control, 59*, 461–467. https://doi.org/10.1016/j.foodcont.2015.06.020.
93. Wang, L., Liao, Y., Peng, Z., Chen, L., & Zhang, W. (2019). Food raw materials and food production occurrences of deoxynivalenol in different regions. *Trends in Food Science & Technology, 83*(October 2018), 41–52. https://doi.org/10.1016/j.tifs.2018.11.006.
94. Wang, S. Q., Huang, G. Q., Li, Y. P., Xiao, J. X., Zhang, Y., & Jiang, W. L. (2015). Degradation of aflatoxin B1 by low-temperature radio frequency plasma and degradation product elucidation. *European Food Research and Technology, 241*(1), 103–113. https://doi.org/10.1007/s00217-015-2496-1.
95. Wu, Y., Cheng, J. H., & Sun, D. W. (2021). Blocking and degradation of aflatoxins by cold plasma treatments: Applications and mechanisms. *Trends in Food Science & Technology, 109*, 647–661. https://doi.org/10.1016/j.tifs.2021.08.003.
96. Zheng, Z., Huang, Y., Liu, L., Chen, Y., Wang, Y., & Li, C. (2022). Zearalenone degradation by dielectric barrier discharge cold plasma: The kinetics and mechanism. *Foods, 11*(10), Article ID 1494–1499. https://doi.org/10.3390/foods11101494.

PART III

Non-Food Applications of Cold Plasma Technology

CHAPTER 9

COLD PLASMA TO CONTROL BIOFILMS ON FOOD AND FOOD PRODUCTS

E. J. RIFNA

ABSTRACT

The surrounding conditions around food products greatly permit microbial cells to proliferate and develop on surfaces, resulting in the formation of resistant and robust microbial biofilms. Indeed, the progressively hazardous conditions triggered through biofilm formation have unquestionably turned out to be the highest concern in food quality and safety sectors, leading to various foodborne illnesses and hospital-acquired infections. However, due to the constructive nature of biofilm building, the microbial cells enclosed inside the biofilms display improved tolerance to conventional decontamination techniques as well as to synthetic antibiotics when compared to free-living microbial cells. Consequently, to tackle this communal health risk and to effectively eliminate the development of biofilms from food product surfaces, cold plasma, an emerging non-thermal technique, is applied in the food industry. The successful elimination of developed biofilms through plasma technology has resulted in a boost of concerns regarding this matter. The chapter provides a comprehensive review of the development of biofilms on food surfaces, as well as studying the mode of action, resulting affecting factors, associated advantages, and limitations of the cold plasma technique to avert and eliminate detrimental biofilms from foods and food products.

Nonthermal Plasma Technology for Food and Food Products.
R. Mahendran, Preeti Birwal, and Megh R. Goyal (Eds.)

9.1 INTRODUCTION

Contamination of food products through microorganisms is a dreadful concern faced by numerous consumers, food processors, as well as industries and is particularly an alarming issue in fresh food and partially processed food goods such as ready-to-eat products owing to lack of supplementary processing [2]. Infestation by foodborne microorganisms is unavoidable since any process in the manufacturing line of preparation of food products possibly results in interaction with foodborne pathogens, resulting in dreadful food contaminations [15]. In this context, there is a necessity for decontamination of harmful microbial load present in the food products. Various technologies involving thermal technologies, the use of synthetic antibiotics, and so forth have been adopted to preserve the quality and safety of food. Even if the deactivation mechanism of these techniques was identified as effective on a wide range of harmful microorganisms, unexpectedly, through extended works, it was identified that the majority of harmful microorganisms occur chiefly in biofilm form, replacing their existence in abiotic and biotic environments. Various studies also suggest that microbes developed in biofilm matrix are highly resilient compared to their biotic and abiotic form when facing antibacterial action, leading to a novel threat to food processing industries and producers [14, 36]. Biofilms are immobile structures robustly fixed on a medium. The biofilm structures consist of solitary organisms or various kinds of bacteria and are architected with an outer membrane made of protein, exopolysaccharides, and DNA concealed by pathogens [3]. The establishment of biofilm on food product surfaces is a normal incidence in a period of consistent situations and it has been described that about 90% of fresh products, particularly fruits and vegetables, comprise biofilms [30]. In addition, biofilms are assessed to account for 70% of all microbial infections happening in the USA [18]. Therefore, the investigation of a real and innovative methodology for the control of biofilms is on the plan.

Cold plasma is a novel technology also recognized as non-thermal plasma and is the main focus in research owing to its modest apparatus, reduced energy use, and nonexistence of poisonous leftovers. Plasma is produced through excitation of a targeted composition of gas using electrical ejections at reduced pressure, vacuum conditions, or atmospheric pressure at ambient temperature [32]. Depending on the gas composition utilized for the generation of plasma, numerous methods of vigorous agents including UV rays, ions, charged species, and free radicals could be present in the composition of plasma. The flexibility of the aforementioned components has driven the application of plasma in numerous applications [19].

Furthermore, the vigorous components in the mixture of plasma are supposed to be included in the intervention of components occurring within cells, metabolic pathways, and the veracity of microbial cell membranes [38]. A considerable amount of previous works have been studied to investigate the antibacterial properties of plasma against harmful microorganisms. However, the majority of these works simply evaluated the microorganisms that survived in a water medium, without fully replicating the chief environmental conditions and existence of pathogenic organisms as biofilms. The planktonic form of microorganisms present within the matrix of a biofilm shows varying vulnerability to the application of cold plasma. Jahid et al. [44] demonstrated that a reduction of greater than five logs was achieved in the case of *Aeromonas hydrophila* after plasma treatment for 15 seconds. Furthermore, the vigorous components, such as reactive nitrogen and oxygen species produced during plasma treatment, were also shown to restrict the membrane potential of microbial cells, thereby demonstrating the potential of plasma technology in the field of anti-biofilms [38].

Furthermore, there are few works that have been focused on improving plasma technology as an anti-biofilm method in food processing industries. However, the mechanism of action of cold plasma on degrading biofilm growth is still unclear.

This chapter offers a detailed review of the formation of biofilms on food surfaces, the mechanism of action, influencing factors, advantages, and limitations of using cold plasma to prevent and eliminate harmful biofilms from foods and food products.

9.2 BIOFILM DEVELOPMENT ON FOOD

Development of the biofilm is an active procedure that, in a consecutive mode, comprises attachment, maturation, and dispersion. In the foremost process of development of biofilm, microorganisms transfer into the near vicinity of the food product surface, which could possess various chemical as well as physical properties that are vital to assess the level and degree of attachment. Furthermore, the medium exhibited in aqueous food products could get covered with compounds that might lead to physicochemical adjustments that could encourage bacteriological attachment [11]. Transfer of pathogenic cells to the external area is usually attained through a number of entrenched processes, such as gravitational and Brownian effects [13].

In the initial phase of the process, the adhesion process is temporary. However, as inaccurately attached pathogenic organisms amalgamate with the process of adhesion and the production of exopolymers, the adhesion of microorganisms to the food surface turns out to be permanent. The connection of bacterial cells to the medium is trailed by the growth of microbial cells, formation of colonies, and attachment of extra microbial cells existing within the native surroundings (Figure 9.1) [48]. As microorganism attachment occurs, the cells propagate with the release and accretion of polymers, ultimately leading to being fixed and arrested in the hydrated matrix of the food surface. Biofilms developed at the interface (solid or liquid) of food are highly non-uniform and are designed through microcolonies enclosed in a matrix, which are disconnected from other formed colonies through voids and act as materials of a viscoelastic nature [27]. The last process of biofilm development is dispersion or diffusion, which directs the greatest conveyance of solutes inside the formed biofilm and between the outside environment and the biofilm. As the process of diffusion is gradual with respect to the metabolism at the cellular level, the biochemical surroundings frequently vary significantly at diverse heights of biofilms [28]. The aforementioned chemical inclines produce ecological micro-niches that permit the cohabitation of various species.

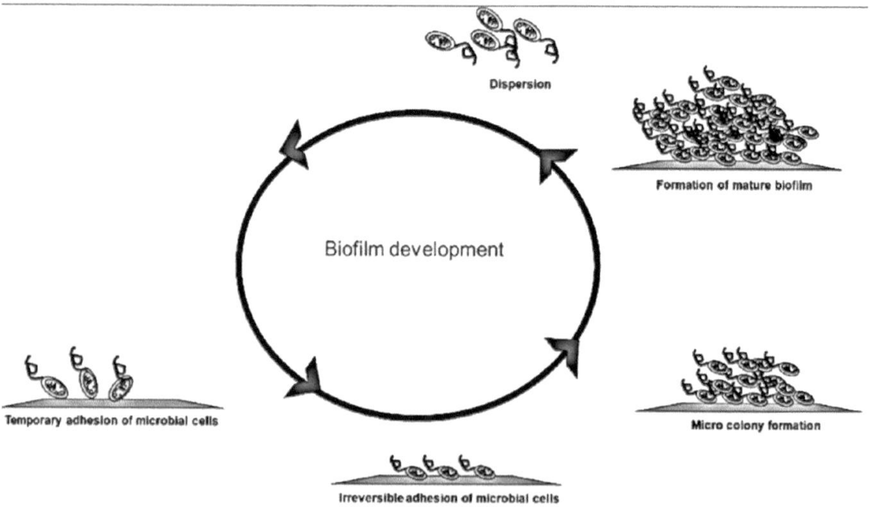

FIGURE 9.1 Various phases of development of biofilm on a food surface.

9.3 FACTORS INFLUENCING BIOFILM ATTACHMENT ON FOOD SURFACE

The biofilm attachment is determined through physical and chemical attributes of substrata of cells and is altered by acclimatizing biofilms which initiate through the adjacent where the cell chemistry modifies in reaction to environmental changes. Permanent adhesion is a biological method within genetic guidelines. Before working on *E. coli, C. sakazakii, S. aureus*, validate that genetic factors accountable for the expression of proteins present within the cell surface get triggered in response to exterior inducements such as nutrient restriction and density of population. [1, 6, 33, 64].

9.3.1 FOOD CONTACT SURFACE PROPERTIES

It was reported by Bryers et al. [10] that biofilm adhesion enhances with a perilous value of surface tension. The greatest adhesion of microbial cells relies on a great free energy of the surface. Surfaces with high surface energy are predominantly hydrophilic in nature. Hence, such surfaces usually permit better attachment of bacterial biofilm than food surfaces that possess a hydrophobic nature [66]. The preliminary attachment of *L. monocytogenes* on stainless steel (SS) surfaces was highly swifter compared to rubber [66]. Supplementary proof for the prominence of surface energy regarding adhesion was described by Bos et al. [7], who revealed that biofilm attachment happened frequently in the hydrophilic area of the hydrophobic-hydrophilic boundary on SS surfaces. Furthermore, a report by Bos et al. [7] illustrated that the cell membrane pressure of bacterial cells also affected adhesion; for example, the interaction of gram-positive bacteria with SS resulted in reduced attachment. Sinde et al. [66] identified that washing SS conditions resulted in a reversible change in surface properties. A more hydrophilic nature was found to be developed when cleaning the equipment surfaces with a strong alkali, whereas cleaning using a weak acid resulted in a hydrophobic nature.

9.3.2 FOOD CONTACT SURFACE TOPOGRAPHY

Stainless steel material is typically utilized for manufacturing utensils. Nylon, Teflon, and rubber are also utilized for manufacturing numerous utensils, as well as food processing apparatus parts. The surfaces of these utensils are grazed through frequent use, enhancing their potential to trap soil as well as microorganisms [39]. The condition mentioned above develops a harborage for the development and proliferation of microorganisms. Faille et al. [31] reported that the value of surface roughness of stainless steel does not associate with the adherence of cells; however, other dimensions of roughness which point out surface imperfections [68] connect more precisely with the retention of soil and its detachment. Works by Myszka et al. [55] elucidate that surface faults were connected with noteworthy surges in the attachment of bacterial cells.

9.3.3 CONTACT TIME

The time of contact between the attaching medium and bacterial cells is compulsory for permanent adhesion. In a study, the authors confirmed that the greatest predominant *L. monocytogenes* strain identified in food processing equipment possessed strong attachment potential and required only a small time period for adhesion [17].

9.3.4 ADHESIVE NATURE OF BACTERIAL CELLS

The attachment of microbial cells is influenced by the chemical and physical attributes of microbial cells, which in turn are influenced by various parameters such as the temperature of the growth matrix and the medium used for growth. Microbial cells usually possess a negative charge on their surface and often behave as particles with a water-repelling nature; however, the level of hydrophobicity changes with the development stage. Hydrophobicity typically decreases as the growth level increases [8]. In another study, it was established that *Listeria* displayed a high level of attachment when the microbial cells were at their maximum metabolic stage [50]. In the case of *Listeria*, the optimal conditions for strong attachment are a pH of 7 and a temperature of 30°C. Nitschke et al. [57] found that undernourishment of certain bacterial cells also resulted in an increase in the hydrophobicity of cells. Several other studies have linked attachment to hydrophobicity and surface charge. The attachment of *Staphylococcus* strains is directly associated with hydrophobicity and electronegativity of cells, while the attachment of some other cells is inversely related to the level of negative charge but has no effect on hydrophobicity [5].

9.4 EFFECT OF COLD PLASMA AS AN ANTI-BIOFILM FOR FOOD SAFETY

The foods that fall into the fresh food category (such as fruits and vegetables) recurrently nurture biofilms. Fresh produce has been conventionally disinfected by means of chlorine water [24]. Nevertheless, the use of chlorine does not considerably decrease the occurrence of a few harmful microbial cells, such as *E. coli, Salmonella,* and *S. aureus*, and also poses some risks to human health. Cold plasma has been identified as an effective method to treat fresh produce that may not be preserved with the help of other techniques without causing deviations in color, food quality, texture, and taste. In most cases, cold plasma has been used against bacterial cells. However, there are few studies on the application of plasma for the treatment or elimination of biofilms that develop on fresh produce [45]. Atmospheric plasma discharge has been shown to deactivate biofilms formed by *Pantoea agglomerans* on *Capsicum annuum* (bell peppers) without triggering the thermal effects that could occur in food. Researchers described a log reduction of two, which is a strong benchmark for food safety regulators as food does not require the same level of decontamination as medical equipment [45]. One drawback of this work is that investigators did not explore the effects on bell peppers at the cellular level. Decrement of value higher than 6 logs was found for microorganisms such as *E. coli* and *Salmonella* in biofilms on food products. A crucial restriction for non-thermal plasma is the principally unknown influences of plasma technique on nutritional, quality, and sensorial aspects of treated sustenance [56]. The inactivation potential of non-thermal plasma has also been investigated on egg shells by inoculating them with strains of *Salmonella* at ambient temperature. It was revealed that the biofilm was significantly destroyed with no change in the quality characteristics of eggs [45]. In another study, Patange et al. [60] used cold plasma under atmospheric conditions to eliminate a biofilm of *P. fluorescens* developed on lettuce leaves. The outcomes revealed that the amount of enduring microbial cells in biofilms reduced to untraceable limits after six minutes of treatment using cold plasma.

In another study, Gabriel et al. [34] found that *P. aeruginosa* developed on food utensils made of stainless steel were eliminated using plasma generated under atmospheric conditions. The influence of cold plasma against biofilms on stainless steel surfaces under atmospheric conditions was studied in detail in this work. Moreover, to enhance the activity of anti-biofilms on food products that possess uneven surfaces, plasma techniques are frequently combined with other constituents to mutually address biofilms. For instance, Cui et al. [19] inspected the mutual influence between cold plasma (produced using nitrogen gas) on biofilms of *E. coli* developed on carrots and cucumbers. The authors stated that bacteriophages might considerably enhance the anti-biofilm influence of plasma without adverse effects on the food surface color of treated food products compared to untreated food products. The influence of cold plasma on biofilms of *Candida albicans* was evaluated in both *in vivo* and *in vitro* conditions [40]. The *in vitro* work indicated that biofilms formed by *Candida albicans* were considerably deactivated through cold plasma technology. The *in vivo* work specified that treatment with cold plasma achieves the finest antifungal properties under ideal conditions, with a treatment time of 4 minutes, as extreme exposure to plasma harms healthy cells and tissues, leading to an intensification of contamination compared to the 4-minute treatment condition. Treatment using cold plasma also offers a potential alternative method for inactivating packaged food, as it may reduce biofilm growth on polymeric packaging materials [9, 63]. To decontaminate packaging materials used for food packaging, the application of cold plasma methodology could be considered either indirect or direct. The appropriate materials are exposed to the plasma discharge zone and decontaminated through exposure to intense plasma discharges containing photons, free radicals, and ions [61].

9.4.1 DECONTAMINATION OF BIOFILMS FROM FOOD PACKAGING SURFACES USING COLD PLASMA

The researchers inspected the antimicrobial properties of cold plasma on the food-contact surfaces that exist within the food-processing units. These chiefly comprise polymers, and metals packaging materials (Table 9.1) As these are commonly dried areas, various outcomes are estimated associated with earlier sections (food surfaces), owing to the alterations in nature of microbes in the dried surroundings and subsequent chemistry of plasma.

9.5 MODE OF ACTION OF COLD PLASMA FOR BIOFILM CONTROL

Plasma technology, which plays a significant role in surface alteration, is a capable technique to prevent or hinder the formation of biofilms. As

TABLE 9.1 Effect of Cold Plasma on Eliminating Biofilms from Food Packaging Surfaces

	Cold Plasma System	Treatment Parameters	Gas Used	Microorganisms	Reduction	Ref.
Metallic surfaces	Dielectric barrier discharge	1 kHz, 10 Kv, 35 mW/cm^2	Air	*B. atrophaeus* spores	2.6 log and 5.3 log for 1 minute and 3 minute	[16]
	Atmospheric plasma jet	30 W	Oxygen, nitrogen	*B. atrophaeus* spores	5 log for 2 minute	[53]
	Dielectric barrier discharge	8 kV, 20 kHz	Air	*C. sakazakii*	6.88 log for 1.5 hours	[46]
	Gliding arc	15 kV, 16 kHz	Nitrogen	*E. coli*	1.44 log reduction after for 1 min N2	[22]
	Surface barrier discharge	400 W	Nitrogen	*E. coli*	4 log reduction	[20]
	Dielectric barrier discharge	30 kV and 150 mW/cm^2	Air	*C. jejuni*	No visible growth after 1 minute	[26]
	Surface micro discharge	10 kV 0 and 0.1 W/cm^2	Air	*E. coli*	5 log reduction after 30 s	[65]
	Gliding arc	15 kV, 16 kHz	Nitrogen	*E. coli*	2.46 log reduction after plasma exposure of 1 minute	[22]
Polymer surfaces	Dielectric barrier discharge	170 W	Air	*B. atrophaeus*	Less than 6 log reduction	[54]
	Dielectric barrier discharge	130 W	Air	*B. atrophaeus*	5 log reduction after exposure to 1 s	[41]
	Atmospheric plasma jet	200 W	Air	*B. atrophaeus*	1.46 log reduction for 10 min	[49]
	Atmospheric plasma jet	10–40 kHz	Argon	*B. subtilis*	4.0 log reduction after 1 minute	[25]
	Dielectric barrier discharge	5 kHz	Air	*E. coli*	Less than 5log after exposure to 24 s	[35]
	Plasma jet	43.5 kHz	Helium	*E. coli*	2.5log for 2.5 min	[62]
	Dielectric barrier discharge	7 W/cm^2	Argon	*B. subtilis*	4.0log for 4 s	[43]
	Gliding arc	15 kV	Air	*E. coli*	2.63 log reduction after exposure to 1 min	[21]

energy is delivered to a gas species, it gets ionized and produces plasma, which is encompassed by electrons, ions, and free radicals. Depending on the environments in which the gas ionization occurs, plasmas with various properties and appropriate for numerous uses might be produced [4, 37, 67]. The matrix of biofilms includes constituents such as polysaccharides, DNA, and RNA, which get damaged during the treatment with cold plasma. The bacterial cells consist of alginate and polysaccharides components that interrelate with reactive oxygen-nitrogen species produced by cold plasma and hence lead to disintegration and damage of cellular components of biofilms. The alginate and polysaccharides components are the chief components of a bacterial cell responsible for the adhesion of biofilm to food surface, important properties of biofilms, and their maturation and dispersion [69]. Prominently, the alginate layer in the cell is also responsible for scavenging free radicals in a bacterial cell. So, due to exposure to cold plasma, when the reactive nitrogen-oxygen species of plasma interact with these cellular components, the macrophages and neutrophils of bacterial biofilms burst out [69]. Additionally, the damage to alginate walls or layers alginate binds, confines diffusion which leads to damage to bacterial cells comprising the peptide layer of bacteria in biofilms [12], which ultimately results in decreased diffusion of extracellular polymeric substances (EPS). The vital role exhibited by developed biofilm in the extracellular polymeric matrix in convening reduces the antimicrobial lenience of bacteria both through damaging the physiological protection which is present over biofilms and dehydrate the cells which lead to the final destruction of biofilm architecture and small environments around the biofilm [23, 37].

9.6 INVESTIGATION OF COLD PLASMA TECHNOLOGY AS AN ANTI-BIOFILMS AGENT FOR FOOD PRODUCTS

9.6.1 *ADVANTAGES*

Despite the fact that eradicating the biofilms on surfaces of food, the effect of plasma technology on food quality cannot be disregarded either. The prospect of plasma technology on the quality of food relies on the kind of food, time of exposure, and species generated during plasma exposure. Few studies have inspected the physical, chemical, organoleptic attributes, and phytochemical components of food products treated with plasma for inactivation of biofilms, and the majority of outcomes displayed that no negative impacts were observed on studied parameters of food after exposure to cold plasma conditions. For instance, Hertwig et al. [42] revealed that the color of bell pepper remains unaltered during the period of plasma treatment to attain a 4-log reduction for bacterial biofilm using the plasma technique. Whereas in other studies, the enhanced visual appearance of treated food products after cold plasma treatment has been reported [51]. Some authors also demonstrated that cold plasma had no impact on food texture [52, 59]. Furthermore, the findings in current studies even described a positive enhancement concerning the rheology of wheat flour dough that contained biofilm over it subsequent to plasma exposure [52]. In conclusion, without compromising the normal functionality, cold plasma technology displays negligible confrontational influence on the organoleptic attributes of food such as texture, color, and other sensory properties of food products.

Furthermore, in food processing units, the preservation of phytochemical components during the process of removing biofilms from food surfaces has been a significant concern for scientists for an extended period of time. Phytochemical components in food products are often sensitive to treatment conditions involving heat [58]. Alongside thermal methods for biofilm removal, the use of cold plasma technology shows great potential in preserving phytochemical compounds such as phenols, tannins, flavonoids, antioxidant activity, and more. A study demonstrated that cold plasma was more effective than traditional heat methods in maintaining phenolic components and anthocyanin molecules found in fresh sour cherries and pomegranate fruits [29].

9.6.2 DISADVANTAGES

The application of plasma as an anti-biofilms agent greatly depends on the produced vigorous species, importantly reactive nitrogen-oxygen species. Unpredictably, the energetic species, particularly the ions, could effortlessly break the lipid bonds present within the food matrix, which could eventually lead to oxidation of lipids and result in disapproving attributes on the quality of food. The degree of oxidation of lipids on exposure to plasma is in linear association with the double bonds present in the lipids owing to the energy required for the perception of a hydrogen atom present in the CH- bonds of lipids. Distinctive lipid compounds that are prone to free radicals and plasma-generated ions include α-Linolenic acid and linoleic acid comprising double bonds. Furthermore, the oxidation of lipid products that are developed on exposure to plasma leads to unacceptable food quality, particularly in the case of foods that contain high-fat content. In a study by Kim et al. [47], cold plasma was used to control biofilm development on the surface of pork;

however, oxidation of lipid molecules was perceived owing to the creation of malondialdehyde, therefore distressing the sensory and trophism properties of the treated pork meat [70]. To diminish the effect of lipid oxidation on exposure to cold plasma, a few permissible ingredients with outstanding antioxidant characteristics, such as essential oils, are advised to be incorporated into the food matrix to preserve against the oxidation of lipids.

9.7 SUMMARY

The recurrent incidences of food security occurrences produced through biofilm development have encouraged the food industries and scientific communities to use cold plasma as a real and effective tool against the developed anti-biofilms for the preservation of safety and quality of food. The assortment of vigorous species with regards to the composition of gas used, which comprises ions, radicals, vigorous oxygen species, and nitrogen radicals, are estimated as chief influencing parameters accountable for the regulation of biofilms after exposure to the plasma treatment process. Furthermore, etching influence, as well as the intervention of quorum sensing, could be strongly described as the properties which offer anti-biofilm activity to the cold plasma technique. The competence of plasma technology for the elimination of biofilms is controlled by numerous parameters, comprising the composition of the gas, surface adhesion, multiple species biofilm or single-species biofilm, process independent variables, kind of microorganism, the thickness of the biofilm, and so forth. Though with regards to high fatty foods, the technology of cold plasma is expected to result in the oxidation of lipids, the advancement of cold plasma furthermore offers a novel and effective solution to control or completely discard the development of biofilms in food products and food contact surfaces, which would pave the way for auxiliary progress of food safety in food industries.

KEYWORDS

- ***Aeromonas hydrophila***
- **cold plasma**
- **DNA**
- **extracellular polymeric substances**
- ***L. monocytogenes***

REFERENCES

1. Adams, J. L., & McLean, R. J. C. (1999). Impact of rpoS deletion on *Escherichia coli* biofilms. *Applied and Environmental Microbiology, 65*(9), 4285–4287. https://doi.org/10.1128/AEM.65.9.4285-4287.1999.
2. Alegbeleye, O. O., Singleton, I., & Sant'Ana, A. S. (2018). Sources and contamination routes of microbial pathogens to fresh produce during field cultivation: A review. *Food Microbiology, 73*, 177–208. https://doi.org/10.1016/j.fm.2018.01.017.
3. Alvarez-Ordóñez, A., Coughlan, L. M., Briandet, R., & Cotter, P. D. (2019). Biofilms in food processing environments: Challenges and opportunities. *Annual Review of Food Science and Technology, 10*, 173–195. https://doi.org/10.1146/annurev-food-032818-121223.
4. Amulya, P. R., Sudheer, K. P., & Rifna, E. J. (2016). Effect of edible wax coating and MAP on the quality of mango during storage. *International Journal of Agricultural Sciences Research, 6*(4), 13–18.
5. Araújo, E. A., de Andrade, N. J., da Silva, L. H. M., de Carvalho, A. F., de Silva, C. A. S., & Ramos, A. M. (2010). Control of microbial adhesion as a strategy for food and bioprocess technology. *Food and Bioprocess Technology, 3*(3), 321–332. https://doi.org/10.1007/s11947-009-0264-7.
6. Bagge, D., Hjelm, M., Johansen, C., Huber, I., & Gram, L. (2001). Shewanella putrefaciens adhesion and biofilm formation on food processing surfaces. *Applied and Environmental Microbiology, 67*(5), 2319–2325. https://doi.org/10.1128/AEM.67.5.2319-2325.2001.
7. Bos, R., Van Der Mei, H. C., Gold, J., & Busscher, H. J. (2000). Retention of bacteria on a substratum surface with micro-patterned hydrophobicity. *FEMS Microbiology Letters, 189*(2), 311–315. https://doi.org/10.1111/j.1574-6968.2000.tb09269.x.
8. Boulangé-Petermann, L., Rault, J., & Bellon-Fontaine, M. N. (1997). Adhesion of Streptococcus thermophilus to stainless steel with different surface topography and roughness. *Biofouling, 11*(3), 201–216. https://doi.org/10.1080/08927019709378443.
9. Bourke, P., Ziuzina, D., Han, L., Cullen, P. J., & Gilmore, B. F. (2017). Microbiological interactions with cold plasma. *Journal of Applied Microbiology, 123*(2), 308–224. https://doi.org/10.1111/jam.13455.
10. Bryers, J. D. (1987). Biologically active surfaces: Processes governing the formation and persistence of biofilms. *Biotechnology Progress, 3*(2), 57–68. https://doi.org/10.1002/btpr.5420030202.
11. Cappitelli, F., Polo, A., & Villa, F. (2014). Biofilm formation in food processing environments is still poorly understood and controlled. *Food Engineering Reviews, 6*(1–2), 29–42. https://doi.org/10.1007/s12393-013-9078-5.
12. Chan, C., Burrows, L. L., & Deber, C. M. (2005). Alginate as an auxiliary bacterial membrane: Binding of membrane-active peptides by polysaccharides. *Journal of Peptide Research, 65*(3), 343–351. https://doi.org/10.1111/j.1399-3011.2005.00263.x.
13. Characklis, W. G. (1981). Bioengineering report: Fouling biofilm development – A process analysis. *Biotechnology and Bioengineering, 23*(9), 1923–1960. https://doi.org/10.1002/bit.260230904.
14. Chmielewski, R. A. N., & Frank, J. F. (2003). Biofilm formation and control in food processing facilities. *Comprehensive Reviews in Food Science and Food Safety, 2*(1), 22–32. https://doi.org/10.1111/j.1541-4337.2003.tb00012.x.

15. Cho, G. L., & Ha, J. W. (2019). Application of X-ray for inactivation of foodborne pathogens in ready-to-eat sliced ham and mechanism of the bactericidal action. *Food Control, 96*, 343–350. https://doi.org/10.1016/j.foodcont.2018.08.034.
16. Clarke, D., Tyuftin, A. A., Cruz-Romero, M. C., Bolton, D., & Fanning, S. (2017). Surface attachment of active antimicrobial coatings onto conventional plastic-based laminates and performance assessment of these materials on the storage life of vacuum packaged beef sub-primals. *Food Microbiology, 62*, 196–201. https://doi.org/10.1016/j.fm.2016.10.013.
17. Corcoran, M., Morris, D., De Lappe, N., O'Connor, J., & Lalor, P. (2014). Commonly used disinfectants fail to eradicate Salmonella enterica biofilms from food contact surface materials. *Applied and Environmental Microbiology, 80*(4), 1507–1514. https://doi.org/10.1128/AEM.02995-13.
18. Coutinho, N. M., Silveira, M. R., Rocha, R. S., Moraes, J., & Ferreira, M. V. S. (2018). Cold plasma processing of milk and dairy products. *Trends in Food Science & Technology, 74*, 56–68. https://doi.org/10.1016/j.tifs.2018.02.010.
19. Cui, H., Bai, M., Yuan, L., Surendhiran, D., & Lin, L. (2018). Sequential effect of phages and cold nitrogen plasma against *Escherichia coli* O157 biofilms on different vegetables. *International Journal of Food Microbiology, 268*, 1–9. https://doi.org/10.1016/j.ijfoodmicro.2018.01.014.
20. Cui, H., Ma, C., & Lin, L. (2016). Synergistic antibacterial efficacy of cold nitrogen plasma and clove oil against *Escherichia coli* O157 biofilms on lettuce. *Food Control, 66*, 8–16. https://doi.org/10.1016/j.foodcont.2016.01.002.
21. Cullen, P. J., Lalor, J., Scally, L., Boehm, D., & Milosavljević, V. (2018). Translation of plasma technology from the lab to the food industry. *Plasma Processes and Polymers, 15*(2), Article ID 201700085. https://doi.org/10.1002/ppap.201700085.
22. Dasan, B. G., Onal-Ulusoy, B., Pawlat, J., Diatczyk, J., Sen, Y., & Mutlu, M. (2017). A new and simple approach for decontamination of food contact surfaces with gliding arc discharge atmospheric non-thermal plasma. *Food and Bioprocess Technology, 10*(4), 650–661. https://doi.org/10.1007/s11947-017-1898-5.
23. Deacon, J., Abdelghany, S. M., Quinn, D. J., Schmid, D., & Megaw, J. (2015). Antimicrobial efficacy of tobramycin polymeric nanoparticles for Pseudomonas aeruginosa infections in cystic fibrosis: Formulation, characterization and functionalization with dornase alfa (DNase). *Journal of Controlled Release, 198*, 55–61. https://doi.org/10.1016/j.jconrel.2014.12.024.
24. Delaquis, P. J., Stewart, S., Toivonen, P. M. A., & Moyls, A. L. (1999). Effect of warm, chlorinated water on the microbial flora of shredded iceberg lettuce. *Food Research International, 32*(1), 7–14. https://doi.org/10.1016/S0963-9969(99)00042-7.
25. Deng, S. X., Cheng, C., Ni, G. H., Meng, Y. D., & Chen, H. (2010). The interaction of an atmospheric pressure plasma jet using argon or argon plus hydrogen peroxide vapor addition with Bacillus subtilis. *Chinese Physics B, 19*(10), Article ID 105203. https://doi.org/10.1088/1674-1056/19/10/105203.
26. Dirks, B. P., Dobrynin, D., Fridman, G., Mukhin, Y., Fridman, A., & Quinlan, J. J. (2012). Treatment of raw poultry with nonthermal dielectric barrier discharge plasma to reduce *Campylobacter jejuni* and *Salmonella enterica*. *Journal of Food Protection, 75*(1), 22–28. https://doi.org/10.4315/0362-028X.JFP-11-108.
27. Donlan, R. M. (2002). Biofilms: Microbial life on surfaces. *Emerging Infectious Diseases, 8*(9), 881–890. https://doi.org/10.3201/eid0809.020063.

28. Dunne, W. M. (2002). Bacterial adhesion: Seen any good biofilms lately? *Clinical Microbiology Reviews, 15*(2), 155–166. https://doi.org/10.1128/CMR.15.2.155-166.2002.
29. Elez Garofulić, I., Režek Jambrak, A., Milošević, S., Dragović-Uzelac, V., Zorić, Z., & Herceg, Z. (2015). The effect of gas phase plasma treatment on the anthocyanin and phenolic acid content of sour cherry Marasca (*Prunus cerasus* var. Marasca) juice. *LWT – Food Science and Technology, 62*(1), 894–900. https://doi.org/10.1016/j.lwt.2014.10.045.
30. Fagerlund, A., Langsrud, S., & Møretrø, T. (2021). Microbial diversity and ecology of biofilms in food industry environments associated with Listeria monocytogenes persistence. *Current Opinion in Food Science, 37*, 171–178. https://doi.org/10.1016/j.cofs.2021.07.009.
31. Faïlle, C., & Carpentier, B. (2009). Food contact surfaces, surface soiling and biofilm formation. In *Biofilms in the Food and Beverage Industries* (pp. 303–330). Elsevier. https://doi.org/10.1533/9781845697167.3.303.
32. Farber, R., Dabush-Busheri, I., Chaniel, G., Rozenfeld, S., & Bormashenko, E. (2019). Biofilm grown on wood waste pretreated with cold low-pressure nitrogen plasma: Utilization for toluene remediation. *International Biodeterioration & Biodegradation, 139*, 62–69. https://doi.org/10.1016/j.ibiod.2018.12.012.
33. Fysun, O., Kern, H., Wilke, B., & Langowski, H. C. (2019). Evaluation of factors influencing dairy biofilm formation in filling hoses of food-processing equipment. *Food and Bioproducts Processing, 113*, 39–48. https://doi.org/10.1016/j.fbp.2018.11.008.
34. Gabriel, A. A., Ugay, M. C. C. F., Siringan, M. A. T., Rosario, L. M. D., Tumlos, R. B., & Ramos, H. J. (2016). Atmospheric pressure plasma jet inactivation of Pseudomonas aeruginosa biofilms on stainless steel surfaces. *Innovative Food Science & Emerging Technologies, 36*, 311–319. https://doi.org/10.1016/j.ifset.2016.06.004.
35. Gadri, R. B., Roth, J. R., Montie, T. C., Kelly-Wintenberg, K., & Tsai, P. P. Y. (2000). Sterilization and plasma processing of room temperature surfaces with a one atmosphere uniform glow discharge plasma. *Surface and Coatings Technology, 131*(1–3), 528–541. https://doi.org/10.1016/S0257-8972(00)00820-3.
36. Giaouris, E., Heir, E., Hébraud, M., Chorianopoulos, N., & Langsrud, S. (2014). Attachment and biofilm formation by foodborne bacteria in meat processing environments: Causes, implications, role of bacterial interactions and control by alternative novel methods. *Meat Science, 97*(3), 298–309. https://doi.org/10.1016/j.meatsci.2013.07.026.
37. Gordon, C. A., Hodges, N. A., & Marriott, C. (1988). Antibiotic interaction and diffusion through alginate and exopolysaccharide of cystic fibrosis-derived Pseudomonas aeruginosa. *Journal of Antimicrobial Chemotherapy, 22*(5), 667–674. https://doi.org/10.1093/jac/22.5.667.
38. Govaert, M., Smet, C., Vergauwen, L., Ećimović, B., & Walsh, J. L. (2019). Influence of plasma characteristics on the efficacy of Cold Atmospheric Plasma (CAP) for inactivation of Listeria monocytogenes and Salmonella Typhimurium biofilms. *Innovative Food Science & Emerging Technologies, 52*, 376–386. https://doi.org/10.1016/j.ifset.2019.02.008.
39. Gu, T., Meesrisom, A., Luo, Y., Dinh, Q. N., & Lin, S. (2021). *Listeria monocytogenes* biofilm formation as affected by stainless steel surface topography and coating

composition. *Food Control, 130*, Article ID 108275. https://doi.org/10.1016/j.foodcont.2021.108275.

40. He, M., Duan, J., Xu, J., Ma, M., & Chai, B. (2020). *Candida albicans* biofilm inactivated by cold plasma treatment in vitro and in vivo. *Plasma Processes and Polymers, 17*(4), Article ID 201900068. https://doi.org/10.1002/ppap.201900068.
41. Heise, M., Neff, W., Franken, O., Muranyi, P., & Wunderlich, J. (2004). Sterilization of polymer foils with dielectric barrier discharges at atmospheric pressure. *Plasmas and Polymers, 9*(1), 23–33.
42. Hertwig, C., Reineke, K., Ehlbeck, J., Knorr, D., & Schlüter, O. (2015). Decontamination of whole black pepper using different cold atmospheric pressure plasma applications. *Food Control, 55*, 221–229. https://doi.org/10.1016/j.foodcont.2015.02.019.
43. Hessing, A., Schneider, R. M. G., Gutierrez, A., Silverberg, R., Gutter, M. S., & Schneider, K. R. (2020). The cost of food safety: FSHN15-07/FS270-10/2015. *EDIS, 2016*(1), 5–11.
44. Jahid, I. K., Han, N., Zhang, C. Y., & Ha, D. (2015). Mixed culture biofilms of *Salmonella typhimurium* and cultivable indigenous microorganisms on lettuce show enhanced resistance of their sessile cells to cold oxygen plasma. *Food Microbiology, 46*, 383–394. https://doi.org/10.1016/j.fm.2014.08.013.
45. Kamgang-Youbi, G., Herry, J.-M., Brisset, J.-L., Bellon-Fontaine, M.-N., Doubla, A., & Naïtali, M. (2008). Impact on disinfection efficiency of cell load and of planktonic/adherent/detached state: Case of *Hafnia alvei* inactivation by plasma activated water. *Applied Microbiology and Biotechnology, 81*(3), 449–457. https://doi.org/10.1007/s00253-008-1721-0.
46. Khan, M. S. I., Lee, E. J., & Kim, Y. J. (2016). A submerged dielectric barrier discharge plasma inactivation mechanism of biofilms produced by *Escherichia coli* O157, *Cronobacter sakazakii*, and *Staphylococcus aureus*. *Scientific Reports, 6*, Article 370–379. https://doi.org/10.1038/srep37079.
47. Kim, H. J., Yong, H. I., Park, S., Choe, W., & Jo, C. (2013). Effects of dielectric barrier discharge plasma on pathogen inactivation and the physicochemical and sensory characteristics of pork loin. *Current Applied Physics, 13*(7), 1420–1425. https://doi.org/10.1016/j.cap.2013.03.020.
48. Klapper, I., Rupp, C. J., Cargo, R., Purvedorj, B., & Stoodley, P. (2002). Viscoelastic fluid description of bacterial biofilm material properties. *Biotechnology and Bioengineering, 80*(3), 289–296. https://doi.org/10.1002/bit.10384.
49. Kramer, B., Hasse, D., Guist, S., Schmitt-John, T., & Muranyi, P. (2020). Inactivation of bacterial endospores on surfaces by plasma processed air. *Journal of Applied Microbiology, 128*(4), 920–933. https://doi.org/10.1111/jam.14445.
50. Kurincic, M., Jersek, B., Klancnik, A., Mozina, S. S., & Fink, R. (2016). Effects of natural antimicrobials on bacterial cell hydrophobicity, adhesion, and zeta potential. *Archives of Industrial Hygiene and Toxicology, 67*(1), 39–45. https://doi.org/10.1515/aiht-2016-67-2731.
51. Lacombe, A., Niemira, B. A., Gurtler, J. B., Fan, X., & Sites, J. (2015). Atmospheric cold plasma inactivation of aerobic microorganisms on blueberries and effects on quality attributes. *Food Microbiology, 46*, 479–484. https://doi.org/10.1016/j.fm.2014.08.014.
52. Misra, N. N., Pankaj, S. K., & Segat, A. (2016). Cold plasma interactions with enzymes in foods and model systems. *Trends in Food Science & Technology, 55*, 39–47. https://doi.org/10.1016/j.tifs.2016.06.008.

53. Muranyi, P., Wunderlich, J., & Heise, M. (2008). Influence of relative gas humidity on the inactivation efficiency of a low temperature gas plasma. *Journal of Applied Microbiology, 104*(6), 1659–1666. https://doi.org/10.1111/j.1365-2672.2007.03736.x.
54. Muranyi, P., Wunderlich, J., & Langowski, H. C. (2010). Modification of bacterial structures by a low-temperature gas plasma and influence on packaging material. *Journal of Applied Microbiology, 109*(6), 1875–1885. https://doi.org/10.1111/j.1365-2672.2010.04822.x.
55. Myszka, K., & Czaczyk, K. (2011). Bacterial biofilms on food contact surfaces – A review. *Polish Journal of Food and Nutrition Sciences, 61*(3), 173–180. https://doi.org/10.2478/v10222-011-0005-6.
56. Niemira, B. A. (2012). Cold Plasma Decontamination of Foods. *Annual Review of Food Science and Technology, 3*(1), 125–142. https://doi.org/10.1146/annurev-food-022811-101141.
57. Nitschke, M., Araújo, L. V., Costa, S. G. V. A. O., Pires, R. C., & Zeraik, A. E. (2009). Surfactin reduces the adhesion of food-borne pathogenic bacteria to solid surfaces. *Letters in Applied Microbiology, 49*(2), 241–247. https://doi.org/10.1111/j.1472-765X.2009.02656.x.
58. Pankaj, S. K., Misra, N. N., & Cullen, P. J. (2013). Kinetics of tomato peroxidase inactivation by atmospheric pressure cold plasma based on dielectric barrier discharge. *Innovative Food Science & Emerging Technologies, 19*, 153–157. https://doi.org/10.1016/j.ifset.2013.04.005.
59. Pasquali, F., Stratakos, A. C., Koidis, A., Berardinelli, A., & Cevoli, C. (2016). Atmospheric cold plasma process for vegetable leaf decontamination: A feasibility study on red chicory (*Cichorium intybus* L.). *Food Control, 60*, 552–559. https://doi.org/10.1016/j.foodcont.2015.08.023.
60. Patange, A., Boehm, D., Ziuzina, D., Cullen, P. J., Gilmore, B., & Bourke, P. (2019). High voltage atmospheric cold air plasma control of bacterial biofilms on fresh produce. *International Journal of Food Microbiology, 293*, 137–145. https://doi.org/10.1016/j.ijfoodmicro.2019.01.008.
61. Peng, P., Chen, P., Zhou, N., Schiappacasse, C., & Cheng, Y. (2019). Packed food and packaging materials disinfected by cold plasma. In C. J. Cullen, P. J. Cullen, & V. Patil (Eds.), *Advances in Cold Plasma Applications for Food Safety and Preservation* (pp. 269–286). Elsevier. https://doi.org/10.1016/B978-0-12-814921-8.00009-8.
62. Perni, S., Shama, G., & Kong, M. G. (2008). Cold atmospheric plasma disinfection of cut fruit surfaces contaminated with migrating microorganisms. *Journal of Food Protection, 71*(8), 1619–1625. https://doi.org/10.4315/0362-028X-71.8.1619.
63. Sen, Y., & Mutlu, M. (2013). Sterilization of food contacting surfaces via non-thermal plasma treatment: A model study with *Escherichia coli*-contaminated stainless steel and polyethylene surfaces. *Food and Bioprocess Technology, 6*(12), 3295–3304. https://doi.org/10.1007/s11947-012-0895-0.
64. Shi, X., & Zhu, X. (2009). Biofilm formation and food safety in food industries. *Trends in Food Science & Technology, 20*(9), 407–413. https://doi.org/10.1016/j.tifs.2009.04.010.
65. Shimizu, T., Sakiyama, Y., Graves, D. B., Zimmermann, J. L., & Morfill, G. E. (2012). The dynamics of ozone generation and mode transition in air surface micro-discharge plasma at atmospheric pressure. *New Journal of Physics, 14*(10), Article 103028. https://doi.org/10.1088/1367-2630/14/10/103028.

66. Sinde, E., & Carballo, J. (2000). Attachment of Salmonella spp. and Listeria monocytogenes to stainless steel, rubber and polytetrafluorehtylene: The influence of free energy and the effect of commercial sanitizers. *Food Microbiology, 17*(4), 439–447. https://doi.org/10.1006/fmic.2000.0345.
67. Sousa, J. S., Niemi, K., Cox, L. J., Algwari, Q. T., Gans, T., & O'Connell, D. (2011). Cold atmospheric pressure plasma jets as sources of singlet delta oxygen for biomedical applications. *Journal of Applied Physics, 109*(12), Article 3601347. https://doi.org/10.1063/1.3601347.
68. Verran, J., Airey, P., Packer, A., & Whitehead, K. A. (2008). Microbial retention on open food contact surfaces and implications for food contamination. *Advances in Applied Microbiology, 64*, 223–246. https://doi.org/10.1016/S0065-2164(08)00408-5.
69. Vu, B., Chen, M., Crawford, R. J., & Ivanova, E. P. (2009). Bacterial extracellular polysaccharides involved in biofilm formation. *Molecules, 14*(7), 2535–2554. https://doi.org/10.3390/molecules14072535.
70. Ziuzina, D., Boehm, D., Patil, S., Cullen, P. J., & Bourke, P. (2015). Cold plasma inactivation of bacterial biofilms and reduction of quorum sensing regulated virulence factors. *PLoS One, 10*(9), Article e0138209. https://doi.org/10.1371/journal.pone.0138209.

CHAPTER 10

IMPACT OF COLD PLASMA ON APPLICATIONS IN FOOD PACKAGING

CHEEYATTIL SUBITH, M. DHARINI, and R. MAHENDRAN

ABSTRACT

Packaging materials have a sole purpose of protecting the product that is to be contained. However, the property and the quality of the packaging material differs based on the material used for packaging and the various modifications that has undergone. The modifications in packaging materials can be done by treating it using various methods, such that it can perform better and maintain a higher level of quality and safety. Potential applications for cold plasma also comprise of modifying the packaging materials; in case of food packaging materials, the cold plasma induced structural transformation become a powerful tool. Cold plasma assisted modification methods are safe, low cost, and pollution-free; it also replaces traditional means by decreasing environmental pollution and energy consumption. Decontamination of packed food (without heating) with plasma becomes trending in modern food industry. Surface roughness is increased by plasma treatment, which improves, ink printability, surface adhesion and minimizes the contact angle (CA). This chapter discusses the impact of nonthermal cold plasma on the packages made from polysaccharides, proteins and their mixtures, as well as the various plasma producing devices utilized for film modification. The potential applications and future perspectives of plasma treated films are also discussed.

10.1 INTRODUCTION

Packaging is utilized for protection, distribution, presentation, and retention of the quality and safety of foods. The packaging sector is an emerging part

Nonthermal Plasma Technology for Food and Food Products.
R. Mahendran, Preeti Birwal, and Megh R. Goyal (Eds.)

of the food industry; more technological innovation has been underway. Biodegradable, active, and antimicrobial packaging materials with better barrier (water vapor transmission rate (WVTR) and oxygen transmission rate (OTR)) and printable properties are leading areas of research [53].

Food sterilization, decontamination, and pasteurization process start from the beginning of food processing lines; pre-sterilization or decontamination of food materials before packaging without treating the packaging material expedites the deterioration of foods. Thus, the stage of packaging, sterilization, and decontamination play a critical role in preserving the quality during food storage. The functional characteristics and effectiveness of the packaging materials can be enhanced by different physicochemical treatments, nonthermal treatments, and the addition of antimicrobial agents [51]. Modified atmospheric packaging, active and smart packaging are getting more significant in the current food industry due to improved properties and maintenance of a better-quality standard. The criterion for selecting packaging material varies according to the food material to be packed and the purpose of packaging. Numerous properties decide the selectivity of the desired package, including its availability, cost, and properties of packaging, including barrier properties, transparency, printability, thickness, strength, biodegradability, and reliability of the packages [3].

Apart from synthetic polymers, an increasing interest in packaging regulation and environmental concerns, biodegradable packages are getting more attention and consumer acceptance [35]. However, synthetic polymers are having better barrier properties and chemically inert, which leads to the industrial-scale higher demand for synthetic packaging materials compared to biodegradable packaging material. The barrier properties can be enhanced by lamination, coating, coextrusion and addition of other substances with the packaging material. Nevertheless, the lamination and coating efficiency of synthetic and biodegradable packaging materials vary according to the chemical nature of packages. The surface morphology, contact angle (CA), and hydrophilicity of the packaging materials determine the printability, coating efficiency and adhesion properties during lamination and coextrusion of packages [21].

The biopolymeric materials shows weak oxygen and water vapor transmission rate as compared to synthetic polymers; chemical crosslinking and surface coating/lamination have been implemented to improve these properties [65]. Yet, there are other methods to increase the properties of these synthetic and biodegradable packages.

Modification of the morphological and physicochemical characteristics can improve properties of packages [43]. Irradiation, laser treatment, UV rays, X-rays, IR radiation and cold plasma, are the standard physical treatments that alter the characteristics of packaging material; among them, cold plasma treatment shows desirable properties like adhesion properties, improved printability, enhanced barrier properties and decontamination of the packaging materials [62].

Cold plasma has drawn a tremendous significance in food processing for over a decade. The quality standards of foods are retained during the cold plasma treatment. Packaging film sterilization, in-package sterilization, and surface modification plays vital roles in retaining the safety and to meet international standards [57]. Unlike other non-thermal technologies, cold plasma is effectively used to change the packaging film's physicochemical properties [46].

Cold plasma is becoming a trending technology as it has demonstrated to have a variety of applications in active packaging and improved antimicrobial packaging. Enhanced adhesion properties, and surface roughness has improved the coating capability of packaging materials with antimicrobial and active substances [31]. Plasma treatments are classified based on the process's temperature, including cold plasma and thermal plasma. Thermal plasma expedites thermal damage, whereas cold plasma operates at atmospheric temperature and produces desirable physicochemical changes to the packages [33].

Cold plasma technology has been claimed to increase the barrier functions and surface modifications of both synthetic and biodegradable packing materials [50]. Apart from physicochemical changes, active species and high-energy photons in the form of UV rays are produced within the region of plasma discharge provides sterilization effect, antimicrobial effect, pesticide reduction and biodegradable properties to the packages [2, 44]. This property of plasma in sterilizing the packaging material along with improving the barrier properties makes it an outstanding technology in the food packaging sector.

This chapter discusses principle behind plasma in improving the packaging materials, their applications and prospective ideas that can be studies in the future for maintaining the quality of foods by means of packaging.

10.2 COLD PLASMA DEVICES

The fourth state of matter, also known as plasma, comprised of numerous charged particles, including, electrons, reactive species, gaseous molecules and ions (in ground and excited state). The higher temperature generated in a gaseous environment aid in creating thermal plasma. In contrast, the electrical energy applied to gas, generate non thermal plasma, which gets absorbed by electrons and produces a non-equilibrium state between electrons and ions. The energized electron collides with feed gas (neutral atom or molecules), creating impact ionization (collision of the electron) and photoionization at atmospheric temperature [37].

The most commonly used cold plasma discharges methods are, Corona discharge, Dielectric barrier discharge (DBD), and Atmospheric pressure plasma jets (APPJ) [55]. Corona discharge plasma apparatus consists of a high voltage probe and a metal electrode; the high voltage power supply creates a non-uniform stream of corona discharge between the electrode and probe, where the sample is kept between them [66].

DBD plasma system comprise of two electrodes (planar/cylindrical) arranged parallelly along with an insulating material (dielectric barrier) wrapped over one electrode, and the other electrode is earthed. Filamentary discharge (non-uniform) and glow discharge (uniform) are the two discharge patterns in DBD systems according to the design of the electrode arrangement and the pressure used for generating plasma [58].

Atmospheric pressure plasma jet systems are either corona discharge or DBD discharge with an addition of continuous gas supply within the conducting electrodes to carry the reactive species produced inside the corona or DBD chamber in the form of jet. Plasma jets are conventional and can be used for direct treatment of biological tissues [11]. Air, CO_2, N_2, Ar and He are commonly used working gasses for all three types of discharges [64].

10.3 NEED FOR FOOD PACKAGING

The food safety and quality control sectors are getting more significant in the advanced era of food processing, along with the safety concern, the field of packaging in food processing receives extreme importance. New innovative packaging technologies and packaging materials acquired a characteristic role in the food processing sector (Figure 10.1). Synthetic polymers, metals, glass, paper and biopolymers are now introduced with significant modification and improved properties [68]. The increasing interest in polymer packaging materials is still prevalent due to the availability, cost, lightweight, barrier properties, chemically inert nature and accessibility of these packages [26].

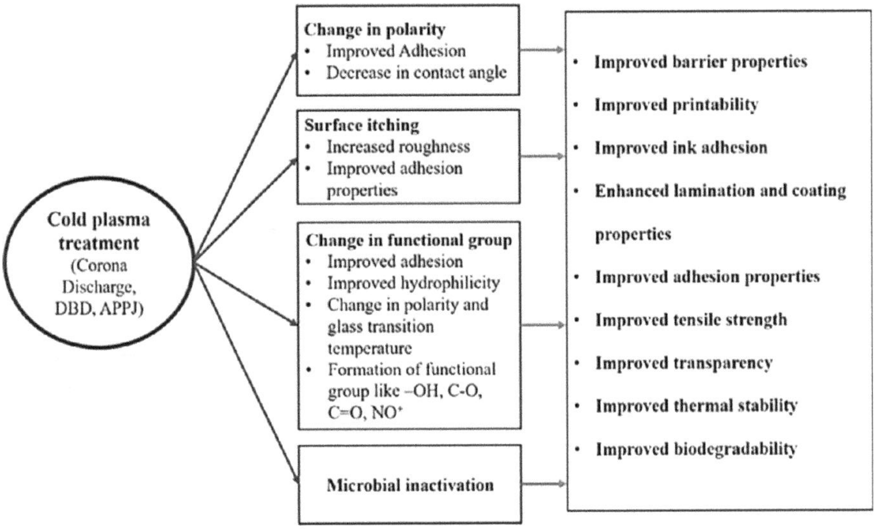

FIGURE 10.1 The applications of cold plasma treatment in food packaging.

Plastic packages are getting more prominence due to their versatile properties, availability, and ability to modify according to the requirements. However, the inherent properties such as barrier properties of an individual material limit the wide applicability of a package. At the same time, various physical modifications, lamination and coating are applied to heighten the utility and functionality of the packages [34].

In this current century, demand for the use of green technology, biodegradable and biopolymers are gaining attention; moreover, the chemical cross-linking using citric acid, calcium chloride, plasticizers, and other aldehyde were also found to improve the barrier and functional properties [1]. With an increasing demand for green technology and biodegradable packages, biopolymers are gaining more importance and acceptance [69]. Biodegradable biopolymer limits its barrier properties, printability, thermal stability, and coating properties compared to synthetic polymers. Simultaneously the it minimizes the limitations of plasma treatment by cross-linking reaction, surface itching, and the development of the polar functional group on the surface of the package [23].

The formation of hydrophobic layer by coating oil nanoparticles after plasma treatments in order to reduce the microbial contamination and spoilage of packaged foods [16]. Conventional plastic of thin cross-sectional area provides better barrier properties, whereas the biopolymers of the same thickness show comparatively lower barrier properties; these issues were resolved up to a level by cross-linking of biopolymers by plasma treatment [61].

The LDPE film treated with cold plasma (30 W, 13.56 MHz) provide enhanced and uniform coating (Gallic acid monohydrate) as compared to untreated film [65]. The coating and lamination capacities improved the surface roughness and adhesion characteristics of plasma-treated biopolymer. These characteristic features provide the integration of active components like plant extract, antimicrobial agents, bacteriocins, and oxygen, carbon dioxide, ethylene, and moisture scavengers to enhance the functionality of the packages [59].

10.4 EFFECTS OF COLD PLASMA ON PACKAGING FILMS

Surface structure and characteristics of the packaging material can be altered by plasma treatment. Bombardment of reactive species while the exposure of plasma resulted in roughness of the surface and a reduction in the sharp edges of the packaging material [44]. The highly energized plasma species like electrons, protons, ions, and free-radicles bombard the surface of the packaging materials causing surface modification and etching effect over the surface [13]. The free radicles of plasma create a breakdown of chemical bonds, fragmentation of polymeric chain, and degradation of certain functional groups, which directly interfere with the adhesiveness and barrier functions of the film [20].

The rate of surface etching and roughness created over the polymer surfaces is attributed to the types of plasma treatment, intensity, time of exposure, the working gas, and the type of the polymer used [54]. The surface etching on PET was reported to be mainly caused by the reactive species; the rate of etching formed by different gasses is in the order oxygen > hydrogen > nitrogen > argon > ammonia, whereas the degree of roughness created on the surface by different gasses is in the order ammonia > nitrogen > hydrogen > argon > oxygen [22].

Table 10.2 illustrates the impact of plasma on surface modification. However, the uniform etching over the surface is created by DBD plasma; RF plasma microwave-driven plasma was found to produce non-uniform etching, yet it also yielded a desirable effect [17]. Plasma treatment on polypropylene films created proton abstraction from the -CH group, resulting in the cleavage of the backbone chain. As a result, the polymer properties were found to be different compared to the untreated polymer; it constituted better coating and printability of polymeric films [20] (Table 10.1).

TABLE 10.1 Impact of Cold Plasma Technology on the Morphological and Structural Characteristics of Packaging Materials

Packaging Material	Type	Treatment Conditions	Results	Applications	References
BOPP/LDPE laminates	S	Cold plasma (1,600 Pa using an O_2), 84 W, 120 seconds	Hydrophilicity increased, to provide better coating efficiency	Improved coating/lamination efficiency Improved printability of packaging	[32]
CMC films	B	Atmospheric pressure plasma, 14 kV, 20 kHz	Water vapor permeability reduced, Improved coating properties and adhesive properties	Improving coating/adhesive properties for bilayer formation Improving barrier properties	[20]
Corn starch/poly (caprolactone) films.	B	Air plasma, 13.56 MHz, 40 W, 2 min	Tensile strength reduced, dilation at break increased, CA increased, surface coarseness increased, biodegradability also increased	Enhanced biodegradability and printability of packages	[5]
Fish protein films	B	Glow discharge plasm (Alternating current), 60 Hz, 4.4 kV, 5 min	Improved transparency, improved thermal stability, water vapor permeability reduced, improved tensile strength, roughness increased	Improving barrier properties and physicochemical properties	[56]
Gelatin films	B	DBD, 60, 70 and 80 kV for 1, 2, 3, 4 and 5 min	Increased topography and roughness of the film, increased surface polarity and hydrophilicity	Used for coating and lamination application Improved printability of the film	[48]
High amylose corn starch films	B	DBD plasma 60–80 kV for 1 to 5 minutes.	Raised the outside roughness and increased hydrophilicity	Improved surface properties of packing film enhance printability	[47]
Momordica charantia polysaccharide (MCP)	B	Cold plasma, 350 W, 30s	Increased antimicrobial activity and antioxidant property	Improved coating efficiency Improved stability of coating	[10]

TABLE 10.1 *(Continued)*

Packaging Material	Type	Treatment Conditions	Results	Applications	References
PLA film	B	Cold plasma treatment, 900 W, 2.45 GHz, 667 Pa, 40 min	Increased roughness, printability, CA and biodegradability	Used as printable biodegradable packaging material	[62]
PLA films	B	DBD, 70 and 80 kV for 0.5, 1.5, 2.5 and 3.5 min	Increased the roughness, increased water vapor permeability	Used for better printability of the package.	[45]
Polycaprolactone or Poly (lactic acid) (PLA)	B	DBD, 50 Hz and 31 kV for 1 to 20 min	Increased surface erosion and decrease in CA	Used for making multi-layer film making Improved printability	[19]
polythene oxide nanofibers with Beta cyclodextrin and tea tree oil	S	Air jet plasma, 300 W, 0.5–5 min	Improved coating efficiency and antimicrobial activity	Used for improving coating efficiency	[9]
Sodium caseinate	B	DBD, 60 and 70 kV for 1 to 5 min	Increased surface roughness, WVTR, OTR was noticed, and decreased glass transition temperature	Used for in-package plasma decontamination of foods	[46]
Stainless steel material PET	S S	Gliding Arc Discharge 15 kV, 40 mA, 16 kHz	Hydrophobicity and CA reduced	Reducing surface microbial load by modifying the surface properties of packaging material (hydrophobic surface)	[12]
Starch film (rice, cassava, maize)	B	DBD, 80 kV for 5 min	Surface roughness and glass transition temperature also increases	Used for printable and biodegradable packages	[49]
Zein films	B	Airglow discharge plasma (ADP), 5 kV, 40 kHz, 30s, 60s, 90s, and 120s	Mechanical strength increased, wettability is increased, increased crosslinking of protein (increased barrier properties)	Mechanical properties and barrier of biofilm can be improved by using this method	[30]

The slight degradation of the polymer network from the surface (up to a few nanometers deep) caused by surface etching during plasma treatment had a negative impact on the tensile strength and the general bulk characteristics of BOPP/ LDPE polymeric packaging materials. [32]. Plasma is an advanced technology used in food packaging to efficiently optimize the packaging conditions and enhance the properties like adsorption, coating properties, printability, and decontamination of the packages. The following are the applications of plasma processing of food packaging materials.

10.5 APPLICATIONS OF COLD PLASMA TREATMENT ON FOOD PACKAGING

10.5.1 BARRIER PROPERTIES OF FILMS

The major components of barrier properties of a material are WVTR and OTR. The WVTR depends on the combination of migration and soluble nature over the void space across the polymer network. The plasma treatment has both desirable and adverse effects on the barrier functions of the food packages; it mostly attributed as a result of surface etching, cross-linking, and surface modification of the packaging film [63].

Lower OTR and WVTR are essential for an ideal package to retain the sensory and nutritional qualities of food. Characteristic reductions in OTR were observed in plasma-treated BOPP/LDPE film; in contrast, the plasma-treated nitrocellulose-coated film created cracks on the material. A reduction in the strength was due to the formation of the cracks over the surfaces as observed after SEM analysis [32]. This drawbacks can be rectified by lamination or coating of plasma treated biopolymers with cellulose and other bio-nanocomposite [67].

The hydrophilic nature of biopolymers has more tendency to react with oxygen-free radicles at higher voltage, delivering the formation of polar function group and hydrophilic nature of polymer [7]. The elevation of WVTR due to increased hydrophilicity of the material is the major drawback of some bio-polymers during the plasma treatment. It is due to the reaction of free radicles with the packaging material and the formation of a polar functional group [48]. The DBD plasma had a significant effect on thin film surface characteristics, although their bulk properties were unaffected [21].

Similarly, the WVTR of polylactic acid (PLA) was also increased after the application of DBD plasma exposure (80 kV). The plasma treatment had both desirable and adverse effect on the barrier functions of the packages.

The treatment conditions, types of treatment, and types of packages determine the effectiveness; barrier properties of biopolymers have enhanced by crosslinking, whereas the formation of cracks reduces barrier properties of the synthetic polymer due to surface etching.

10.5.2 PRINTING

The principal parameters that regulate the printability of packaging material are the adhesion properties, wettability, dye-uptake, CA, and surface roughness. Cold plasma-treated polymer shows a higher rate of ink adhesion due to the improved roughness of the sample. The printability of PLA film is enhanced by the application cold plasma due to the higher adhesion and reduced CA [62]. The lower surface energy of polymer like PET and PP is enhanced by the plasma treatment to improve the printability, adhesion power and ink uptake. The polymerization of materials while the plasma treatment leads to the creation of rough surface and thus enhances printability [14]. Plasma treatment also has impact on the printability of paper board coated with polyethylene; the ink adhesion, leveling, and stability of the ink have been reported to be improved after the treatment [36].

Morsy et al. [32] revealed that DBD plasma (15s) treated polypropylene film shows improved print density, ink adhesion, and reduced gloss of printing due to the formation of the roughness of the package. Paper coated with LDPE material after plasma treatment exhibit enhanced rub-off resistance to the surface, which improves printability. The surface modification is constituted by enhanced bonding ability, surface energy, and surface oxidation [27]. The enhanced adoption power and microstructural surface modification, during the plasma treatment is efficiently employed for high pixel printing on glass and metal cans without color variation and clarity loss [47]. The printability and ink intake of packaging materials is improved by the change in surface roughness, adhesion properties, and polarity during plasma treatment. In addition to it, plasma treatment creates abrasion-resistant and scratch-resistant polymer surface, which enhance the uniformity, quality and clarity of printing.

10.5.3 CONTACT ANGLE (CA)

The CA of packaging material is used to represent the hydrophilic/hydrophobic nature, wettability, and ink adhesion. The small CA represents the wettable surface and hydrophilic nature of the package[24]. The time of exposure during plasma treatment was inversely proportional to the CA of PET bottles, while the surface polarity got increased [49].

The DBD plasma treatment on polyethylene film forms polar hydrophilic groups (C-O, C=O, and -OH), which improve polarity of the polymeric surface. The impact of plasma on CA is shown in Table 10.2.

The polar functional group creates a hydrophilic surface and reduces CA [60]. However, upon storage, the surface free energy gets reduced, and the recovered hydrophobic group is called the aging phenomenon after plasma treatment [70]. In oxygen gas plasma treatment, the oxygen reactive species were found to react with the polymeric packaging material to form a polar functional group on the surface, and reduce the CA of the package [18]. The reduction in contact angle provides better printability, adhesion properties, ink intake, surface coating, and lamination properties [8]. The development of a polar functional group causes a change in contact angle, which influences the water and microbial adhesion properties. Coating of hydrophobic materials and fillers reduces these drawbacks. The treatment condition and type of polymer used during plasma treatment determine whether it induces the formation of a polar functional group or the surface roughness, which determines the contact angle. It is essential to scrutinize the treatment condition and the type of polymer to achieve the desired functional characteristics of the packaging film.

10.5.4 ADHESION

Surface interaction of packaging material and chemical substance is known as the adhesive property of the packaging material. The plasma treatment converts the free energy of the surface to the adhesion force [42]. Airglow discharge plasma using oxygen as working gas creates an oxygen-containing functional group that improves the polarity and adhesion power of zein film. Apart from this, the formation of cross-linking of polymers occurred, which destroyed the crystalline structure of zein film [30].

The cold plasma (20 kV) treatment imparts chemical transformation to polypropylene film, which enhances wettability and improves adhesion properties, resulting in improved coating efficiency of the polypropylene film with carboxymethyl cellulose [36]. The cellulose nanofiber (CNFs) treated with cold plasma observed under SEM image shows a significant increment in adhesion properties of CNF matrices [15]. The improved adhesion properties during plasma can be effectively used to enhance the printability,

TABLE 10.2 Impact of Cold Plasma on Surface Sterilization of Packaging Materials

Packaging Material	Type	Treatment Conditions	Reported Sterilization Effect on Packaging Materials	Applications	References
Collagen casing	B	APP, 0.5 kW, 10 minute	3.87 log_{10} reduction of *Salmonella*	In order to reduce the production of biofilm on food container	[25]
PP	S		4.48 log_{10} reduction of *Salmonella*		
PET	S		4.32 log_{10} reduction of *Salmonella*		
Defatted soybean meal (DSM) edible film	B	Cold plasma, 2.45 GHz, 400–900 W	6.2–6.7 log_{10} reduction of total aerobic plate count	Improving properties of the DSM edible film along with inactivation of microorganisms	[42]
LDPE	S	Low-pressure plasma 5 minutes	4 log_{10} reduction of vegetative cells of *E. coli* O157:H7 and *Staphylococcus aureus*	Used for reduction of food-borne pathogens in different packaging materials	[28]
PE, PP, Nylon	S	10 kV, 1.25 A, 10 minutes	4.5 log_{10} reduction for *Escherichia. coli* O157:H7 3.5 log_{10} reduction for *Salmonella. Typhimurium*	Disinfection of pathogenic bacteria on packages without affecting physiochemical properties	[52]
PE, PP, Paper foil	S	CDPJ, 20 kV, 1.5 A 120 seconds	4.5 to 5 log_{10} decrease in *E. coli* O157:H7 3 1 log_{10} decrease in *Salmonella typhimurium*	Food should meet microbial requirements set by regulatory bodies	[29]
PET bottles	S	CDBD, 10–50 kHz, 170 and 180 W 3 minutes	3 log_{10} reduction for *Escherichia coli*	Plasma effect on DNA of spores and cells of microbes in the package. Inactivation of microbes on the package without changes in the package	[41]
PET film	S	CDBD, 10–15 kHz 180 W for 25 second CDBD, 10–15 kHz 180 W for 10 second	7 log_{10} reduction for *Bacillus atrophaeus* spores 4 log_{10} reduction for *Bacillus atrophaeus* vegetative cells	Proper inactivation of spores and vegetative cells without affecting packaging material	[41]

TABLE 10.2 *(Continued)*

Packaging Material	Type	Treatment Conditions	Reported Sterilization Effect on Packaging Materials	Applications	References
Polyethylene terephthalate (PET) foils	B	CDBD 130 W 5 minutes	5 $\log_{10}$ reduction for *Aspergillus Niger*	Industrial sterilization of packaging material	[40]
		1 minute	6.6 $\log_{10}$ reduction for *Clostridium botulinum*	Disinfection of heat-sensitive material in a few minutes	
Stainless steel	S	Gliding Arc Discharge 15 kV, 40 mA, 16 kHz	5.18 $\log_{10}$ reduction for *S. epidermidis*	Decontamination of packages effectively	[12]

adsorption of active components, coating, and lamination purpose to enhance the barrier properties.

10.5.5 SURFACE STERILIZATION

Aseptic packaging and antimicrobial packaging systems are receiving increasing interest in the food processing sector. In aseptic packaging, both the packaging system and the food product are sterilized before packaging. However, the packaging materials are sterilized before packaging using different nonthermal technologies like UV radiation, irradiation, and cold plasma technologies. It is difficult to sterilize packages using irradiation due to higher operating costs and handling difficulties. Compared to UV treatment, cold plasma has improved decontamination properties and enhanced barrier and printability properties of the packaging material [52].

The UV rays produced along with the reactive species during plasma generation create the antimicrobial effect. In addition, the penetration of plasma on the cytoplasmic membrane and oxidation of cell components also destroy microorganisms [4]. The application of plasma on surface sterilization is provided in Table 10.2. DBD Plasma treatment of polythene material for 5 minutes effectively reduces Clostridium botulinum (5 log reduction) over the surface of the packaging material [40]. Plasma treatment shows a 7 $\log_{10}$ reduction in *Bacillus atrophaeus* spores, effectively sterilizing PET films [42]. Besides the modification of polymeric materials, plasma treatments are effectively used for the surface sterilization and decontamination of packaging materials, reducing complications during the packaging and processing of foods.

10.6 AVENUES FOR FUTURE RESEARCH

The trend of using nonthermal and green technologies is becoming more prominent in the current world. The decontamination and sterilizing properties of plasma treatment can effectively substitute chemical sterilizers; apart from that, synthetic polymers can be effectively substituted with biopolymers by improving the physicochemical properties using plasma treatment [67].

Cold plasma is one of the green technologies that is efficient in producing both surface sterilization and modification of packaging material [6]. According to current trends in the area of food processing, the versatility of food products is higher, and the requirement for cheap and efficient packages is also essential. In order to fulfill the demand and requirement, it is essential to improve barrier properties, biodegradability, and printability of packages [38]. The future involves improving coating, lamination, printability, surface modification, and biodegradability using plasma treatment [21]. Plasma treatment increased tensile strength (TS) and elongation at break (EAB) values, which improves the biodegradability of biopolymers [8]. The future of food packaging aims at biodegradable and green packaging technologies with advanced printability, adhesion properties, barrier properties, and antimicrobial properties, to create an active and intelligent package. Cold plasma could be a potential technology in improving these properties in biopolymers. Further research should be undertaken in the field to replace synthetic polymers with biopolymers using plasma technology.

10.7 SUMMARY

The commercialization of biodegradable packages is limited due to their barrier properties, printability, and stability. These drawbacks can be efficiently eliminated by cold plasma technology, owing to the improvement in surface sterilization, printability, barrier properties, CA, adhesion property, ink intake, lamination, and coating efficiency of different packaging materials after plasma application. Cold plasma technology productively substitutes the application of subsequent conventional technologies for decontamination, sterilization, as well as physicochemical changes to the packages by a single treatment. The critical parameters (barrier property, adsorption, and diffusivity) of biopolymers (PCL, PLA, cellulose, CMC, zein, and other protein films) show desirable changes after plasma treatment. The absorptivity and adhesion properties of plasma-treated biopolymers open an excellent opportunity for antimicrobial, active, and intelligent packaging sectors. It also contributes to the adhesion of lamination between two polymers to enable the formation of a composite film and improve printability. Cold plasma is an affordable, sustainable, and green technology to modify the properties of synthetic and biopolymeric films. However, the possibility of diffusion of reactive species (formed after reacting with the package) from packaging films to the food in the case of packed foods poses a threat of causing toxic substances. This may cause health issues; therefore, scientific research is essential to study the toxicity of plasma technology.

KEYWORDS

- **atmospheric pressure plasma jets**
- **cellulose nanofiber**
- **dielectric barrier discharge**
- **elongation at break**
- **oxygen transmission rate**
- **surface etching**
- **tensile strength**
- **water vapor transmission rate**

REFERENCES

1. Ahmed, J., Mulla, M. Z., Vahora, A., Bher, A., & Auras, R. (2021). Polylactide/graphene nanoplatelets composite films: Impact of high-pressure on topography, barrier, thermal, and mechanical properties. *Polymer Composites*, *42*(6), 2898–2909.
2. Anbarasan, R., Jaspin, S., Bhavadharini, B., Pare, A., Pandiselvam, R., & Mahendran, R. (2022). Chlorpyrifos pesticide reduction in soybean using cold plasma and ozone treatments. *LWT*, *159*, 113193. https://doi.org/10.1016/j.lwt.2022.113193.
3. Antón, N., González-Fernández, Á., & Villarino, A. (2020). Reliability and mechanical properties of materials recycled from multilayer flexible packages. *Materials (Basel)*, *13*(18), 3183992. https://doi.org/10.3390/ma13183992.
4. Aparajhitha, S., & Mahendran, R. (2019). Effect of plasma bubbling on free radical production and its subsequent effect on the microbial and physicochemical properties of Coconut Neera. *Innovative Food Science & Emerging Technologies*, *58*, 102230. https://doi.org/10.1016/j.ifset.2019.102230
5. Arolkar, G. A., Salgo, M. J., Kelkar-Mane, V., & Deshmukh, R. R. (2015). The study of air-plasma treatment on corn starch/poly(ε-caprolactone) films. *Polymer Degradation and Stability*, *120*, 262–272.
6. Bahrami, R., Zibaei, R., Hashami, Z., Hasanvand, S., & Garavand, F. (2022). Modification and improvement of biodegradable packaging films by cold plasma; a critical review. *Critical Reviews in Food Science and Nutrition*, *62*(7), 1936–1950.
7. Chaiwong, C., Rachtanapun, P., Wongchaiya, P., Auras, R., & Boonyawan, D. (2010). Effect of plasma treatment on hydrophobicity and barrier property of polylactic acid. *Surface and Coatings Technology*, *204*(18–19), 2933–2939.
8. Chen, G., Chen, Y., Jin, N., Li, J., & Dong, S. (2020). Zein films with porous polylactic acid coatings via cold plasma pre-treatment. *Industrial Crops and Products*, *150*, 112382. https://doi.org/10.1016/j.indcrop.2020.112382.

9. Cui, H., Bai, M., & Lin, L. (2018). Plasma-treated poly (ethylene oxide) nanofibers containing tea tree oil/beta-cyclodextrin inclusion complex for antibacterial packaging. *Carbohydrate Polymers*, *179*, 360–369.
10. Cui, H., Yang, X., Abdel-Samie, M. A., & Lin, L. (2020). Cold plasma treated phlorotannin/Momordica charantia polysaccharide nanofiber for active food packaging. *Carbohydrate Polymers*, *239*, 116214. https://doi.org/10.1016/j.carbpol.2020.116214.
11. Das, B., Chakrabarty, D., Guha, C., & Bose, S. (2021). Effects of corona treatment on surface properties of co-extruded transparent polyethylene film. *Polymer Engineering and Science*, *61*(5), 1449–1462.
12. Dasan, B. G., Onal-Ulusoy, B., Pawlat, J., Diatczyk, J., Sen, Y., & Mutlu, M. (2017). A new and simple approach for decontamination of food contact surfaces with gliding arc discharge atmospheric non-thermal plasma. *Food and Bioprocess Technology*, *10*(4), 650–661.
13. De Vietro, N., Conte, A., Incoronato, A. L., Del Nobile, M. A., & Fracassi, F. (2017). Aerosol-assisted low pressure plasma deposition of antimicrobial hybrid organic-inorganic Cu-composite thin films for food packaging applications. *Innovative Food Science & Emerging Technologies*, *41*, 130–134.
14. Dey, A., Rasane, P., Choudhury, A., Singh, J., Maisnam, D., & Rasane, P. (2016). Cold plasma processing: A review. *Journal of Chemical and Pharmaceutical Sciences*, *9*(4), 2980–2984.
15. Fazeli, M., & Simão, R. A. (2019). Preparation and characterization of starch composites with cellulose nanofibers obtained by plasma treatment and ultrasonication. *Plasma Processes and Polymers*, *16*(6), 1800167. https://doi.org/10.1002/ppap.201800167.
16. Ghiasi, F., Golmakani, M. T., Eskandari, M. H., & Hosseini, S. M. H. (2020). A new approach in the hydrophobic modification of polysaccharide-based edible films using structured oil nanoparticles. *Industrial Crops and Products*, *154*, 112679. https://doi.org/10.1016/j.indcrop.2020.112679.
17. Goiana, M. L., de Brito, E. S., Alves Filho, E. G., Miguel, E. de C., & Fernandes, F. A. N. (2021). Corn starch-based films treated by dielectric barrier discharge plasma. *International Journal of Biological Macromolecules*, *183*, 2009–2016.
18. Guillard, V., Mauricio-Iglesias, M., & Gontard, N. (2010). Effect of novel food processing methods on packaging: Structure, composition, and migration properties. *Critical Reviews in Food Science and Nutrition*, *50*(10), 969–988.
19. Heidemann, H. M., Dotto, M. E. R., Laurindo, J. B., Carciofi, B. A. M., & Costa, C. (2019). Cold plasma treatment to improve the adhesion of cassava starch films onto PCL and PLA surface. *Colloids and Surfaces A: Physicochemical and Engineering Aspects*, *580*, 123739. https://doi.org/10.1016/j.colsurfa.2019.123739.
20. Honarvar, Z., Farhoodi, M., Khani, M. R., Mohammadi, A., & Shokri, B. (2017). Application of cold plasma to develop carboxymethyl cellulose-coated polypropylene films containing essential oil. *Carbohydrate Polymers*, *176*, 1–10.
21. Hoque, M., McDonagh, C., Tiwari, B. K., Kerry, J. P., & Pathania, S. (2022). Effect of cold plasma treatment on the packaging properties of biopolymer-based films: A review. *Applied Sciences*, *12*(3), 2031346. https://doi.org/10.3390/app12031346.
22. Inagaki, N., Narushim, K., Tuchida, N., & Miyazaki, K. (2004). Surface characterization of plasma-modified poly (ethylene terephthalate) film surfaces. *Journal of Polymer Science Part B: Polymer Physics*, *42*(20), 3727–3740.

23. Kabir, E., Kaur, R., Lee, J., Kim, K. H., & Kwon, E. E. (2020). Prospects of biopolymer technology as an alternative option for non-degradable plastics and sustainable management of plastic wastes. *Journal of Cleaner Production, 258*, 120536. https://doi.org/10.1002/polb.20234.
24. Karam, L., Jama, C., Dhulster, P., & Chihib, N. E. (2013). Study of surface interactions between peptides, materials and bacteria for setting up antimicrobial surfaces and active food packaging. *Journal of Materials and Environmental Science*, *4*(5), 798–821.
25. Kim, H. J., Jayasena, D. D., Yong, H. I., Alahakoon, A. U., & Park, S. (2015). Effect of atmospheric pressure plasma jet on the foodborne pathogens attached to commercial food containers. *Journal of Food Science and Technology*, *52*(12), 8410–8415.
26. Kostov, K. G., Nishime, T. M. C., Castro, A. H. R., Toth, A., & Hein, L. R. O. (2014). Surface modification of polymeric materials by cold atmospheric plasma jet. *Applied Surface Science*, *314*, 367–375.
27. Lahti, J., Tuominen, M., & Kuusipalo, J. (2008). The influence of atmospheric plasma treatment on digital print quality of extrusion coated paper. *TAPPI 2008 PLACE Conference Innovative Flexibles and Consumer Packaging*, *24*(3), 471–492.
28. Lee, T., Puligundla, P., & Mok, C. (2015). Inactivation of foodborne pathogens on the surfaces of different packaging materials using low-pressure air plasma. *Food Control*, *51*, 149–155.
29. Lee, T., Puligundla, P., & Mok, C. (2017). Corona discharge plasma jet inactivates food-borne pathogens adsorbed onto packaging material surfaces. *Packaging Technology and Science*, *30*(10), 681–690.
30. Li, Y., Bai, Y., Huang, J., Yuan, C., & Ding, T. (2020). Airglow discharge plasma treatment affects the surface structure and physical properties of zein films. *Journal of Food Engineering*, *273*, 109813. https://doi.org/10.1016/j.jfoodeng.2019.109813.
31. Lin, L., Liao, X., & Cui, H. (2019). Cold plasma treated thyme essential oil/silk fibroin nanofibers against Salmonella Typhimurium in poultry meat. *Food Packaging and Shelf Life*, *21*, 100337. https://doi.org/10.1016/j.fpsl.2019.100337.
32. Lu, P., Guo, M., Xu, Z., & Wu, M. (2018). Application of nano fibrillated cellulose on BOPP/LDPE film as oxygen barrier and antimicrobial coating based on cold plasma treatment. *Coatings*, *8*(6), 8060207. https://doi.org/10.3390/coatings8060207.
33. Manoharan, D., Stephen, J., & Radhakrishnan, M. (2021). Study on low-pressure plasma system for continuous decontamination of milk and its quality evaluation. *Journal of Food Processing and Preservation*, *45*(2), 15138. https://doi.org/10.1111/jfpp.15138.
34. Mehta, D., & Yadav, S. K. (2022). Recent advances in cold plasma technology for food processing. *Food Engineering Reviews*, *14*, 555–578. https://doi.org/10.1007/s12393-022-09317-z
35. Mendes, A. C., & Pedersen, G. A. (2021). Perspectives on sustainable food packaging: Is bio-based plastics a solution? *Trends in Food Science & Technology*, *112*, 839–846.
36. Mesic, B. (2006). Printability of polyethylene-coated paper and paperboard. *Karlstad University Studies*, ISBN: 9170630828, Vol. 1, pp. 1–51. Retrived from: https://www.diva-portal.org/smash/record.jsf?pid=diva2%3A596318&dswid=-3459 (accessed on 16 June 2024).
37. Mishra, R., Bhatia, S., Pal, R., Visen, A., & Trivedi, H. (2016). Cold plasma: Emerging as the new standard in food safety. *Research and Investigations in International Journal of Engineering and Science*, *6*(2), 15–20.

38. Moosavi, M. H., Khani, M. R., Shokri, B., Hosseini, S. M., Shojaee-Aliabadi, S., & Mirmoghtadaie, L. (2020). Modifications of protein-based films using cold plasma. *International Journal of Biological Macromolecules*, *142*, 769–777.
39. Morsy, F. A., Elsayad, S. Y., Bakry, A., & Eid, M. A. (2006). Surface properties and printability of polypropylene film treated by an air dielectric barrier discharge plasma. *Surface Coatings International Part B: Coatings Transactions*, *89*(1), 49–55.
40. Muranyi, P., Wunderlich, J., & Heise, M. (2007). Sterilization efficiency of a cascaded dielectric barrier discharge. *Journal of Applied Microbiology*, *103*(5), 1535–1544.
41. Muranyi, P., Wunderlich, J., & Langowski, H. C. (2010). Modification of bacterial structures by a low-temperature gas plasma and influence on packaging material. *Journal of Applied Microbiology*, *109*(6), 1875–1885.
42. Oh, Y. A., Roh, S. H., & Min, S. C. (2016). Cold plasma treatments for improvement of the applicability of defatted soybean meal-based edible film in food packaging. *Food Hydrocolloids*, *58*, 150–159.
43. Paneru, R., Lamichhane, P., Chandra Adhikari, B., Ki, S. H., Choi, J., et al. (2019). Surface modification of PVA thin film by nonthermal atmospheric pressure plasma for antifogging property. *AIP Advances*, *9*(7), 75008. https://doi.org/10.1063/1.5100776.
44. Pankaj, S. K., Bueno-Ferrer, C., Misra, N. N., Milosavljević, V., & O'Donnell, C. P. (2014). Applications of cold plasma technology in food packaging. *Trends in Food Science & Technology*, *35*(1), 5–17.
45. Pankaj, S. K., Bueno-Ferrer, C., Misra, N. N., O'Neill, L., & Jiménez, A. (2014). Characterization of polylactic acid films for food packaging as affected by dielectric barrier discharge atmospheric plasma. *Innovative Food Science & Emerging Technologies*, *21*, 107–113.
46. Pankaj, S. K., Bueno-Ferrer, C., Misra, N. N., O'Neill, L., & Tiwari, B. K. (2014). Physicochemical characterization of plasma-treated sodium caseinate film. *Food Research International*, *66*, 438–444.
47. Pankaj, S. K., Bueno-Ferrer, C., Misra, N. N., O'Neill, L., & Tiwari, B. K. (2015). Dielectric barrier discharges atmospheric air plasma treatment of high amylose corn starch films. *LWT – Food Science and Technology*, *63*(2), 1076–1082.
48. Pankaj, S. K., Bueno-Ferrer, C., O'Neill, L., Tiwari, B. K., Bourke, P., & Cullen, P. J. (2017). Characterization of dielectric barrier discharge atmospheric air plasma treated chitosan films. *Journal of Food Processing and Preservation*, *41*(1), 61–67.
49. Pankaj, S. K., Wan, Z., De León, J. E., Mosher, C., Colonna, W., & Keener, K. M. (2017). High-voltage atmospheric cold plasma treatment of different types of starch films. *Starch-Stärke*, *69*(11–12), 00009. https://doi.org/10.1002/star.201700009.
50. Peng, P., Chen, P., Zhou, N., Schiappacasse, C., Cheng, Y., et al. (2019). Packed food and packaging materials disinfected by cold plasma. In *Advances in Cold Plasma Applications for Food Safety and Preservation*, pp. 269–286, ISBN: 978-0-12-814921-8, Academic Press. https://doi.org/10.1016/B978-0-12-814921-8.00009-8.
51. Perera, K. Y., Sharma, S., Pradhan, D., Jaiswal, A. K., & Jaiswal, S. (2021). Seaweed polysaccharide in food contact materials (active packaging, intelligent packaging, edible films, and coatings). *Foods*, *10*(9), 2088. https://doi.org/10.3390/foods10092088.
52. Puligundla, P., Lee, T., & Mok, C. (2016). Inactivation effect of dielectric barrier discharge plasma against foodborne pathogens on the surfaces of different packaging materials. *Innovative Food Science & Emerging Technologies*, *36*, 221–227.

53. R, R., Philip, E., Madhavan, A., Sindhu, R., & Pugazhendhi, A. (2021). Advanced biomaterials for sustainable applications in the food industry: Updates and challenges. *Environmental Pollution, 283*, 117071. https://doi.org/10.1016/j.envpol.2021.117071.
54. Ranjitha Gracy, T. K., Gupta, V., & Mahendran, R. (2019). Influence of low-pressure nonthermal dielectric barrier discharge plasma on chlorpyrifos reduction in tomatoes. *Journal of Food Process Engineering, 42*(6), 13242. https://doi.org/10.1111/jfpe.13242.
55. Ratish Ramanan, K., Sarumathi, R., & Mahendran, R. (2018). Influence of cold plasma on mortality rate of different life stages of *Tribolium castaneum* on refined wheat flour. *Journal of Stored Products Research, 77*, 126–134.
56. Romani, V. P., Olsen, B., Pinto Collares, M., Meireles Oliveira, J. R., Prentice-Hernández, C., & Guimarães Martins, V. (2019). Improvement of fish protein films properties for food packaging through glow discharge plasma application. *Food Hydrocolloids, 87*, 970–976.
57. Romani, V. P., Olsen, B., Pinto Collares, M., Meireles Oliveira, J. R., Prentice, C., & Martins, V. G. (2020). Cold plasma and carnauba wax as strategies to produce improved bi-layer films for sustainable food packaging. *Food Hydrocolloids, 108*, 106087. https://doi.org/10.1016/j.foodhyd.2020.106087.
58. Sasmazel, H. T., Alazzawi, M., & Alsahib, N. K. A. (2021). Atmospheric pressure plasma surface treatment of polymers and influence on cell cultivation. *Molecules, 26*(6), 1665. https://doi.org/10.3390/molecules26061665.
59. Shaikh, S., Yaqoob, M., & Aggarwal, P. (2021). An overview of biodegradable packaging in food industry. *Current Research in Food Science, 4*, 503–520.
60. Sharmin, N., Sone, I., Walsh, J. L., Sivertsvik, M., & Fernández, E. N. (2021). Effect of citric acid and plasma activated water on the functional properties of sodium alginate for potential food packaging applications. *Food Packaging and Shelf Life, 29*, 100733. https://doi.org/10.1016/j.fpsl.2021.100733.
61. Sifuentes-Nieves, I., Hernández-Hernández, E., Neira-Velázquez, G., Morales-Sánchez, E., Mendez-Montealvo, G., & Velazquez, G. (2019). Hexamethyldisilane cold plasma treatment and amylose content determine the structural, barrier, and mechanical properties of starch-based films. *International Journal of Biological Macromolecules, 124*, 651–658.
62. Song, A. Y., Oh, Y. A., Roh, S. H., Kim, J. H., & Min, S. C. (2016). Cold oxygen plasma treatments for the improvement of the physicochemical and biodegradable properties of polylactic acid films for food packaging. *Journal of Food Science, 81*(1), E86–E96.
63. Soni, B., Hassan, E. B., Schilling, M. W., & Mahmoud, B. (2016). Transparent bio nanocomposite films based on chitosan and TEMPO-oxidized cellulose nanofibers with enhanced mechanical and barrier properties. *Carbohydrate Polymers, 151*, 779–789.
64. Tolouie, H., Mohammadifar, M. A., Ghomi, H., & Hashemi, M. (2018). Cold atmospheric plasma manipulation of proteins in food systems. *Critical Reviews in Food Science and Nutrition, 58*(15), 2583–2597.
65. Wong, L. W., Hou, C. Y., Hsieh, C. C., Chang, C. K., Wu, Y. S., & Hsieh, C. W. (2020). Preparation of antimicrobial active packaging film by capacitively coupled plasma treatment. *LWT – Food Science and Technology, 117*, 108612. https://doi.org/10.1016/j.lwt.2019.108612.
66. Wu, T.-Y., Sun, N.-N., & Chau, C.-F. (2018). Application of corona electrical discharge plasma on modifying the physicochemical properties of banana starch indigenous to Taiwan. *Journal of Food and Drug Analysis, 26*(1), 244–251.

67. Yadav, A., Mangaraj, S., Singh, R., Das, K., Kumar, N., & Arora, S. (2018). Biopolymers as packaging material in food and allied industry. *International Journal of Chemical Studies*, *6*(2), 2411–2418.
68. Yan, M. R., Hsieh, S., & Ricacho, N. (2022). Innovative food packaging, food quality and safety, and consumer perspectives. *Processes*, *10*(4), 747–755. https://doi.org/10.3390/pr10040747.
69. Yaradoddi, J. S., Banapurmath, N. R., Ganachari, S. V., Soudagar, M. E. M., & Mubarak, N. M. (2020). Biodegradable carboxymethyl cellulose based material for sustainable packaging application. *Scientific Reports*, *10*(1), 1–13.
70. Zhang, Z. H., Wang, L. H., Zeng, X. A., Han, Z., & Brennan, C. S. (2019). Non-thermal technologies and its current and future application in the food industry: a review. *International Journal of Food Science and Technology*, *54*(1), 1–13.

PART IV

Emerging Trends in Cold Plasma Applications

CHAPTER 11

COLD PLASMA IN COMBINATION WITH OTHER NOVEL TECHNOLOGIES

APARAJHITHA SUDARSAN and ANJALY PAUL

ABSTRACT

Within the food industry, a relatively new technology called cold plasma is increasingly taking precedence as an ideal non-thermal intervention in the food processing sector. Cold plasma, which involves a cocktail of reactive species, has increasingly been implemented as a stand-alone technology, as well as in combination with a number of other thermal, non-thermal, and chemical-based interventions to improve both the safety and quality characteristics of various foods. The advantages associated with cold plasma in terms of the design of electrodes, flexibility to implement on different substrates, as well as its scale-up specific to each application, make it a versatile technology in the food industry. While there's an incentive in exploring the potential of cold plasma, traditional interventions are still highly effective and reliable at a commercial scale. A number of researchers have thus explored the dual benefits of cold plasma and traditional interventions to mitigate challenges associated with each technology while simultaneously exploiting the advantages associated with each. This approach is increasingly becoming significant in certain key areas such as improved food safety, drying, and extraction of bioactives. This chapter aims to cover these recent studies on different combinations of traditional technologies and cold plasma on various foods.

11.1 INTRODUCTION

The importance of nutrition and a healthy lifestyle is gaining increasing awareness among the public. The role of vital food components such

Nonthermal Plasma Technology for Food and Food Products.
R. Mahendran, Preeti Birwal, and Megh R. Goyal (Eds.)

as polyphenols, antioxidants, dietary fibers, and their importance in preventing chronic diseases and ailments has now shifted the focus of food processing towards sustainable interventions that ensure food safety as well as quality [70]. Commonly used commercial technologies involve thermal methods to increase shelf life but end up altering product characteristics. Although these technologies are effective and result in maximal microbial decontamination, they also cause nutritional losses since many micronutrients in food are heat-sensitive. Non-thermal technologies have been explored as potential options to produce safer and more nutritious foods wherein processing is carried out at or below room temperature [63, 67]. Although promising, their apparent lack of commercialization is due to the cost of operation, complexity, and lack of more conducive research for some technologies. A more holistic approach would be to use a combination of thermal and non-thermal interventions to optimize maximum quality while also ensuring safe food [10]. This would not only ensure a minimal detrimental effect of thermal treatment but also minimize the cost of operation of non-thermal treatment [47]. This multi-hurdle approach can also help successfully eliminate certain pathogenic, resistant strains that show resistance to certain technologies. This combination approach can either be done sequentially or simultaneously to achieve an additive or synergistic effect.

Cold plasma is one of the innovative non-thermal technologies that is either used as a stand-alone or in combination with other food processing technologies to achieve the target effect. A number of books and review articles have previously been published that emphasize the range of applications of cold plasma within the food industry [14, 52, 56, 59, 72]. Cold plasma, also commonly called non-equilibrium plasma, is often defined as a cocktail of a number of reactive gas species (RGS) arising from the ionization of a gas between two electrodes. The extent of ionization for cold plasma is usually around 2%, while the rest of the gas is in its unionized form, causing the difference in bulk and electron temperatures [21, 52]. The kind of plasma generation source plays a significant role in the extent of ionization as well as the distribution density of reactive species produced.

This chapter focuses on common plasma sources used in the food industry and recent studies that have explored the combination of cold plasma with one or more thermal and non-thermal technologies to achieve improved food preservation.

11.2 PLASMA SOURCES IN THE FOOD INDUSTRY

11.2.1 PLASMA JET

This configuration is made up of two electrodes that are placed concentric to each other wherein the inner electrode is powered by either a high frequency (usually radio-frequency), microwave, direct current power source [45, 58]. A pressurized gas flows between these electrodes that gets ionized as it passes through the potential difference and exits in the form of a jet [58]. A jet configuration is commonly used in applications requiring a targeted action of RGS produced for specific target sites. Hence, within the medical industry, plasma jets find great use for tissue and wound treatments, blood coagulation, and targeted surface sterilization of tools [16, 54, 61]. But this targeted delivery is also one of the limitations of this type of plasma generation source. However, this has been overcome by researchers by designing multi-array systems that have been successful in achieving targeted results [7, 23].

11.2.2 CORONA DISCHARGE

These are a type of localized sparks formed around sharp edges. This can either be a pin type or a thin wire that is connected to a high voltage power source. When an electric field passes through this wire, the air around the sharp edge gets ionized resulting in plasma production that is highly confined to this small region [22, 71]. This type of localized plasma discharge is called corona. It is a weak luminant zone that does not extend all the way to the ground electrode; hence it is also sometimes called a partial discharge. This type of discharge is commonly powered by a pulsed generator, alternating current (AC), or direct current (DC), wherein the ground electrode is usually a flat plate or cylinder [12, 76]. It is an unstable discharge with intense streamers produced in this local region. It finds great application for the production of ozone and treatment of residues from wastewater [3, 8, 46].

11.2.3 DIELECTRIC BARRIER DISCHARGE (DBD)

Also called a silent discharge, this mode of plasma generation was first discovered by Siemens in 1857. It consists of one or more dielectric materials placed between the electrodes to prevent arcing as well as enable a more uniform discharge. Some common dielectric materials used are glass, ceramic, enamel, quartz, or polymers [12, 69]. The thickness of the dielectric layer determines the homogeneity of the plasma and insulation of the discharge current [44, 76]. This discharge in a DBD is usually in the form of microdischarges that help in achieving a higher power density as well as higher electron energies [21]. A plasma is produced when the supplied voltage surpasses the ionization potential of the working gas. A DBD setup geometry is quite flexible with a variety of electrode shapes and setups. They can also be operated under a wide range of gas pressure ranging from 0.1–1.0 atm and are primarily driven by an AC power source or pulsed voltage source. The discharge gap in a DBD system typically varies between 0.1 mm to several centimeters [69].

11.3 COLD PLASMA WITH CONVENTIONAL TECHNOLOGIES

11.3.1 HEAT TREATMENT

Thermal food processing and preservation techniques are prominent and mostly employed in the food manufacturing industries. However, these techniques raise various concerns including nutrient losses, high energy consumption, and carbon emissions [60]. Recently, several novel nonthermal processing techniques like high-pressure processing (HPP), acoustic treatments, pulsed electric field (PEF), ultraviolet radiation (UV), and cold plasma (CP) are being scrutinized for mitigating the disadvantages of high-temperature treatments [88]. Combined interventions of novel and conventional techniques are also explored (Figure 11.1). Cold or nonthermal plasma is a proven technique widely recognized for its antimicrobial and property modification effects mostly by the chemical reactions of reactive gas species with food components. Lately, simultaneous pre- or post-treatments of CP with heat have also found various applications. This hurdle approach improves process efficiency and avoids temperature stress response activation in microorganisms [42]. It also helps in the reduction of heat treatment time, energy requirements, and quality deterioration.

Mild heat treatment of plasma-activated water (PAW) was effective in controlling microorganisms with minor changes in the product quality. For instance, heating of PAW to 55°C for 30 min reduced *Saccharomyces cerevisiae* to undetectable levels, without adversely affecting pH, color, and TSS of grapes. The bioactive components like phenolics, antioxidants, and vitamin C were retained as that of fresh grapes and the sterilization effect was higher than sole PAW and heat applications [82].

Similar results were found by Xiang et al. [81] in the inactivation of *Escherichia coli* O157:H7 suspended in PAW placed in a heated water bath. Microstructural changes were observed in the bacterial cell, and the cell integrity was disrupted because of the interaction of reactive gas species. Choi et al. [9] observed synergistic microbial reductions of *Listeria monocytogenes* (3.4 log CFU/g) and *Staphylococcus aureus* (3.7 log CFU/g) in salted cabbage after sequential PAW (10 min) and mild heat (60°C for 5 min) treatment with a significant reduction in peroxidase activity. The PAW (15 min) treatment followed by hot water blanching (60°C for 5 min) was also beneficial in reducing *Klebsiella pneumoniae* and other viable microorganisms in tiger nuts. The study observed that the color, pH, flavonoid content, and FRAP radical scavenging activity of tiger nuts were altered by the sequential treatment, but the phenolic content, DPPH scavenging, lipid oxidation, and sensory attributes were comparable to the control [55]. However, in a study on the application of dry heat (80°C for 25 min) sequentially with CP, the combination did not improve the deoxynivalenol mycotoxin reduction compared to CP applied alone [20]. Heat treatments above 140°C were required for deoxynivalenol reduction [78].

Ge et al. [25] showed that the combination of heat pre-treatment (130°C) followed by CP was more effective in modifying the properties of red adzuki bean starch than the treatments applied individually. These studies prove the synergistic effect of CP (PAW in particular), combined with heat in enhancing microbial inactivation and property modifications.

One of the recent applications that have gained interest among researchers is CP pre-treatment to augment the drying kinetics and final product quality in hot air drying. Ashtiani et al. [49] investigated a gliding arc plasma pre-treatment for grapes for 10–60 s before thermal drying at 60°C and 1.5 m/s air convection. The study found that CP was effective in improving the moisture transfer rates, with higher effective diffusivity. This improvement in drying kinetics and reduction in drying time was attributed to the microstructural changes on the grape surface caused by the plasma etching by reactive gas species. The nutritional value was enhanced with a rise in antioxidant activity and phenolic content with better vitamin C retention. The authors optimized 50 s CP pre-treatment because of the lower color changes and low energy consumption. Similar works on CP pretreated drying were found effective on tucuma pulp [43], jujube slices [6], and golden berry [50] as presented in Table 11.1. The most effective exposure time and frequency of CP treatment varied with the raw material and usually, a 30-second exposure could facilitate better drying kinetics without considerable changes in quality [73, 87].

As the phenolic content and antioxidant activity are increased upon intense plasma exposures but the drying kinetics and other quality aspects may be adversely affected due to microstructural changes, the CP pre-treatment can be finely adjusted to obtain dried products of desired quality. The industrial adaptation of CP-assisted drying operations can reduce energy consumption and deliver products of better quality.

11.3.2 ORGANIC ACIDS

Organic acids are naturally occurring compounds with acidic properties and have great potential as food preservatives. These compounds are known for their antimicrobial and anti-browning effects, especially in fresh produce [74]. Yadav & Roopesh [83] investigated the synergistic effect that food-grade organic acids, namely gallic acid (GA) and lactic acid (LA) have, in combination with CP for the inactivation of *Salmonella typhimurium* from the surface of poultry meat. Each of the acids was sprayed separately onto poultry meat surfaces and allowed to act for 5 minutes before subjecting to 30 seconds and 2 minutes of CP treatment. It was observed that a higher concentration (50 mM) of the acids combined with 30 seconds of cold plasma exposure was necessary for a 2.5 log CFU/m^2 reduction in *Salmonella.* Treatment only with the acids resulted in less than 1.0 log CFU/m^2 reduction, regardless of the acid type. The study also observed a significantly higher reduction in the bacterial count when the acid concentration was increased from 10 mM to 50 mM and a simultaneous increase in reduction when the plasma exposure time was increased from 30 seconds to 2 minutes, with up to 3.1 log CFU/cm^2 reduction after 2 minutes with LA. Tests on cell permeability found that LA was more effective than GA and when combined with CP, it increased the cell permeability to >80%, resulting in significant membrane damage and a subsequent drop in metabolic activity by 85%. The authors attribute the antimicrobial activity of these organic acids to the ability to passively diffuse through the bacterial cell membrane. Organic acids do this by compromising crucial mitochondrial enzymes such as NADH oxidase and esterase while simultaneously inducing sublethal acid stress, increasing their susceptibility to CP.

Similarly, Adam et al. [1] reported the inactivation of polyphenol oxidase by 78% in the sequential application of citric acid (5%) and CP integrated low-pressure cooling method. The authors also found a 5.68 log CFU/g reduction of *Salmonella typhimurium* on fresh-cut apples in the combined treatment with no significant changes in the color of apple slices during 7 days of refrigerated storage.

Organic acids have also been found practical in developing biodegradable food packaging materials. A recent study explored the potential of citric acid (CA) in improving the film-forming capacity of sodium alginate – a common biodegradable food packaging material [65]. The simultaneous application of PAW and CA (1%) along with glycerol increased plasticizing effect and reduced water vapor transmission rates of the film. The crosslinking of the alginate chain by reactive gas species resulted in the improvement of mechanical and barrier properties. The outcomes of these studies are promising in further exploring the benefits of organic acids combined with CP for several food-based applications.

11.3.3 ADDITIVES

11.3.3.1 CHEMICALS

Cold plasma combined with chemicals or sanitizers has been reported to show synergistic functions. For example, CP-treated aerosolized hydrogen peroxide was more effective in microbial inactivation than spraying untreated hydrogen peroxide [68]. CP exposure commenced advanced oxidation processes in the H_2O_2 mist to produce hydroxyl radicals and further secondary reactive species. The study investigated the efficiency of six different fruit and vegetable surfaces like smooth apple skin, upper leaves and midribs of romaine lettuce, cantaloupe rinds with a thin layer of flesh, and stem scars and smooth surfaces of grape tomatoes. On smooth surfaces, 8 s of spraying CP activated H_2O_2 aerosols with 30 min of dwell time resulted in a >5 log reduction of *Salmonella typhimurium*. However, on the rougher and corrugated surfaces, 30 s of spraying time resulted in only 2–3 log reductions. On the other hand, *Listeria innocua* was inactivated by greater than 5 log reduction except in the case of tomato stem scars. This indicates that the Gram-positive bacteria were more susceptible to the combined action of CP and H_2O_2. It should be noted that the deleterious effect of H_2O_2 by the peroxidation of membrane lipids, nucleic acids, and other essential cell components is more prominent here and the interaction effect of other plasma species is limited. Similar results on decontamination of *E. coli, S. Typhimurium,* and *L. innocua* were obtained on grape tomato, cantaloupe, and spinach by CP-activated H_2O_2 aerosols, without affecting the quality [31]. These studies demonstrate the potential of simple and scalable activated aerosol spraying treatments on microbial inactivation of fresh produce.

11.3.3.2 NATURALLY DERIVED ADDITIVES

Interventions of CP with naturally derived botanical extracts have also been a widely researched area. The essential oils and plant extracts are a group of commonly employed antimicrobial agents even in ancient cultures [80]. However, microbes develop resistance to these extracts eventually, and the complete eradication of microorganisms to safe and acceptable limits is often cumbersome [19]. Furthermore, overdosage of essential oils affects the sensory attributes of the foods. Hence, the use of botanical extracts is usually combined with other mild food preservation techniques to have either an additive or synergistic effect [64].

Yoo et al. [84] reported that the combined application of clove oil and CP treatment enhanced the microbial deactivation of *E. coli* and *S. aureus* from about 3 log reductions in individual treatments to greater than 7.5 log reductions when applied together. The chromatographic analysis of clove oil revealed that the CP treatment did not affect the composition of phenolic compounds responsible for the bactericidal effect in the oil. Yet, the combined treatment disrupted the cell membrane integrity, induced morphological changes, and increased leakage of cellular components to a greater extent.

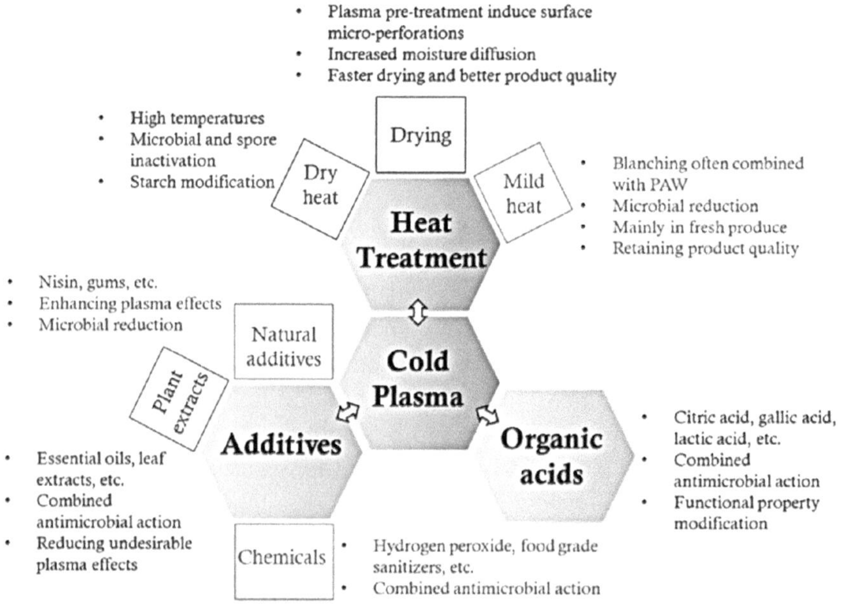

FIGURE 11.1 Effects of cold plasma with conventional technologies.

Other naturally derived compounds were also applied simultaneously with CP to investigate the combined benefits (Figure 11.1). Nisin is a natural antimicrobial agent derived as a by-product of fermentation. Costello et al. [13] assessed the synergistic effect of CP and nisin on the inactivation of *Listeria innocua*. The study recommended that CP combined with nisin is more effective in the inactivation of planktonically grown cells in a liquid-like system compared to microbial colonies on a solid-like surface. Moreover, the order in which nisin was applied also affected the microbial reduction rates, i.e., nisin applied before CP treatment had higher *Listeria* inactivation than CP treatment followed by nisin.

Olatunde et al. [57] found that the fish (Asian sea bass) fillets dipped in liposomal-encapsulated ethanolic coconut husk extract followed by CP treatment and stored in modified atmospheric packages at 4°C had a better shelf life (18 days) than untreated samples (6 days). The extract also helped in reducing the lipid oxidation caused by CP and resulted in products of better sensory profile. In another study, CP treatment improved hydration properties, gel structure, and free radical scavenging activity of peanut protein isolate by facilitating covalent bond formation with sesbania gum [85]. It can be concluded from these studies that CP treatment could synergistically improve the individual effects of additives and the adverse effects of CP can be alleviated by some of these applications. Further, the antimicrobial and functional property-enhancing applications of cold plasma-assisted treatments are summarized in Table 11.1.

In a study by González-González et al. [27], the authors combined linalool, a monoterpenoid essential oil, with cold plasma to inactivate pathogenic microorganisms from RTE chicken meat. Linalool has the potential to damage and penetrate microbial cell membranes, altering crucial cell functions by limiting the availability of ions and other functional molecules. Its application in nanoparticle form (average particle size: 103.24 nm) helps increase its solubility and enables enhanced diffusion. The smaller particle size also helps in effective electrostatic interaction of the positively charged essential oil emulsion with the negatively charged microbial cells, resulting in disruption of cellular osmotic balance. The authors generated plasma through a piezoelectric system, followed by linalool nanoemulsion washing to eliminate *E. coli* O157:H7 and *Salmonella* (*Typhimurium* and *Senftenberg*). A combination of five-minute CP and 25 minutes of linalool-loaded nanoemulsion washing (LW) was sufficient to achieve a 2.76 and greater than 3.2 log CFU/g reduction in *E. coli* and *Salmonella*, respectively, from the surface of chicken meat. Four hours post-treatment, the population of E. coli and Salmonella reduced by up to 4.0 log CFU/g when inoculated on a

TABLE 11.1 Combination of Cold Plasma with Conventional Technologies and Additives

Sample	Conventional Treatment	Treatment Parameters	Significant Result(s)	References
Grapes	Mild heat	• Cold Plasma parameters: ▪ Source: atmospheric plasma jet ▪ Treatment time: 90 s Volume: 300 mL sterile distilled water (SDW) • Thermal treatment parameters: ▪ Temperature: 25, 50, 52.5, and 55°C ▪ Treatment time: 30 min	• PAW with mild heat at 55°C reduced *Saccharomyces cerevisiae* to undetectable levels • No significant difference in the firmness of grapes • pH, color, and TSS were not affected • Phenolics, antioxidant activity, and vitamin C were not significantly different compared to fresh grapes.	[82]
Red pepper	Mild heat	• Thermal treatment parameters: ▪ Temperature: 60°C ▪ Treatment time: 5, 10, 15, and 20 min • Cold Plasma parameters: ▪ Source: DBD-nitrogen ▪ Gas: ▪ Voltage: 3 kV ▪ Discharge gap: 3 mm ▪ Frequency: 43 kHz ▪ Gas flowrate: 1.5 L/min ▪ Treatment time: 5, 10, 15, and 20 min	• 6.43 log CFU/g reduction of *Bacillus cereus* in combined 20 min heat and 20 min CP treatment. • A synergistic effect was observed in all combinations except for 5 min – 5 min heat – CP treatment. • No significant change in color and pH of red pepper powder	[30]
Red adzuki bean starch	Dry heat	• Thermal treatment parameters: ▪ Temperature: 130°C ▪ Treatment time: 1, 3, and 9 h • Cold Plasma parameters: ▪ Source: DBD ▪ Gas: atmospheric air ▪ Voltage: 40 V ▪ Discharge gap: 2 mm ▪ Treatment time: 1, 5, and 10 min	• Cracks appeared on the starch granules due to high temperature and plasma etching • Reduced amylose content and reduction in molecular weight due to starch depolymerization • Significantly higher reduction in relative crystallinity compared to individual treatments • Solubility increased with treatment time and swelling power decreased for both CP and dry heat. • CP-dry heat combination had higher modifications in starch with low viscosity and digestibility	[25]

TABLE 11.1 *(Continued)*

Sample	Conventional Treatment	Treatment Parameters	Significant Result(s)	References
Tucuma pulp	Drying	• Cold Plasma parameters: ▪ Source: DBD ▪ Gas: atmospheric air ▪ Voltage: 20 kV ▪ Discharge gap: 1.5 cm ▪ Treatment time: 10 min ▪ Frequency: 200, 500, 800 Hz • Thermal treatment parameters: ▪ Temperature: 60°C ▪ Airflow rate: 0.5 m/s ▪ Treatment time: up to 3 h	• CP treatment significantly reduced drying time. • Higher drying rate at 200 Hz, which decreased at higher frequencies due to the structural damage to moisture diffusion channels. • Deformation of cell microstructure was observed at 200 and 800 Hz. • Higher rehydration in CP treated. • Color changes were lower at 200 Hz compared to untreated dried control • Bioactive compounds were higher than sole drying but lower than fresh samples. • Carotenoids were higher at 200 Hz, while total phenolics and antioxidants increased at higher frequencies.	[43]
Shiitake mushrooms	Drying	• Cold Plasma parameters: ▪ Source: DBD ▪ Gas: atmospheric air ▪ Power: 650 W ▪ Discharge gap: 5 cm ▪ Gas flow rate: 135 L/min ▪ Treatment time: 1 min • PAW parameters: ▪ Source: jet from DBD ▪ Treatment time: 10 minutes ▪ Sample volume: 2 L • Thermal treatment parameters: ▪ Temperature: 50, 60, and 70°C ▪ Treatment time: up to 3 h	• Microstructural changes in both PAW and CP; with higher shrinkage and intracellular space in CP. • Drying was faster for CP at 70°C, followed by PAW compared to control. • Higher bulk density and swelling index for CP pretreatment at 70°C. • Reduced water retention in CP pretreated at 70°C • Relative vitamin content (C and B) increased with CP and temperature; this might be due to lower drying times • Reduced browning in CP • No change in total sugar content • Higher phenolics and antioxidants	[33]

TABLE 11.1 *(Continued)*

Sample	Conventional Treatment	Treatment Parameters	Significant Result(s)	References
Jujube slices	Drying	• Cold Plasma parameters: ▪ Source: DBD ▪ Source: atmospheric air ▪ Voltage: 5 kV ▪ Discharge gap: 5 cm ▪ Treatment time: 15, 30, and 60 s ▪ Frequency: 40 kHz • Thermal treatment parameters: ▪ Temperature: 50, 60, and 70°C ▪ Treatment time: up to 3 h	• CP reduced initial moisture content after 30 and 60 s pre-treatment. • Smaller cavities and dense microstructure at higher CP pre-treatment times; reduced effective moisture diffusivity • CP for 15 s was effective with larger cavities and higher moisture diffusivity: higher drying rate at 60 and 70°C. • CP increased procyanidins, flavonoids, phenolics, and antioxidant activity. • Browning index was higher for CP treated but reduced Hydroxymethyl furfural content after 15 s pre-treatment.	[6]
Goldenberry	Drying	• Cold Plasma parameters: ▪ Source: Gliding arc ▪ Gas: atmospheric air ▪ Gas flowrate: 5 L/min ▪ Treatment time: 15, 30, 45, and 60 s • Thermal treatment parameters: ▪ Temperature: 50°C ▪ Airflow rate: 1 m/s Convective dryer and Ultrasound-assisted convective dryer	• Micro-perforations on the goldenberry surface increased in size with exposure time. • The water contact angle was reduced from 77.5 to 30 degrees. • Better color retention, shelf stability, and rehydration in CP treated • Ultrasound-assisted drying after 60 s CP treatment had higher phenolics, vitamin C retention, and antioxidant activity. Also, higher moisture diffusivity, significantly lower drying time, and specific energy consumption.	[50]
Juice from blanched carrot slices	Citric acid blanching	• Acid blanching (AB) parameters: ▪ Citric acid (35 g/L) ▪ Blanching temperature: 100°C ▪ Treatment time: 5 min • Cold Plasma parameters: ▪ Source: DBD ▪ Gas: atmospheric air ▪ Voltage: 80 kV ▪ Treatment time: 4 min	• Higher lycopene, lutein, chlorogenic acid, sugar content, and carotenoids in AB-CP treatment than control and individual treatments. • Total plate count and total yeast and mold count were reduced to undetectable levels after combined AB-CP treatment.	[77]

TABLE 11.1 *(Continued)*

Sample	Conventional Treatment	Treatment Parameters	Significant Result(s)	References
Poultry meat Polycarbonate membrane filter paper	Gallic acid Lactic acid	• Organic acid treatments: ▪ Gallic acid (GA) and Lactic acid (LA) spray ▪ Concentration: 5, 10, and 15 mm ▪ Treatment time: 5 min • Cold Plasma parameters: ▪ Source: DBD ▪ Gas: atmospheric air ▪ Frequency: 3.5 kHz ▪ Discharge gap: 5 mm ▪ Treatment time: 15, 30, and 60 s ▪ Incubation time: 10 min	• 3.5 log CFU/cm^2 reduction of *Salmonella typhimurium* on filter paper at 10 mM GA or LA. • 0.4, 0.3, and 1.2 log CFU/cm^2 reductions when LA, GA, and CP were applied alone. • 50 mM acid was required for 2.5 log CFU/cm^2 reduction on the meat after 30 s CP. • The combined treatment enhanced membrane lipid oxidation and reduced metabolic activity in *Salmonella*. • Cell membrane permeability was highest for 10 mM LA and 30 s CP. • CP alone (30 s) had the highest intracellular reactive oxygen levels	[83]
Fresh cut apples	Citric acid	• Citric acid (CA) parameters: ▪ Concentration: 1, 2.5, 5, and 10% w/v ▪ Treatment time: 3 min • Cold Plasma integrated low-pressure cooling parameters: ▪ Vacuum: 200 mbar ▪ Voltage: 30 kV ▪ Discharge gap: 12 mm ▪ Treatment time: 1, 2, and 3 min	• Sequential 5% CA followed by 3 min CP resulted in 5.68 log CFU/g reduction of *Salmonella typhimurium.* Highest poly-phenol oxidase inactivation (78%) at the same combination. • The color of apple slices was retained during 7 days of cold storage (4°C). • CA increased cooling time and reduced moisture loss, which was attributed to moisture replacement by dipping in CA.	[1]
Sodium alginate film	Citric acid	• Citric acid (CA) ▪ Concentration: 1% w/v • PAW parameters: ▪ Source: DBD ▪ Gas: atmospheric air ▪ Activation time: 30 min • Glycerin concentration: 30%	• Tensile strength and tensile modulus decreased with the combination of CA, PAW, and glycerol compared to PAW alone. While elongation at break and storage modulus increased. • 44% reduction in water vapor transmission rate of sodium alginate film compared to 34% in CA alone. • Higher viscosity, which decreased under shear stress • Improved barrier properties and shear-thinning effects due to crosslinking of alginate chains by reactive species.	[65]

TABLE 11.1 *(Continued)*

Sample	Conventional Treatment	Treatment Parameters	Significant Result(s)	References
Chicken drumsticks	Lactic acid	• Plasma activated Lactic acid: ▪ Source: Plasma jet ▪ Gas: atmospheric air ▪ Voltage: 19 kV ▪ Frequency: 20 kHz ▪ Activation time: 80 s ▪ Discharge gap: 10 mm ▪ Immersion time: 10 min • Lactic acid ▪ Volume: 300 mL ▪ Concentration: 0.05, 0.1, 0.15, and 0.2% w/v	• The total microbial count was lowest at PALA (0.2%) compared to controls (distilled water, PAW, and sodium hypochlorite) • Slowed down the growth of *Escherichia coli* and *Pseudomonas* on storage (8 days). • The color of the chicken was maintained, but the pH and malondialdehyde content decreased. • The residual nitrite content increased but was within the acceptable limits • PALA could reduce lipid oxidation, off-odor development, and changes to myofibrillar protein.	[62]
Grape tomatoes, Apples, Cantaloupe, Romaine lettuce	Hydrogen peroxide	• H_2O_2 parameters: ▪ Concentration: 7.8% ▪ Form: Ionized aerosol ▪ Air pressure: 7 psi • Cold Plasma parameters: ▪ Voltage: 17 kV ▪ Discharge gap: 9 mm ▪ Gas flow rate: 5.0 mL/min ▪ Treatment time: 8–30 s ▪ Incubation time: 30 min	• 1.54–3.17 log CFU/piece reduction of *Salmonella typhimurium* when H_2O_2 aerosol applied alone and 2.35–5.5 log CFU/piece for CP- H_2O_2 combination • *Listeria innocua* was also better reduced with combination treatment with greater than 5 log CFU/piece reduction on all surfaces studied. • Higher log reduction on smooth surfaces than on rough surfaces	[68]
Food model systems: Liquid broth and Xanthan gum gel system	Nisin	• Nisin conditions: ▪ Concentration: 35 IU/mL ▪ Treatment time: 30 min • Cold Plasma parameters: ▪ Source: DBD ▪ Gas: helium: oxygen (99:1) ▪ Voltage: 7 V ▪ Power: 7 W ▪ Frequency: 15 kHz ▪ Discharge time:0–30 min	• Higher *Listeria innocua* inactivation (2 log reduction) in nisin applied before CP • Lower antimicrobial effect when nisin applied after CP treatment, especially for planktonic growth. • The combined treatment was more effective on liquid broth than solid surface when compared to individual treatments • Cells grown as surface colonies were less sensitive to combined treatments	[13]

TABLE 11.1 *(Continued)*

Sample	Conventional Treatment	Treatment Parameters	Significant Result(s)	References
Fish (Asian sea bass) slices	Liposomal encapsulated ethanolic coconut husk extract	• Coconut Husk Extract concentration: 400 mg/kg • Cold Plasma parameters: ▪ Source: DBD in package ▪ Gas: Argon: CO_2: O_2 (30:60:10) ▪ Voltage: 80 kV ▪ Discharge gap: 4 cm ▪ Discharge time: 5 min ▪ Air flow rate: 5 mL/min	• The combined treatment could increase the shelf life of fish from 6 days to 18 days when stored at 4°C. • The total viable mesophilic and psychrophilic bacteria was lower in the combination and within the acceptable limits till the 18th day. • Total volatile nitrogen base and trimethylamine content were also the lowest, indicating lower microbial spoilage • The lipid oxidation caused by CP was reduced with the application of husk extract. Better sensory profile in combination • pH was lower in combined treatment than individual CP and husk extract applications	[57]
High denatured peanut protein isolate	Sesbania gum	• Sesbania gum concentration: 1: 10 w/w • Cold Plasma parameters: ▪ Source: DBD ▪ Gas: atmospheric air ▪ Voltage: 35 V ▪ Power: 70 W ▪ Frequency: 15 kHz ▪ Discharge time: 1–5 min	• The combination of sesbania gum and CP treatment of HPPI improved water binding properties. Solubility increased from 610 g/L to 1,580 g/L. • Hydrophilic surface development in glycosylation products and improved antioxidant activity • Development of gel network structure and enhanced gel property along with free radical scavenging activity	[86]
Beef jerky	Clove oil	• Clove oil conditions: ▪ Concentration: 0.05% ▪ Time: 3 h ▪ Temperature: 37°C • Cold Plasma parameters: ▪ Source: encapsulated DBD ▪ Gas: atmospheric air ▪ Voltage: 8.4 kV ▪ Frequency: 2.2 kHz ▪ Discharge time: 2 and 4 min	• Combined treatment resulted in greater than 7.5 log reduction of *Escherichia coli* and *Staphylococcus aureus*: Only 3 log reductions in individual treatments • The bacterial reduction of *E. coli* was higher than *S. aureus* on beef jerky. • Higher bacterial cell damage was observed in TEM images for combined treatment • EAP treatment did not increase the antibacterial activity of clove oil, but a synergistic effect on microbial reduction was observed.	[84]

TABLE 11.1 *(Continued)*

Sample	Conventional Treatment	Treatment Parameters	Significant Result(s)	References
Bacteria sample	Olive leaf extract	• Olive leaf extract concentration: 10 mg/mL • Cold Plasma parameters: ▪ Source: DBD ▪ Gas: atmospheric air ▪ Voltage: 8 kV ▪ Frequency: 20 kHz ▪ Discharge time: 1 min ▪ Air flow rate: 5 mL/min	• *Escherichia coli* was reduced to non-detectable levels after 6 h, but viable cells reappeared after 24 h in OLE and CP combined treatment. Similar was the case of Listeria innocua. • The highest reduction was observed for *Staphylococcus aureus* with the number of viable cells at ~3 log CFU mg/mL compared to 10 log CFU mg/mL in individual and control treatments.	[17]

smoother surface, i.e., a petri dish, indicating the effect of surface features on the inactivation efficacy of the interventions.

11.4 COLD PLASMA IN COMBINATION WITH NON-THERMAL TECHNOLOGIES

11.4.1 ULTRASOUND

Ultrasound-assisted applications have seen increasing use in the past decade. It has found a variety of applications within the food industry such as texture modification, improved mass transfer properties, aiding in thermal treatment of thermally-sensitive foods, as well as improved food safety. Ultrasounds are electromagnetic waves of low frequency, typically in the 20 kHz range. An ultrasound (US) generator releases sound energy in the form of longitudinal waves causing constant compression and rarefaction of constituents present within the medium. This results in a number of biochemical, chemical, and physical changes in the food matrix that are dose-dependent on incident sound energy. Depending on the frequency range employed in food processing, US is divided into low energy (low intensity, low power) and high energy (high intensity, high power).

- Low energy US is usually characterized by frequencies in MHz (<1 W/cm^2) while high energy US is employed at frequencies between 20 and 500 kHz (>1 W/cm^2).
- High energy US is typically employed to induce physical and/or biochemical changes for texture modification while low energy.
- US helps with non-destructive analysis and processing of foods. The mechanism behind US-induced microbial inactivation is the aggressive formation and breaking of cavitation bubbles causing localized high temperatures and pressures, free radical production, as well as microstreaming within the food matrix.

The first antibacterial effect of US was observed by Harvey & Loomis in 1929, following which Kinsloe et al., in 1954 elucidated the physical damage induced upon biological cells. Studies further show a synergistic or an additive effect when US is combined with other technologies such as those of high-temperature treatments, high-pressure processes, or other non-thermal technologies.

Wang & Wu [79] experimented with US (25 kHz), free chlorine (10 ppm), peracetic acid (PAA) (80 ppm), and a DBD cold plasma for in-package disinfection of *E. coli* O17:H7 and *Salmonella typhimurium* from blueberries. The authors subjected fresh blueberries to a 1-minute wash cycle with the chemical disinfectants combined with US. The packaged product was then subjected to cold plasma exposure for 1 minute at 400 Hz. Owing to their nature and limitations of thermal technologies from a nutritional perspective, fresh produce such as blueberries are highly susceptible to contamination by pathogenic microbes. Low-cost chemical disinfectants such as free chlorine (FC) or peracetic acid (PAA) are commonly used as disinfectants by food industries to prevent cross-contamination of fresh produce like fruits and vegetables by the recirculation of wash water. These chemicals require constant monitoring of dosage and are only effective to reduce 1.0–2.0 logs of natural microbiota. When combined with other technologies such as US and cold plasma, the disinfection rates have shown to increase and prevent cross-contamination of *E. coli* and *Listeria monocytogenes* during fresh produce washing [29]. While these are effective controls during washing, there is very little control exercised to prevent contamination during or after product packaging. In-package technologies such as cold plasma in combination with others are effective in killing spoilage and pathogenic microorganisms. Cold plasma is highly flexible as a post-packaging intervention and has been previously used by a number of researchers on fresh produce [34, 38, 51]. When combined with other processing interventions, it helps in enhancing the antimicrobial effect compared to their individual application effects. Wang & Wu [79] observed a 2.06 and 1.99 log CFU/g reduction in *E. coli* and *Salmonella* for the US-FC+CP treatment. A similar reduction profile was observed for the US-PAA+CP treatment too, and this reduction was sustained even after 3 days of post-treatment storage at 4°C. When the interventions were applied individually, only a 0.73, 0.83, and 0.63 log CFU/g reduction for *E. coli* and 0.64, 0.74, and 0.68 log CFU/g reductions were observed for FC, PAA, and US, respectively. Without cold plasma intervention, a maximum reduction of 1.38–1.43 log CFU/g was observed for the US+FC and US+PAA treatments for both *E. coli* and *Salmonella*. Despite multiple interventions, there was no change in the anthocyanin content and firmness of the berries. The authors also observed lower ROS activity and higher antioxidant enzyme activity. This can arise from cold plasma activating the antioxidant system in the blueberries to scavenge the produced ROS, thus ensuring lower quality loss. The bacterial cells undergo instantaneous high pressure and shear forces induced by US as a pre-treatment that damage microbial cell membranes, while FC and PAA cause membrane oxidation and intracellular damage that enables easy action of RGS produced by cold plasma. The interventions were comparatively lesser for aerobic mesophiles and yeast and mold, possibly due to the presence of extracellular polymeric substances (EPS) offering physical protection for the microbes, different layers of microbial cells, or the development of resistant psychrophilic microbial species post-treatment.

Miraei Ashtiani et al. [50] observed that when US was combined with traditional convective drying, it helped reduce the drying time and improved vitamin C content in ripe goldenberries by 175%. When the authors used cold plasma as a pre-treatment, they observed that it helped bring down the drying temperature from 90°C to 50°C and helped improve the nutritional value of the berries by retaining a higher percentage of Vitamin C and polyphenols that are thermally sensitive and are usually lost during hot air drying. Upon observing the surface of the berries, the authors observed the formation of micropores from the CP pre-treatment that enables increased mass diffusion enabling higher moisture loss during drying in a shorter time. It also served as a pathway for easy penetration for the US waves and changed the surface functionality that helped preserve the product for up to 3 months without any quality changes.

Zhao et al. [89] studied different combinations of US (25 kHz, 550 W), plasma-activated water (PAW), and PAA (200 ppm) on the native microflora (mesophilic and psychrophilic), *Listeria monocytogenes*, *E. coli,* and *Pseudomonas fluorescens* in mackerel fillets. The study employed the use of a plasma jet system operated at 300 W and 20 kHz frequency with an atmospheric air flow rate of 11 L/min. 60 mL of deionized water was subjected to the jet for 10 mins to generate PAW. All treatment interventions were applied for 10 mins. The authors observed that PAA-involved treatments showed the greatest reduction in *Listeria* with up to 0.72 log CFU/g reductions compared to the initial load of 5.45 log CFU/g with no significant change between the PAA-involved interventions, while the PAW-PA-US combination was more effective for *E. coli* with a 0.61 log CFU/g reduction. A similar trend was observed for *E. coli* inactivation with PAA-involved interventions achieving a higher log reduction than other combination treatments. *P. fluorescens* showed a comparatively lower log reduction for all treatment combinations when compared to *E. coli* and *Listeria*, with the highest of 0.5 log CFU/g reduction. The authors observed a synergistic outcome in terms of pH lowering when PAA and PAW treatments were combined, but this effect was not significant when compared to the individual treatments. The pH dropped to 3.11 while ORP and conductivity showed a significant increase after plasma exposure. The combination of PAW and PAA also helped in a 10-fold increase in the concentration of peroxide species, which are established oxidizers of microbial cell membranes.

A similar study was done by Esua et al. [32] on green carp – a kind of freshwater fish to eliminate *E. coli* and *Shewanella putrefaciens*. The authors elucidated the effect of US as a pre-treatment before subjecting the fish tissue to PAW or plasma functionalized buffer (PFB). A DBD system with atmospheric air as the working gas and a discharge gap of 5 mm was used to generate PAW and PFB. A bath-type reactor operating at 40 kHz frequency, 500 W power was used for 4 minutes to generate the ultrasound waves. The compression and mechanical shock waves and rarefaction generated produced hydrogen atoms, hydroxyl radicals, and peroxides in the tissue that damage the functional and cellular constituents of microbial cells, including causing cell lysis and ultimately, microbial cell death. This also affects the myofibrillar tissues and muscle structure of the meat, helping in increased water holding capacity and decreased water losses, thus helping in improved meat quality. This can also result in enzyme and non-enzyme-based reactions within the myofibrils and cause degradation of the pigmented myofibrillar proteins. The US-PAW and US-PFB treated fish samples showed an increased hardness and springiness when compared to just the PAW or PFB interventions. FTIR analysis showed a decreased intensity with increasing voltages indicating protein denaturation and unfolding as a result of cavitation-induced shearing and the RONS from the plasma. The implosion of bubbles during cavitation ruptures the cytoplasmic membrane of microbial cells which are then more susceptible to the RGS from PAW. US+PAW was the most effective in reducing the pathogens at all voltage levels (30–70 V) tested with the highest of 1.39 and 1.49 log CFU/g reduction for US-PAW and 1.31 and 1.39 log CFU/g reduction for US-PFB, respectively for *E. coli* and *S. putrefaciens*. The authors observed an increasing trend in the concentration of NO_3, NO_2, H_2O_2, and O_3 with an increase in voltage confirming the role of RGS in the inactivation efficiency. There was a significant increase in peroxide value after the treatments. This can be attributed to fish meat that is usually rich in polyunsaturated fatty acids (PUFA) that are not stable and get readily oxidized in the presence of free radicals and peroxides produced by the treatment interventions employed in the study. Although the TBARS value as an indication of lipid peroxidation showed a significant increase post-treatment, this increase was still well under the limit of acceptability of 0.5 mg MDA/kg representative of off-flavor. The tissues also showed bleaching possibly due to oxidation of color pigments within the fish meat such as carotenoids, hemoglobin, and myoglobin by the RONS generated.

11.4.2 PULSED ELECTRIC FIELD (PEF)

PEF technology is one of the novel non-thermal food processing technologies commonly used for food preservation. It traces its roots back to the Electro-Pure process, which was based on Joule heating of milk for shelf-life extension. Further development of this process by using a pulsed discharge system with voltages reaching up to 30 kV gained traction in the 1950s, resulting in the development of the Elsteril® process by Krupp Maschinentechnik in 1988.

PEF involves the application of current in the form of pulses (1–100 μs) at fields of 0.1–1 kV/cm for transient electroporation, and 15–40 kV/cm for permanent electroporation [35]. When applied, it causes cell membranes to develop pores that may be transient or permanent in nature, depending on the intensity and treatment parameters employed. This, in turn, results in cell content leakage and subsequent lysis of microbial cells. PEF is also shown to affect some key enzymes and proteins in microorganisms, impairing crucial metabolic processes and ultimately leading to cell death [2]. Some authors have previously explored the use of PEF to induce the production of secondary metabolites by stressing plant cells [18, 24, 26], while others have explored its potential application for structural modifications of food tissues [28, 39], as well as improved water binding properties in meat tissues [75]. More research is required in these applications to further develop an understanding of the underlying mechanisms to target desired applications in different food matrices.

Some of the important factors that influence microbial inactivation through PEF include the pulse width and shape, electrical field intensity, exposure time, microorganism type and growth stage, external media factors such as presence or absence of ionic compounds, salts, ionic strength, pH, and other antimicrobial agents. The selective ability of PEF for microbial inactivation is an advantage wherein larger cells such as those of yeasts require lesser intensity of electric field potential to cause inactivation. It also offers a significant edge over thermophilic and chemical-resistant microorganisms that can be targeted through PEF or in combination with other hurdle technologies to achieve a synergistic effect. A recent study explored the combination of PEF pre-treated *Litopenaeus vannamei*, a type of popular crustacean that was later immersed in Chamuang leaf extract (CLE) – a type of antimicrobial agent before treating with HVACP. The PEF treatment was set at 15 kV/cm with two different pulse intensities tested – 600 (PEFT1) and 800 pulses (PEFT2). It was observed that treatment combinations involving more pulses, i.e., PEFT2, had higher microbial reduction as well as a lower melanosis score, which was used as a criterion for visual appeal. When tested with different concentrations of CLE (1% or 2%), it was noted that the polyphenol oxidase (PPO) activity was dose-dependent as well as PEF energy-dependent. When combined with DBD-HVACP treatment

(80:20 Argon: Air mixture) for 10 minutes at 16 kV, it was observed that the combination of PEFT2-CLE2-HVACP exhibited the least melanosis score, the lowest microbial count throughout 18 days of refrigerated storage, as well as lower pH and microbial lipase activity throughout, indicating lower microbial enzyme inactivation. The authors postulate cell wall electroporation and disintegration caused by PEF, which enables enhanced diffusion of CLE into microbial cells causing death and slower cell recovery. The authors also measured a higher TBARS value with HVACP-treated samples, indicating the formation of hydroperoxides which further help maintain the antimicrobial effect. Despite HVACP-induced oxidation, the PEFT2-CLE2-HVACP treated samples retained the highest content of fatty acids after 18 days of post-treatment storage, with MUFA and PUFA still higher than control or other treated samples while retaining the highest visual appeal.

11.4.3 ULTRAVIOLET (UV) RADIATION

UV (190–400 nm) is yet another non-thermal intervention that makes use of non-ionizing wavelengths of the electromagnetic spectrum. Non-ionizing radiation results in electronic excitations at atomic and molecular levels and depending on the gas used, has a characteristic color [36]. The ultraviolet region is further divided into three zones based on the wavelengths and their relevance to the food industry. UV-C encompasses the region between 190 and 280 nm and is of utmost importance in the food industry owing to its germicidal effect, capable of inactivating bacteria and viruses [36]. It is a well-established technique for air and water purification as well as surface disinfection of food matrices and biomedical devices. Its application in liquid foods is quite complex owing to different food constituents and their varied interaction with the UV spectrum. Within the emission spectrum of the most common mercury-based source used for UV generation, 253.7 nm and 185 nm are the most significant [48]. The 253.7 nm emission line represents 85% of the total emitted UV intensity and is prominent for germicidal activity since microbial DNA absorbs this wavelength in particular resulting in DNA mutations and lesions [36].

Microbial cell wall characteristics such as thickness and structure, composition, presence of certain UV-susceptible proteins, and other repair and defense mechanisms within the microorganism influence the germicidal effect of UV. Other factors concerned with the food matrix itself such as the presence of suspended matter, food composition, viscosity, density, thickness of the food layer, pH, and water activity have a major effect on UV effectiveness [11, 37]. Bang, Kim, et al. [4] carried out a study on the effect of UV-C and cold plasma on *Salmonella Montevideo, Typhimurium*, and *Enteritidis* from black peppercorns. A cold plasma system with Helium as the working gas flowing at 1 L/min operating at 9.7 kV and 15 kHz frequency was used after UV pre-treatment. With a UV dose of 616 $\mu W/cm^2$ intensity, each of the treatment interventions was applied for 22.1 minutes to assess the antibacterial effectiveness. When applied individually, UV and CP achieved a 3.2 and 2.5 log CFU/g reduction, respectively, while their combination achieved a higher reduction of 3.7 log CFU/g. The authors postulate that this could potentially be due to the increased sensitivity of *Salmonella* species to cold plasma following a UV exposure that partially damaged the microbial cells. Subsequent application of cold plasma rendered these partially damaged cells dead, impairing their ability to persist and regrow. The authors observed only a 0.4 log CFU/g of sub-lethally injured cells. *Salmonella* species have the ability to repair UV damage through photolyases that absorb light at 300–600 nm for their repair mechanism. With cold plasma as the further hurdle step, these enzymes are inactivated, thus impairing *Salmonella*'s ability to repair itself. The authors also observed a lower antioxidant activity and a decrease in piperine concentration arising from the reaction between cold plasma-produced RGS with the peppercorns.

Another example of multiple hurdle application using UV was demonstrated by Lee et al. [40], wherein the authors used DBD-cold plasma with UV-activated zinc oxide (ZnO) to induce photocatalysis. They observed the effect of this combination on the *Salmonella* inactivation in steamed fish cake. The authors noted that when ZnO and the CP were applied simultaneously (ZnO & CP), the production of RGS was more than the individual treatments. Although more, the difference was less significant, potentially due to the multiple reactions between the free radicals generated through ZnO and CP. The treatment led to a 3 log CFU/g reduction and was ascertained as primarily due to CP owing to the fact that CP treatment alone also achieved a similar log reduction with no significant difference. The authors observed that the headspace to food volume ratio played a significant role in microbial inactivation, with 66:1 as the optimized ratio achieving the highest 2.3 log CFU/g reduction for one piece of cake. The microbial reduction was also sustained throughout 21 days of storage at 4 and 10°C in comparison to the control. The combined treatments did not significantly affect the moisture, color, gel strength, or water activity of the sample. No significant change was observed in sample pH, possibly due to the fact that there was not enough RGS produced for any significant reaction with water molecules.

Lee et al. [41], demonstrated decontamination of rice germ with an integrated large-scale plasma jet pulsed light UV-C system. The samples were first treated with pulsed light of 3 kV at 2 Hz followed by a plasma jet operating at 1 kV, 30 Hz, 2 kW with a dry air flow rate of 840 L/min. The plasma treatment was followed by UV exposure wherein the samples were placed on a 120 cm long conveyor belt that passed through a UV chamber with 8 W lamps at a speed of 4.6 cm/s. A higher microbial reduction was observed when there was either an increase in treatment time or decrease in sample size/amount (0.2–3.0 kg). A maximum reduction of 1.0 log CFU/g was observed in bacteria while yeast and mold showed up to 2.0 log CFU/g reduction. The authors postulate that this might arise from the fact that the initial load of yeasts and mold was higher which can result in a higher reduction. Some other studies have found that the shape and size of microbial cells also influence the effectiveness of RGS to diffuse and inactivate them. Some microbial species also have defense mechanisms that enable them to recover under favorable environments such as temperature, pH, light, and so on. Although there was a small increase in sample temperature due to pulsed light treatment as well as a drop in moisture content due to the flow of dry air during plasma exposure, there was no change observed in the rice germ color nor in the lipid profile. RGS have the ability to increase the antioxidant activity of certain foods by causing the dissociation of agglomerates. Pulsed light and UV are also known to induce photo-protective antioxidant defense reactions in foods, but no such effects were observed by the authors in the rice germs.

11.4.4 COLD PLASMA FOR SUSTAINABLE PACKAGING

Food packaging is an integral part in preserving the effect of processing of food until consumption. Packaging plays multiple roles such as physical protection, enhanced preservation, containment as well as traceability. Development of biodegradable and sustainable packaging materials with enhanced mechanical and gas barrier characteristics has seen an increasing trend owing to the global plastic pollution. In recent years, cold plasma has been increasingly explored for modification of biodegradable polymers with improved characteristics that can maintain food product's integrity. One such study was done by Cui et al. [15] where the authors studied the potential of cold plasma to improve the mechanical properties of nano-bio-polymer made from a polysaccharide from bitter gourd (*Momordica charantia*) (MCP) and a bioactive agent from brown algae – Phlorotannin (PT). The authors tested different ratios of MCP and PT and observed their effect on film thickness and their mechanical strength. They also observed that with an increase in the polysaccharide content, the moisture content of the resultant nanofibers increased due to the water-soluble nature of MCP. This also resulted in an opaque film formed potentially due to agglomeration of the polysaccharide chains and subsequent blocking of light. The agglomeration also resulted in a lower tensile strength of the fibers leading to an increased chance of breakage. When these nanofibers were subjected to a nitrogen-based cold plasma, the RGS helped in improving the release rate of PT by 23.5–25.0% from the nanofibers resulting in an enhanced antibacterial and antioxidant effect.

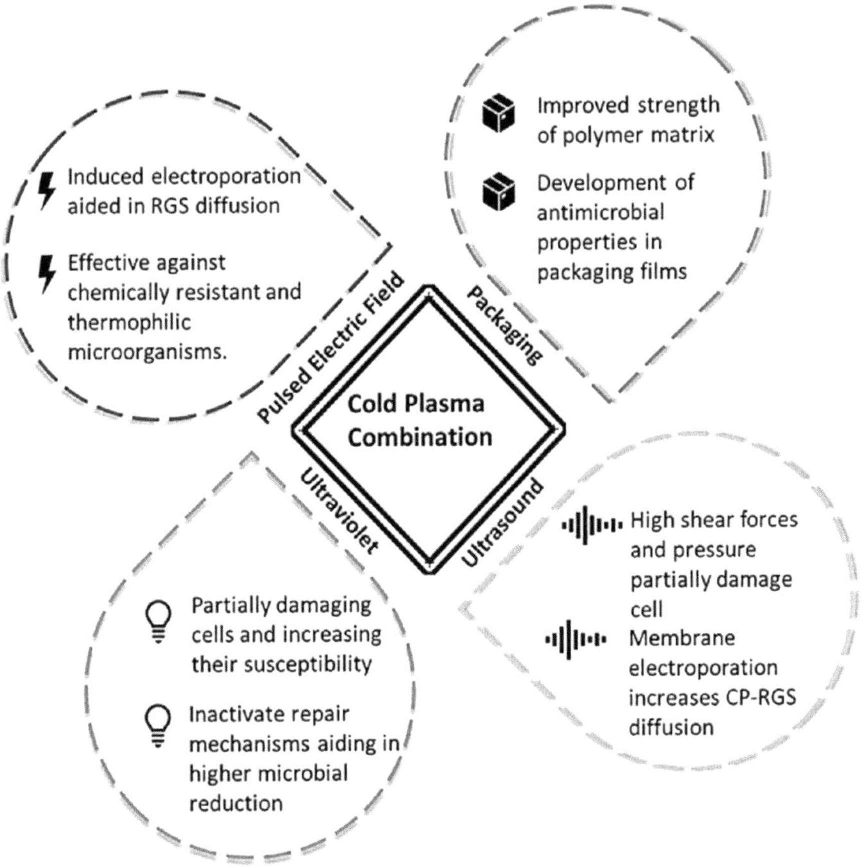

FIGURE 11.2 Effect of cold plasma with non-thermal technologies.

Another study indicated similar film modifications in the development of biodegradable packaging material from protein sources such as whey and gluten [53]. An argon-based cold plasma was shown to improve the tensile strength of the protein films, but exposure of more than 10 minutes resulted in the loss of film integrity. This might be due to the intense bombardment and etching caused by Ar^+ ions, as well as the higher ionization yield for Argon compared to air, which can compromise the structure with longer processing times. Despite an increase in tensile strength, air-based plasma decreased the elongation level of whey proteins, potentially due to the development of peroxide and hydroxyl functional groups within the film matrix that can alter polymer flexibility. An opposite effect was observed for the gluten film, wherein an increase in flexibility was observed for both Ar and air-based plasma, with no significant difference between them. The authors attribute it to the difference in amino acid composition between the two protein matrices that can lead to different dipole-dipole interactions, hydrogen bonds, and cross linkages that can influence the mechanical properties. This also resulted in an increase in oxygen barrier property for both gas-based plasmas. Despite the formation of polar functional groups and changes in surface roughness, the films showed no change in water vapor permeability, confirming that other thermodynamic properties also influence the permeability of moisture. More research is crucial to understand the effect of different gas-based plasmas on different polymers that can then help target a suitable packaging matrix to help preserve food.

A summary of cold plasma applications with non-thermal technologies has been shown in Figures 11.2 and Table 11.2.

11.5 FUTURE CONSIDERATIONS

The incidence of foodborne pathogens and consumption risks is only increasing each year. The shift in awareness towards safer and healthier foods is a key deciding factor in the development and application of processing technologies for food. Cold plasma has shown promising potential to improve the shelf stability of foods while preventing detrimental effects on the chemical, physical, and sensory aspects of food. However, some of the complexities associated with it, such as the influence of food surface characteristics on the penetration of RGS, the influence of the RGS cocktail on food quality, as well as the difficulty in precisely defining the reactive chemistry. Some of these have been overcome with the combination of CP

TABLE 11.2 Cold Plasma in Combination with Other Nonthermal Technologies

Application Goals	Treatment Parameters	Significant Result(s)	References
• Cold plasma pre-treated US assisted convective drying of goldenberries. • Improved quality and nutritional value	• Cold plasma parameters: ▪ Gas: Air at 35–40% RH. ▪ Flow rate: 5 L/min. ▪ Source: Gliding arc with divergent stainless-steel electrodes. ▪ Discharge gap: 7.5 mm ▪ Power: 300 W ▪ Voltage: 10 kV ▪ Frequency: 50 Hz ▪ Treatment time(s): 15, 30, 45 and 60 seconds ▪ Sample distance: 5 cm. • Ultrasound parameters: ▪ Power: 100 W ▪ Frequency: 20 kHz ▪ Intensity: 1,780.3 W/m^2 • Drying parameters: ▪ Temperature: 50°C ▪ Air flow rate: 1 m/s ▪ Sample size: 70 g. ▪ Final M.C: 12% (w.b)	• Increased porosity after CP treatment – proportional to treatment time. • RGS from CP pre-treatment ruptures hydrogen and other non-covalent bonds in the waxy cuticle of the berries leading to increased porosity. • Increase in hydrophilicity due to production of new surface functional groups by ROS – C=O to C-OH and C-O and break of C-H and C-C bonds by RNS. • Decrease in contact angle from 77.5° to 30° after 60s of CP exposure. • US reduced drying time by 10% from 890 mins (only hot air drying) while CP + drying reduced time by 19.33%; CP + US + drying reduced drying time by 29%. • Reduced drying time due to increased porosity facilitating increased mass transfer channels from inner cell layers. • CP pre-treatment facilitates easy penetration of US due to breakdown of waxy layer on berries' skin by RGS. • CP pre-treated samples showed higher rehydration rate (85.2%) due to disruption of surface microstructures and increased surface hydrophilicity. • Upto 80.87% retention in Vitamin-C after CP+US+drying – faster drying results in shorter exposure to heat and oxygen resulting in lesser nutritional loss. • Higher retention of phenolic content due to decreased thermal stress; highest retention for CP treated samples due to release of phenolic compounds attached to cell walls getting released due to rupture of bonds and cell membranes. • Significant reduction in CO_2, SO_2 and NO_x emissions as a result of CP pre-treatment.	[50]

TABLE 11.2 *(Continued)*

Application Goals	Treatment Parameters	Significant Result(s)	References
PAA/Chlorine assisted US-washing followed by in-package cold plasma disinfection of *E. coli* O157:H7, *Salmonella typhimurium* and native microflora (mesophiles, yeasts and molds) from blueberries.	• PAA concentration: 80 ppm • Free chlorine concentration: 10 ppm • Sample: disinfectant volume – 1:20 (w/v) • Washing time: 1 minute • Sample size: 100 g • Ultrasound parameters: ▪ Power: 400 W ▪ Frequency: 25 kHz. • Cold plasma parameters: ▪ Source: DBD with 10 cm diameter electrodes ▪ Frequency: 400 Hz ▪ Voltage: 78.80 kV ▪ Gas: Atmospheric air ▪ Power: 33.91 W ▪ Discharge gap: 20 mm	• US-FC + CP (log CFU/g) reductions – ▪ 0.86 – *E. coli* ▪ 0.71 – *Salmonella* ▪ 0.42 – aerobic mesophilic ▪ 0.29 – yeasts and mold. • US-PAA + CP (log CFU/g) reductions – ▪ 0.71 – *E. coli* ▪ 0.59 – *Salmonella* ▪ 0.32 – aerobic mesophilic ▪ 0.21 – yeasts and mold. • With just FC, PAA and US, ▪ *E. coli* reduction: 0.73, 0.83 and 0.63 log CFU/g; ▪ *Salmonella* reduction: 0.64, 0.74 and 0.68 log CFU/g ▪ No significant difference between the treatments. • With just US+FC, *E. coli* and *Salmonella* reduced by 1.32 and 1.29 log CFU/g, respectively. • With just US+PA, *E. coli* and *Salmonella* reduced by 1.38 and 1.43 for *E. coli* and *Salmonella* – no significant difference between these two combinations. • US-FC+CP resulted in a 2.06, 1.99.105 and 0.77 log CFU/g reduction in E. coli, *Salmonella*, aerobic mesophilic count and yeast & mold. • Similar for US-PA+CP treatment – no significant difference. • Sustained microbial reduction even after 3 days post treatment. • No change in anthocyanin content, firmness post treatment, even after 3 days. • CP combination treatments resulted in lower ROS activity with higher antioxidant enzyme activity – CP activating the antioxidant system in the blueberries to scavenge ROS, lowering quality loss. • US induced pores on bacterial membrane – collapsing cavitation bubbles and localized shear force and high pressures. • FC and PAA cause cell membrane, DNA and intracellular damage through oxidation as well as enzyme inactivation.	[79]

TABLE 11.2 *(Continued)*

Application Goals	Treatment Parameters	Significant Result(s)	References
US, PAW, PAA and their combination on native microflora (mesophiles and psychrophiles) and inoculated *E.coli, Listeria monocytogenes* and *Pseudomonas fluoroscens* on mackerel filets.	• PAA concentration: 200 ppm for 10 minutes. • Ultrasound parameters: ▪ Frequency: 25 kHz ▪ Power: 550 W ▪ Time: 10 minutes. • Cold plasma parameters for PAW generation: ▪ Gas: Atmospheric air ▪ Flow rate: 11 L/min ▪ Source: Plasma jet ▪ Power: 300 W ▪ Frequency: 20 kHz ▪ Treatment time(s): 10 mins	• Combination treatments achieved 0.72, 0.62 and 0.5 log CFU/g reduction of *Listeria, E. coli* and *Pseudomonas resp.* • A 0.41 and 0.70 log CFU/g reduction observed for mesophile and psychrophile reduction for PAW-PAA+US combination, respectively. • PAW+PAA achieved lower pH demonstrating synergistic effect even though it had no significantly higher reduction than the individual treatments. • No change in LAB color values except for decrease in L value after US treatment potentially due to denaturation of muscle proteins resulting in aggregation and change in optical properties. • pH dropped to 3.11, ORP and conductivity significantly increased after plasma treatment. • Peroxide conc at 140.6 micromoles for PAW-PAA – 10 times more than just PAW treatment.	[89]
Application of PAW and PFB in combination with US for inactivation of *E. coli* and *Shewanella putrefaciens* from grass carp (freshwater fish)	• Cold plasma parameters: ▪ Gas: Atmospheric air ▪ Source: DBD with 50 mm diameter electrodes ▪ Discharge gap: 5 mm ▪ Power: 73.5 W ▪ Voltage: 30–70 V ▪ Frequency: 10 kHz at 1.05 A ▪ Treatment time(s): 8 mins • US treatment parameters: ▪ Frequency: 40 kHz ▪ Power: 500 W ▪ Time: 4 minutes. ▪ Volumetric power: 17.45 W/L	• pH drops to 2.60 from 6.67 for PAW; 2.59 from 3.75 for PFB. • O_3 content increases to 5.0 ppm in PAW – showed an increasing trend with increasing voltage. • O_3 content decreasing with increasing voltage – 0.38 ppm from an initial 0.68 ppm. • PAW and PFB showed increasing trend for H_2O_2 and reached a max of 2.26 micromolar. • NO_3, NO_2 increasing with increasing voltage – PFB showed highest with 2.99 and 3.38 mM for NO_2 and 6.87 and 11.94 mM for NO_3 for PAW and PFB resp. • Electrical conductivity: 0.82–2.73 mS/cm and 7.04–8.72 mS/cm for PAW and PFB, resp. • ORP: 564.50–598.43 mV and 582/33–630.79 mV for PAW and PFB, respectively. • Residual NO_2 and NO_3 increasing with increasing voltage: 19.01, 21.94, 20.26 and 23.77 mg/kg for NO_2, and 34.17, 37.60, 35.87 and 39.01 mg/kg for NO3 for PAW, PFB, USPAW and USPFB.	[32]

TABLE 11.2 *(Continued)*

Application Goals	Treatment Parameters	Significant Result(s)	References
		• Microbial reductions: 0.22 and 0.23 log CFU/g for *E. coli* and *S. putrefaciens* for US; 0.88 and 0.92 log CFU/g for PAW, 1.15 and 1.18 log CFU/g for PFB resp. • PAW+US was most effective in reducing pathogens at all voltage levels for *E. coli;* only at 50–70 V for *S. putrefaciens.* • Highest reduction as a combo was 1.39 and 1.49 log CFU/g for USPAW and 1.31 and 1.39 log CFU/g for USPFB resp. for *E. coli* and *S. putrefaciens.* • Acidified environment weakens bacterial resistance to penetration of ROS. • Significant rise in Peroxide Value (PV) with increasing voltage – 2.35 and 2.4 mEq peroxide/kg for USPAW and USPFB. • Increase in TBARS value probably due to oxidizing ability of PAW and US – only minimal increase from 0.081 mg MDA/kg for DI water to 0.128 and 0.134 mg MDA/kg for PAW and PFB, resp – with increasing voltages. • Slightly higher TBARS for PFB due to more RONS in PFB. • Drop in hardness and cohesiveness and small increase in adhesiveness, springiness and chewiness for all voltage levels – hardness was significantly reduced from 515.65 g when compared to initial of 594.48 g. • Beaching and oxidation of carotenoid, hemoglobin and myoglobin pigments by RONS generated in the PAW, in addition to enzymatic and non-enzymatic reactions resulting in disorganization of myofibrils and degradation of myofibrillar proteins. • FTIR peak height decreased with increasing voltage – protein denaturation due to RONS action in PAW and US cavitation-based shear actions. • Overall, combination treatments induced increased lipid oxidation and protein degradation compared to individual applications but still were within acceptable limits. • Better biomedical index and nutritional value, improved functional properties and samples from combined treatment exhibited distinct color change and increased brightness.	

TABLE 11.2 *(Continued)*

Application Goals	Treatment Parameters	Significant Result(s)	References
• PEF pre-treated *Litopenaeus vannamei* (a popular crustacean) immersed in Chamuang leaf extract (CLE) (1 or 2%) followed by in-package HVACP exposure.	• Cold plasma parameters: ▪ Gas: Argon/Air mix (80:20) ▪ Source: DBD with 15 cm diameter electrodes ▪ Discharge gap: 2 cm ▪ Voltage: 16 kVrms ▪ Treatment time(s): 10 mins • PEF parameters: ▪ Field: 15 kV/cm ▪ Pulses: 600 (PEFT1) and 800 pulses (PEFT2) ▪ Energy: 483 kJ/kg and 697 kJ/kg. ▪ *L. vannamei* was pre-treated in either PEFT1 or PEFT2 for 30 mins in 1 or 2% CLE solution.	• PEFT2-CLE2-HVACP exhibited least melanosis. • Melanosis scores: PEFT1-CLE2-HVACP > PEFT2-CLE1-HVACP > PEFT1-CLE1-HVACP. • PPO was inhibited by CLE in a dose dependent manner while it was inactivated by PEF in an energy dependent manner. • PEFT2-CLE2-HVACP had the lowest microbial count (total viable count) at the end of 18 days of storage – 5.03 logs when compared to 8.17 log CFU/g in control. • PEF previously reported to inactivate mesophilic bacteria in *L. vannamei* due to cell wall disintegration due to electroporation. • Psychrophilic count continued to stay low for PEFT2-CLE2-HVACP throughout storage – enhanced CLE passage through PEF generated pores, especially at 800 pulses. • *Pseudomonas* was the dominant bacterial species present. • H_2S producing bacteria were the lowest in the PEFT2-CLE2-HVACP treated samples – even when PEF intervention was not present. • Lowest enterobacteria in PEFT2-CLE2-HVACP – PEF with lower pulses with 2% CLE could also inactivate it – attributed to passage of CLE 2% through PEF induced pores – so is the LAB count. • For LAB, the count was highest for plasma treated samples after 18 days – MAP with less oxygen with LAB as facultative anaerobes – no prior intervention resulted in them growing more. • Lowest pH for PEFT2-CLE2-HVACP – no change through storage – more basic pH in control due to increased activity of microbial enzymes – inactivation in the treated sample helped maintain lower pH. • Highest TBARS value in control for upto 18 days of storage – PEFT2-CLE2-HVACP had the lowest value – microbial lipase activity higher in control. • Higher TBARS in only HVACP treated sample suggested hydroperoxides formation and further decomposition into secondary products. • PEF + 2% CLE alone alleviated lipid oxidation while HVACP alone induced oxidation.	[66]

TABLE 11.2 *(Continued)*

Application Goals	Treatment Parameters	Significant Result(s)	References
		• PEFT2-CLE2-HVACP had retained the highest content of fatty acids – after 18 day, EPA and DHA were depleted while myristic, oleic and linoleic acid were augmented – lower PUFA contents possibly due to oxidation by microbial growth induced lipases. • Higher visual appeal and odor likeness of PEFT2-CLE2-HVACP than other treatments possibly due to lower oxidation and melanosis.	
• UV-C + Cold plasma on inactivation of *Salmonella Montevideo, Typhimurium* and *Enteritidis* from black peppercorns.	• Cold Plasma parameters: ▪ Gas: Helium ▪ Flow rate: 15 L/min. ▪ Source: electrode formed out of copper plates wrapped around outer surface of treatment chamber at 3 cm intervals. ▪ Voltage: 9.7 kV ▪ Frequency: 15 kHz ▪ Treatment time(s): 22.1 minutes • UV treatment parameters: ▪ Intensity: 616 μW/cm^2 ▪ Treatment time: 22.1 minutes	• 3.7 log CFU/g reduction with UV+CP; 3.2 logs for UV and 2.5 logs for CP. • Combination treatment more effective potentially due to sensitivity of *Salmonella* to UV followed by CP which kills the partially damaged cells that could persist and regrow. • 0.7 and 0.4 log CFU/g sub-lethally injured after CP and UV-CP treatments. • Photolyases in *Salmonella* repair UV damaged cells by absorbing light at 300–600 nm range. • Bacterial enzymes inactivated by CP – cannot induce photoreactivation to enable damaged cells to grow. • Lowered antioxidant activity of lipophilic antioxidants. • Decrease in piperine concentration – a lipophilic antioxidant.	[4]
• CP + UV-activated ZnO (photocatalysis) for inactivation of *Salmonella* in steamed fish cake.	• Cold plasma parameters: ▪ Gas: Air ▪ Source: DBD with upper electrode as pins – 3,352 steel pins ▪ Discharge gap: 3.5 cm	• ZnO + CP More effective than ZnO or CP alone – higher number of RGS in ZnO + CP including those from ZnO such as H_2O_2, OH + the ROS and RNS from CP. • No difference between ZnO & CP and CP only treatment – low UV light intensity and shorter UV treatment in ZnO & CP (12.2 uW/cm^2, 3 mins) compared to those in ZnO + CP (490 uW/cm^2, 10 mins) – limited production of RGS during ZnO treatment.	[40]

TABLE 11.2 *(Continued)*

Application Goals	Treatment Parameters	Significant Result(s)	References
• Inside of lid coated with Whey protein isolate solution (5% w/w + 5% glycerol) having ZnO particles. • 0.25 g ZnO in 100 g WPI glycerol solution ? sonicated for 10 mins ? 2 mL of this cast inside lid.	▪ Voltage: 23 kV ▪ Treatment time(s): 3 mins. • CP simultaneously with ZnO (ZnO & CP) and CP applied after ZnO (ZnO + CP). • CP treatment chamber placed between UV lamps (350 nm). • UV treatment parameters: ▪ Distance between lamps: 40 cm. ▪ Power: 12.2 μW/cm^2 • For ZnO + CP – sample placed in plywood box with UV lamp (40 W; 490 uW/cm^2) for 10 mins followed by CP for 1 min. • Stored for 0, 7, 14 and 21 days at 4 and 10°C at 99% RH.	• The less significant difference may also be due to the decreasing number of radicals due to potential reactions between radicals generated by ZnO and CP treatments. • 10 mins of ZnO chosen based on treatment times tested between 5–20 mins – no synergistic significant effect of ZnO observed beyond 10 mins – 2.3 log CFU/g. • 3 mins of CP led to 3.0 log CFU/g survivors – lowest among all tested – no significant difference between ZnO+CP and CP alone – so primary effect due to CP. • Headspace volume to food volume ratio played a significant role in inactivation efficacy with a 66:1 ratio with the highest reduction of 2.3 log CFU/g for one piece. • No growth below 10 C for ZnO+CP – 1.3–1.6 log CFU/g lower than control at 21 days for both 4 and 10°C. • No significant difference in moisture content, color and water activity between treated (ZnO+CP) and untreated at both temperatures. • No pH changes due to RGS and OH radicals – maybe not enough to diffuse and react with water inside to shift pH. • No moisture change, no gel strength change.	
• Decontamination of rice germ using a large-scale plasma jet pulsed light UV-C integrated system (PPU)	• Max production rate of system – 226 kg/h. • Sample amount: 200, 500, 1,000 and 3,000 g. • Initial microbial load:	• Higher treatment time and lesser sample amount had maximum reduction – 7 mins and 200 g best combo. • Increase in treatment time ? increase in IPL fluence ? increase in physical damage to cell membranes and walls of microbes. • 22% lipids in rice germ – mainly oryzanol and tocopherol which did not get affected by the treatments.	[41]

TABLE 11.2 *(Continued)*

Application Goals	Treatment Parameters	Significant Result(s)	References
• IPL? plasma jet? UV	▪ Bacteria: 5.7 log CFU/g; ▪ Yeast and mold (Y&M): 3.7 log CFU/g. • Cold plasma parameters: ▪ Gas: Dry air ▪ Flow rate: 840 L/min ▪ Source: Plasma jet with 3.8 mm nozzles ▪ Length of plasma discharge: 5 mm x 18 mm ▪ Voltage: 1 kV ▪ Power: 2 kW ▪ Treatment time(s): 3, 5, 7 and 10 mins. ▪ Frequency: 30 Hz • Pulsed light parameters: ▪ Chamber volume: 19,460 cm^3, ▪ Energy supply: FP squared ▪ Voltage: 3 kV ▪ Frequency: 2 Hz • UV-C parameters: ▪ Intensity: 5 lamps, each 8 W at 253.7 nm ▪ Approximately 120 cm distance of conveyor at a speed of 4.6 cm/s for UV exposure.	• M.C of the germ, reduced from 8.67% by 46.5% after PPU – heat generated + dry air circulation caused drop in M.C even for the individual treatments. • No change in antioxidant activity – DPPH and ABTS radical scavenging assays – not enough RGS to cause changes.	

TABLE 11.2 *(Continued)*

Application Goals	Treatment Parameters	Significant Result(s)	References
Low voltage piezoelectric direct discharge CP and linalool loaded nanoemulsion washing (LW) against *E. coli* O157:H7 and *Salmonella* on RTE chicken	• Cold plasma parameters: ▪ Voltage: 15 V ▪ Frequency: 50 Hz. ▪ Gas: air at 60% RH ▪ Discharge gap: 10 mm ▪ Treatment time: 0–5 mins • LW parameters: ▪ Treatment time: 0–25 mins ▪ Concentration: 8,000 ppm	• Bacterial reductions: ▪ *Salmonella*: > 3.2 log CFU/g ▪ *E. coli:* 2.76 log CFU/g • Linalool caused increase in electrical conductivity after 6 h – leakage of ions compared to control – confirmed that increased cell membrane permeability is one of the mechanisms. • Up to 1.83 and 1.67 log CFU/g reduction observed for *E. coli* and *Salmonella* when treatment time increased to 15 mins. • No significant increase in reduction when treatment time increased up to 25 mins. • With CP – 1 log reduction after 1 min – highest was 1.8 and 1.47 logs after 5 mins for *Salmonella* and *E. coli,* respectively. • 2.38 log reduction in *E. coli* (LW?CP) but 2.76 log reduction when CP?LW. • 2.96 reduction in *Salmonella* when LW?CP but below LOD when CP?LW. • Antibacterial effect more pronounced when CP was applied first. • Surface etching resulting in pores or sublethal injuries making cells more susceptible for linalool penetration. • Higher lipid oxidation levels cause of CP, but not significantly higher.	[27]
Cold plasma to improve antimicrobial and material properties of nanofibers from bitter gourd (*Momordica charantia*) polysaccharide (MCP) and Phlorotannin (PT).	• Cold plasma parameters: ▪ Source: Commercial system ATV Electronic Technology ▪ Gas: Nitrogen ▪ Flow rate: 100 sccm ▪ Power: 350 W ▪ Treatment time: 30 seconds. • MCP:PT ratios tested: 2:1, 4:1, 6:1, 8:1 and 10:1 with PT concentration set at 20 mg/mL	• MCP:PT ratio of 2:1 showed maximum film thickness of 0.063 m while a ratio of 10:1 showed the minimum of 0.027 mm. • Moisture content of nanofibers formed increased with increase in MCP content – probably due to hygroscopic nature of MCP. • Formed nanofibers showed maximum tensile strength when MCP:PT ratio was at 6:1 – higher MCP concentration led to agglomeration and breakage. • CP helped an enhanced release of PT in-vitro – surface modification of nanofibers promoted higher release rate. • FTIR analysis revealed successful integration of PT into MCP nanofibers. • Oxidation resistance of nanofibers improved by CP treatment.	

TABLE 11.2 *(Continued)*

Application Goals	Treatment Parameters	Significant Result(s)	References
Low pressure cold plasma for improving mechanical properties of protein films (whey and gluten)	• Cold plasma parameters: ▪ Source: Low pressure glow discharge ▪ Frequency: 20 kHz ▪ Power: 50 W ▪ Treatment time(s): 5, 10 and 15 mins. ▪ Gas: Air and Argon	• ATR-FTIR analysis revealed enhanced β-sheet structure and removal of H atoms from polymer structure after cold plasma treatment. • Surface etching and creation of active sites by Argon based plasma resulted in development of functional groups upon film's exposure to ambient atmosphere. • Amino acid composition of protein films shown to affect bond linkages causing an increase in tensile strength of the films. • Improved oxygen permeability for both films, irrespective of gas – cross linkages and intermolecular bonds reduce free volume between polymeric chains.	[53]

with other technologies such as PEF or ultrasound, while other underexplored areas include technologies such as modified atmosphere packaging or nanotechnology that can exploit the hurdle approach to sustain shelf stability throughout the supply chain. Foods high in fat have often not been well suited for treatment with CP; however, more research is required to explore the potential of combining other novel interventions as pre-treatments to mitigate the detrimental effects.

11.6 SUMMARY

Cold plasma as a non-thermal intervention is an ideal technology that can either be used as a stand-alone or in combination with other interventions to process food matrices to enhance food security and ensure quality retention. The use of other technologies with cold plasma utilizes a hurdle approach that offers an advantage to overcome the multiple resistance mechanisms developed by microorganisms. This chapter summarizes recent studies that have explored the potential of cold plasma to work with other thermal and non-thermal technologies as a pre-treatment or an additional hurdle step to improve safety and quality of the food matrix. It is also being explored more for the development of biodegradable polymers wherein the plasma reactive species have shown potential to improve polymerization and enhance certain crucial properties of the polymer matrix that result in improved mechanical properties.

KEYWORDS

- **cold plasma combination**
- **decontamination**
- **food quality**
- **food safety**
- **hurdle technology**
- **pre-treatment**
- **shelf-life**

REFERENCES

1. Adam, A. M., Jeganathan, B., Vasanthan, T., & Roopesh, M. S. (2022). Dipping fresh-cut apples in citric acid before plasma-integrated low-pressure cooling improves Salmonella and polyphenol oxidase inactivation. *Journal of the Science of Food and Agriculture*, *102*(8), 3425–3434.
2. Aguiló-Aguayo, I., Soliva-Fortuny, R., & Martín-Belloso, O. (2010). Impact of high-intensity pulsed electric field variables affecting peroxidase and lipoxygenase activities of watermelon juice. *LWT – Food Science and Technology*, *43*(6), 897–902.
3. Banaschik, R., Lukes, P., Jablonowski, H., Hammer, M. U., Weltmann, K. D., & Kolb, J. F. (2015). Potential of pulsed corona discharges generated in water for the degradation of persistent pharmaceutical residues. *Water Research*, *84*, 127–135.
4. Bang, I. H., In, J., & Min, S. C. (2021). Inactivation of Salmonella on black peppercorns using an integrated ultraviolet-C and cold plasma intervention. *Food Control*, *119*, 107498.
5. Bang, I. H., Kim, Y. E., & Min, S. C. (2021). Preservation of mandarins using a microbial decontamination system integrating calcium oxide solution washing, modified atmosphere packaging, and dielectric barrier discharge cold plasma treatment. *Food Packaging and Shelf Life*, *29*, 100682. https://doi.org/10.1016/j.fpsl.2021.100682.
6. Bao, T., Hao, X., Shishir, M. R. I., Karim, N., & Chen, W. (2021). Cold plasma: An emerging pretreatment technology for the drying of jujube slices. *Food Chemistry*, *337*, 127783. https://doi.org/10.1016/j.foodchem.2020.127783.
7. Cao, Z., Nie, Q., Bayliss, D. L., Walsh, J. L., & Ren, C. S. (2010). Spatially extended atmospheric plasma arrays. *Plasma Sources Science and Technology*, *19*(2), 025003. https://doi.org/10.1088/0963-0252/19/2/025003.
8. Chizoba Ekezie, F. G., Sun, D. W., & Cheng, J. H. (2017). A review on recent advances in cold plasma technology for the food industry: Current applications and future trends. *Trends in Food Science & Technology*, *69*, 46–58.
9. Choi, E. J., Park, H. W., Kim, S. B., Ryu, S., Lim, J., et al. (2019). Sequential application of plasma-activated water and mild heating improves microbiological quality of ready-to-use shredded salted kimchi cabbage (*Brassica pekinensis* L.). *Food Control*, *98*, 501–509.
10. Choudhury, N., Iftikhar, F., & Makroo, H. A. (2022). Non-thermal processing of foods. (pp. 181–196). CRC Press.
11. Christensen, J., & Linden, K. (2001). How particles affect UV light in the UV disinfection of unfiltered drinking water. *Journal-American Water Works Association, 95*(4), 179–189.
12. Conrads, H., & Schmidt, M. (2000). Plasma generation and plasma sources. *Plasma Sources Science and Technology*, *9*(4), 441–454. https://doi.org/10.1088/0963-0252/9/4/301.
13. Costello, K. M., Smet, C., Gutierrez-Merino, J., Bussemaker, M., Van Impe, J. F., & Velliou, E. G. (2021). The impact of food model system structure on the inactivation of *Listeria innocua* by cold atmospheric plasma and nisin combined treatments. *International Journal of Food Microbiology*, *337*, 108948. https://doi.org/10.1016/j.ijfoodmicro.2020.108948.
14. Coutinho, N. M., Silveira, M. R., Rocha, R. S., Moraes, J., Ferreira, M. V. S., et al. (2018). Cold plasma processing of milk and dairy products. *Trends in Food Science & Technology*, *74*, 56–68.

15. Cui, H., Yang, X., Abdel-Samie, M. A., & Lin, L. (2020). Cold plasma treated phlorotannin/Momordica charantia polysaccharide nanofiber for active food packaging. *Carbohydrate Polymers*, *239*, 116214. https://doi.org/10.1016/j.carbpol.2020.116214.
16. Daeschlein, G., Von Woedtke, T., Kindel, E., Brandenburg, R., Weltmann, K. D., & Jünger, M. (2010). Antibacterial activity of an atmospheric pressure plasma jet against relevant wound pathogens in vitro on a simulated wound environment. *Plasma Processes and Polymers*, *7*(3–4), 224–230.
17. De la Ossa, J. G., Kadri, H. El, Gutierrez-Merino, J., Wantock, T., & Harle, T. (2021). Combined antimicrobial effect of bio-waste olive leaf extract and remote cold atmospheric plasma effluent. *Molecules*, *26*(7), 1890. https://doi.org/10.3390/molecules26071890.
18. Dörnenburg, H., & Knorr, D. (1995). Strategies for the improvement of secondary metabolite production in plant cell cultures. *Enzyme and Microbial Technology*, *17*(8), 674–684.
19. El-Wakeil, N. E. (2018). Retracted article: Botanical pesticides and their mode of action (Gesunde Pflanzen, (2013), 65, 4, (125–149), 10.1007/s10343-013-0308-3). *Gesunde Pflanzen*, *70*(2), 11–15. https://doi.org/10.1007/s10343-018-0415-2.
20. Feizollahi, E., Arshad, M., Yadav, B., Ullah, A., & Roopesh, M. S. (2021). Degradation of Deoxynivalenol by Atmospheric-Pressure Cold Plasma and Sequential Treatments with Heat and UV Light. *Food Engineering Reviews*, *13*(3), 696–705.
21. Fridman, A. (2008). *Plasma Chemistry*. Cambridge University Press, p. 978.
22. Fridman, A., Kennedy, L. A. (2004). *Plasma Physics and Engineering*. DOI: https://doi.org/10.1201/9781482293630, pp. 1–884, eBook ISBN: 9780429190766, CRC Press.
23. Furmanski, J., Kim, J. Y., Kim, S. O. (2011). Triple-coupled intense atmospheric pressure plasma jet from honeycomb structural plasma device. *IEEE Transactions on Plasma Science*, *39*(11), 2338–2339.
24. Galindo, F. G., Dejmek, P., Lundgren, K., Rasmusson, A. G., Vicente, A., Moritz, T. (2009). Metabolomic evaluation of pulsed electric field-induced stress on potato tissue. *Planta*, *230*(3), 469–479.
25. Ge, X., Shen, H., Su, C., Zhang, B., Zhang, Q. (2021). The improving effects of cold plasma on multi-scale structure, physicochemical and digestive properties of dry heated red adzuki bean starch. *Food Chemistry*, *349*, 129–159.
26. Gómez Galindo, F., Wadsö, L., Vicente, A., Dejmek, P. (2008). Exploring metabolic responses of potato tissue induced by electric pulses. *Food Biophysics*, *3*(4), 352–360.
27. González-González, C. R., Labo-Popoola, O., Delgado-Pando, G., Theodoridou, K., Doran, O., Stratakos, A. C. (2021). The effect of cold atmospheric plasma and linalool nano emulsions against *Escherichia coli* O157 and *Salmonella* on ready-to-eat chicken meat. *LWT*, *149*, 111898. https://doi.org/10.1016/j.lwt.2021.111898.
28. Gudmundsson, M., Hafsteinsson, H. (2001). Effect of electric field pulses on microstructure of muscle foods and roes. *Trends in Food Science & Technology*, *12*(3–4), 122–128.
29. Huang, R., de Vries, D., Chen, H. (2018). Strategies to enhance fresh produce decontamination using combined treatments of ultraviolet, washing and disinfectants. *International Journal of Food Microbiology*, *283*, 37–44.
30. Jeon, E. B., Choi, M. S., Kim, J. Y., Park, S. Y. (2020). Synergistic effects of mild heating and dielectric barrier discharge plasma on the reduction of *Bacillus cereus* in red pepper powder. *Foods*, *9*(2), 171. https://doi.org/10.3390/foods9020171.

31. Jiang, Y., Sokorai, K., Pyrgiotakis, G., Demokritou, P., Li, X. (2017). Cold plasma-activated hydrogen peroxide aerosol inactivates *Escherichia coli* O157, *Salmonella typhimurium*, and *Listeria innocua* and maintains quality of grape tomato, spinach and cantaloupe. *International Journal of Food Microbiology*, *249*, 53–60.
32. Johnson Esua, O., Cheng, J. H., Sun, D. W. (2021). Novel technique for treating grass carp (*Ctenopharyngodon idella*) by combining plasma functionalized liquids and ultrasound: Effects on bacterial inactivation and quality attributes. *Ultrasonics Sonochemistry*, *76*, 105660. https://doi.org/10.1016/j.ultsonch.2021.105660.
33. Karim, N., Shishir, M. R. I., Bao, T., Chen, W. (2021). Effect of cold plasma pretreated hot-air drying on the physicochemical characteristics, nutritional values and antioxidant activity of shiitake mushroom. *Journal of the Science of Food and Agriculture*, *101*(15), 6271–6280.
34. Klockow, P. A., Keener, K. M. (2008). Quality and safety assessment of packaged spinach treated with a novel atmospheric, non-equilibrium plasma system. *American Society of Agricultural and Biological Engineers Annual International Meeting ASABE*, *8*(6), 4625–4635.
35. Knorr, D., Froehling, A., Jaeger, H., Reineke, K., Schlueter, O., Schoessler, K. (2011). Emerging technologies in food processing. *Annual Review of Food Science and Technology*, *2*, 203–235.
36. Koutchma, T. (2019). *Ultraviolet Light in Food Technology: Principles and Applications, Vol. 2*, p. 344. CRC Press.
37. Koutchma, T., Parisi, B., Patazca, E. (2007). Validation of UV coiled tube reactor for fresh juices. *Journal of Environmental Engineering and Science*, *6*(3), 319–28.
38. Kumar Mahnot, N., Siyu, L.-P., Wan, Z., Keener, K. M., Misra, N. N. (2020). In-package cold plasma decontamination of fresh-cut carrots: microbial and quality aspects. *Journal of Physics D: Applied Physics*, *53*(15), 154002. https://doi.org/10.1016/j.ifset.2003.12.001.
39. Lebovka, N. I., Praporscic, I., Vorobiev, E. (2004). Effect of moderate thermal and pulsed electric field treatments on textural properties of carrots, potatoes and apples. *Innovative Food Science & Emerging Technologies*, *5*(1), 9–16.
40. Lee, H. S., Kim, N., Min, S. C. (2022). Inactivation of *Salmonella* in steamed fish cake using an in-package combined treatment of cold plasma and ultraviolet-activated zinc oxide. *Food Control*, *135*, 108772. https://doi.org/10.1016/j.foodcont.2021.108772.
41. Lee, S. Y., Lee, W. K., Lee, J. W., Chung, M. S., Oh, S. W. (2021). Microbial decontamination of rice germ using a large-scale plasma jet-pulsed light-ultraviolet-c integrated treatment system. *Food and Bioprocess Technology*, *14*(3), 542–553.
42. Liao, X., Cullen, P. J., Muhammad, A. I., Jiang, Z., Ye, X. (2020). Cold plasma–based hurdle interventions: New strategies for improving food safety. *Food Engineering Reviews*, *12*(3), 321–332.
43. Loureiro, A. da C., Souza, F. das C. d. A., Sanches, E. A., Bezerra, J. de A., Lamarão, C. V., et al. (2021). Cold plasma technique as a pretreatment for drying fruits: Evaluation of the excitation frequency on drying process and bioactive compounds. *Food Research International*, *147*, 110462. https://doi.org/10.1016/B978-0-12-801365-6.00004-4.
44. Lu, P., Cullen, P. J., Ostrikov, K. (2016). Atmospheric pressure nonthermal plasma sources. In *Cold Plasma in Food and Agriculture: Fundamentals and Applications* (pp. 83–116). Elsevier.

45. Lu, X., Laroussi, M., Puech, V. (2012). On atmospheric-pressure non-equilibrium plasma jets and plasma bullets. *Plasma Sources Science and Technology*, *21*(3), 034005. https://doi.org/10.1088/0963-0252/21/3/034005.
46. Magureanu, M., Bradu, C., Piroi, D., Mandache, N. B., Parvulescu, V. (2013). Pulsed corona discharge for degradation of methylene blue in water. *Plasma Chemistry and Plasma Processing*, *33*(1), 51–64.
47. Manoharan, D., Radhakrishnan, M., Tiwari, B. K., Schlueter, O. K. (2020). Plasma-assisted combination processes. *Innovative Food Processing Technologies: A Comprehensive Review*, *1*, 667–681. https://doi.org/10.1016/B978-0-12-815781-7.00007-X.
48. Masschelein, W. J., Rice, R. G. (2017). *Ultraviolet Light in Water and Wastewater Sanitation.* (p. 174). CRC Press.
49. Miraei Ashtiani, S. H., Rafiee, M., Mohebi Morad, M., Khojastehpour, M., Khani, M. R. (2020). Impact of gliding arc plasma pretreatment on drying efficiency and physicochemical properties of grape. *Innovative Food Science & Emerging Technologies*, *63*, 102381. https://doi.org/10.1016/j.ifset.2020.102381.
50. Miraei Ashtiani, S. H., Rafiee, M., Mohebi Morad, M., Martynenko, A. (2022). Cold plasma pretreatment improves the quality and nutritional value of ultrasound-assisted convective drying: The case of goldenberry. *Drying Technology*, *40*(8), 1639–1657.
51. Misra, N. N., Keener, K. M., Bourke, P., Mosnier, J. P., Cullen, P. J. (2014). In-package atmospheric pressure cold plasma treatment of cherry tomatoes. *Journal of Bioscience and Bioengineering*, *118*(2), 177–182.
52. Misra, N. N., Tiwari, B. K., Raghavarao, K. S. M. S., Cullen, P. J. (2011). Nonthermal plasma inactivation of food-borne pathogens. *Food Engineering Reviews*, *3*(3–4), 159–170.
53. Moosavi, M. H., Khani, M. R., Shokri, B., Hosseini, S. M., Shojaee-Aliabadi, S., Mirmoghtadaie, L. (2020). Modifications of protein-based films using cold plasma. *International Journal of Biological Macromolecules*, *142*, 769–777.
54. Morent, R., De, N. (2011). Inactivation of bacteria by non-thermal plasmas. *Biomedical Engineering – Frontiers and Challenges*, ISBN: 978-953-51-4472-4, Vol. 1, pp. 25–54. https://doi.org/10.5772/18610.
55. Muhammad, A. I., Chen, W., Liao, X., Xiang, Q., Liu, D. (2019). Effects of plasma-activated water and blanching on microbial and physicochemical properties of tiger nuts. *Food and Bioprocess Technology*, *12*(10), 1721–1732.
56. Muhammad, A. I., Liao, X., Cullen, P. J., Liu, D., Xiang, Q. (2018). Effects of nonthermal plasma technology on functional food components. *Comprehensive Reviews in Food Science and Food Safety*, *17*(5), 1379–1394.
57. Olatunde, O. O., Benjakul, S., Vongkamjan, K. (2020). Microbial diversity, shelf-life and sensory properties of Asian sea bass slices with combined treatment of liposomal encapsulated ethanolic coconut husk extract and high voltage cold plasma. *LWT*, *134*, 110232. https://doi.org/10.1111/1541-4337.12379.
58. Pankaj, S., Wan, Z., Keener, K. (2018). Effects of cold plasma on food quality: A review. *Foods*, *7*(1), 4–11.
59. Pankaj, S. K., Bueno-Ferrer, C., Misra, N. N., Milosavljević, V., O'Donnell, C. P. (2014). Applications of cold plasma technology in food packaging. *Trends in Food Science & Technology*, *35*(1), 5–17.

60. Pereira, R. N., Vicente, A. A. (2010). Environmental impact of novel thermal and non-thermal technologies in food processing. *Food Research International*, *43*(7), 1936–1943.
61. Perni, S., Shama, G., Kong, M. G. (2008). Cold atmospheric plasma disinfection of cut fruit surfaces contaminated with migrating microorganisms. *Journal of Food Protection*, *71*(8), 1619–1625.
62. Qian, J., Wang, C., Zhuang, H., Nasiru, M. M., Zhang, J., Yan, W. (2021). Evaluation of meat-quality and myofibrillar protein of chicken drumsticks treated with plasma-activated lactic acid as a novel sanitizer. *LWT*, *138*, 110642.https://doi.org/10.4315/0362-028X-71.8.1619.
63. Ross, A. I. V., Griffiths, M. W., Mittal, G. S., Deeth, H. C. (2003). Combining nonthermal technologies to control foodborne microorganisms. *International Journal of Food Microbiology*, *89*(2–3), 125–138.
64. Seow, Y. X., Yeo, C. R., Chung, H. L., Yuk, H. G. (2014). Plant Essential Oils as Active Antimicrobial Agents. *Critical Reviews in Food Science and Nutrition*, *54*(5), 625–644.
65. Sharmin, N., Sone, I., Walsh, J. L., Sivertsvik, M., Fernández, E. N. (2021). Effect of citric acid and plasma activated water on the functional properties of sodium alginate for potential food packaging applications. *Food Packaging and Shelf Life*, *29*, 100733. https://doi.org/10.1080/10408398.2011.599504.
66. Shiekh, K. A., Zhou, P., Benjakul, S. (2021). Combined effects of pulsed electric field, Chamuang leaf extract and cold plasma on quality and shelf-life of *Litopenaeus vannamei*. *Food Bioscience*, *41*, 100975.https://doi.org/10.1016/j.fpsl.2021.100733.
67. Singh, A. K., Sarkar, P., Janaswamy, S., Yao, Y., Bhunia, A. K. (2014). Encapsulation and delivery of antimicrobial compounds. *Novel Food Preservation and Microbial Assessment Techniques*, Vol. 1, pp. 1–468, ISBN: 9780429166785, 1st Edition. https://doi.org/10.1201/b16758.
68. Song, Y., Fan, X. (2020). Cold plasma enhances the efficacy of aerosolized hydrogen peroxide in reducing populations of *Salmonella Typhimurium* and *Listeria innocua* on grape tomatoes, apples, cantaloupe and romaine lettuce. *Food Microbiology*, *87*, 103391.https://doi.org/10.1016/j.fm.2019.103391.
69. Souza, V. R., Illera, A. E., Keener, K. M. (2020). Plasma Sources. *Innovative Food Processing Technologies: A Comprehensive Review*, pp. 1–8, 1st Edition, ISBN: 9781315271217. https://doi.org/10.1201/b21952-6.
70. Sudarsan, A., Keener, K. M. (2022). Inactivation of spoilage organisms on baby spinach leaves using high voltage atmospheric cold plasma (HVACP) and assessment of quality. *Innovative Food Science and Emerging Technologies*, *79*, 103023. https://doi.org/10.1016/j.ifset.2022.103023.
71. Sudarsan, A., Keener, K. M. (2022). Inactivation of *Salmonella enterica* serovars and *Escherichia coli* O157 surrogate from baby spinach leaves using high voltage atmospheric cold plasma (HVACP). *University of Guelph*, Volume 155, Article ID: 112903. https://doi.org/10.1016/j.lwt.2021.112903.
72. Surowsky, B., Schlüter, O., Knorr, D. (2015). Interactions of non-thermal atmospheric pressure plasma with solid and liquid food systems: A review. *Food Engineering Reviews*, *7*(2), 82–108.
73. Tabibian, S. A., Labbafi, M., Askari, G. H., Rezaeinezhad, A. R., Ghomi, H. (2020). Effect of gliding arc discharge plasma pretreatment on drying kinetic, energy

consumption and physico-chemical properties of saffron (*Crocus sativus* L.). *Journal of Food Engineering, 270*, 109766. https://doi.org/10.1016/j.jfoodeng.2019.109766.
74. Theron, M. M., Lues, J. F. R. (2007). Organic acids and meat preservation: A review. *Food Reviews International, 23*(2), 141–158.
75. Topfl, S., Heinz, V. (2007). Application of pulsed electric fields to improve mass transfer in dry cured meat products. *Fleischwirtschaft International Journal of Meat Production and Meat Processing, 22*(1), 62–64.
76. Uhm, H. S., Hong, Y. C., Shin, D. H. (2006). A Microwave Plasma Torch and Its Applications, *Plasma Sources Science and Technology, 15*(2), S26–F29.
77. Umair, M., Jabbar, S., Nasiru, M. M., Senan, A. M., Zhuang, H., Zhang, J. (2020). Sequential application of high-voltage electric field cold plasma treatment and acid blanching improves the quality of fresh carrot juice (*Daucus carota* L.). *Journal of Agricultural and Food Chemistry, 68*(51), 15311–15318.
78. Vidal, A., Sanchis, V., Ramos, A. J., Marín, S. (2015). Thermal stability and kinetics of degradation of deoxynivalenol, deoxynivalenol conjugates and ochratoxin A during baking of wheat bakery products. *Food Chemistry, 178*, 276–286.
79. Wang, J., Wu, Z. (2022). Combined use of ultrasound-assisted washing with in-package atmospheric cold plasma processing as a novel non-thermal hurdle technology for ready-to-eat blueberry disinfection. *Ultrasonics Sonochemistry, 84*, 105960. https://doi.org/10.1016/j.ultsonch.2022.105960.
80. Wińska, K., Mączka, W., Łyczko, J., Grabarczyk, M., Czubaszek, A., Szumny, A. (2019). Essential oils as antimicrobial agents—myth or real alternative? *Molecules, 24*(11), 2130. https://doi.org/10.3390/molecules24112130.
81. Xiang, Q., Wang, W., Zhao, D., Niu, L., Li, K., Bai, Y. (2019). Synergistic inactivation of Escherichia coli O157 by plasma-activated water and mild heat. *Food Control, 106*, 106741. https://doi.org/10.1016/j.foodcont.2019.106741.
82. Xiang, Q., Zhang, R., Fan, L., Ma, Y., Wu, D., et al. (2020). Microbial inactivation and quality of grapes treated by plasma-activated water combined with mild heat. *LWT, 126*, 109336. https://doi.org/10.1016/j.lwt.2020.109336.
83. Yadav, B., Roopesh, M. S. (2022). Synergistically enhanced *Salmonella typhimurium* reduction by sequential treatment of organic acids and atmospheric cold plasma and the mechanism study. *Food Microbiology, 104*, 103976. https://doi.org/10.1016/j.fm.2021.103976.
84. Yoo, J. H., Baek, K. H., Heo, Y. S., Yong, H. I., Jo, C. (2021). Synergistic bactericidal effect of clove oil and encapsulated atmospheric pressure plasma against *Escherichia coli* O157 and *Staphylococcus aureus* and its mechanism of action. *Food Microbiology, 93*, 103611. https://doi.org/10.1016/j.fm.2020.103611.
85. Yu, J., Chen, G. Y., Zhang, Y. F., Zheng, X. C., Jiang, P. (2021). Enhanced hydration properties and antioxidant activity of peanut protein by covalently binding with sesbania gum via cold plasma treatment. *Innovative Food Science & Emerging Technologies, 68*, 102632. https://doi.org/10.1016/j.ifset.2021.102632.
86. Yu, J., Wang, G., Wang, X. (2018). Improving the freeze-thaw stability of soy protein emulsions via combining limited hydrolysis and Maillard-induced glycation. *LWT, 91*, 63–69.
87. Zhang, X. L., Zhong, C. S., Mujumdar, A. S., Yang, X. H., Deng, L. Z. (2019). Cold plasma pretreatment enhances drying kinetics and quality attributes of chili pepper (*Capsicum annuum* L.). *Journal of Food Engineering, 241*, 51–57.

88. Zhang, Z. H., Wang, L. H., Zeng, X. A., Han, Z., Brennan, C. S. (2019). Non-thermal technologies and its current and future application in the food industry: a review. *International Journal of Food Science & Technology*, *54*(1), 1–13.
89. Zhao, Y. M., Oliveira, M., Burgess, C. M., Cropotova, J., Rustad, T., et al. (2021). Combined effects of ultrasound, plasma-activated water, and peracetic acid on decontamination of mackerel fillets. *LWT*, *150*, 111957. https://doi.org/10.1016/j.lwt.2021.111957.

CHAPTER 12

NOVEL AND ADVANCED TRENDS IN COLD PLASMA TECHNOLOGY FOR THE FOOD INDUSTRY

G. NAGAMANIAMMAI, R. SMRUTHI, NISAR S. NIKITHA, and S. NIRANJAN

ABSTRACT

Nowadays, people are looking for sophisticated, reliable, and healthy food products that must be sterilized to reduce chemical alterations in food and extend the product's shelf life. In the past, thermal processing such as pasteurization and sterilization was used to reduce microbial growth, but there were many losses in quality and nutritional attributes. Non-thermal preservation technology plays a powerful and significant role in preventing microbial deterioration. Cold plasma (CP) technology is an emerging non-thermal plasma technology used in the food sector to inactivate microbes by utilizing ionized gas. This ionized gas interacts with biological materials and contributes to the fatality of those microbes. Cold plasma technology is considered an emerging trend in food preservation technology because it is energy-saving and easy to use. Cold plasma can prevent the growth of spoilage microbes while retaining the physicochemical properties of food products. This chapter discusses the interaction of plasma species in food, the impact of cold plasma on the quality and safety of various food products, the combination of plasma technology with other emerging technologies, as well as the advantages and disadvantages of cold plasma, applications of cold plasma in the food sector, and scale-up challenges.

Nonthermal Plasma Technology for Food and Food Products.
R. Mahendran, Preeti Birwal, and Megh R. Goyal (Eds.)

12.1 INTRODUCTION

Thermal processing is a coalescence of temperature and time required to preserve food from spoilage microorganisms. There are many traditional and novel, cutting-edge thermal processes commonly utilized in food industries, as mentioned in Figure 12.1, such as sterilization, pasteurization, ohmic heating, radio frequency heating, etc. Regardless of various advantages in thermal processing, there are many limitations like damage in texture, color, loss of nutritional compounds, overcooking, etc. These limitations compromise the food quality and increase the customer demand for high-quality and safe food [48]. The development of new disruptive technology or non-thermal technology has been studied mainly to slow down the thermal effect on food, decontaminate any microbes on food products and instruments present on the surface, and infection control [64]. The alternatives to non-thermal/cold plasma processing (Figure 12.1) are ionizing radiation, high hydrostatic pressure (HHP), cold plasma technology, and ohmic heating [21]. Non-thermal processing technology can preserve taste, odor, and flavor and extend microbial food safety without any loss. The alternative new generation non-thermal preservation technology is plasma technology [14].

Plasma technology is developed for bonding and curing polymers, mainly using ionized gas for sterilization. Cold Plasma is an emerging technology, relatively used in the food industry to inactivate bacteria, spores, and viruses. Cold plasma can be split into two classifications based on the pressure factor: low and high-pressure plasma. The low-pressure plasma generates low pressure or vacuum and inactivates the microbes. The high-pressure plasma operates at radio frequency, creating ionizations that use fast electrical stimulations at different levels [61].

The primary principle behind cold plasma technology is the fragmentary ionization of charged ions. Cations or anions and other charged particles such as electrons and free radicals are discharged from several gases such as helium, argon, oxygen, and atmospheric air, and photons are some of them in use. Helium is readily ionized among all the feed gases (Figure 12.2) as it requires low electrical consumption [27, 54].

In the cold plasma process, the food, dielectric barrier, and voltage play a significant part in the food industry. The food acts as an insulating material; a dielectric barrier such as quartz allows the voltage to pass through the food inside the container. The sensitive species present in the food interact with different chemical reactions and thus impart food composition [5].

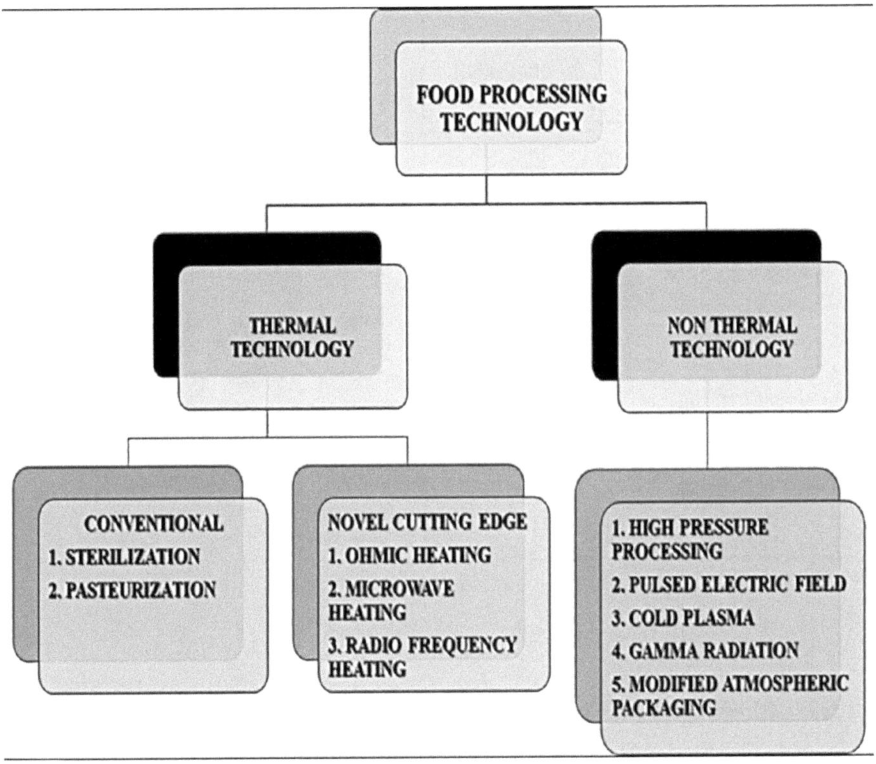

FIGURE 12.1 Flowchart for food processing technology.

This technology in the food industry has many pros and cons. The techniques are mainly employed in processing food packaging materials to increase obstruction properties and improve antimicrobial activity [62]. Cold plasma has many research outputs but has not been completely explored in the food industry. One of the latest research projects showed that cold plasma applications could be expanded in vegetable oils processing by hydrogenation. This process yields trans-free edible oil by inactivating antinutritional factors and preventing the invasion of a food allergen [84].

In this chapter, authors have discussed the interaction between plasma species and food, and the sequel of cold plasma on the qualitative and quantitative effects of food products. Also, combining this technology with other emerging innovative technologies highlights the challenges of cold plasma, current applications of cold plasma, and scale-up challenges.

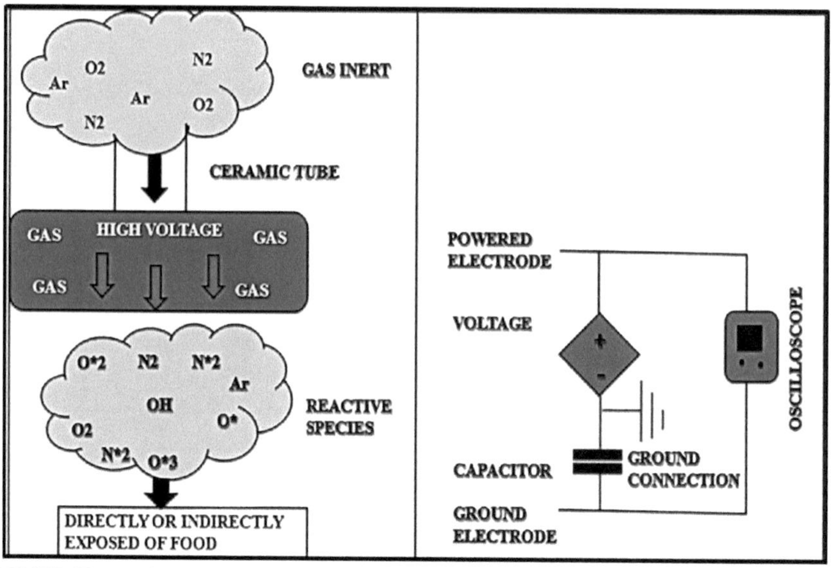

FIGURE 12.2 Physical description of cold plasma set-up.

Source: Adapted from Ref. [27,54]

12.2 PLASMA CHEMISTRY

The gas ionization led to the evolution of reactive species that strongly relate to the gas passed into the plasma system. The principle behind this process is that when air acts as a carrier gas, the oxygen and nitrogen mixture is set up, including the ozone and NOx. The reactive species interact with the food matrix, challenging its actions due to its synergistic effect on the food component [60]. The versatility of cold plasma is that it offers unique designs concerning the food industry equipment. The most commonly used devices are dielectric barrier discharge and jet plasma. The dielectric barrier system consists of two metallic electrodes, where one electrode has been completely covered with the dielectric barrier. The jet plasma consists of two electrodes (concentric) in which the high ionization is caused by connecting the inner electrode with high radio frequency and just looks like a jet-like appearance [11, 20, 79].

Based on the varying temperature and pressure, plasma is categorized as thermal and non-thermal plasma. Thermal plasma needs extreme pressures for its propagation and uniform temperature for all the constituents. Non-thermal plasma has low pressure and power and non-uniform thermal equilibrium, also known as non-equilibrium plasma [73].

12.2.1 DESIGN OF PLASMA

The atmospheric cold plasma treatments are nowadays concerned with establishing the process condition which can minimize the damage to nutritional and sensorial quality of food and increase the microbial decontamination. There are many issues addressed in designing cold plasma sources for food [37].

- The first issue in designing the cold plasma sources was air plasmas. A simple air plasma can emit 75 different and unique chemical species and involve more than 500 simultaneous reactions.
- The next issue in this plasma technology is standardizing and stabilizing the plasma analytics. To harmonize the research and accelerate the development of plasma sources in the food and medical industries.
- The final issue in designing the cold plasma source is that these processes must not affect the safety and health issues in the work environment.

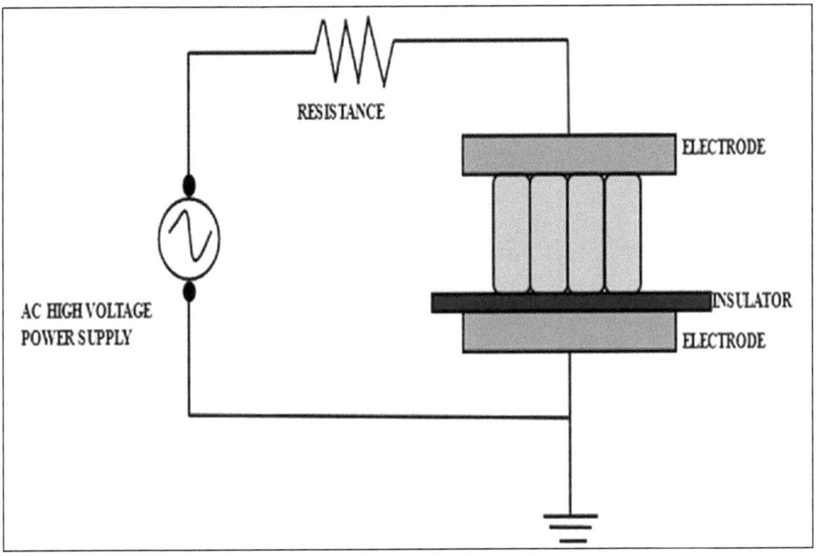

FIGURE 12.3 Schematic diagram of dielectric barrier discharge.

Source: Reprinted with permission from Ref. [16]. Copyright 2017. American Chemical Society.

12.2.2 SOURCES OF PLASMA

The plasma technology brought innovations in the plasma sources, which can be operated at different atmospheric pressures. The plasma source comprises dielectric barrier discharges (DBD), gliding arc discharges (GAD), corona discharges (CD), and radiofrequency discharges (RFD).

12.2.2.1 DIELECTRIC BARRIER DISCHARGES

The dielectric barrier discharge plasma sources are initiated by an alternating current (A/C) released by two electrodes (made of metals). They are separated using some dielectric barrier materials such as quartz or plastics at a gap distance varying up to several centimeters (Figure 12.3). These dielectrics will form sparks due to the charge fluctuation. The main applications of these sources are ozone generation lamps, UV generation in excimer lamps, and CO_2 lasers [16].

12.2.2.2 GLIDING ARC DISCHARGES

The gliding arc releases are framed in a reactor containing at least two separating metallic electrodes, operated at a high potential difference in optimum conditions. The principle behind this discharge is that the inlet gas with humid air is discharged into the release hole between the electrodes, which leads to an arc formation (Figure 12.4). GAD usually forms both thermal and non-thermal plasma. This source has been utilized to degrade chemical contaminants and bacterial decontamination [16].

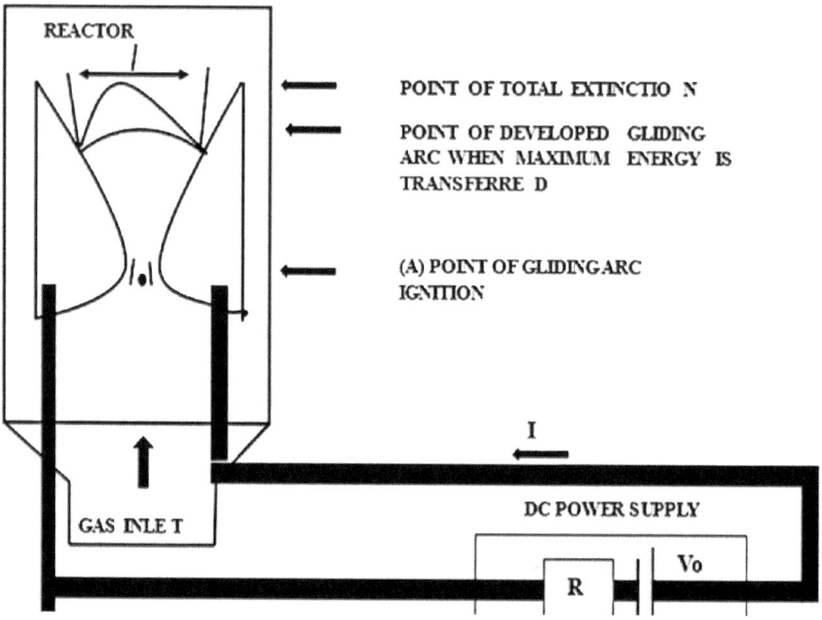

FIGURE 12.4 Schematic diagram of gliding arc discharge.

Source: Reprinted with permission from Ref. [17]. Copyright 2022. Elsevier.

12.2.2.3 CORONA DISCHARGE PLASMA

The most crucial plasma source for plasma ignition is created around sharp-pointed electrodes. These electrodes contain substantial electric fields for accelerating the ionizing energy of the electrons formed under the arbitrary state of any gas or molecules [73]. These are mainly generated at high voltage, occurring only on one electrode (Figure 12.5). These discharges are used for industrial applications such as microbial decontamination and electro-precipitation.

12.3 INTERACTION OF PLASMA WITH FOOD

The composition of cold plasma is very intricate. Many reactive agents play an important role in inactivating targeted microbes; most bacteria, especially anaerobes, are susceptible to ROS [77]. The diffusion of radicals which contain oxygen, through their cell wall results in local damage to the DNA strands and protein-membrane [29]. The charged particles bombard the cell wall and break the chemical bonds, which causes erosion. The creation of openings and lesions in the membrane can cause the penetration of toxic compounds into the bacteria [29]. Despite much research in cold plasma on antimicrobial activity, this technology must explain the mechanisms and mode of delivery must be optimized to give an effective substitute to microbial reduction.

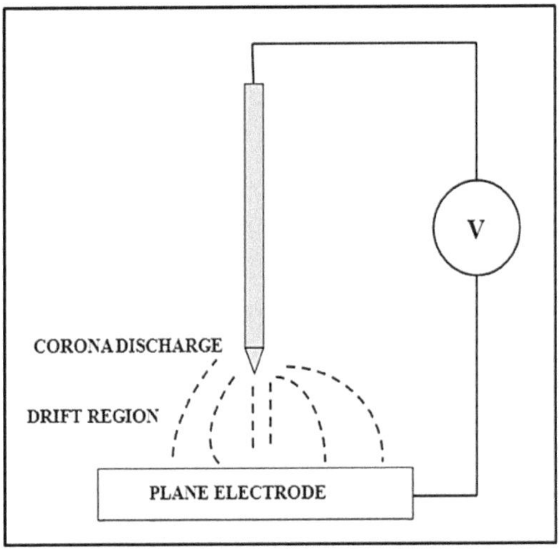

FIGURE 12.5 Schematic diagram of corona discharge plasma.

Source: Adapted from Ref. [73]

12.3.1 SAFETY AND QUALITY ASPECTS OF FOOD BY COLD PLASMA

For decades, surface modifications in the packaging industry have used cold plasma technology. However, food products are exposed to high and robust electric fields during processing and subjected to various reactive gas species. This process affects the chemical and physical attributes of food. The results of cold plasma on different foods are displayed in Table 12.1.

12.3.2 MICROBIAL INACTIVATION

12.3.2.1 BACTERIAL INACTIVATION

The vegetative and bacterial spores showed various elements, including process parameters, microbial properties, and environmental conditions, and microbial properties based on their inactivation efficiency. A higher frequency of input voltage has been analogous to an excessive increase in the inactivation potential of cold plasma. Recent research papers reported that both gram-negative and gram-positive bacteria are inactivated by cold plasma resulting from plasma intensities that were competent enough to destroy the microorganisms [36].

12.3.2.2 FUNGAL INACTIVATION

In eukaryotic cells, the inactivation becomes less intense when CAP is applied than in prokaryotic cells. The structural complexity of eukaryotic cells protects them from the stress caused by CAP. Fungi are the most common eukaryotes that can taint food in different mediums like air, soil, and water. Fungi are mainly classified into three categories based on their shape, size, morphology, and mobility: single-celled microscopic yeasts, multicellular microscopic molds, and filamentous mushrooms [32]. Gaunt et al have reported that the inactivation led to the breach in the cell wall against the fungal spore caused by RONS and charged particles [31].

12.4 BIOACTIVE COMPOUNDS

Vitamins and phenolic compounds have significant scope in the food industry.

TABLE 12.1 Outcomes of Plasma on Different Food Products.

Product Name	Plasma Used	Observations on the Quality Attributes	Microbial Count
Black gram	• Plasma: Radiofrequency plasma barrier discharge • Feed gas rate: The pressure of 2 Pa, air, the frequency at the rate of ~13.56 MHz, 30–50 W, 5–15 min	• Black gram showed impression and hydrophilization of surface • There was a reduction in the hardness, moisture content, ash content, and cooking time	N/A [70]
Cherry tomato	• Plasma: DBD • Feed gas rate: 100 kV, 150 s, air	• Color, firmness pH or TSS were observed, and there were no significant differences	• Sample reduction of >5 and 3.5 $\log_{10}$ cfu/ in *E. coli* and *Listeria innocua* [86]
Fresh mackerel filets	• Plasma: DBD • Feed gas rate: 70–80 kV, frequency of ~50 Hz, air, 1–5 min	• There is an increase in polyunsaturated fatty acids such as oleic and eicosapentaenoic acid in plasma-treated samples • TBARS values were observed with no significant differences	• The total aerobic mesophilic count showed no depletion. • There was a significant reduction in psychrotrophic LAB, other bacterial species, *Pseudomonas* [1]
Golden apples	• Plasma: Gliding arc • Feed gas rate: The frequency at 60 Hz, the gas used is air, and the time interval is 1–3 min	• It has been notified that the texture and color of the apples were not changed	• There was 3.5 $\log_{10}$ depletion in *Salmonella* and *E. coli* O157:H7 [61]
Orange juice	• Plasma: Dielectric barrier discharge (DBD) • Feed gas rate: Air/MA65, 90 kV, 30–120s	• Brix or pH has no significant change • Reduced Vitamin C by 22% in air	• Up to 5 $\log_{10}$ reduction of *Salmonella enterica* [83]
Pork	• Plasma: Pulsed plasma • Feed gas rate: ~ 0.8 MPa, 20–100 kHz, 1.2 kVa, N_2, He, Ar	• Color and pH have no significant differences.	• Up to 3 $\log_{10}$ cfu/cm^3 reduction of psychrotrophic bacteria, yeast, and mold [81]
Radish sprouts	• Plasma: Microwave plasma • Feed gas rate: 2.45 GHz, 900 W, 669 Pa, 1–20 min, N_2, 1 L/min	• There have been less significant changes in color, water activity, antioxidant activity, and vitamin c concentration. • Moisture content was low.	• There was a reduction of 2.6 $\log_{10}$ in *Salmonella typhimurium* • And also 0.8 $\log_{10}$ reduction in mesophilic aerobes [62]
Refined wheat flour	• Plasma: DBD plasma • Feed gas rate: 1–2.5 kV, 50 Hz, 1–5 min	• Refined wheat flour showed no depletion in the color	• *Tribolium castaneum* showed increased growth [67]

12.4.1 ASCORBIC ACID

Ascorbic acid is also referred to as vitamin C. It is usually used as a functional ingredient in food. Ascorbic acid results in the reduction of some free radicals and ROS, thus damaging the DNA and lipids. Mehta et al. compared the effects of different thermal treatments with non-thermal treatments, such as ultraviolet, ultrasonication, and atmospheric cold plasma, on 100 ml of a beverage based on tomato [50]. Atmospheric cold plasma technology showed the highest retention of 95% ascorbic acid. The loss of ascorbic acid may be due to the generation of free radicals or ROS during cold plasma processing.

12.4.2 PHENOLIC COMPOUNDS

Phenolic compounds are the major bioactive compounds in fresh foods. Cold plasma has been observed to be beneficial in improving functional activity. This treatment also helps in maintaining the antioxidant concentration in food [48]. Illera et al. [34] observed a significant increase in the total phenolic content with an increase in the processing time of cold plasma. However, during the 28-day storage period, there was a decrease in the total phenolic content. Research has shown that the increase in phenolic content may be due to the breakdown of cell membranes by reactive plasma species.

12.5 PHYSICAL AND CHEMICAL PROPERTIES

12.5.1 TSS, pH, and TITRATABLE ACIDITY

Recent literature has indicated that acidity, pH, and total soluble solids (TSS) showed minor changes when cold plasma was applied to orange juice. Almeida et al. [2] reported that the low pH value of prebiotic orange juice was observed after DBD treatment. According to Pankaj et al. [65], the titratable acidity of grape juice was significantly reduced after exposure to high-voltage atmospheric cold plasma at 80 kV for 4 minutes.

12.5.2 COLOR

Nowadays, studies express a noticeable color change after plasma treatment. There is a visual difference in the color when the total color difference value is 2 [17]. Amini et al have observed that the loss in saffron quality is due to the increase in voltage input and additional oxygen in the working gas [6].

12.5.3 TEXTURE

It has been observed that there are no significant changes in the texture retention of food products like vegetables and fruits after cold plasma treatment. In grains and legumes, the CP treatment decreased the hardness and chewiness. Plasma-treated food products decreased soaking/cooking time [71].

12.6 SURFACE MODIFICATIONS ON POLYMERS

In polymers, the development of any relevant technology, such as polymer processing and surface treatments, should impart modification without inducing any undesirable changes [82].

12.6.1 CRYSTAL STRUCTURE

Researchers found that the cold plasma treatment has changed the crystal structures in the polymer. The fluctuations in energy source, plasma gas, discharge systems, and voltage levels make it hard to draw definite conclusions. The plasma technology is found to be leading to the degradation of polymers that are treated, which causes the creation of radical chains and alters polymer crystallinity [8].

12.6.2 BARRIER PROPERTIES

The polymer modification technology or in-package food processing methods do not severely affect the packaging materials' barrier properties. In a recent literature survey, treatment of the plasma with gases such as oxygen, carbon dioxide, and air has less effect on the polymer's oxygen and water vapor permeability [65]. Tenn et al. [80] investigated that the films coated with carbon tetrafluoride and tetramethyl silane gave rise to the hydrophobicity of the films that are treated and reduced the permeability of water on the surface of the films.

12.7 FACTORS AFFECTING COLD PLASMA TECHNOLOGY

12.7.1 INTRINSIC AND EXTRINSIC ATTRIBUTES

The internal characteristics of microorganisms also influence CP efficiency. The inherent properties of microorganisms play an essential role in microbial decontamination of food products since sensitivity may vary within a species or strain. According to Chizoba Ekezie et al. [16], bacteria in the stationary phase generally exhibit greater inactivation sensitivity than those in the growth phase. In contrast, bacteria that are gram-positive exhibit lower plasma resistance than bacteria that are gram-negative. This is due to the peptidoglycan thickness and features of their lipopolysaccharide membranes. Spores of bacteria are more resistant to plasma treatment, which reduces their ability to penetrate reactive species. Spores are clustered together and are more resistant to plasma treatment than vegetative cells. The intricate rigidity of fungi's cell walls makes them tolerable to plasma treatment, as their walls are composed principally of chitin [15].

12.7.2 PROCESSING VARIABLES

The effectiveness of cold plasma is highly dependent on a few factors. They are majorly influenced by the fact that different plasmas and the methods used for their induction have their characteristics. As an example, the process gas which is employed primarily determines the number and type of reactive species created in the discharge and the efficiency of the procedures involved in treating these discharges. Similarly, the energy density increases with increased frequency and input voltage associated with the active species generated. In addition, the mode of exposure is a process variable affecting CP efficiency. Direct exposure is preferable to indirect or remote exposure due to the cross-relaxation nature of charged particles and their ability to rejoin before reaching the matrix [16].

12.7.3 MILIEU FACTOR

The efficiency of the CP treatment is highly influenced by environmental parameters such as pH, relative humidity, and sample type. A solid food matrix varies with reactive species from a liquid food matrix, as liquids can vaporize at the time of treatment and participate in reactions that are ensuing.

Muranyi et al. [59] showed that foods with a lower pH are more reactive to stress, heat, and other environmental factors. They found that when plasma treatment was completed at pH 5, Bacillus cereus was reduced by 4.7 logs, as opposed to 2.1 logs at pH 7. Similarly, as the composition of hydroxyl radical changes, increased relative humidity accelerates the inactivation of microorganisms. Matrix properties also influence process efficiency. Microbes can be effectively decontaminated on an agar plate or filter membrane, but not with a surface of a fruit or even a cheese slice, due to a definite rate of migration of microbes from the external to inner tissues [16].

12.8 MERGING OF COLD PLASMA WITH DIFFERENT EMERGENT TECHNOLOGIES

Emerging trends in cold plasma technology have allowed combining different technology processes to produce practical applications and have been listed in Table 12.2.

12.9 APPLICATIONS OF COLD PLASMA ON POLYMERS

Plasma technology offers many applications, including remediation of gaseous pollutants, treatment of volatile organic compounds, lasers, reduction of greenhouse gases, excimer lamps, and display panels. Cold plasma and its utilization in packaging traditionally focus on modifications of the polymer surface. However, more recently, the applications have expanded into innovative food product in-package plasma decontamination [19, 66].

Cold plasma (CP) causes a number of physiochemical changes to the characteristics of the plasma surface and the plasma polymer interface. This phenomenon is used to functionalize the surface to impart the surface energies for package polymer to enhance adhesion, printability, sealability, anti-mist characteristics, and boost resistance to prevent mechanical failure. Cold plasma can quickly sterilize the packaging materials, such as lids, films, etc., without compromising their bulk qualities or leaving residues [64]. This section discusses the physical and chemical aspects of cold plasmas. This study discusses the present condition of cold plasma in varying the surface polymers and also surface sterilization of polymers relevant to food packaging.

TABLE 12.2 Combinations of Different Technologies with Cold Plasma Technology

Application	Target Materials	Combination Method	Process Conditions	Observations
Active packaging using nanotechnology	Phlorotannin/ *Momordica charantia* polysaccharide-based nanofiber	The CP reaction chamber was processed in a nanofiber membrane with the power of 350 W	Nitrogen was applied at 100 (standard cubic cm)/ min of flowrate	The radical scavenging activity increased from 80.64% to 91.74%. Improvement of the nanofiber structures. The antibacterial and antioxidant activity in the nanofiber has been enhanced [23]
Antimicrobial packaging using nanotechnology	Polypropylene/ Low density polyethylene (BOPP/LDP) laminates	The pretreatment of 80 × 120 mm films with the DBD system. Then nano fibrillated coating solution (cellulose/nisin) was tested along the CP treated BOPP/LDPE	The pressure was applied at 2 minutes interval using an O_2/ air mixture at 200 ml/min of flowrate.	The CP tempered films have improved the attachment of coatings due to increased surface hydrophobicity. The concentration of nisin increased to 0.005 g/g, causing a 94% decline in *Listeria monocytogenes* [45]
Antimicrobial packaging using nanotechnology	Nanofibers made from thyme silk/oil fibrion (TO/SF) against *S. typhimurium*	The exposure of TO/SF in the plasma along with nanofiber membrane at the power of 400 W	Nitrogen was applied at the rate of 100 cubic cm)/ min at 4 min interval	*S. typhimurium* in chicken and duck meat at the temperature of 25°C declined to ~ 6 log CFU/g [43]
Modification of structure using nanotechnology	Starch nanoparticle (SNP)	The Starch nanoparticles were treated with a vacuum cold plasma system of 30–50 kPa pressure and power of 1 kW.	O_2 and NH_3 were repetitively used three times in the intervals of 3, 5, or 7 min, and a flow rate of 300 mL/min was applied	There is better dispersibility of O_2 than NH_3 High polyphenols in tea were observed There was a decrease in relative crystallinity from 44.13% to 33.80% [12]
Nanotechnology for the for biofilm inhibition	Electrospun Nanofibers made of chitosan	The process starts with the pretreatment of chitosan powders in a free rotation glass chamber and dissolved in acetic acid. Then, the powders were electrospun with polyvinyl alcohol.	Time interval of 30 min and 13.56 MHz power was given 5–6 × 10^2 Torr 50 W energy	The polyvinyl alcohol chitosan furan nanofiber showed the highest prevention value of about 80.4% [49]

TABLE 12.2 *(Continued)*

Application	Target Materials	Combination Method	Process Conditions	Observations
Nanotechnology forstructure modification	Grass Pea Protein Nano Particle	0.3 g of GPPIs were dispersed over the surface of the electrode as a thin layer and exposed directly to the DBD plasma.	The time interval between 5 min and 10 min and feed gas rate of air (RH 42 ± 5%) at a temperature of 25 ± 2°C	A solid interfacial layer is formed due to complex tertiary structure and disulfide bonds. The size of NP decreases the non-uniformity, and hydrophobicity in the morphology is increased due to voltage and time [47]
Quality Improvement using Ultrasound	Italian salami applied with sea asparagus and *Salicornia neei* extract	400 mg of dried *Salicornia neei* were preprocessed with an AC glow discharge plasma, and extraction assisted by ultrasound was done.	Time interval: 5–60 minutes Feed gas rate: 0–15 kV atmospheric air 60 Hz 2×10^{-2} bar pressure and 14 W	*Salicornia neei* showed scarce lipid oxidation and texture deterioration. The sensory quality inhibited against *E. coli* has a negative impact [26]
Quality Improvement using Pulsed Light	Pulse light treatment in powdered red pepper	Treatment of red pepper powder was performed by amalgamating cold plasma and intense pulse light treatment	Plasma system rate: 0–10 kV voltage ~ 6,400–40,000 Hz He-PL system rate: Input 0.22 kV voltage AC supply source at 26,000–30,000 V voltage and 0.38 μF capacitor	Treatment by plasma reduced the aerobic plasma at the time interval of 1, 3, and 5 min and 0.9 logs CFU/g Capsaicin showed no adverse effects on vitamin C, color, and antioxidant activity [40]

TABLE 12.2 *(Continued)*

Application	Target Materials	Combination Method	Process Conditions	Observations
Quality improvement using PEF treatment	PEF and cold plasma treatment of shrimp (*Litopenaeus vannamei*)	Shrimps were conditioned with PEF by immersing them in 1–2% chamuang leaf extract solution for 30 minutes. Then the samples were packed with MAP along with Ar/Air mixture and then applied with high voltage cold atmospheric plasma.	The process condition includes transformer input voltage of 230 V, 0–120 kV output voltage, and 50 Hz electrodes at the time interval of 10 mins.	The longevity of the product increased by 9 days The microbial spoilage and count in the meat by PEF treated with 2% CLE was low [76]
Quality Improvement with VI and PEF	PWS treated with PEF and VI along with CP	The pacific white shrimp samples were treated earlier with PEF and CLE solutions. Then for 15 minutes, vacuum impregnation was conducted under a vacuum pressure of 5 kPa. The samples were packed using MAP and vanquished with HVCAP using the DBD system.	The conditions include a time interval of 10 min and a voltage of 16 kV at atmospheric pressure.	This combined process of PEF and VI delayed the growth of spoilage microbes such as *Entero bactericeae*, *TVC*, *LAB*, etc. [75]
Quality Improvement using Ultrasound	Ultrasound treatment for *S. aureus* and *E. coli* in the muscle and skin of the chicken.	Pretreatment of chicken tissues was done by soaking them in PAW. The PAW can be discharged below the water at a distance of 5 mm between the ground and the electrode. Ultrasonication was performed at 4, 25, 60°C at a time interval of 30, 45, and 60 mins	Argon plasma jet worked at a flow rate of 3 s/m at 6.5 mins. The ultrasonication operation was conducted at 40 Hz and outpower power of 220 W	Both *S. aureus* (0.83 log CFU/ml) and *E. coli* (1.33 log CFU/ml) with a sample thickness of 4 mm were inactivated at 40°C for 60 min. The tissue samples' protein, hardness, and lipid values were maintained. [69]

TABLE 12.2 *(Continued)*

Application	Target Materials	Combination Method	Process Conditions	Observations
Quality Improvement using IPL – ultraviolet	Rice germ was treated with IPL-ultrasound along with cold plasma	There were 3 main operations performed; 1. Atmospheric pressure plasma 2. IPL 3. UV-C	The process conditions for APPJ: 7.4 kW Air compressor, pressure: 800 kPa, 14 L/h air production) IPL: 3,000 V voltage, 2 Hz UV- C: 5 lamps, 8 Watts each with a distance of 5 cm, size: 253.7 nm	The bacteria, mold, and yeast were reduced by 1.0 and 2.0 log CFU/g [41]
Ultrasound for quality improvement	Vegetable juice has been treated with cuminaldehyde/β-cyclodextrin inclusion complex (CUM/βCD-IC)	Preparation of inclusion complex was performed by ultrasound technique. The powders were treated with cold nitrogen plasma	Nitrogen flow rate: 100 cubic cm/min Time: 6 mins Power: 50 W	The combination of cold plasma with ultrasound revealed some antibacterial and antioxidant activity effects against *E. coli* in tomato juice (2.58 log CFU/ml) and cucumber juice (2.51 log CFU/ml) for 3 days at 4°C. The vegetable juices did not show any color changes. [44]

12.9.1 ACTIVATION OF POLYMER SURFACE

Functionalizing polymeric materials and surface activation refer to physicochemical or biological changes in the polymer's surface that allow certain functional groups to be attached to the surface for desired functionality. Surface activation is commonly used to change or improve a surface's adherence and printing qualities. Surface activation with cold plasma showed polar or cross-linked molecules' formation by increasing the polymer's total surface free energy. The energy which is free can also produce etching on the surface due to particles that have energy and bombard the surface, such as ions (cations and anions), high radical particles, electrons (positive and negative), excited atoms, and in improving adhesion for inert, biocompatible materials. Chemical groups mainly interact with plasma species forming functional groups such as ester, carboxyl, hydroxyl, aldehyde, amino, carbonyl, and amide, increasing adhesion, wettability, and biocompatibility by changing the polymer characteristics from hydrophobic to hydrophilic surfaces, which can be challenging to achieve with traditional activation methods [66].

12.9.1.1 ADHESION

Surface bonding qualities can be changed using cold plasma without changing the physiochemical properties of the treated polymers [25]. Adhesion is the tendency of various surfaces to stick together due to intermolecular interface interaction. Plasma treatments are mainly used to change polymers' surface free energy (total) to increase or decrease anti-adhesion, depending on the application. In the case of food packaging, changes in adhesive qualities caused by plasma can be used to create composites of polymer with desirable water vapor or oxygen barrier characteristics.

12.9.1.2 PRINTING

Functionalization by atmospheric pressure plasma (APP) and surface activation (SA) allows the synthesis of transparent materials and coatings used in the manufacture of composite packaging of food. The significant problem in labeling jars, printing on glass containers, or package sealing is done to process materials consistently at a reasonable cost.

Impressions in the food packaging and pharmaceutical sectors vary (e.g., European Article Number (EAN) codes- or best-before dates), and such imprints must resist abrasion. While decorating the glass bottles or jars, this treatment can meet the needs for exact color matching and high pixel precision.

The surface contact angle is a very important and common technique used to look into surface energy and hydrophilicity (wettability) is related to the adhesion of ink on the surface [25]. Moreover, the most important factor for future applications is that surface wetting is driven by the size of the surface energy and free energy of adhesion in between the liquid and solid.

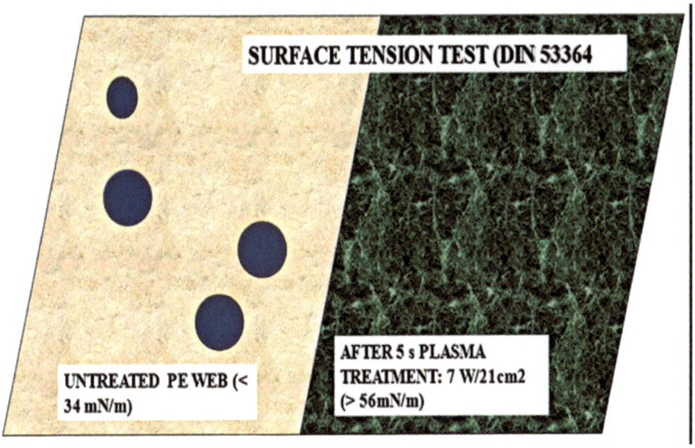

FIGURE 12.6 Effect of fused hollow cathode plasma exposure on spreadability of ink in a polyethylene web surface.

Source: Adapted from Ref. [11]

Effects of plasma treatment on printability, as shown by Bárdos et al. [11], exhibited a significant improvement in the dispersibility of ink on a polyethylene surface at a time interval of 5 seconds (Figure 12.6). This treatment has created abrasion-resistant imprints, high pixel precision in decorating the glass bottles or jars, accurate color matching, and homogenous coatings with great scratch resistance with no collateral negative effects [63, 66].

12.9.1.3 SURFACE STERILIZATION

The most important functions of food packaging are to preserve the food supplies enclosed within it. After the food has been processed, the packaging should protect it against contamination from the environment or the package. Because the intrinsic properties of the polymers do not enable microbial development; however, this is not always the case. As a result, surface sterilization is essential to eliminate any health concerns for customers and potential economic losses [54].

The regulatory guidelines, as per FSSAI, classify microbial specifications for food packaging materials and the packaging process [55]. However, some limitations led to the search for fresh approaches. Cold plasma technology has exhibited significant potential as an alternative to traditional surface sterilization procedures [63, 66].

Plasma-related food substances are used in the food packaging industry, including antifouling and printable surface sterilization, and a decrease in polymer permeability for carbon dioxide (CO_2) and oxygen (O_2). Gas plasma sterilization at low temperatures is a quick and safe method for packaging materials without changing the product qualities. Muranyi et al. [59] utilized polyethylene terephthalate (PET) as the medium for the treatment, resulting in the decrease of $2\log_{10}$ with just minor modifications to the packaging functionality. Table 12.3 provides a review of several significant research in plasma-aided sterilizing of packaging materials [66].

12.9.1.4 IN-PACKAGE DECONTAMINATION

Decontamination of in-package is one of the important and new applications of cold plasma in the food industry (Figure 12.7) [85]. Here, the plasma species are generated into the food materials that are prepackaged. The sensitive species that react will perform the sterilization of packages and food decontamination and return to its previous gas composition during the storage duration, preventing the contamination of post-processing [54, 66].

In-package plasma decontamination of food materials is based on the use of dielectric, which is the polymer package and has been researched using a variety of packing materials, including LDPE (low density), HDPE (high density), polystyrene (PS), polypropylene, and cellulosic polymers [66].

12.10 UTILIZATION OF COLD PLASMA IN THE FOOD INDUSTRY

Non-thermal preservation technology has vast and diverse scope in the processing and packaging industry. This technology is mainly used for surface sterilization in food products like cooked meat, cheese, and milk packing materials for microbial inactivation from raw materials [61]. Also, cold plasma treatment could change hydrophobic surfaces to hydrophilic surfaces or incorporate a hydrophilic coating on surfaces. The applications of cold plasma in various fields have been illustrated in Figure 12.8.

TABLE 12.3 Results of Different Plasma Processes for Wide Range of Microbes

Type of Polymer	Microbes	The Process Involved in Plasma	Results
Bottles and foils made from polyethylene terephthalate	Spores of *Bacillus atrophaeus*	The plasma utilized is low-pressure microwave plasma with a power of 1/42,000 W. The reactor consists of a gas mixture connected with hydrogen, nitrogen, and oxygen cylinders to mix gases in the ratio of 2:1:0.8 at a pressure of 1/430 Pa.	*Bacillus atrophaeus* showed a reduction of 6 $\log_{10}$ after 5 s exposure to plasma [24]
Foil made from polyethylene terephthalate	Spores of *Bacillus subtilis* and *Aspergillus Niger*	The process involves atmospheric pressure dielectric barrier system with the power of 1/47 W/cm^2 and 20–50 kHz frequency	There was a significant reduction of *Bacillus sps.* and *Aspergillus spores* of at least four orders of magnitude of plasma treatment within seconds [33]
Foil made of glass, polypropylene, polyethylene, paper and nylon	Microorganisms like *Salmonella typhimurium*, *Escherichia coli* O157:H7, and *Staphylococcus aureus*	The plasma utilized was low-pressure air plasma at a pressure of 1/40 Pa and 5–5.0 Torr power. The density of the plasma is 1/412.4–54.1 mW/cm^3	After 300s treatment with this plasma, there was a reduction of up to 4 log CFU [42]
Foils made of polyethylene terephthalate	Spore of *Bacillus subtilis*	The process starts with the use of low-pressure microwave driven plasma with the frequency of 1/42.45 GHz	*Bacillus subtilis* showed reduction up to 4 orders of magnitude with the treatment range of 0.5–4s. [72]
Foils of polyethylene terephthalate	Different microorganisms such as *Salmonella* serotype mons, *Escherichia coli*, and *Staphylococcus aureus*	The plasma uses a cascaded DBD system with the power input of excimer lamps was ¼, 130 W. The absorbance to detect microbes was 282 nm	Bacterial cells reduction of 5–6 log within 1–3s [59]
Nitro cellulose	*Bacillus stearothermophilus* endospores	The process involved was OAUGDP	After plasma treatment at 5.5 mins, there was 5 log CFU reduction [28]
Polypropylene	*E. coli* K12	The process involved was OAUGDP	Exposure of OAUGDP after 24 s, there was 5 log CFU reduction observed [28]
	P. aeruginosa	The process involved was OAUGDP	Exposure of OAUGDP after 22s, *P. aeruginosa* showed a reduction to 6 log CFU [28]
	S. aureus	The process involved was OAUGDP	The *S. aureua* showed depletion of 6 log CFU after plasma treatment exposure at a time interval of 22 s [28]

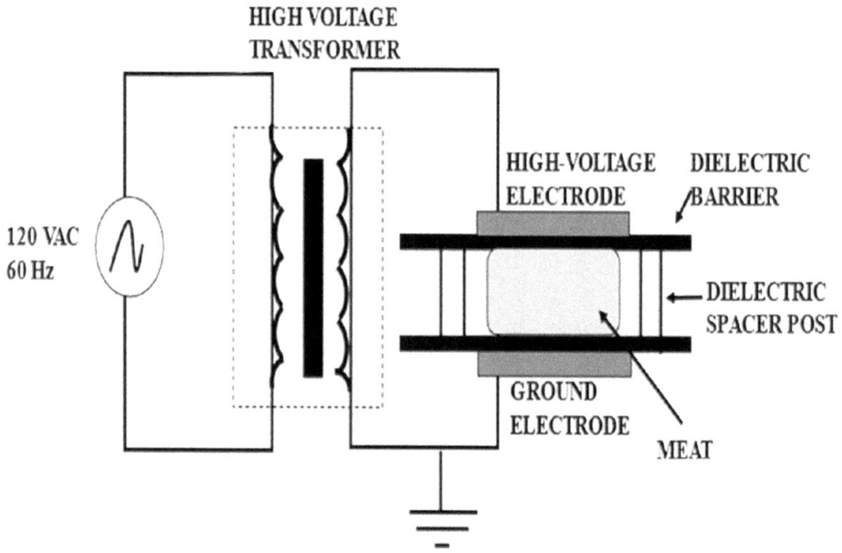

FIGURE 12.7 Schematic representation of in-package contamination of plasma treatment system.

Source: Adapted from Ref. [85]. Copyright © 2019 Hong Zhuang et al. https://creativecommons.org/licenses/by/4.0/

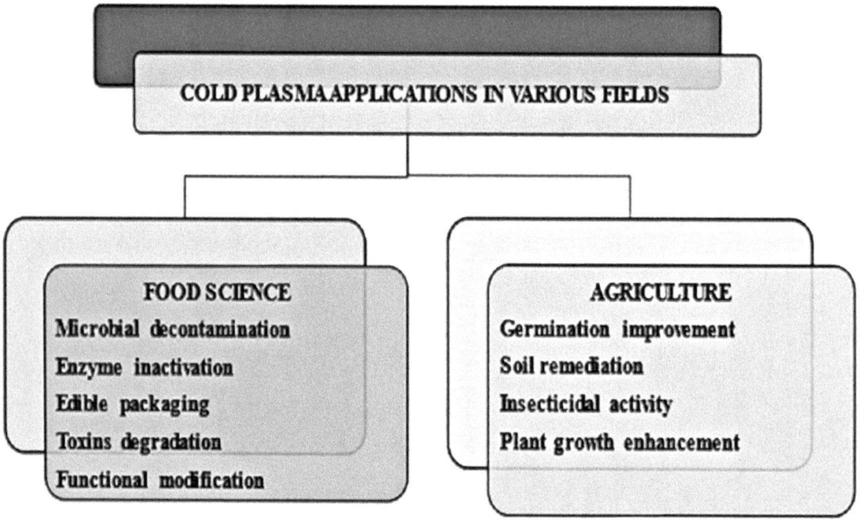

FIGURE 12.8 Applications of cold plasma in food technology.

12.10.1 COLD PLASMA IN FRUITS AND VEGETABLE PROCESSING

Orange juice was treated by atmospheric cold plasma for 15, 30, 45, and 60 seconds using two different types of exposure with quantitative Nuclear Magnetic Resonance (qNMR) combined with chemometrics [3]. Applications of different amino acids such as alanine, arginine, histidine, glutamic acid, proline, valine, and aromatic amino acids like phenylalanine, tryptophan, and tyrosine; organic acids such as citric, acetic, lactic, malic, GABA, formic, succinic; and disaccharides like sucrose, lactose, and monosaccharides like fructose, α-glucose, β-glucose, and other components like ethanol were evaluated thermometrically. There were no substantial changes in the orange juice, showing that the ACP is one of the important techniques that can be used for orange juice without causing major changes in its physiochemical composition [4].

In certain trials, the antioxidant activity in plasma-treated fruits remained unchanged [18]. According to reports, CAP treatment on fresh-cut apples reduces O_2 fixation while increasing CO_2 concentration in the gaseous environment, producing alterations in the cell respiratory pathway [78]. Endogenous enzymes in fresh fruit and vegetables are inactivated; as a result of this alteration, the enzymatic browning and maturation process is delayed [30].

12.10.2 COLD PLASMA IN CEREALS AND GRAINS

Starch is an important food component with drawbacks, such as low solubility, stability due to retrogradation, non-stable viscosity, and low freezing/thawing stability. This process is used to change the functional properties of starches to improve the food additives' efficiency. The mechanisms for starch modification are cross-linking of starch granules and depolymerization that occurs at the amylopectin side chains between two starch molecules.

Roy et al. [68] explored the change of reaction in plasma technology on seed germination and quality. Plasma jet- DBD therapy processed wheat seed showed 20–30% more seedling growth compared to untreated samples. Low glow discharge plasmas applied to wheat seeds at medium pressure (10 Torrs) improved water absorption and provided 95–100% seed germination. Plasma-treated seed produced a plant with increased germinal growth, high dry matter, and high chlorophyll content in the leaves [30].

Misra et al. [51] depicted that the outcome of ACP on rheological properties such as elasticity, plasticity, and tensile strength between strong and weak flour. They found that enhanced dough elasticity, strength, and reduced mixing time improved the rheological properties of dough.

12.10.3 COLD PLASMA IN DAIRY PRODUCTS

Milk is extremely sensitive to various processing procedures due to its complicated structure. As a result, the modern dairy business is looking for novel ways to retain functional dairy proteins/peptides, lactose, lipids, and minerals in order to develop new milk products. Pasteurization and sterilization are the most common techniques employed to enhance the product's safety and extend the lifespan of dairy products. However, this technique can result in protein denaturation, vitamin and mineral loss, and flavor compounds [10].

The physicochemical and textural qualities of milk are influenced by CAP treatment; the changes mentioned always depend on the time and flow rate (ml/min). For example, chocolate milk treated at a reduced flow rate (0.1666 ml/hr) for a 5-minute interval with N_2 plasma has a small particle size of 14–87 m, while operating at an intermediate flow rate (0.333 ml/h) for a 15-minute interval escalates the size of the particle to 17–103 m [22]. The creaminess of the milk is inversely proportional to the particle size. The presence of a higher number of tiny particles will make the milk creamier.

Segat et al. [74] explored the contact of APP with WPI model solutions throughout the range of time intervals (1–60 minutes). According to this study, this technology can alter the structure of proteins, increasing their functional properties and accessibility. It was also discovered that after 15 minutes of exposure to cold plasma, the protein's 3D structure changed, reducing foaming and emulsifying product while increasing foam stability.

12.10.4 COLD PLASMA IN MEAT PRODUCTS

Many researchers have reported that CAP treatment improves the quality of life [13, 46]. Nitrite (NO_2) is a common additive used in curing meat products, causing color and flavor changes in the finished product [30].

CAP is made up of various reactive nitrogen species (stable species) that dissolve into nitrogen oxide, and nitric acid with water can decompose into nitrate/nitrite. Plasma-activated brine (PAB) was utilized by Inguglia et al. [35] to cure beef jerky. The discovery in the generation of nitrites in brine water was higher in air plasma than in N_2 gas plasma, with nitrite residue ranging from 25 to 181 ppm. The nitrite reduction is caused by nitrite reacting directly with the meat components. However, they did note that the nitrite reaction did not affect the lipid oxidation or texture of beef jerky. Similarly, air-plasma treated jerky has reduced the overall population of

Listeria innocua to 0.85 log_{10} CFU/g after CAP application, which mainly disinfects the meat surface.

Gas mixtures of CO_2: Ar: O_2 in the ratio 60:30:10 and Ar: O_2 in the ratio 10:90 are tested in bag packages of Asian sea bass slices (ASBS), which have shown an increase in 9- and 12-days shelf-life. Regardless of the gas mixture utilized in plasma, there was a notable depletion in the overall viable cell count of ASBS [30].

12.10.5 MICROBIAL DECONTAMINATION

Currently, the food industry faces the challenging task of providing consumers with safe and shelf-stable food products while avoiding microbial contamination. Microorganism invasion can occur during processing or post-harvest processes, including machine operation, human handling, and transportation. Thermal/Heat processing is typically employed to create microbiologically stable and safe food. Moreover, it is not a favored practice because of the reduction in sensory and nutritional properties. However, food supplies containing artificial preservatives are now disapproved of by most customers. Non-thermal plasma, HPP, irradiation, and PEF are some of the examples of reasonable non-thermal decontamination procedures [16].

12.10.6 DEGRADATION OF AGROCHEMICAL WASTES

Pesticides are a class of chemical substances used to manage crop pests and weeds in modern farming. They are used regularly to reduce crop losses and boost agricultural output. They are typically regarded as potentially harmful to other creatures, including humans, and as a result, extreme caution should be exercised when using them. Over decades, it has been found that they possibly contaminate the drinking water supplies by mixing with the groundwater after being runoff from the crops. Because of the overtime bio-accumulation and the risk of long-term harmful consequences, toxicity may still be a worry even when present in low amounts. Prolonged exposure to certain chemicals like dichlorvos, dimethoate, etc., causes disturbance to glucose metabolism, induces hyperglycemia, and causes oxidative stress and DNA damage [16].

Some studies implied non-thermal plasma treatment to reduce or diminish the pesticide content in food and drinking supplies. When dichlorvos coated on glass surfaces were exposed to an inductively linked plasma source, it was successfully decomposed. However, an increment in the power had no influence on the effectiveness of dissipation [9]. When exposed to plasma therapy, fungicides such as boscalid, pyrimethanil, and iprodione on strawberries have no effect [52]. The results indicated that while CP can eradicate a wide range of pesticides [7], its efficiency is highly dependent on the gas composition and natural/structural properties of the pesticide in question.

12.10.7 PROCESSING OF FOOD PACKAGING MATERIALS

CAP can alter material surface properties and increase the antioxidant and antibacterial effects displaced the traditional packing materials. According to several studies, CAP increases the mechanical and functional properties of packaging film [30]. Oh et al. [62] recently evaluated smoked salmon coated with edible films prepared from defatted soybean meal (DSM) which was treated with the CP. They observed that the elongation, tensile strength, and moisture preventive performance of the DSM-based film increased by 6.8%, 13.4%, and 24.4% respectively, after 15 minutes of argon-induced CP at 400 W. During storage at 4°C, lipid oxidation was also reduced.

Heise et al. [33] found that CP can maintain the sealing qualities of polymer laminates or foils. Alternatively, using layers of deposition barrier, the approach is employed to cover coatings on the surface of polymers. Consequently, gas permeation into packaging material (oxygen and carbon dioxide) may be decreased, and unwanted pollutants could be eliminated without leaving residues [16].

12.11 PROS AND CONS OF COLD PLASMA TECHNOLOGY

Recent research showed that cold plasma has technological applications at pilot scale levels. The larger the plasma generative electrode dimensions, the greater the plasma coverage. These improvements are labor-intensive and cost-effective. Jacofsky et al., discovered that the MHRCP can unveil a wide surface area with an increased infiltration effect on the food matrix [30]. MHRCP is considered a replacement for large-scale purposes, but no research has been conducted on its application in the food industry due to its high cost.

Meat products such as beef, pork, chicken, and seafood showed adverse effects due to their increase in 2-thiobarbituric acid reactive substances (TBARS) and peroxide value [39].

12.12 FUTURE INNOVATIONS

Cognizance of the consequences of cold plasma technology is lacking in numerous sectors within the food industry. There are some broad perspectives about the future innovations of cold plasma.

12.12.1 MASS TRANSFER ENHANCEMENT

The technology has led to surface etching in food materials. It is observed that the plasma discharge from DBD on the lemon peels damages the oil glands, resulting in increased diffusion of oil extraction [38]. The enhancement of mass transfer operations and modification of food property using cold plasma are the two trending sectors where possible innovations in the food industry can occur.

12.12.2 FOOD STRUCTURE-PROPERTY MODIFICATION

The polymer processing and electronic sectors have more advantages of cold plasma applications for structural modification. There are two types of structural modification: Surface property modification and functional property modification.

- **Surface Property Modification:** Misra et al. [53] demonstrated the hydrophobic or hydrophilic modulations of the food surface for less oil seepage. An increase in the hydrophobicity of the biscuits is increased the spreadness of oil/butter used for spraying. With the help of plasma processing, the spreading of oil is increased by 50%.
- **Functional Property Modification:** The interesting application of plasma is explored in protein modification and its functionality. Misra et al. [51] demonstrated that the weak wheat flour with gluten having a secondary structure has more stability in CAP treatment. There is a noticeable change in dough rheological properties.

12.13 FOOD ALLERGEN MITIGATION AND ANTINUTRITIONAL FACTORS

Allergy from food products is a serious threat when the immune system responds abnormally to specific meals. Food processors and researchers are still interested in developing hypoallergenic foods and speeding up the development of oral tolerance. CAP enhances certain processes that influence linear epitopes and allergenic characteristics because food allergens are basically proteins. Different techniques are applied for the disparity in food allergenicity on exposure to CAP, while the exact mechanical pathway involved is still uncertain. The protein solubility loss induces the emergence of crosslinking of proteins or insoluble aggregates, which changes the isotopic conformation while fragmentation modifies linear epitopes. It is also possible that intrinsic species in plasma gas, like hydroxyl radicals, can rupture peptide bonds, destroy side chains by burning, and cause fragmentation. Disulfide bonds inside peptide chains possibly cleave existing hydroxyl radicals, producing RSH and ROS at the cleavage point [16].

The reactive oxygen species such as nitrogen, oxygen, and hydroxyl radicals that strike the amino acids are prone to oxidation, damaging the antibody binding sites. The ANF, such as glycosides, lecithins, tannins, etc., can significantly impact the gastrointestinal system and body metabolism. For example, tannins reduce protein digestibility, and lecithin inhibits hydrolysis within the intestinal absorptive cells. Heat degradation, soaking, germination, fermentation, irradiation, and other methods can be used to achieve this. High-temperature treatment for lengthy periods, on the other hand, causes undesirable changes in protein physicochemical characteristics and also the loss of heat-labile molecules. These research studies revealed CP technology as a feasible alternative to thermal techniques for allergenic management and ANF inactivation in foods. As a result, conclusions cannot be made until further research is done to understand the precise mechanisms of ANF and study its effects on nutritional and organoleptic profiles after post-treatment of food substances [16].

12.14 MANAGEMENT OF WASTEWATER AND EFFLUENT

In food processing operations, water is an important raw material used for cleaning equipment, processing raw materials, and generating steam, among several other things. The organic loads in wastewater from food processing plants are too high to oxidize or decontaminate using traditional, biological, or physicochemical methods. It is essential that adequate technologies for treating waste discharge from the food industry be developed. When diffused into liquid waste, oxidative species of ROS were shown to cause the degradation of pollutants. Mohamed et al. [56] showed that CP could be used to treat effluents from the processing of fruits and vegetables. This study used a plasma jet operating at a voltage of 25 kV. Results indicated that efficient wastewater decontamination was obtained after 120 and 150 seconds, respectively, for the dates and tomato processing facilities. After 180 seconds of exposure, the TBC dropped by 2.24 and 0.41 logs, respectively, for beetroot and blackberry wastewater [56]. According to Mohamed et al. [56], among the bacterial species, *E. coli* was the most susceptible, having up to 90.22% of endotoxins removed. Likewise, CP from GAD was utilized to eliminate AG25 dye from water in 10 minutes instead of the 20 minutes required by traditional salt coagulation ($FeCl_3$, $6H_2O$).

12.15 SAFETY CONSIDERATION OF COLD PLASMA TECHNOLOGIES

Apart from engineering design, operator health and safety considerations in an industrial manufacturing environment cannot be neglected while prospecting for large-scale applications. As a result, engineers should think about their well-being and safety considerations for cold plasma device designs and methodologies. A variety of issues come into play when assessing the fitness and safety implications of CP in the food processing industry.

Higher absorption of reactive gas species and the higher voltage are the two key safety concerns with cold plasma devices and procedures. Depending on the generating conditions, plasma will typically contain significant quantities of sensitive gas species such as ozone, peroxides, nitrogen oxides, etc. Reactive and excited species of plasma, like any other hazardous gas or powerful oxidizing agent, can be harmful to people if they are exposed to dangerous amounts over long periods [37].

Cold plasma RONS have been linked to various physical/chemical changes in food, including the structural changes in amino acids, the high molecular-weight compound oxidation to organic acids, and secondary metabolites formed during lipid peroxidation, leading to toxic compounds like short-chain aldehydes [58]. Some of these substances can be innocuous to cells at first sight, but they can be converted into toxins during cell metabolism by some enzyme known as cytochrome P450 (mammalian enzymes) [27]. In addition, food allergy is mediated by immunoglobulin E (IgE) due to protein activity, which is a serious threat to the public's well-being [57].

12.16 SUMMARY

Cold plasma is developing a dependable technology for decontamination, preservation, and sterilization of food materials. As seen in some foods, it enhances the bioactive profile, physicochemical properties, rheology, and shelf-life. In addition, the CAP increases nutritional quality in some food products by increasing total phenolic compounds, amino acids, and sugars. However, these changes depend on the gas mixture used to make plasma and the way of exposure/penetration over the food. Factors such as consumer preference for healthier foods and increased consumer knowledge, demand for non-heat-treated foods or raw is rising. The food industry must progress in technology, food safety measures, and environmental sustainability practices to meet the world's expanding need for food. Hence, this demands a commitment to innovative technology research and development.

Several studies have shown that cold plasma technology is extremely effective at ensuring food safety while preserving crucial organoleptic features. Without a doubt, cold plasma technology offers a new way to ensure food safety. The thoughts and ideas discussed in this chapter can assist in overcoming current barriers to plasma technology acceptance in foods.

KEYWORDS

- **cold plasma**
- **cold plasma safety evaluation**
- **food quality**
- **ionized gas**
- **microbial deterioration**
- **non-thermal**
- **pasteurization**

REFERENCES

1. Albertos, I., Martín-Diana, A. B., Cullen, P. J., Tiwari, B. K., Ojha, S. K. (2017). Effects of dielectric barrier discharge (DBD) generated plasma on microbial reduction and quality parameters of fresh mackerel (*Scomber scombrus*) fillets. *Innovative Food Science & Emerging Technologies*, *44*, 117–122.
2. Almeida, F. D. L., Cavalcante, R. S., Cullen, P. J., Frias, J. M., Bourke, P. (2015). Effects of atmospheric cold plasma and ozone on prebiotic orange juice. *Innovative Food Science & Emerging Technologies*, *32*, 127–135.
3. Alves Filho, E. G., Almeida, F. D. L., Cavalcante, R. S., De Brito, E. S., Cullen, P. J. (2016). 1H NMR spectroscopy and chemometrics evaluation of non-thermal processing of orange juice. *Food Chemistry*, *204*, 102–107.

4. Alves Filho, E. G., de Brito, E. S., Rodrigues, S. (2019). Effects of cold plasma processing in food components (pp. 253–268). Elsevier Inc. *Advances in Cold Plasma Applications for Food Safety and Preservation,* DOI: https://doi.org/10.1016/B978-0-12-814921-8.00008-6, ISBN: 978-0-12-814921-8.
5. Alves Filho, E. G., Rodrigues, T. H. S., Fernandes, F. A. N., de Brito, E. S., Cullen, P. J. (2019). An untargeted chemometric evaluation of plasma and ozone processing effect on volatile compounds in orange juice. *Innovative Food Science & Emerging Technologies, 53,* 63–69.
6. Amini, M., Ghoranneviss, M., Abdijadid, S. (2017). Effect of cold plasma on crocin esters and volatile compounds of saffron. *Food Chemistry, 235,* 290–293.
7. Anbarasan, R., Jaspin, S., Bhavadharini, B., Pare, A., Pandiselvam, R., Mahendran, R. (2022). Chlorpyrifos pesticide reduction in soybean using cold plasma and ozone treatments. *LWT, 159,* 113193. https://doi.org/10.1016/j.lwt.2022.113193.
8. Ataeefard, M., Moradian, S., Mirabedini, M., Ebrahimi, M., Asiaban, S. (2009). Investigating the effect of power/time in the wettability of Ar and O_2 gas plasma-treated low-density polyethylene. *Progress in Organic Coatings, 64*(4), 482–488.
9. Bai, Y., Chen, J., Yang, Y., Guo, L., Zhang, C. (2010). Degradation of organophosphorus pesticide induced by oxygen plasma: Effects of operating parameters and reaction mechanisms. *Chemosphere, 81*(3), 408–414.
10. Barba, F. J., Esteve, M. J., Frígola, A. (2012). High pressure treatment effect on physicochemical and nutritional properties of fluid foods during storage: A Review. *Comprehensive Reviews in Food Science and Food Safety, 11*(3), 307–322.
11. Bárdos, L., Baránková, H. (2010). Cold atmospheric plasma: Sources, processes, and applications. *Thin Solid Films, 518*(23), 6705–6513.
12. Chang, R., Lu, H., Tian, Y., Li, H., Wang, J., Jin, Z. (2020). Structural modification and functional improvement of starch nanoparticles using vacuum cold plasma. *International Journal of Biological Macromolecules, 145,* 197–206.
13. Chaplot, S., Yadav, B., Jeon, B., Roopesh, M. S. (2019). Atmospheric cold plasma and peracetic acid-based hurdle intervention to reduce Salmonella on raw poultry meat. *Journal of Food Protection, 82*(5), 878–888.
14. Charoux, C. M. G., Patange, A., Lamba, S., O'Donnell, C. P., Tiwari, B. K., Scannell, A. G. M. (2021). Applications of nonthermal plasma technology on safety and quality of dried food ingredients. *Food Research International, 130,* 325–340.
15. Chizoba Ekezie, F.-G.,Cheng, J. H., & Sun, D. W. (2018). Effects of nonthermal food processing technologies on food allergens: A review of recent research advances. *Trends in Food Science & Technology, 74,* 12–25.
16. Chizoba Ekezie, F. -G., Sun, D. -W., Cheng, J. -H. (2017). A review on recent advances in cold plasma technology for the food industry: Current applications and future trends. *Trends in Food Science & Technology, 69,* 46–58.
17. Choi, M. H., Kim, G. H., Lee, H. S. (2002). Effects of ascorbic acid retention on juice color and pigment stability in blood orange (*Citrus sinensis*) juice during refrigerated storage. *Food Research International, 35*(8), 753–759.
18. Chutia, H., Mahanta, C. L., Ojah, N., Choudhury, A. J. (2020). Fuzzy logic approach for optimization of blended beverage of cold plasma treated TCW and orange juice. *Journal of Food Measurement and Characterization, 14*(4), 1926–1938.
19. Connolly, J., Valdramidis, V. P., Byrne, E., Karatzas, K. A., Cullen, P. J. (2013). Characterization and antimicrobial efficacy against *E. coli* of a helium/air plasma

at atmospheric pressure created in a plastic package. *Journal of Physics D: Applied Physics, 46*(3), 035041. https://doi.org/10.1088/0022-3727/46/3/035401.
20. Conrads, H., Schmidt, M. (2000). Plasma generation and plasma sources. *Plasma Sources Science and Technology, 9*(4), 441–454.
21. Coutinho, N. M., Silveira, M. R., Fernandes, L. M., Moraes, J., Pimentel, T. C. (2019). Processing chocolate milk drink by low-pressure cold plasma technology. *Food Chemistry, 278*, 276–283.
22. Coutinho, N. M., Silveira, M. R., Pimentel, T. C., Freitas, M. Q., Moraes, J. (2019). Chocolate milk drink processed by cold plasma technology: Physical characteristics, thermal behavior and microstructure. *LWT, 102*(November 2018), 324–329.
23. Cui, H., Yang, X., Abdel-Samie, M. A., Lin, L. (2020). Cold plasma treated phlorotannin/Momordica charantia polysaccharide nanofiber for active food packaging. *Carbohydrate Polymers, 239*, 116214. https://doi.org/10.1016/j.carbpol.2020.116214.
24. Deilmann, M., Mann, H. H., Steves, S., Bibinov, N., Awakowicz, P. (2009). Silicon oxide permeation barrier coating and plasma sterilization of PET bottles and foils. *Plasma Processes and Polymers, 6*(Suppl. 1), 695–699.
25. Dixon, D., Meenan, B. J. (2012). Atmospheric dielectric barrier discharge treatments of polyethylene, polypropylene, polystyrene and poly (ethylene terephthalate) for enhanced adhesion. *Journal of Adhesion Science and Technology, 26*(20–21), 2325–2337.
26. Faria, G. Y. Y., Souza, M. M., Oliveira, J. R. M., Costa, C. S. B., Collares, M. P., Prentice, C. (2020). Effect of ultrasound-assisted cold plasma pretreatment to obtain sea asparagus extract and its application in Italian salami. *Food Research International, 137*, 109435. https://doi.org/10.1016/j.foodres.2020.109435.
27. Fernández, A., Noriega, E., Thompson, A. (2013). Inactivation of Salmonella enterica serovar Typhimurium on fresh produce by cold atmospheric gas plasma technology. *Food Microbiology, 33*(1), 24–29.
28. Gadri, R. B., Roth, J. R., Montie, T. C., Kelly-Wintenberg, K., Tsai, P. P. Y. (2000). Sterilization and plasma processing of room temperature surfaces with a one atmosphere uniform glow discharge plasma (OAUGDP). *Surface and Coatings Technology, 131*(1–3), 528–541.
29. Gallagher, M. J., Vaze, N., Gangoli, S., Vasilets, V. N., Gutsol, A. F. (2007). Rapid inactivation of airborne bacteria using atmospheric pressure dielectric barrier grating discharge. *IEEE Transactions on Plasma Science, 35*(5), 1501–1510.
30. Ganesan, A. R., Tiwari, U., Ezhilarasi, P. N., Rajauria, G. (2021). Application of cold plasma on food matrices: A review on current and future prospects. *Journal of Food Process Engineering, 45*(1), 15070. https://doi.org/10.1111/jfpp.15070.
31. Gaunt, L. F., Beggs, C. B., Georghiou, G. E. (2006). Bactericidal action of the reactive species produced by gas-discharge nonthermal plasma at atmospheric pressure: A review. *IEEE Transactions on Plasma Science, 34*(4 II), 1257–1269.
32. Walker, G. M., White, N. A., Whiteway, M., Fitzpatrick, D., Curran, B., Doyle, S., Baars, J. (2018). *Fungi: Biology and Applications.* Publisher: Wiley, pp. 1–416, Edition: 2, ISBN: 9781119374169, https://www.researchgate.net/profile/Kevin-Kavanagh-4/publication/228030934_Fungal_Fermentation_Systems_and_Products/links/5a82c7aaa6fdcc6f3eadd318/Fungal-Fermentation-Systems-and-Products.pdf.
33. Heise, M., Neff, W., Franken, O., Muranyi, P., Wunderlich, J. (2004). Sterilization of polymer foils with dielectric barrier discharges at atmospheric pressure. *Plasmas and Polymers, 9*(1), 23–33.

34. Illera, A. E., Chaple, S., Sanz, M. T., Ng, S., Lu, P. (2019). Effect of cold plasma on polyphenol oxidase inactivation in cloudy apple juice and on the quality parameters of the juice during storage. *Food Chemistry X, 3*, 100049. https://doi.org/10.1016/j.fochx.2019.100049.
35. Inguglia, E. S., Oliveira, M., Burgess, C. M., Kerry, J. P., Tiwari, B. K. (2020). Plasma-activated water as an alternative nitrite source for the curing of beef jerky: Influence on quality and inactivation of *Listeria innocua. Innovative Food Science & Emerging Technologies, 59*, 102276. https://doi.org/10.1016/j.ifset.2019.102276.
36. Kayes, M. M., Critzer, F. J., Kelly-Wintenberg, K., Roth, J. R., Montie, T. C., Golden, D. A. (2007). Inactivation of foodborne pathogens using a one atmosphere uniform glow discharge plasma. *Foodborne Pathogens and Disease, 4*(1), 50–59.
37. Keener, K. M., Misra, N. N. (2016). Future of Cold Plasma in Food Processing. Elsevier Inc., pp. 343–360.
38. Kodama, S., Thawatchaipracha, B., Sekiguchi, H. (2014). Enhancement of Essential Oil Extraction for Steam Distillation by DBD Surface Treatment. *Plasma Processes and Polymers, 11*(2), 126–132.
39. Kulawik, P., Alvarez, C., Cullen, P. J., Aznar-Roca, R., Mullen, A. M., Tiwari, B. (2018). The effect of non-thermal plasma on the lipid oxidation and microbiological quality of sushi. *Innovative Food Science & Emerging Technologies, 45*(December 2017), 412–417.
40. Lee, H. S., Park, H. H., Min, S. C. (2020). Microbial decontamination of red pepper powder using pulsed light plasma. *Journal of Food Engineering, 284*(January), 110075. https://doi.org/10.1016/j.jfoodeng.2020.110075.
41. Lee, S. Y., Lee, W. K., Lee, J. W., Chung, M. S., Oh, S. W. (2021). Microbial decontamination of rice germ using a large-scale plasma jet-pulsed light-ultraviolet-C integrated treatment system. *Food and Bioprocess Technology, 14*(3), 542–553. https://doi.org/10.1007/s11947-020-02490-4.
42. Lee, T., Puligundla, P., Mok, C. (2015). Inactivation of foodborne pathogens on the surfaces of different packaging materials using low-pressure air plasma. *Food Control, 51*, 149–155. https://doi.org/10.1016/j.foodcont.2014.10.030.
43. Lin, L., Liao, X., Cui, H. (2019). Cold plasma treated thyme essential oil/silk fibroin nanofibers against *Salmonella Typhimurium* in poultry meat. *Food Packaging and Shelf Life, 21*(February), 100337. https://doi.org/10.1016/j.fpsl.2019.100337.
44. Lin, L., Liao, X., Li, C., Abdel-Samie, M. A., Siva, S., Cui, H. (2021). Cold nitrogen plasma modified cuminaldehyde/β-cyclodextrin inclusion complex and its application in vegetable juices preservation. *Food Research International, 141*(August 2020), 110132. https://doi.org/10.1016/j.foodres.2021.110132.
45. Lu, P., Guo, M., Xu, Z., Wu, M. (2018). Application of nanofibrillated cellulose on BOPP/LDPE film as oxygen barrier and antimicrobial coating based on cold plasma treatment. *Coatings, 8*(6), 207. https://doi.org/10.3390/coatings8060207.
46. Luo, J., Muhammad Nasiru, M., Yan, W., Zhuang, H., Zhou, G., Zhang, J. (2020). Effects of dielectric barrier discharge cold plasma treatment on the structure and binding capacity of aroma compounds of myofibrillar proteins from dry-cured bacon. *LWT, 117*(July 2019), 108606. https://doi.org/10.1016/j.lwt.2019.108606.
47. Mahdavian Mehr, H., Koocheki, A. (2021). Physicochemical properties of grass pea (*Lathyrus sativus* L.) protein nanoparticles fabricated by cold atmospheric-pressure

plasma. *Food Hydrocolloids, 112*(August 2020), 106328. https://doi.org/10.1016/j.foodhyd.2020.106328.
48. Mandal, R., Singh, A., Pratap Singh, A. (2018). Recent developments in cold plasma decontamination technology in the food industry. *Trends in Food Science & Technology, 80*, 93–103. https://doi.org/10.1016/j.tifs.2018.07.009.
49. Maslakci, N. N., Ulusoy, S., Oksuz, A. U. (2017). Investigation of the effects of plasma-treated chitosan electrospun fibers onto biofilm formation. *Sensors and Actuators B: Chemical, 246*, 887–895. https://doi.org/10.1016/j.snb.2017.01.154.
50. Mehta, D., Sharma, N., Bansal, V., Sangwan, R. S., Yadav, S. K. (2019). Impact of ultrasonication, ultraviolet and atmospheric cold plasma processing on quality parameters of tomato-based beverage in comparison with thermal processing. *Innovative Food Science & Emerging Technologies, 52*, 343–349. https://doi.org/10.1016/j.ifset.2019.01.007.
51. Misra, N. N., Kaur, S., Tiwari, B. K., Kaur, A., Singh, N., Cullen, P. J. (2015). Atmospheric pressure cold plasma (ACP) treatment of wheat flour. *Food Hydrocolloids, 44*, 115–121. https://doi.org/10.1016/j.foodhyd.2014.09.019.
52. Misra, N. N., Pankaj, S. K., Walsh, T., O'Regan, F., Bourke, P., Cullen, P. J. (2014). In-package nonthermal plasma degradation of pesticides on fresh produce. *Journal of Hazardous Materials, 271*, 33–40. https://doi.org/10.1016/j.jhazmat.2014.02.002.
53. Misra, N. N., Sullivan, C., Pankaj, S. K., Alvarez-Jubete, L., Cama, R. (2014). Enhancement of oil spreadability of biscuit surface by nonthermal barrier discharge plasma. *Innovative Food Science & Emerging Technologies, 26*, 456–461. https://doi.org/10.1016/j.ifset.2014.09.005.
54. Misra, N. N., Tiwari, B. K., Raghavarao, K. S. M. S., Cullen, P. J. (2011). Nonthermal plasma inactivation of food-borne pathogens. *Food Engineering Reviews, 3*(3–4), 159–170. https://doi.org/10.1007/s12393-011-9042-4.
55. Mittendorfer, J., Bierbaumer, H. P., Gratzl, F., Kellauer, E. (2002). Decontamination of food packaging using electron beam – Status and prospects. *Radiation Physics and Chemistry, 63*(3–6), 833–836. https://doi.org/10.1016/S0969-806X(01)00649-8.
56. Mohamed, A. A. H., Al Shariff, S. M., Ouf, S. A., Benghanem, M. (2016). Atmospheric pressure plasma jet for bacterial decontamination and property improvement of fruit and vegetable processing wastewater. *Journal of Physics D: Applied Physics, 49*(19), 195401. https://doi.org/10.1088/0022-3727/49/19/195401.
57. Muhammad, A. I., Ding, T., Liu, D. (2022). Safety evaluation of cold plasma technology. In *Applications of Cold Plasma in Food Safety*, pp. 341–349, ISBN: 978-981-16-1827-7. Springer Singapore. https://doi.org/10.1007/978-981-16-1827-7_15.
58. Muhammad, A. I., Xiang, Q., Liao, X., Liu, D., Ding, T. (2018). Understanding the impact of nonthermal plasma on food constituents and microstructure—A review. *Food and Bioprocess Technology, 11*(3), 463–486. https://doi.org/10.1007/s11947-017-2024-0.
59. Muranyi, P., Wunderlich, J., Heise, M. (2007). Sterilization efficiency of a cascaded dielectric barrier discharge. *Journal of Applied Microbiology, 103*(5), 1535–1544. https://doi.org/10.1111/j.1365-2672.2007.03377.x.
60. Niemira, B. A. (2012). Cold plasma decontamination of foods. *Annual Review of Food Science and Technology, 3*(1), 125–142. https://doi.org/10.1146/annurev-food-022811-101202.

61. Niemira, B. A., & Sites, J. (2008). Cold plasma inactivates *Salmonella Stanley* and *Escherichia coli* O157 inoculated on golden delicious apples. *Journal of Food Protection, 71*(7), 1357–1365. https://doi.org/10.4315/0362-028X-71.7.1357.
62. Oh, Y. J., Song, A. Y., & Min, S. C. (2017). Inhibition of *Salmonella typhimurium* on radish sprouts using nitrogen-cold plasma. *International Journal of Food Microbiology, 249*, 66–71. https://doi.org/10.1016/j.ijfoodmicro.2017.03.012.
63. Pankaj, S. K., Bueno-Ferrer, C., Misra, N. N., Milosavljević, V., & O'Donnell, C. P. (2014). Applications of cold plasma technology in food packaging. *Trends in Food Science & Technology, 35*(1), 5–17. https://doi.org/10.1016/j.tifs.2013.11.007.
64. Pankaj, S. K., Bueno-Ferrer, C., Misra, N. N., O'Neill, L., & Jiménez, A. (2014). Characterization of polylactic acid films for food packaging as affected by dielectric barrier discharge atmospheric plasma. *Innovative Food Science & Emerging Technologies, 21*, 107–113. https://doi.org/10.1016/j.ifset.2013.11.001.
65. Pankaj, S. K., & Keener, K. M. (2017). Cold plasma: Background, applications and current trends. *Current Opinion in Food Science, 16*, 49–52. https://doi.org/10.1016/j.cofs.2017.06.009.
66. Pankaj, S. K., & Thomas, S. (2016). Cold plasma applications in food packaging. In S. K. Pankaj & S. Thomas (Eds.), *Cold Plasma in Food and Agriculture: Fundamentals and Applications* (pp. 293–307). Elsevier Inc.
67. Radhakrishnan, M., Ramanan, K. R., Sargunam, R., & Sarumathi, R. (2016). Effect of cold plasma on mortality of *Tribolium castaneum* on refined wheat flour. In *Proceedings of the 10th International Conference on Controlled Atmosphere and Fumigation in Stored Products (CAF 2016)*, pp. 37–44, Publisher: CAF.
68. Roy, N. C., Hasan, M. M., Kabir, A. H., Reza, M. A., Talukder, M. R., & Chowdhury, A. N. (2018). Atmospheric pressure gliding arc discharge plasma treatments for improving germination, growth and yield of wheat. *Plasma Science and Technology, 20*(11), 2058–6272. https://doi.org/10.1088/2058-6272/aac647.
69. Royintarat, T., Choi, E. H., Boonyawan, D., Seesuriyachan, P., & Wattanutchariya, W. (2020). Chemical-free and synergistic interaction of ultrasound combined with plasma-activated water (PAW) to enhance microbial inactivation in chicken meat and skin. *Scientific Reports, 10*(1), 1–14. https://doi.org/10.1038/s41598-020-66964-1.
70. Sarangapani, C., Thirumdas, R., Devi, Y., Trimukhe, A., Rajendra, R., Deshmukh, R. R., & Annapure, U. S. (2016). Effect of low-pressure plasma on physicochemical and functional properties of parboiled rice flour. *LWT – Food Science and Technology, 69*, 482–489. https://doi.org/10.1016/j.lwt.2016.01.065.
71. Sarangapani, C., Yamuna Devi, R., Thirumdas, R., Trimukhe, A. M., Deshmukh, R. R., & Annapure, U. S. (2017). Physicochemical properties of low-pressure plasma treated black gram. *LWT – Food Science and Technology, 79*, 102–110. https://doi.org/10.1016/j.lwt.2017.01.036.
72. Schneider, J., Baumgärtner, K. M., Feichtinger, J., Krüger, J., & Muranyi, P. (2005). Investigation of the practicability of low-pressure microwave plasmas in the sterilization of food packaging materials at industrial level. *Surface and Coatings Technology, 200*, 962–966. https://doi.org/10.1016/j.surfcoat.2005.01.109.
73. Scholtz, V., Pazlarova, J., Souskova, H., Khun, J., & Julak, J. (2015). Nonthermal plasma – A tool for decontamination and disinfection. *Biotechnology Advances, 33*(6), 1108–1119. https://doi.org/10.1016/j.biotechadv.2015.08.010.

74. Segat, A., Misra, N. N., Cullen, P. J., & Innocente, N. (2015). Atmospheric pressure cold plasma (ACP) treatment of whey protein isolate model solution. *Innovative Food Science & Emerging Technologies, 29*, 247–254. https://doi.org/10.1016/j.ifset.2015.03.014.
75. Shiekh, K. A., Benjakul, S., Qi, H., Zhang, B., & Deng, S. (2021). Combined hurdle effects of pulsed electric field and vacuum impregnation of Chamuang leaf extract on quality and shelf-life of Pacific white shrimp subjected to high voltage cold atmospheric plasma. *Food Packaging and Shelf Life, 28*(October 2020), 100660. https://doi.org/10.1016/j.fpsl.2021.100660.
76. Shiekh, K. A., Zhou, P., & Benjakul, S. (2021). Combined effects of pulsed electric field, Chamuang leaf extract and cold plasma on quality and shelf-life of *Litopenaeus vannamei*. *Food Bioscience, 41*, 100975. https://doi.org/10.1016/j.fbio.2021.100975.
77. Stoffels, E., Sakiyama, Y., & Graves, D. B. (2008). Cold atmospheric plasma: Charged species and their interactions with cells and tissues. *IEEE Transactions on Plasma Science, 36*(4), 1441–1457. https://doi.org/10.1109/TPS.2008.923362.
78. Tappi, S., Ragni, L., Tylewicz, U., Romani, S., Ramazzina, I., & Rocculi, P. (2019). Browning response of fresh-cut apples of different cultivars to cold gas plasma treatment. *Innovative Food Science & Emerging Technologies, 53*, 56–62. https://doi.org/10.1016/j.ifset.2019.01.008.
79. Tendero, C., Tixier, C., Tristant, P., Desmaison, J., & Leprince, P. (2006). Atmospheric pressure plasmas: A review. *Spectrochimica Acta Part B: Atomic Spectroscopy, 61*(1), 2–30. https://doi.org/10.1016/j.sab.2005.10.001.
80. Tenn, N., Follain, N., Fatyeyeva, K., Valleton, J. M., & Poncin-Epaillard, F. (2012). Improvement of water barrier properties of Poly (ethylene-co-vinyl alcohol) films by hydrophobic plasma surface treatments. *Journal of Physical Chemistry C, 116*(23), 12599–12612. https://doi.org/10.1021/jp211401p.
81. Ulbin-Figlewicz, N., Brychcy, E., & Jarmoluk, A. (2015). Effect of low-pressure cold plasma on surface microflora of meat and quality attributes. *Journal of Food Science and Technology, 52*(2), 1228–1232. https://doi.org/10.1007/s13197-013-1127-y.
82. Whittle, J. D., Short, R. D., Steele, D. A., Bradley, J. W., & Bryant, P. M. (2013). Variability in plasma polymerization processes – An international round-robin study. *Plasma Processes and Polymers, 10*(9), 767–778. https://doi.org/10.1002/ppap.201300044.
83. Xu, L., Garner, A. L., Tao, B., & Keener, K. M. (2017). Microbial inactivation and quality changes in orange juice treated by high voltage atmospheric cold plasma. *Food and Bioprocess Technology, 10*(10), 1778–1791. https://doi.org/10.1007/s11947-017-1981-7.
84. Yepez, X. V., & Keener, K. M. (2016). High-voltage atmospheric cold plasma (HVACP) hydrogenation of soybean oil without trans-fatty acids. *Innovative Food Science and Emerging Technologies, 38*, 169–174. https://doi.org/10.1016/j.ifset.2016.10.012.
85. Zhuang, H., Rothrock, M. J., Hiett, K. L., Lawrence, K. C., & Gamble, G. R. (2019). In-package air cold plasma treatment of chicken breast meat: Treatment time effect. *Journal of Food Quality, 2019*, Article ID 1837351. https://doi.org/10.1155/2019/1837351.
86. Ziuzina, D., Misra, N. N., Cullen, P. J., Keener, K., & Mosnier, J. P. (2016). Demonstrating the potential of industrial scale in-package atmospheric cold plasma for decontamination of cherry tomatoes. *Plasma Medicine, 6*(3–4), 397–412. https://doi.org/10.1615/PlasmaMed.2018026356.

CHAPTER 13

COLD PLASMA FOR SPACEFLIGHT-GROWN PRODUCES: DISINFECTION AND OTHER APPLICATIONS

ANBARASAN RAJAN and R. MAHENDRAN

ABSTRACT

Safe, nutritious, and fresh food products are essential for astronauts to maintain their physical and mental health. However, considering the space mission duration and distance, it is not possible to supply food from Earth. Meanwhile, astronauts also cannot eat freeze-dried and pre-packed food for longer durations. Furthermore, accommodating food that lasts for the entire space mission demands more space in the spacecraft, which is practically and economically not feasible. These problems can be addressed by cultivating food products (ready to eat/use) in space using artificial plant production systems. However, the microbial risk associated with spaceflight-grown products (SFGP) limits their usage during missions. Unlike on Earth, microbial decontamination in spaceflight cannot be performed using conventional processing equipment due to high resource consumption (i.e., water, space, mass, cost). In addition, most of the existing processing methods cannot be used for multi-purpose operations. Therefore, researchers have decided to experiment with cold plasma (CP) equipment in a microgravity environment to discover its potential applications in SFGP processing. Since cold plasma can be used under both ambient and vacuum conditions, it can sustain the microgravity environment without requiring water or oxygen. In this way, CP can be used for processing plants at different life stages. Furthermore, this technique can be used for other applications such as wound healing, surface sterilization, seed drought, and saline-tolerant enhancement. This

Nonthermal Plasma Technology for Food and Food Products.
R. Mahendran, Preeti Birwal, and Megh R. Goyal (Eds.)

chapter provides an overview of the significance of CP in SFGP processing and its future scopes.

13.1 INTRODUCTION

The concept of "Space food" began almost in 1926, when the Russian scientist named Konstantin penned his "Plan of Space Exploration." The main proposal of Konstantin's publication was to colonize the universe and utilize the sun's natural radiation to cultivate plants for food and energy. Later, the importance of space-grown produce was understood by the scientists as it not only provides food but also oxygen for human survival [6].

Nutritious foods are essential for astronauts' physical and mental health and work efficiency [18, 33]. Thus, the food products required for astronauts are provided in prepacked condition to meet the quality standards of NASA or any respective space research center. The foods supplied in this form are highly nutritive, stable, safe, and of supreme quality. However, the food quality deteriorates during extended space missions (i.e., Mars mission: 1–3 years). Thus, it is essential that food needs to have a minimum of 3–5 years stability at room temperature for space missions [7]. As a result, astronauts are forced to eat dried foods instead of juice, dried fruit, or supplements for months considering the health and performance consequences. However, these processed/packed goods are not capable of providing some essential minerals and nutrients such as Ca, K, vitamin D, and K; furthermore, they contain highly unstable vitamins such as A, B1, B6, and C which deteriorate during the storage period [6].

Furthermore, it is also not possible for a spacecraft to carry a huge volume of food that is adequate for longer space missions [45]. In this regard, food supply for the ISS or any other future space stations (or missions) requires an unmanned spaceflight shipment from Earth to address the deficiency of nutritive food supply and space requirements. However, for a journey like a Mars mission, it will take around 8 months (225 million km travel) to reach from Earth. Hence, it is difficult to invest extra money and time for another journey to provide resources to the crew members [6]. Thus, it is impossible to depend on Earth for resources. Meanwhile, a planet like Mars does not have the high-tech technical and infrastructural support available on Earth to process and preserve food. At the same time, astronauts need food, water, and other essentials to continue their mission [8, 31, 32, 35, 38].

Considering the mission duration of 3–5 years, it is possible to cultivate plants in the spaceflight itself for sustainable resource management. Thus, researchers worked hard and succeeded in designing independent spaceflight plant production units with a suitable atmosphere to support plant growth. The key objective of this process is to utilize wastewater and carbon dioxide for plant growth; subsequently, obtaining food and water (as water vapor condensation) [6]. Plants are typically regarded as safe to eat with minimal or no preparation. Salad crops, coupled with various horticultural operations, can provide a diversity of tastes, textures, and colors which may contribute to psychological advantages in spaceflight. Hence, spaceflight crop production is considered the sustainable future approach for astronauts' food and energy resources [8]. Though onions, leeks, and cabbages were cultivated in spaceflight in 1971, the consumption of space-grown produce (onion) started in 1975 only. However, even today, astronauts consume mostly freeze-dried or canned foods due to various environmental and technical reasons [6].

On Earth, food and agricultural products are washed and processed before consumption to prevent human exposure to microbes and other toxic contaminants. However, the spaceflight plant production units do not have any such processing facilities (including water) to wash or disinfect the plant products. This is due to the fact that the preservation techniques require certain energy resources like heat, water, pressure, or chemicals. Among these resources, heat and chemical applications are not preferred due to safety and quality reasons. In this regard, CP could be the possible technology for disinfecting SFGP.

CP has the potential to disinfect, decontaminate (i.e., toxins and chemicals), and preserve food products under both atmospheric and vacuum conditions [1, 2, 4].

This chapter discusses contamination risks and challenges associated with SFGP; challenges in disinfecting food products in space; CP processing of SFGP; advantages of CP technology in space applications; future scope for CP applications in SFGP.

13.2 RISKS OF CONTAMINATION IN SPACE FLIGHT GROWN PRODUCE

Cultivation of plants in spaceflights started almost 50 years ago to experiment with the SFGP eating quality (aspects include safety and nutritional quality), physical nature, and reproductive ability. Since plants are grown under different environments (microgravity, artificial light, and soil), earlier studies (ISS Russian segment: 2003–2005) focused on the changes in the physical nature and reproductive capacity of (pea) seeds [39]. However, recent studies have started focusing on the nutritional and microbiological aspects of SFGP. The increasing concern over the microbiological safety of SFGP is to avoid crew members' exposure to certain life-threatening pathogens through these products [24].

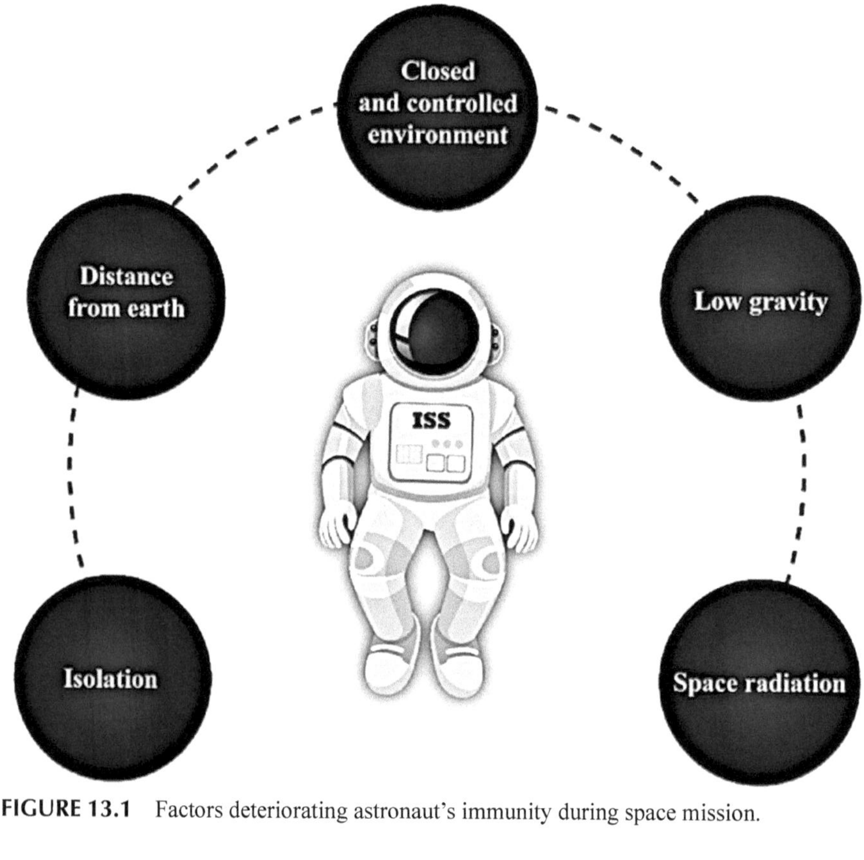

FIGURE 13.1 Factors deteriorating astronaut's immunity during space mission.

When it comes to spaceflight or the ISS, the entire system is composed of enclosed chambers where the astronauts perform their day-to-day tasks. When the human body is exposed to high pressure, sun rays, low gravity, defined food and hygiene practices, it affects their immune system. Furthermore, this new environment also affects the nature of microbes present in human respiratory and gastrointestinal tracts and leads to the risk of infection (contamination speed between astronauts, astronauts to spaceflight surface and then to astronauts, contaminated surface/soil/astronauts to plant and then to crew members). Though all these microbes are not harmful to healthy humans under ideal conditions (on Earth), space mission-induced weakened immunity (Figure 13.1) of astronauts makes them susceptible to infection risks. Evidently, data collected during US space missions (1989–98) reported around 26 cases within the span of 10 years [40]. As a result, contamination levels and types must be regulated to ensure sufficient living conditions and decrease the possibility of adverse effects on the astronauts. It is apparent that opportunistic organisms that are able to produce infections and sickness play a significant role in these preventative efforts. Hence, predefined microbial quality standards for food, air, surfaces, and water have been established for space agencies to prevent, monitor, and mitigate contamination risks in spaceflights [42].

However, the problem arises when microbes contaminate SGFP and are transferred from plant produce to humans through their diet. In space, VEGGIE, APH, and LADA-GH are different hardware systems that were used for the cultivation of plant produce. Among these systems, VEGGIE was used on the ISS for the successful production of leafy vegetables. However, these cultivated leafy vegetables are prone to microbial contamination. The experiment (VEG-01C) conducted in VEGGIE showed *Fusarium oxysporum* infection in Zinnia hybrid plants [15].

Fusarium oxysporum is a pervasive genus of soil-borne fungus that adapts and grows in different habitats. These strains are saprophytic, so they can persist for long periods on organic materials in soil and the rhizosphere of many plant species. Furthermore, certain strains of *F. oxysporum* are harmful to several plant species; when they infect the vascular system, they reach the roots and cause root rots or tracheomycosis [10]. The major concern is that they can cause human diseases. Harmful strains of *F. oxysporum* can cause *Fusarium* wilt in practically all agriculturally valuable crops. These multi-host isolates from plants are also capable of causing invasive fungal infections in immunocompromised animals. Despite the fact that the fungal infection on the Zinnia plants was most likely caused by a fan fault in the Veggie hardware, it emphasized the need to describe the healthy basal microbiome of all plants grown in space. Hence, methodologies are developed to analyze the risk of contamination in ISS-produced plant products. With this method, the interaction between agricultural produce and microorganisms can be analyzed by sending the material (frozen plant produce) to Earth. However, the current method of monitoring the safety of SFGP through Earth-dependent analysis will not be a sustainable solution for space missions involving long-distance (high orbit) travels [15, 16]. Thus, scientists are in search of technology that can be used in space to prevent the infection risks that come from the SGFP.

13.3 ISSUES ASSOCIATED WITH SPACE FLIGHT GROWN PRODUCES

The section 13.2 discussed the various possible causes of microbial infections to astronauts (sources include: humans, contact surfaces, and plants due to environmental and technical faults). The current section will discuss the changes in interaction of microbes with SFGP and the microbial risk associated with SFGP throughout its growth stages.

In general, there are two categories of agricultural products available for astronauts to cater to their food needs. The first category comprises plants that can be cultivated in spacecraft for ready-to-eat purposes (i.e., salad crops). The second category comprises products that can either be cultivated in space or shipped from Earth to space to fulfill food requirements [32]. The list of plants under each of these categories is shown in Figure 13.2.

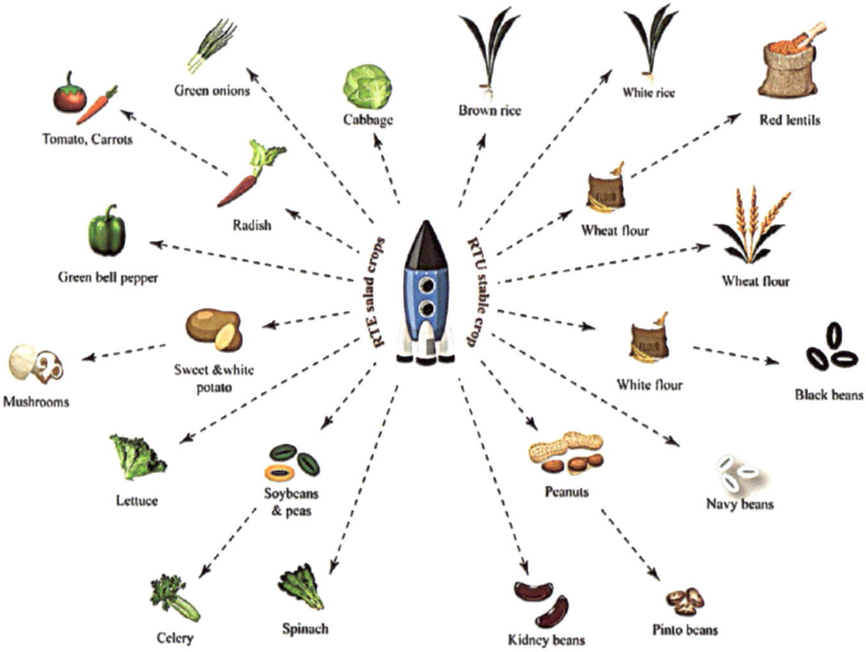

FIGURE 13.2 Agriculture produce that can be cultivated or shipped to space station.

Plants and their associated microbes have intricate and dynamic connections that are often affected by various stress and environmental conditions (i.e., spaceflight). These changes influence the plant-microbe interaction both at genetic and physiological levels and make them vulnerable to diseases. Evidently, soybean seedlings cultivated in space have gone through more fungal *(Phytophthora sojae)* attacks than the control (earth-grown) samples. Furthermore, the growth and development of legumes (*Medicago truncatula*) are also affected by microgravity and clino-rotation conditions, which in turn affect the plants' association with nitrogen-fixing fungi. Similarly, the bacteria (e.g., *Pseudomonas, Paenibacillus,* and *Bacillus*) that produce wheat seedling growth substances are also found to be affected due to the microgravity environment [15]. However, all these stated changes are associated with either SFGP growth or its nutritional changes. But it is more important to understand the problems that arise due to the microbes associated with these SFGP. Though SFGP have the potential to augment the pre-packaged system with improved nutritional quality, astronauts' food choice and acceptance, they may also bring microbial risk. Hence, both seeds and plant growth support systems are disinfected prior to launch to fulfill safety standards. However, the space shuttle and its components will remain untreated. Since plants are cultivated in the same spacecraft where astronauts are living, microbes might get transferred from humans to plants through the air. Evidently, various studies have suggested that fresh produce from SFGP may exceed existing NASA microbiological guidelines for non-commercially sterile space goods [7].

For example, when lettuce leaves were cultivated in space (VEG-01B), the aerobic microbial count on the leaves was found to be higher than the samples that were cultivated on Earth. In addition, performing repeated harvesting in space further increases the microbial populations of SFGP [24]. However, the risk of foodborne disease on a space mission is unacceptable since it would have a higher impact in space than on Earth owing to restricted medical skills, the problems of microgravity, and resource constraints [7].

13.4 CHALLENGES ASSOCIATED WITH DISINFECTION OF FOOD PRODUCES IN SPACE

The load added to the spaceflight/craft plays a crucial role in deciding the mission cost. Hence, when considering food equipment for a space mission, it should justify its mass by providing a valuable contribution in processing the food products. Nevertheless, the equipment should not occupy more volume of the spaceflight as the vehicle itself is designed in such a way that it can support other essential parts and resources within limited space to facilitate the travel. Thus, it is always appropriate to carry compact lightweight processing equipment for essential needs. In addition, it is also preferable to carry a processing system/equipment that is capable of performing multiple tasks (i.e., knife). Therefore, plans made for long-distance travels like Mars mission will permit only specific equipment that justifies the following requirement: "Electronic and mechanical equipment must be extremely autonomous, self-maintaining or crew-maintained, and perhaps self-repairing. The amount of time required to perform normal procedures must be reduced through system design. The operation of supporting systems (such as electricity, life support, and in-situ resource recovery) should, in theory, be transparent to the crew. In this case, the ideal strategy is to establish the need for technical advancement based on the mission objectives for a certain crew size" [32].

In this regard, selecting processing equipment to disinfect the SFGP (without altering the freshness) becomes complicated as most of the processing (non-thermal) equipment demand more resources (cost, energy, space). For example, the HPP processing system requires high pressure to process the food product. To produce the required pressure, the system (treatment chamber) should be integrated with high-pressure pumps and pressure-transmitting medium supply units [3]. These integrated parts will add extra load to the spaceflight and also occupy more space. Furthermore, the HPP system can only be used for processing food products and does not have any other scope in space. Therefore, it cannot justify the requirement to be added to the list of spaceflight processing equipment.

Similarly, PEF requires a conducting medium, and the Ozone process requires O_2 (also emits O_3), which makes them not suitable for space missions. However, it is necessary to disinfect the SFGP to avoid microbial contamination in food. Thus, ISS uses the PRO-SCAN final wash disinfection solution in the VEGGIE food production system to sanitize the product surfaces. With the application of PRO-SCAN, it is possible to achieve more than a 3-log reduction in the microbial population. Thus, it is possible to use PRO-SCAN – one-time-use wipes to bring the SFGP's microbial population to a safer level as per NASA's standard. Though PRO-SCAN is a citric acid-based safe sanitizer (GRAS) for space missions, the uneven surface characteristics of SFGP (i.e., radish) do not allow complete disinfection of the produce.

Long-duration space travels require regular wipes, which increases PRO-SCAN requirements. It not only affects the mass and volume occupancy of spaceflight but also increases the waste generation inside the space vehicle [7]. Though thermal methods are found to be effective on Earth, they do not fulfill the expectations for being considered as a sustainable processing method in a microenvironment. This is due to the fact that plants might need regular monitoring and surface disinfection procedures to keep the spacecraft environment problem-free. Whereas thermal processes are not suitable to disinfect living cells (i.e., plants, including seeds). Thus, it is appropriate to say that the need for food processing equipment in a spacecraft shuttle is not fulfilled.

13.5 ROLE OF COLD PLASMA IN SPACEFLIGHT FOR DISINFECTION

Currently, the focus of space research organizations is to find suitable processing equipment that can sustain in microgravity conditions without demanding many resources. Currently, CP technology is starting to gain attention in space research due to its wider applications and operational feasibilities. This section discusses the advantages/convenience of using CP in space missions and its potential applications in SFGP cultivation and disinfection.

13.5.1 ADVANTAGES OF ADAPTING CP IN SPACE MISSIONS OVER OTHER DISINFECTION METHODS

Firstly, CP processing does not require water (unlike HPP) to treat the SGFP and saves water resources for humans. Secondly, it can be created in an isolated chamber both at ambient and vacuum conditions without the need for any complex mechanisms. Further, it is possible to obtain uniform plasma exposure throughout produce surfaces by providing microgravity rotation. These factors make CP a microgravity-friendly technique for food processing. Therefore, when plasma is created inside a closed chamber at ambient pressure, the plasma-rich air can be used for disinfecting the SGFP. Nevertheless, the plasma system can also be maintained at vacuum conditions when treating the food products [7]. The use of CP in the Agri and processing sector is not a new approach as its applications (Figure 13.3) are well studied under a normal environment. Further, practical and economical aspects of space missions involving continuous supply of food also suggest cold plasma processing as an alternative for space food applications due to the low resource requirement (e.g., minimal air is sufficient to produce a plasma environment) [34]. However, it is also noteworthy that plasma processing is also associated with a few drawbacks.

The use of a compressed air cylinder in the case of low pressure (approx. 0.3 bar) CP processing might be a threat to spacecraft safety. In addition, vacuum, electromagnetic fields, and free radicals (i.e., O_3) created during plasma processing are also considered as risk factors in CP [7]. Nevertheless, these drawbacks are acceptable, considering the multi-purpose applications provided by CP during space missions. Thus, NASA has recently evaluated two CP systems (low pressure system-0.1–1.0 mbar and atmospheric system) capable of producing a saturated plasma environment using room air. However, these systems are operated under 40°C by utilizing normal air at one atmospheric pressure to keep the operation safe. Furthermore, maintaining these conditions can prevent the toxic oxidant (i.e., O_3) production harmful oxidant production [9]. The following subsections explain various applications of CP processing in space.

13.5.2 PLASMA FOR SEED DISINFECTION

Microbial safety of SFGPs depends on the environmental factors associated with spacecraft and is often influenced by the microorganism carriers such as plants, humans, and spaceflight contact surfaces. However, there are possibilities of seeds carrying the microbes when they are shipped from Earth. To minimize the occurrence of this type of seed-carried microbial infection, seed surfaces are sanitized before the launch of spaceflight. Even though conventional disinfection methods such as alcohol soaking and fumigation are practiced on Earth, they often result in substandard sanitation. Hence, the practice of seed disinfection was performed using CP systems to analyze their efficiency. The positive results of these experiments suggested that plasma disinfection was not only effective against seed microbes but also against other contact surfaces. While the microbial log reduction of CP treatment depends on the plasma source, power level, and gas used during the treatment [34]. Furthermore, most of the produce that is suitable to cultivate in spaceflight (Figure 13.2) has already been experimented with CP on Earth. For example, tomato seeds treated with CP at 80 W are less likely to be affected by bacterial wilt disease as the treatment inhibits the *Ralstonia solanacearum* activity. In addition, the disease severity is also 25% less in the case of plasma-treated seeds than in the control samples [21]. However, for using a plasma system in space, it should be able to reduce the microbial count from 500 CFU to 10 CFU in a 100 cm^2 surface area, only then the system will be considered for the mission by NASA [9].

13.5.3 CP FOR DISINFECTION OF FOOD PRODUCE

Though fruits, vegetables, and other agricultural products are essential for the human diet, they are also carriers of pathogenic microorganisms. Unlike on Earth, sanitizers and disinfectants cannot perform the task of surface disinfection on agricultural produce due to limited resource availability, uneven surface morphology of produce, and the microgravity environment [37]. That's the reason why SFGP are treated using CP systems, and some of the commodities that are treated using CP in a microgravity environment are in this section.

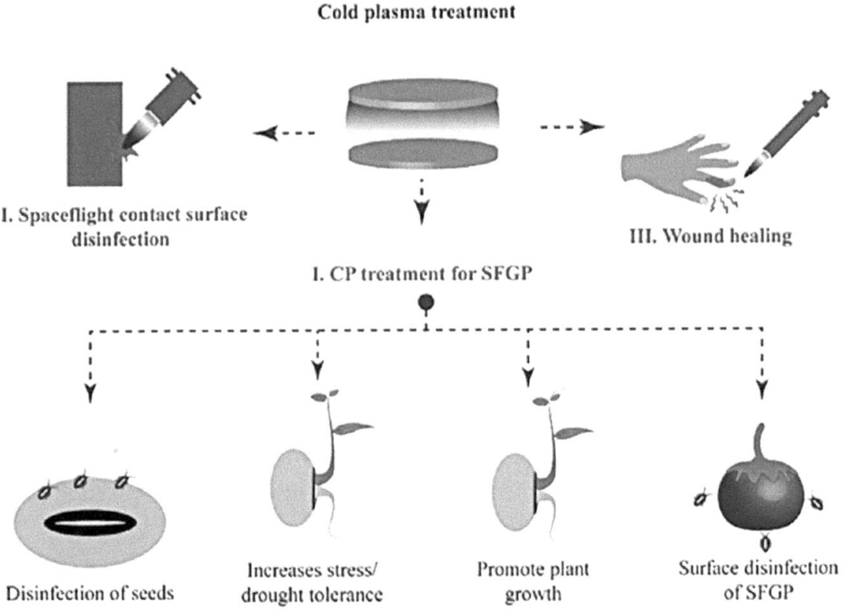

FIGURE 13.3 Potential applications of CP in space missions.

Hintze et al. [17] experimented on the disinfection efficiency of low-pressure CP on space-grown (ISS) agricultural produce. For this purpose, ISS lab-grown cherry tomatoes, radishes, peppers, cabbage, and a metal coupon were selected and inoculated with *B. cereus, E. coli, B. pumilus*, and

A. Niger. These selected commodities were then subjected to cold plasma treatment at low pressure (10–100 pascal) for up to 60 minutes. The initial observation showed that the CP had no significant impact on the microbial load of SFGPs. However, it was able to reduce up to 5 log microbial reduction from the metal surface. Later, when the treatment conditions were optimized for all SFGPs, tomatoes, cabbage, and peppers did not show any conclusive evidence of microbial reduction. However, radishes alone showed more than a 1.7 log reduction in *E. coli* count. The high moisture content of tomatoes, cabbage, and peppers did not allow the plasma to form inside the treatment chamber due to uncontrollable pressure changes. Nevertheless, when atmospheric plasma was used on cherry tomatoes at 100 kV, it was found to reduce the microbial count (*E. coli, Listeria innocua*) by up to 3.5 logs within 2.5 minutes [44].

Microwave-assisted CP treatment on cabbage resulted in more than a 1.5 log reduction in *Salmonella typhimurium* and *Listeria monocytogenes* count within 10 minutes of treatment. Thus, selecting the appropriate environment (vacuum or air) for treatment will increase the efficiency of CP treatment on microbial reduction. However, the quality changes associated with the disinfection process might be an issue when using cold plasma. Evidently, studies also reported changes in food product color, aroma, and nutritional profile after plasma treatment due to the interaction of oxidative reactive species with the produce [7]. However, numerous other studies state negligible changes in product quality during CP treatment. For example, when red chicory was treated in a DBD CP system (~20 V) for 15 minutes, it did not induce any significant changes in product color, texture, or freshness. Similar results were observed for Romaine lettuce, cherry tomatoes, and radish sprouts [43]. Hence, it is clear that both product nature and plasma operating conditions play a major role in deciding the processed product's sensory and microbial quality. Optimizing the operational conditions for all food products before a space mission will reduce the risk of product rejection after CP treatment.

13.6 OTHER APPLICATIONS OF CP IN SPACE MISSIONS

13.6.1 WOUND HEALING

As on Earth, certain human activities in space (spacecraft/station) also cause injuries to the human body during missions. Since the body is exposed to an unusual microgravity environment, these injuries could lead to life-threatening health issues. Considering the medical support available in space, it is necessary to treat the injuries as early as possible to prevent microbial infections.

In this regard, cold plasma (CP) can provide the required medical support through noninvasive, pain-free, and nontoxic therapy to disinfect wounds [41].

The reactive species (nitrogen/oxygen) present in the CP promote wound healing by modulating wound healing-relevant cell characteristics such as viability, proliferation, migration, and inflammation. Furthermore, the reactive species act as transmitters for a wide range of cell responses that aid in wound healing [36]. An additional advantage of CP treatment in the wound healing process is that it permeates and spreads uniformly over the specified area within a short duration. Moreover, it can also inhibit drug-resistant bacteria (i.e., *Staphylococcus aureus*) within a short period of time. CP is not only effective against bacteria but also inhibits fungi and viruses. These conclusions regarding microbial safety are drawn from data obtained from years of clinical studies. Currently, the International Space Station (ISS) uses a DC plasma integrated with the ESA/ROSCOSMOS PK-4 system [22].

13.6.2 SURFACE DISINFECTION

Cleanliness of spaceflight contact surfaces is important to prevent astronauts and SFGP from most of the harmful microbes' contaminants. Furthermore, it is also essential for a rover probe to be sanitized before it starts searching for life's evidence. In this regard, space habitats need to have a provision to disinfect contact surfaces under microgravity environment. Meanwhile, CP systems provide a sustainable solution to disinfection-related issues in space. However, the treatment time to achieve the required microbial reduction varies based on microbes and contact surface. For example, when *E. coli* bacteria are present on a stainless steel surface, the microbial population can be decreased by more than 5 logs within 5 minutes of CP exposure. Whereas, on polymer surfaces, up to 4 log reduction of *E. coli* can be achieved within 3 minutes [23]. Thus, it is essential to experiment with CP treatment combinations on different contact surfaces of spaceflight under artificial microgravity environment to analyze its efficiency during space missions [25].

13.6.3 PLANT GROWTH ENHANCEMENT

There are numerous studies that have reported positive influences of CP treatment on plant growth, including high germination, physical growth of plants, and improved crop yield [12, 14, 26]. However, these studies are performed under normal Earth environment, whereas the problems associated with the space environment might be different. Thus, plant growth in space requires more assistance than on Earth. During spaceflight, the seeds available for sowing are stored for a longer duration due to the complex procedures involved in spaceflight launch preparation. As a result, the seeds will enter a dormant stage, and this time period may be extended up to 2.5 years depending on the mission. Therefore, when the seeds need to be germinated to cultivate crops, they should break their dormancy [28]. In this regard, CP treatments have been proven to be effective in enhancing seed germination by breaking dormancy and increasing seed wettability. Furthermore, it can also enhance plant growth, yield, and produce safety in a similar way as it does on Earth [20, 30]. Hence, plasma treatment can be an all-in-one technique to assist plant growth at different life stages in space.

13.6.4 INCREASE DROUGHT AND SALINE TOLERANCE

Water is a very essential resource for astronauts not only for consumption and cleaning purposes, but also for O_2 generation and space vehicle propellant. Hence, it is essential to utilize the available water with utmost care and recycle it after usage. There are many techniques followed in space stations to recycle used water; however, Water and Urine Processor Assemblies are the two major systems responsible for this process [13]. Nevertheless, SFGP can perform this water recycling process (producing pure water vapor) without any external processing units while producing food for astronauts. In this case, water resources can be effectively utilized for both human and plant growth. Furthermore, the plants produced inside the spacecraft help minimize the CO_2 level of the internal environment. However, for a mission like Mars travel, the available resources are highly limited due to the extended mission period. Thus, it is impossible to supply pure water continuously for plant growth; instead, they can only be irrigated with wastewater (i.e., urine). In addition, regulating the water supply is difficult for SFGP in a microgravity environment [6].

To overcome these situations (water scarcity and saltwater), SFGP needs to have tolerance against drought and salinity. Meanwhile, treating seeds and plants under CP environment increases the drought and saline tolerance. For example, rapeseed treated using CP showed more than a 6% increase in germination rate under drought conditions [27]. While treating the wheat seeding with CP (power intensity: 0.84 W/cm^2) increased the saline tolerance of the plant and also increased the chlorophyll content [19]. Thus, it is evident that CP could be a suitable method for supporting plant growth in space environments shortly.

13.6.5 RESOURCE RECYCLING

As discussed earlier, the CP-treated seeds/plants are capable of growing in drought and saline environments. Hence, the plants grown in spaceflight can utilize the nutrients available in saline-rich wastewater (i.e., urine salt and nitrogen) and recover them for human use (in the form of fruit, vegetables, or leaves). For example, humans require up to two grams of NaCl per day, and 93% of this gets eliminated in urine. Since salt is essential for our human body, it is essential to recover the wasted salt through recycling processes. However, the process of recycling and reintroducing the squandered salt into the diet is a tough task. Nevertheless, plants can perform this process and help in salt recovery.

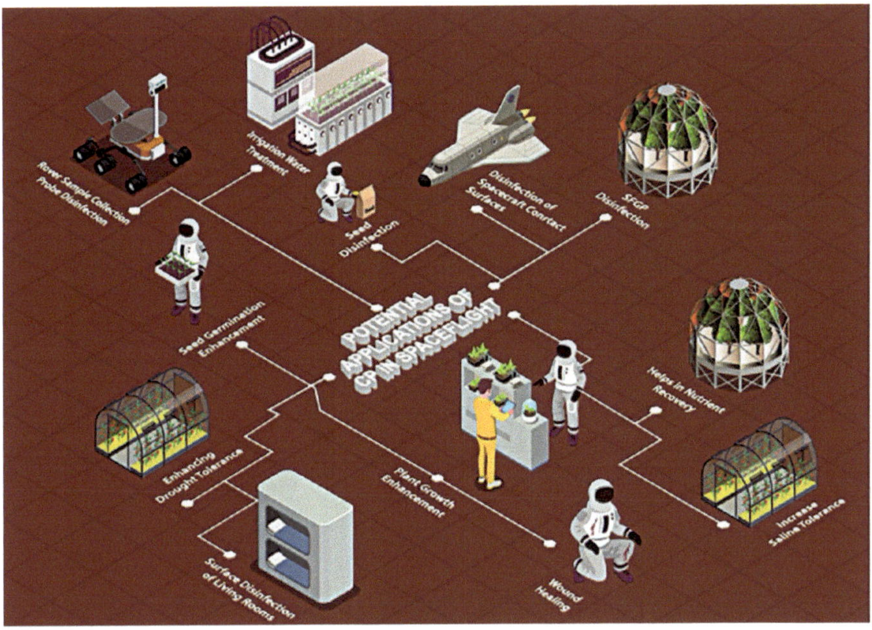

FIGURE 13.4 Potential applications of CP in during space mission program.

Source: Modified from freepik.com

Therefore, it is essential for an SFGP to have saline tolerance to utilize the wastewater for its growth and to facilitate salt recovery for human dietary needs. CP treatments are a proven method to increase the saline tolerance of plants, hence they can be used in space to treat the seeds and plants which will increase the chances of salt recovery. In addition, these plants also recycle the CO_2-containing air and convert it into O_2 during photosynthesis. Furthermore, they also help in recovering the water from wastewater sources in the form of water vapors [5, 11, 29]. Though CP has a wide range of applications (Figure 13.4) in space programs, there are certain areas that need improvement to consider it as a sustainable technique in space missions.

13.7 FUTURE SCOPE

Though cold plasma supports the growth and disinfection of SFGPs, the nutritional profile and sensory acceptance of CP-treated food produce need to be investigated in the future. Furthermore, it is essential for any disinfection treatment methods to provide better (or similar) microbial reduction than the existing PRO-SCAN method to meet NASA safety standards [7]. Agriculture produce with high moisture content will significantly affect the performance of CP systems operating under low pressure. Therefore, future research should focus on developing a CP system that supports both vacuum and air plasma treatment for high and low moisture food produce in space. Additionally, biofilm formation on the contact surfaces of spaceflights poses a serious threat to microbial safety. CP efficiency on biofilm degradation needs to be evaluated during future missions [34]. Although CP can be used for various applications (other than disinfection) in space, studies are yet to be performed and evaluated under microgravity conditions.

13.8 SUMMARY

Considering the CP's wide range of applications and its potential to work under microgravity conditions provides a valid justification for this equipment to be a part of a space mission. Furthermore, this technique is not only useful for treating plants grown in space but also for treating humans in emergency situations. However, designing a cold plasma system that supports all these applications (in terms of configuration and structure) is the major concern. Thus, the emergence of a new portable design for a plasma system and validation of sensory attributes of plasma-treated food will establish a solid place for the plasma system in space exploration missions.

KEYWORDS

- **astronauts**
- **cold plasma**
- **disinfection**
- **space food**
- **space mission**
- **spaceflight grown produces**

REFERENCES

1. Anbarasan, R., Jaspin, S., Bhavadharini, B., Pare, A., Pandiselvam, R., & Mahendran, R. (2022). Chlorpyrifos pesticide reduction in soybean using cold plasma and ozone treatments. *LWT, 159*, Article ID 113193. https://doi.org/10.1016/j.lwt.2022.113193.
2. Aparajhitha, S., & Mahendran, R. (2019). Effect of plasma bubbling on free radical production and its subsequent effect on the microbial and physicochemical properties of Coconut Neera. *Innovative Food Science & Emerging Technologies, 58*, Article ID 102230. https://doi.org/10.1016/j.ifset.2019.102230.
3. Balasubramaniam, V. M., Barbosa-Canovas, G. V., & Lelieveld, H. L. (Eds.). (2016). *High Pressure Processing of Food: Principles, Technology and Application.* ISBN: 978-1-4939-3233-7, Edition: 1, pp. 1–762, Publisher: Springer New York, NY, https://doi.org/10.1007/978-1-4939-3234-4.
4. Baskaran, K., & Radhakrishnan, M. (2021). Surface pasteurization and disinfection of dairy processing equipment using cold plasma techniques. In K. Keener (Ed.), *Non-Thermal Processing Technologies for the Dairy Industry* (pp. 153–166). CRC Press. https://doi.org/10.1201/9781003138716-11.
5. Bormashenko, E., Grynyov, R., Bormashenko, Y., & Drori, E. (2012). Cold radiofrequency plasma treatment modifies wettability and germination speed of plant seeds. *Scientific Reports, 2*, Article 589. https://doi.org/10.1038/srep00589.
6. Carillo, P., Morrone, B., Fusco, G. M., De Pascale, S., & Rouphael, Y. (2020). Challenges for a sustainable food production system on board of the international space station: A technical review. *Agronomy, 10*(5), Article 687. https://doi.org/10.3390/agronomy10050687.
7. Douglas, G. L., Massa, G. D., Hummerick, M. E., & Hintze, P. E. (2019). Cold plasma to disinfect spaceflight grown produce. In P. J. Cullen & S. S. P. Shetty (Eds.), *Advances in Cold Plasma Applications for Food Safety and Preservation* (pp. 225–238). Academic Press. https://doi.org/10.1016/B978-0-12-814921-8.00012-8.
8. Douglas, G. L., Wheeler, R. M., & Fritsche, R. F. (2021). Sustaining astronauts: Resource limitations, technology needs, and parallels between spaceflight food systems and those on Earth. *Sustainability, 13*(16), Article 9424. https://doi.org/10.3390/su13169424.

9. Duda, Z., Gaffney, J., Graves, C., Moore, Q., Watkins, J., & Nagel, J. (2017). Medical sterilization system for NASA Space exploration missions. In *2017 Systems and Information Engineering Design Symposium* (SIEDS), pp. 277–282, Publisher: IEEE. https://doi.org/10.1109/SIEDS.2017.7937731.
10. Fravel, D., Olivain, C., & Alabouvette, C. (2003). *Fusarium oxysporum and Its Biocontrol*. New Phytologist, 157(3), 493–502. https://doi.org/10.1046/j.1469-8137.2003.00700.x.
11. Fu, Y., Li, L., Xie, B., Dong, C., Wang, M., et al. (2016). How to establish a bioregenerative life support system for long-term crewed missions to the Moon or Mars. *Astrobiology, 16*(12), 925–936. https://doi.org/10.1089/ast.2016.1510.
12. Gao, X., Zhang, A., Héroux, P., Sand, W., Sun, Z., et al. (2019). Effect of dielectric barrier discharge cold plasma on pea seed growth. *Journal of Agricultural and Food Chemistry, 67*(39), 10813–10822. https://doi.org/10.1021/acs.jafc.9b03890.
13. Gott, R. P., Miller, S. D., Hodges, J. C., & Xu, K. G. (2021). Plasma-based water purification for crewed space missions: Laboratory experimental comparisons for on-board applicability. *Advances in Space Research, 68*(3), 1591–1600. https://doi.org/10.1016/j.asr.2021.05.010.
14. Hashizume, H., Kitano, H., Mizuno, H., Abe, A., Yuasa, G., et al. (2021). Improvement of yield and grain quality by periodic cold plasma treatment with rice plants in a paddy field. *Plasma Processes and Polymers, 18*(1), Article 00181. https://doi.org/10.1002/ppap.202000181.
15. Haveman, N. J., Khodadad, C. L. M., Dixit, A. R., Louyakis, A. S., Massa, G. D., et al. (2021). Evaluating the lettuce metatranscriptome with MinION sequencing for future spaceflight food production applications. *npj Microgravity, 7*(1), Article 15. https://doi.org/10.1038/s41526-021-00149-7.
16. Heiney, A., & Dunbar, B. (2021). Growing plants in space. National Aeronautics and Space Administration. Publisher: NASA. Retrieved from: https://www.nasa.gov/content/growing-plants-in-space (accessed on 16 June 2024).
17. Hintze, P. E., Franco, C., Hummerick, M. E., Maloney, P. R., Spencer, L. E., et al. (2017). Evaluation of low-pressure cold plasma for disinfection for ISS grown produce and metallic instrumentation. *International Conference on Environmental Systems (ICES)*, pp. 1–10. Conference ID No. KSC-E-DAA-TN39821. Retrieved from: https://ntrs.nasa.gov/api/citations/20170006639/downloads/20170006639.pdf (accessed on 16 June 2024).
18. Houston, C. S. (1998). Polar journeys: The role of food and nutrition in early exploration, by Robert E. Feeney. *Arctic, 51*(4), 386–390. https://www.jstor.org/stable/40512225 (accessed on 16 June 2024).
19. Iranbakhsh, A., Ardebili, N. O., Ardebili, Z. O., Shafaati, M., & Ghoranneviss, M. (2018). Non-thermal plasma induced expression of heat shock factor A4A and improved wheat (*Triticum aestivum* L.) growth and resistance against salt stress. *Plasma Chemistry and Plasma Processing, 38*(1), 29–44. https://doi.org/10.1007/s11090-017-9853-0.
20. Jiang, J., He, X., Li, L., Li, J., & Shao, H. (2014). Effect of cold plasma treatment on seed germination and growth of wheat. *Plasma Science and Technology, 16*(1), 54–58. https://doi.org/10.1088/1009-0630/16/1/10.
21. Jiang, J., Lu, Y., Li, J., Li, L., & He, X. (2014). Effect of seed treatment by cold plasma on the resistance of tomato to *Ralstonia solanacearum* (bacterial wilt). *PLoS One, 9*(5), Article e97753. https://doi.org/10.1371/journal.pone.0097753.

22. Johnson, M. (2019). Cold plasmas assist in wound healing. National Aeronautics and Space Administration. Retrieved from: https://www.dlr.de/en/latest/news/2021/01/20210303_twenty-years-of-plasma-research-on-board-the-iss.
23. Katsigiannis, A. S., Bayliss, D. L., & Walsh, J. L. (2022). Cold plasma for the disinfection of industrial food-contact surfaces: An overview of current status and opportunities. *Comprehensive Reviews in Food Science and Food Safety, 21*(2), 1086–1124. Advance online publication. https://doi.org/10.1111/1541-4337.12885.
24. Khodadad, C. L. M., Hummerick, M. E., Spencer, L. S. E., Dixit, A. R., & Richards, J. T. (2020). Microbiological and nutritional analysis of lettuce crops grown on the International Space Station. *Frontiers in Plant Science, 11*, Article 199. https://doi.org/10.3389/fpls.2020.00199.
25. Laroussi, M. (2022). A brief note on the history of the dielectric barrier discharge and its application for biological decontamination. *IEEE Transactions on Radiation and Plasma Medical Sciences, 6*(1), 121–125.
26. Li, L., Li, J., Shen, M., Hou, J., & Shao, H. (2016). Improving seed germination and peanut yields by cold plasma treatment. *Plasma Science and Technology, 18*(10), 1027–1033. https://doi.org/10.1088/1009-0630/18/10/09.
27. Ling, L., Jiangang, L., Minchong, S., Chunlei, Z., & Yuanhua, D. (2015). Cold plasma treatment enhances oilseed rape seed germination under drought stress. *Scientific Reports, 5*, Article 13033. https://doi.org/10.1038/srep13033.
28. Meier, A., Essumang, D., Hummerick, M., Johnson, C., & Kruger, M. (2021). Reviewing plasma seed treatments for advancing agriculture applications on earth and into the final frontier. *Gravitational Space Research, 9*(1), 133–158.
29. Mortimer, J. C., & Gilliham, M. (2022). SpaceHort: Redesigning plants to support space exploration and on-earth sustainability. *Current Opinion in Biotechnology, 73*, 246–252. https://doi.org/10.1016/j.copbio.2022.01.013.
30. Niemira, B. A. (2012). Cold plasma decontamination of foods. *Annual Review of Food Science and Technology, 3*(1), 125–142. https://doi.org/10.1146/annurev-food-022811-101139.
31. Perchonok, M., Douglas, G., & Cooper, M. (2012). Evidence report: Risk of performance decrement and crew illness due to an inadequate food system. *Human Research Program Space Human Factors and Habitability Element*, 1–33. Retrieved from: https://ntrs.nasa.gov/api/citations/20160011582/downloads/20160011582.pdf (accessed on 16 June 2024).
32. Perchonok, M. H., Cooper, M. R., & Catauro, P. M. (2012). Mission to Mars: Food production and processing for the final frontier. *Annual Review of Food Science and Technology, 3*(1), 311–330. https://doi.org/10.1146/annurev-food-022811-101139.
33. Pimentel, L. (2003). Scurvy: Historical review and current diagnostic approach. *The American Journal of Emergency Medicine, 21*(4), 328–332. https://doi.org/10.1016/S0735-6757(03)00134-8.
34. Poulet, L., Engeling, K., Hatch, T., Stahl-Rommel, S., & Velez Justiniano, Y.-A. (2022). Large-scale crop production for the Moon and Mars: Current gaps and future perspectives. *Frontiers in Astronomy and Space Sciences, 8*, Article 733944. https://doi.org/10.3389/fspas.2021.733944.
35. Smith, S. M., & Zwart, S. R. (2016). Human adaptation to spaceflights: The role of nutrition. National Aeronautics and Space Administration. Retrieved from: https://www.

nasa.gov/sites/default/files/human-adaptation-to-spaceflight-the-role-of-nutrition.pdf (accessed on 16 June 2024).
36. Schmidt, A., Bekeschus, S., Wende, K., Vollmar, B., & von Woedtke, T. (2017). A cold plasma jet accelerates wound healing in a murine model of full-thickness skin wounds. *Experimental Dermatology, 26*(2), 156–162. https://doi.org/10.1111/exd.13111.
37. Sharma, R. R., Reddy, S. V. R., & Sethi, S. (2018). Cold plasma technology for surface disinfection of fruits and vegetables. In S. B. Brul, P. M. Fratamico, & A. C. Oyarzabal (Eds.), *Postharvest Disinfection of Fruits and Vegetables* (pp. 215–232). Elsevier. https://doi.org/10.1016/B978-0-12-812698-1.00010-8.
38. Smith, S. M., & Zwart, S. R. (2008). Nutritional biochemistry of spaceflight. *Advances in Clinical Chemistry, 46*, 87–130. https://doi.org/10.1016/S0065-2423(08)00402-5.
39. Sychev, V. N., Levinskikh, M. A., Gostimsky, S. A., Bingham, G. E., & Podolsky, I. G. (2007). Spaceflight effects on consecutive generations of peas grown onboard the Russian segment of the International Space Station. *Acta Astronautica, 60*(4–7), 426–432. https://doi.org/10.1016/j.actaastro.2006.09.004.
40. Taylor, P. W. (2015). Impact of space flight on bacterial virulence and antibiotic susceptibility. *Infection and Drug Resistance, 8*, 249–262. https://doi.org/10.2147/IDR.S87853.
41. Tipa, R. S., & Kroesen, G. M. W. (2011). Plasma-stimulated wound healing. *IEEE Transactions on Plasma Science, 39*(11), 2978–2979. https://doi.org/10.1109/TPS.2011.2178474.
42. Van Houdt, R., Mijnendonckx, K., & Leys, N. (2012). Microbial contamination monitoring and control during human space missions. *Planetary and Space Science, 60*(1), 115–120. https://doi.org/10.1016/j.pss.2011.09.010.
43. Varilla, C., Marcone, M., & Annor, G. A. (2020). Potential of cold plasma technology in ensuring the safety of foods and agricultural produce: A review. *Foods, 9*(10), Article 1435. https://doi.org/10.3390/foods9101435.
44. Ziuzina, D., Misra, N. N., Cullen, P. J., Keener, K., & Mosnier, J. P. (2016). Demonstrating the potential of industrial scale in-package atmospheric cold plasma for decontamination of cherry tomatoes. *Plasma Medicine, 6*(3–4), 397–412. https://doi.org/10.1016/j.plasmamed.2016.11.001.
45. Zwart, S. R., Kloeris, V. L., Perchonok, M. H., Braby, L., & Smith, S. M. (2009). Assessment of nutrient stability in foods from the space food system after long-duration spaceflight on the ISS. *Journal of Food Science, 74*(7), H209–H217. https://doi.org/10.1111/j.1750-3841.2009.01357.x.

PART V
Research Challenges in Cold Plasma Technology

CHAPTER 14

COMPUTATIONAL FLUID DYNAMICS IN COLD PLASMA PROCESSING: BASICS, SCOPE, AND APPLICATIONS

ANBARASAN RAJAN and R. MAHENDRAN

ABSTRACT

Cold plasma (CP) is an emerging technology in the food sector that provides a wide range of applications in both pre- and post-harvest processing. Although CP technology is used commercially in many fields, it is still primarily used at the laboratory level in food industries. However, the technology has been proven to be effective for improving germination, extraction, surface texture, insect mortality, microbial inhibition, and protein quality. Despite CP not being adapted for industrial-scale applications due to various reasons such as problems associated with the production of product-specific CP devices and their scale-up, defining the process parameters for specific food products, process control, consumer acceptance, and regulatory permission. Computational fluid dynamics (CFD) methods can perform realistic simulations to provide a thorough evaluation of the influences of plasma operating conditions and system designs on reactive species (RS) generation. Thus, the plasma fluid flow behavior and RS transport can be computed for any given plasma applications using CFD. However, when it comes to plasma species interaction with food materials, the data are limited which restricts the efficiency of CFD simulation. Hence, there are only limited CFD studies available for CP-treated food materials. Therefore, the current chapter explains the possibilities of simulating plasma species interaction with food materials.

Nonthermal Plasma Technology for Food and Food Products.
R. Mahendran, Preeti Birwal, and Megh R. Goyal (Eds.)

14.1 INTRODUCTION

Cold plasma (CP) has gained considerable attention as a non-thermal food processing method due to its inexpensive, adaptable, and ecologically safe nature. This technique can be used for food decontamination, disinfestation, enzyme inactivation, toxin removal, food packaging alterations, and wastewater treatment [1, 19]. Technically, CP is an ionized gas containing reactive species, atoms, charged particles (electrons and other positive and negative ions), photons, and ground-level or excited state molecules in a quasi-neutral state at room temperature. In this quasi-neutral state, the temperature of free electrons can reach up to 10,000 K, while the temperature of the heavy species varies only between 300 and 1,000 K. Due to the high-temperature gradient between the plasma reactive species (RS), CP is thermodynamically referred to as non-thermal plasma [16]. Furthermore, the heavy species present inside the plasma chamber have the lowest transport rate (species movement) compared to other species, and because of this, these reactive species (electrons and other species) interact (i.e., collide) within and between them before they interact with the walls of plasma chambers [12].

As a consequence of these complex reactions, understanding plasma processing becomes difficult and it also demands immense knowledge in fluid mechanics, reaction engineering, physical kinetics, heat transfer, mass transfer, and electromagnetics [6]. Hence, the process of controlling and reproducing plasma with desired properties is difficult, and as a result, the process of investigating or analyzing the interaction of plasma species with other compounds becomes challenging at any given treatment condition. As a result, researchers conduct studies using the trial-and-error approach to get the desired plasma characteristic or the requisite treatment outputs. Furthermore, forecasting plasma characteristics becomes more challenging when operating factors such as supply voltage, electrode spacing, pulse switching frequency, gas composition, and air flow rate change. Though previous research findings can be utilized as a reference for future studies, only the quantitative analysis of plasma components can give credible prediction of the processes.

Current breakthroughs in simulation methodologies, combined with the technical assistance of high-end processors (hardware, software), allow for more in-depth modeling of challenging plasma processes. CFD methods can perform realistic simulations at a low cost and in a short amount of time. These simulations provide a thorough evaluation of the influence of plasma operating and design factors on RS generation. Furthermore, using the appropriate reaction kinetics and equations, any treatment outcome might well be predicted before the experiment is performed. The computer simulations can also provide comprehensive space-time data, which is typically difficult or even impossible to gather experimentally but is critical for understanding the physical processes involved in plasma generation [20]. The simulation of plasma at any given conditions through fluid or kinetic approach using CFD simulation software for forecasting its nature and interaction with targeted components is called computational plasma dynamics.

During the computation process, physical processes such as fluid dynamics, thermodynamics, and chemical reactions of plasma species will be considered. Among these, fluid dynamics deals with the plasma flow properties with respect to mass, momentum, and energy conservation. Meanwhile, thermodynamics deals with the heat transfer between the plasma particles along with the work done and volume change caused by pressure (i.e., in the case of low-pressure plasma). Finally, the RS generation and destruction can be determined by chemical reaction equations. All the physical processes involved in CFD are either evolving (at their own time-space scales) or in equilibrium [27].

Plasma is viewed as a continuum in the fluid approach, with the governing equations such as mass, momentum, and energy solved using the Maxwellian equations for the velocity distribution function. Each element in plasma is treated as a separate entity in the kinetic approach; the most common methods for solving the kinetic approaches are the Monte-Carlo collision and particle in cell techniques [12]. In any case, plasma modeling may be used to estimate plasma wastewater pollutant degradation under various operating circumstances.

CFD modeling software includes COMSOL Multiphysics, Autodesk CFD, SimScale, Ansys, OpenFOAM, Simcenter, Flowsquare, ParaView, Altair, and SOLIDWORKS. However, COMSOL Multiphysics offers a distinct module as well as multiphysics support to properly model and simulate plasma under various process conditions.

This chapter discusses the basics of CFD modeling and its potential applications as a predictive tool in different plasma processing.

14.2 COLD PLASMA: BASIC CONCEPTS OF CFD SIMULATION AND MODELLING

One must understand basic concepts of CFD. In CFD, the physical aspects of fluid flow behavior are governed by the fundamental equations such as mass, momentum, and energy conservation. Among these, the first law states that mass can neither be created nor destroyed; hence, mass entering inside the models will always be equal to the mass that went out of the system. This can be explained by the continuity equation (Eqn. 1) for any given CFD model:

$$\frac{\partial \rho}{\partial t} + \nabla.(\rho u) = 0 \tag{1}$$

where, ρ = density, t = time, and u = flow velocity.

Law of Momentum Conservation is described by Newton's Second Law of Motion (Eqn. 2), according to which, the sum of forces acting on a system is equals to the product of mass and acceleration of that system.

$$\rho \frac{dv}{dt} = -\nabla P + \rho f + F \tag{2}$$

where, P = pressure force, f = particle force, and F = viscous forces act together and equalize the density and acceleration of the system.

In CFD, the conservation of energy is solved (in the form of a continuous equation) only when the density and temperature of the fluid vary during processing. To deal with both mass and momentum in a laminar flow interface, plasma modeling uses Navier-Stokes equations, which supply the fluid velocity as well as the system's gas temperature.

Plasma modeling is a multi-physics strategy that incorporates components of reaction engineering, fluid mechanics, physical kinetics, electromagnetics, statistical physics, heat and mass transport, and so on. Hence, the data obtained in CFD simulation are the solutions resulting from the multi-physics coupling of different physics interfaces that were available in the built model.

Software plasma packages have simpler physical interfaces for modeling low-temperature plasma. These interfaces contain all of the necessary tools for modeling plasma discharges by estimating electron transport characteristics and source coefficients (obtained from cross-section data) using Boltzmann and Two-Term Approximation equations. Furthermore, the plasma discharge characteristics data for every given input (operation) state may be acquired using the provided interface capabilities [6]. Figure 14.1 illustrates the computational steps involved in the modeling of the plasma process. Nevertheless, the flow chart shown here is based on the model created in COMSOL; hence, the steps illustrated here need to be considered as an example only.

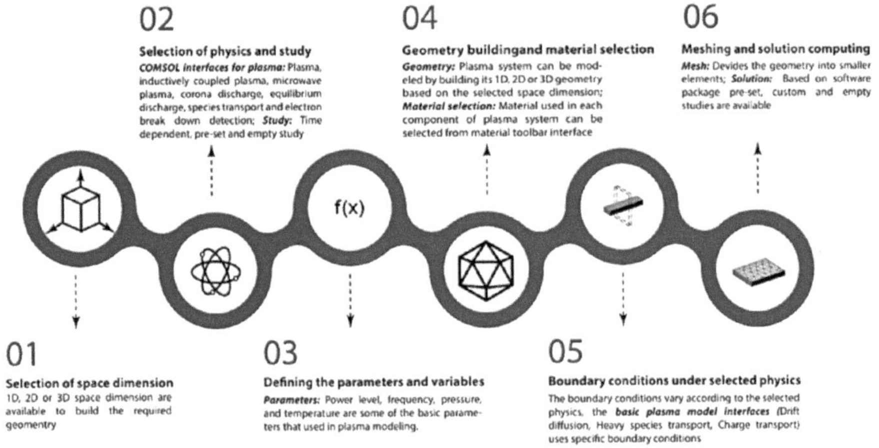

FIGURE 14.1 Steps in Computational fluid dynamics in plasma processing.

14.3 APPLICATIONS OF COMPUTATIONAL FLUID DYNAMICS IN COLD PLASMA

14.3.1 SOIL REMEDIATION

Plasma has the potential to degrade the pollutants present in the soil. However, the degradation efficiency of the treatment depends on various factors such as the pollutant concentration, soil porosity, RS concentration, and other process parameters (i.e., time, voltage). Atrazine is one of the organic pollutants present in the soil that causes health issues in humans. With the help of CFD modeling, it is possible to predict the effect of various operational and soil parameters on the degradation of atrazine for any given plasma treatment.

Bali et al. [3] used microscopic DBD plasma simulation to estimate the RS concentration at the macroscopic level. Initially, the authors predicted the plasma reactions and RS concentration for nanosecond plasma and utilized that data as a source term to find the results at the macroscopic level (minutes). Furthermore, by providing mobility to plasma species in the soil domain, the authors estimated the soil porosity effect on atrazine degradation. The results of the current CFD study identified a reduction in RS concentration (especially O_3) in the soil domain compared to the free discharge space due to the limited mobility (Eqn. 3) of plasma species inside porous soil.

$$\mu_{soil} = \frac{\varepsilon_p}{\tau} \mu_{free} \tag{3}$$

where, μ_{soil} = RS mobility inside the soil domain, μ_{free} = free space mobility, ε_p = soil porosity, and τ = tortuosity.

Once the RS entered inside the soil domain, it starts oxidizing the atrazine located on the surface of soil grains. The Eqn. (4) describes the atrazine degradation behavior inside the soil domain.

$$\frac{\partial C_{air}}{\partial t} = -R_{air}\varepsilon_p^{-1} \tag{4}$$

where, C_{air} and R_{air} are atrazine concentration and reaction rate, respectively.

In addition, an increase in the moisture content (from 5% to 10%) of the soil further reduces the porosity and thus the mobility of RS. However, at 5% soil moisture, the hydroxyl radicals inside the soil domain induce a synergistic effect on the atrazine degradation along with O_3 species. Furthermore, with the application of CFD, authors also predicted the effect of treatment time, air flow rate, and soil permeability on atrazine degradation.

14.3.2 WATER AND WASTEWATER TREATMENTS

14.3.2.1 WATER TREATMENT

Plasma-activated water (PAW) is mainly used for food disinfection and other medical applications. The reactive species (RS) captured inside the water play a major role in these applications. Therefore, understanding the stability and reactivity of these species is important for efficient process control. With the help of CFD modeling, it is possible to compute the species' stability inside the liquid (water) medium.

Heirman et al. [8] developed 0D and 2D CFD models to analyze the RS present inside the buffered water after plasma jet treatment. Essentially, the authors combined both 0-D and 2-D models of CFD to study the chemical reactions and fluid dynamics involved in the plasma-water buffer interaction, respectively, to reduce the obstacles in the computing process. Specifically, the essential RS and reaction data were obtained using the 0D model and used in the 2D model to reduce the computation time. In this manner, the authors computed the gas and liquid flow behavior along with the RS transport and chemical reactions in the liquid phase. Furthermore, the model helped the authors identify the stability of the RS in the liquid medium. Among the RS generated inside the liquid medium, HNO_2/NO_2, HNO_3/NO_3, and H_2O_2 were found to be more stable than RS such as O_3 and HO_2/O_2. The stability of these species depends on the production and loss rate of certain short-lived RS. These results were qualitatively similar to the data collected experimentally.

Therefore, CFD modeling can be adapted for PAW production processes to obtain the required RS concentrations and predict the PAW treatment results.

14.3.2.2 WASTE WATER TREATMENT

Sun et al. [25] studied the reactive blue 19 (RB-19) degradation, showing 59.9%, 49.6%, and 89.8% dye degradation when using Ar, air, and O_2 as carrier gases at 100 kJ/L energy density, respectively. Among the different reactive species produced during plasma treatment, O_3 has a major impact on dye degradation. To determine O_3's contribution to RB-19 oxidation, a 1D model was developed using COMSOL®. Since the plasma discharge was above the water surface in the study by Sun et al. [25], the equations (Eqn. 5) for convection-diffusion were solved to predict the dye and ozone concentrations in the liquid phase.

$$\nabla \bullet (-D_c \nabla C) = R_{cdye} - u \bullet \nabla C \tag{5}$$

To determine the diffusion coefficient (D_c), various assumptions were made (through heptanol dispersion co-efficient model). However, the D_c values for which the simulation results were close to experimental result were presented in the study. In Eqn. (5), the term R_{cdye} indicates the consumption of O_3 (mol m-3s-1) during the dye degradation process that can be obtained from the Eqn. (6) (k = reaction constant, C_{O3} = ozone concentration in liquid, and C_{dye} = dye concentration in liquid). In addition, Fick's Law (Eqn. 7) was used to solve the boundary conditions of the system Eqn. (7).

$$R_{cdye} = -k \bullet C_{O_3} C_{dye} \tag{6}$$

$$\frac{\partial C}{\partial t} = \nabla (D \nabla C) \tag{7}$$

Due to the exclusion of non-ozone RS in the model, the simulation exhibited reduced deterioration for all treatment times (5–15 min) compared to experimental data. According to the simulation model, the ozone contribution to dye degradation was 0.67 during the first 5 minutes of treatment and grew to 0.82 during the next 10 minutes. Apart from predicting the end results, CFD can also be used for optimizing the operational parameters like voltage/potential difference between electrodes.

EL-Tayeb et al. [7] employed a pin-to-plate corona discharge system to treat acid blue-25 dye-contaminated wastewater and to optimize electrode spacing for improved dye degradation. In this regard, a 3D simulation model was created to detect the system's corona onset voltage using software that used Integral Equations Technique (IET) to solve the partial differential equations. The electric field and electric potential of the system were calculated using IET 3D electrostatic design for the given 15 kV input voltage. Based on the results, a 5 mm air gap was shown to be beneficial for acid blue-25 degradation, causing 80% of the dye to be degraded within 10 minutes of treatment. The shaded figure derived from simulation revealed that the electric field distribution was larger near the sharper end of the electrode pins than in the other portions. Hence, the sharper electrode region was decided to be the high corona region (other details are not provided). However, due to complex computational procedures, computing a 3D model is often not achievable. Therefore, adapting a suitable 2D symmetrical geometry can reduce the computational difficulties and simulation time to a greater extent.

Ognier et al. [14] used a 2D symmetrical geometry model (COMSOL) for computing the gas-liquid discharge process and tried to understand the pollutant (1-heptanol, phenol) degradation mechanism in wastewater. To calculate the thickness (Eqn. 8) of the liquid film that comes in contact with the plasma, the wastewater was assumed to be in laminar flow with a parabolic velocity profile.

$$\bar{\delta} = \left(3\frac{\mu}{\rho} \times \frac{R_e}{g}\right)^{1/3} \tag{8}$$

Where, $R_e = \dfrac{\dot{m}}{\pi D \mu}$, μ = viscosity, ρ = density, g = acceleration of gravity (9.81 m/s^2), $\dot{m}$ = mass flow (kg/s), and D = liquid tube diameter.

From Eqn. (8), the film thickness was found to be 0.14 mm (at flow rate of 6 mL/min), while the gas velocity was computed using Navier–Stokes equation. The pollutant and O_3 concentrations at liquid and gas phase were computed using the convection-diffusion equation at both aqueous and gas phase.

$$\Delta.(-D\Delta c) = R - u.\nabla c \tag{9}$$

The RS with shorter life span was considered in the gas phase; and using the Eqn. (10), the oxidation reaction (R) was computed.

$$R = -K_{app} \times C^{P}_{gas} \quad 10$$

Where; C^{P}_{gas} = pollutant concentration in gas, and K_{app} = reaction rate which depends on the selected RS.

However, only O_3 species was considered in liquid phase and its concentration was calculated as the function of discharge energy density. The gas and liquid concentrations were assumed to be thermodynamically equilibrium at the boundary (Eqn. 11).

$$C_{liq} = K_h C_{gas} \quad 11$$

The computed results showed only 4% removal of 1-heptanol while the actual treatment removed up to 90% of the pollutant. Thus, it was found that the RS present in the gas phase alone cannot be considered for modeling as the major part of the pollutant damage was caused by the RS present (mixed) inside the liquid. However, when the assumptions (mass transfer in gas-liquid interface) were considered along with a suitable diffusion coefficient, the computation became effective and showed better results. The computed and calculated results for 1-Heptanol, Ethanol, Phenol & Acetic acid were 91%, 51%, 13% & 7.5% and 88%, 40%, 20% and 7%, respectively. It was also found that the O_3 present in the liquid caused 30% pollutant removal.

14.4 SCOPE OF CFD IN PLASMA MODELLING

Cold plasma has been used for different food processing applications such as food preservation, quality maintenance [13], surface modification [5, 10, 11], pesticide reduction [2, 21], allergen and anti-nutritional factor removal [26]. However, with the help of conventional CFD modeling, we could only compute the plasma parameters (fluid flow behavior, RS concentration and distribution), and not their interaction with food materials. Prediction of plasma parameters and their interaction with food materials will assist the food industries to control the process parameters, improve the treatment efficiency, and help in reactor design and scale-up processes. Though cold plasma-related CFD studies are limited in food applications, yet the studies available in medicinal and engineering fields can be used as a base for food material and cold plasma interaction studies. For example, CFD simulation studies have shown to be effective in predicting the plasma jet-generated OH radical concentration and their transport on the skin [4, 24]. In the same way, plasma species interaction with the food material surfaces can also be modeled using CFD. However, it requires certain data (i.e., species concentration, interaction, and stability as shown in Figure 14.2) to be collected before modeling to obtain the simulation result with a high degree of similarity with the experimental results. The following subsections explain the different possibilities of utilizing CFD modeling in plasma-assisted food applications.

14.4.1 MICROBIAL INACTIVATION

CP treatment generates reactive species that interact with microbial cells and induce cell death. Hence, the amount of reactive species produced during plasma treatment will have a direct impact on the microbial reduction of food samples. When the molar fraction of gas used for plasma treatment is known, the reactive species characteristics can be estimated with the help of other processing parameters using CFD [3, 18]. However, when it comes to liquid foods, the transport of reactive species from the gas phase to the liquid phase plays a major role in microbial reduction. To compute this, the convection-diffusion equation (for the transport of species into the liquid) needs to be solved along with the Navier–Stokes equations (for gas phase flow behavior) in the CFD model [22]. Nevertheless, the modeling becomes complicated if the predominant reactive species data are not input into the model. Despite this, obtaining OES data during plasma processing will reduce the difficulties in computation.

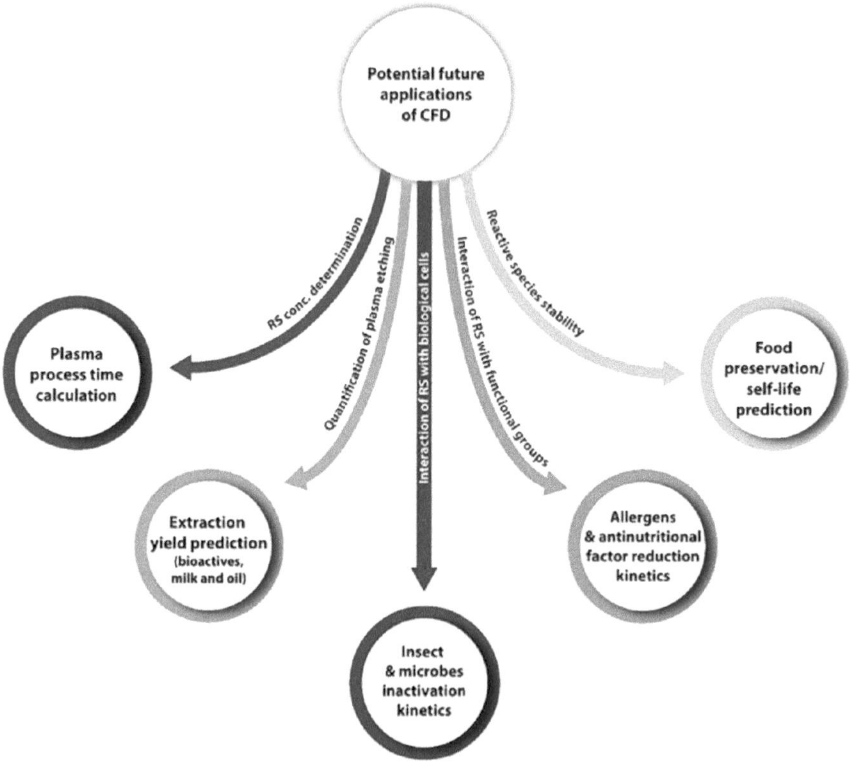

FIGURE 14.2 Possible potential future applications of CFD in CP processing.

At present, the major drawback of CFD modeling of CP-related microbial studies is the unavailability of data such as reactive species' interaction with microbial cells, their reaction rate, and the residence time taken for a known concentration of individual RS to cause cell death. However, a few studies have stated the importance of certain reactive species on microbial inactivation (i.e., O_3 molecules in *Aspergillus* species reduction from hazelnut) [23]. Therefore, by changing the plasma gas source, microbial destruction caused by a specific group of reactive species can be determined, and later these data can be fed to CFD modeling for computing plasma-induced cell death.

14.4.2 SEED GERMINATION

CP treatment has both positive and negative effects on seed germination. Therefore, predicting the surface modification that occurs on seeds can help in setting up the treatment conditions for better germination. Since seed germination depends on their water absorption behavior, predicting mass (water) transport through the modified seed surface is an important task in CFD modeling. However, a quantifiable measure is required to determine the plasma-induced surface modification before it can be modeled, and it can be achieved using chromatic confocal profilometry (CCP). CCP provides data regarding the loss in cross-sectional dimensions of materials that are treated (etched) by the plasma process [9]. Once data such as RS (OES data) concentration [15], surface modification percentage, minimum hydration required for germination, mass (water) transfer rate before and after plasma treatment are collected and calculated; CFD modeling can predict the germination percentage or germination rate for any given seed samples.

14.4.3 IN-PACK PROCESSING OF FOOD MATERIALS

Food commodities such as grains, vegetables, and fruits are currently treated using an in-pack approach, where the food contents within the packaging interact with ionized gas (plasma) immediately and the food surfaces are treated [17]. In this way, the microbial load in the food material is reduced without even introducing the food to the external environment. Since the packaging material properties and physical dimensions are known, plasma species produced inside the isolated food material containing region can be predicted using CFD.

14.4.4 REDUCTION IN USE OF CHEMICAL PESTICIDES

CP operational parameters like input voltage and treatment time directly influence pesticide degradation. Increasing the input voltage during CP treatment induces higher oxidation of the pesticides.

Anbarasan et al. [2] achieved higher chlorpyrifos degradation at 2.0 kV CP than at 1.0 kV. The higher pesticide degradation was due to the increase in RS concentration at higher voltages. The OES data obtained during CP treatment also confirmed the same by showing a considerable increase in the reactive species (i.e., NO, ozone, and oxygen ions) intensities. Since the study already found the major reactive species involved in chlorpyrifos oxidation, it is easier to compute the plasma dynamics for the treatment. However, the major issue in utilizing the CFD for its full potential is the unknown chlorpyrifos oxidation capacity of these reactive species. At the same time, there are also studies available that state the chlorpyrifos oxidation rate of ozone molecules. These data can be utilized in future studies for performing a complete CFD simulation of the plasma-induced pesticide degradation process.

14.5 CHALLENGES AND RESEARCH GAPS ASSOCIATED WITH COLD PLASMA CFD MODELLING

CFD modeling is proven to be valuable in different areas of food processing sectors. However, its effectiveness and practical applications are due to factors such as processing methods, food material properties, precision of the algorithms/equations, CFD programming software, and the computer hardware systems. The physical, rheological, and thermodynamic properties of food materials vary according to their nature; however, these data are applied to CFD models from an existing material library as air, liquid, or any other available material that resembles the given food material. Though some of the data can be fed manually or predicted by a mixed average method, the exact food material property will not be achieved as it varies based on food material size, texture, and composition. Furthermore, physical-chemical, organic-chemical, and micro-biological models are not available in conventional CFD modeling software packages to understand the processing of food materials. Among the above-stated factors, computation time affects the industrial applications of CFD (especially 2D and 3D models), as the computation process takes more than 1,000-fold more time than the actual process. Hence, technological improvements focusing on computational time reduction can increase the practical applications of CFD techniques in food industries [8].

14.6 SUMMARY

Modeling CFD dynamics for food products (solid/liquid) is a complex process due to their heterogeneous nature; hence, food material properties and structure must be taken into consideration while developing a CFD model for plasma processes. Despite the fact that plasma is utilized commercially in a variety of industries, commercialization in food processing is still in its early stages. The five major hurdles that stand in the way of plasma's commercialization in the food sector are the production of product-specific devices and their scale-up, defining the process parameters for specific goods, process control, consumer acceptance, and regulatory permission. All these hurdles can be addressed with the help of CFD technology by studying the plasma dynamics under various process conditions. Furthermore, it will also provide an opportunity to understand the RS behavior and characteristics during and after plasma treatment, respectively. However, all this will be possible only if future research focuses on obtaining in-depth data on reactive species' interaction with biological cells.

KEYWORDS

- **cold plasma**
- **computational fluid dynamics**
- **microbial reduction**
- **pesticide degradation**
- **seed germination**
- **wastewater treatment**
- **water treatment**

REFERENCES

1. Anbarasan, R., Boopathy, B., Stephen, J., & Radhakrishnan, M. (2022). Cold plasma disinfestation of Callosobruchus maculatus infested soybeans: Its subsequent impact on soymilk extraction yield and quality. *Journal of Food Process Engineering, 46*(2), Article e14246. https://doi.org/10.1111/jfpe.14246.
2. Anbarasan, R., Jaspin, S., Bhavadharini, B., Pare, A., Pandiselvam, R., & Mahendran, R. (2022). Chlorpyrifos pesticide reduction in soybean using cold plasma and ozone treatments. *LWT, 159,* Article 113193. https://doi.org/10.1016/j.lwt.2022.113193.
3. Bali, N., Aggelopoulos, C. A., Skouras, E. D., Tsakiroglou, C. D., & Burganos, V. N. (2019). Modeling of a DBD plasma reactor for porous soil remediation. *Chemical Engineering Journal, 373,* 393–405.
4. Beni, M. S., Han, W., & Yu, K. N. (2019). Dispersion of OH radicals in applications related to fear-free dentistry using cold plasma. *Applied Sciences, 9*(10), Article 2119. https://doi.org/10.3390/app9102119.
5. Bulbul, V. J., Bhushette, P. R., Zambare, R. S., Deshmukh, R. R., & Annapure, U. S. (2019). Effect of cold plasma treatment on Xanthan gum properties. *Polymer Testing, 79,* Article 106056. https://doi.org/10.1016/j.polymertesting.2019.106056.
6. COMSOL 5.6. (2013). Introduction to Plasma Module. *Plasma Module User's Guide*, pp. 1–374. Retrieved from: https://doc.comsol.com/5.6/doc/com.comsol.help.plasma/PlasmaModuleUsersGuide.pdf (accessed on 16 June 2024).
7. El-Tayeb, A., El-Shazly, A. H., Elkady, M. F., & Abdel-Rahman, A. (2015). Decolorization of Acid blue 25 dye by non-thermal plasma advanced oxidation process for industrial wastewater treatment. In *2015 IEEE 15th International Conference on Environment and Electrical Engineering,* pp. 807–812, Publisher: IEEE. https://doi.org/10.1109/EEEIC.2015.7165268.
8. Heirman, P., Van Boxem, W., & Bogaerts, A. (2019). Reactivity and stability of plasma-generated oxygen and nitrogen species in buffered water solution: A computational study. *Physical Chemistry Chemical Physics, 21*(24), 12881–12894.
9. Jamali, A., & Evans, P. D. (2011). Etching of wood surfaces by glow discharge plasma. *Wood Science and Technology, 45*(1), 169–182.
10. Jaspin, S., Anbarasan, R., Dharini, M., & Mahendran, R. (2021). Structural analysis of tapioca xerogel and its water and oil triggered shape change. *Food Structure, 30,* Article 100226. https://doi.org/10.1016/j.foostr.2021.100226.
11. Jaspin, S., Anbarasan, R., Dharini, M., & Mahendran, R. (2022). Morphological analysis of corn xerogel and its shape shifting in water. *Journal of Food Engineering, 330,* Article 111107. https://doi.org/10.1016/j.jfoodeng.2022.111107.
12. Manoharan, D., & Radhakrishnan, M. (2022). Computational cold plasma dynamics and its potential application in food processing. *Reviews in Chemical Engineering, 38*(8), 1089–1105.
13. Misra, N. N., Tiwari, B. K., Raghavarao, K. S. M. S., & Cullen, P. J. (2011). Nonthermal Plasma Inactivation of Food-Borne Pathogens. *Food Engineering Reviews, 3*(3–4), 159–170.
14. Ognier, S., Iya-sou, D., Fourmond, C., & Cavadias, S. (2009). Analysis of Mechanisms at the Plasma–Liquid Interface in a Gas–Liquid Discharge Reactor Used for Treatment of Polluted Water. *Plasma Chemistry and Plasma Processing, 29*(4), 261–273.

15. Paatre Shashikanthalu, S., Ramireddy, L., & Radhakrishnan, M. (2020). Stimulation of the germination and seedling growth of *Cuminum cyminum* L. seeds by cold plasma. *Journal of Applied Research on Medicinal and Aromatic Plants, 18,* Article 100259. https://doi.org/10.1016/j.jarmap.2020.100259.
16. Pankaj, S. K., Wan, Z., & Keener, K. M. (2018). Effects of cold plasma on food quality: A review. *Foods, 7*(1), 1–4.
17. Patil, S., Moiseev, T., Misra, N. N., Cullen, P. J., Mosnier, J. P., et al. (2014). Influence of high voltage atmospheric cold plasma process parameters and role of relative humidity on inactivation of *Bacillus atrophaeus* spores inside a sealed package. *Journal of Hospital Infection, 88*(3), 162–169.
18. Perinban, S., Orsat, V., & Raghavan, V. (2019). Nonthermal Plasma–Liquid Interactions in Food Processing: A Review. *Comprehensive Reviews in Food Science and Food Safety, 18*(6), 1985–2008.
19. Rajan, A., Boopathy, B., Radhakrishnan, M., Rao, L., Schlüter, O. K., & Tiwari, B. K. (2023). Plasma processing: A sustainable technology in agri-food processing. *Sustainable Food Technology, 1*(1), 9–49.
20. Ramshaw, J. D., & Chang, C. H. (1992). Computational fluid dynamics modeling of multicomponent thermal plasmas. *Plasma Chemistry and Plasma Processing, 12*(3), 299–325.
21. Ranjitha Gracy, T. K., Gupta, V., & Mahendran, R. (2019). Influence of low-pressure nonthermal dielectric barrier discharge plasma on chlorpyrifos reduction in tomatoes. *Journal of Food Process Engineering, 42*(6), Article e13242. https://doi.org/10.1111/jfpe.13242.
22. Semenov, I. L., Weltmann, K. D., & Loffhagen, D. (2019). Modelling of the transport phenomena for an atmospheric-pressure plasma jet in contact with liquid. *Journal of Physics D: Applied Physics, 52*(31), 1–17. https://doi.org/10.1088/1361-6463/ab208e.
23. Sen, Y., Onal-Ulusoy, B., & Mutlu, M. (2019). Aspergillus decontamination in hazelnuts: Evaluation of atmospheric and low-pressure plasma technology. *Innovative Food Science & Emerging Technologies, 54,* 235–242.
24. Shahmohammadi Beni, M., Han, W., & Yu, K. N. (2019). Modeling OH transport phenomena in cold plasma discharges using the level set method. *Plasma Science and Technology, 21*(5), 3–12. https://doi.org/10.1088/2058-6272/ab008d.
25. Sun, Y., Liu, Y., Li, R., Xue, G., & Ognier, S. (2016). Degradation of reactive blue 19 by needle-plate non-thermal plasma in different gas atmospheres: Kinetics and responsible active species study assisted by CFD calculations. *Chemosphere, 155,* 243–249.
26. Tolouie, H., Mohammadifar, M. A., Ghomi, H., & Hashemi, M. (2018). Cold atmospheric plasma manipulation of proteins in food systems. *Critical Reviews in Food Science and Nutrition, 58*(15), 2583–2597.
27. Yu, S. -T., Chang, S. -C., Jorgenson, P., Park, S. -J., & Lai, M. -C. (1998). Basic equations of chemically reactive flows for computational fluid dynamics. *36th AIAA Aerospace Sciences Meeting and Exhibit, 1,* 1–14, https://doi.org/10.2514/6.1998-1051.

CHAPTER 15

LIMITATIONS AND TOXICOLOGY OF PLASMA TREATMENT

K. SANDHYA, PRAMILA MURUGESAN, and JEYAN A. MOSES

ABSTRACT

Cold plasma (CP) technology has gained significant scientific interest for the inactivation of different microorganisms in food products. The production of reactive (reactive oxygen and reactive nitrogen) species, ultraviolet rays, and shockwaves via plasma discharge could contribute to the lethal action against microbial species. This process is highly productive for temperature-sensitive food products. Nevertheless, the CP process has affected the quality of food products and their toxicological effects on foods. The chapter presents a brief description of the limitations of plasma treatment and its effect on the sensory aspects (odor, flavor, color, and texture), germination behavior, and food packaging material. The effect of such treatment on the environment and the assessment of the mutagenic activity of plasma on food samples are also discussed. In addition, economical aspects related to CP technology in food processing have been demonstrated. Finally, consumer acceptance of choosing plasma-treated food products is also discussed in detail.

15.1 INTRODUCTION

Cold plasma (CP) technology is an advanced oxidation process that holds promise for the sterilization of thermally sensitive food products due to its benefits such as low-temperature processing, not requiring any chemicals, no formation of toxic residues, and environmental compatibility [35]. The use of cold plasma in food processing offers significant advantages, such as preserving color and quality while providing necessary decontamination

Nonthermal Plasma Technology for Food and Food Products.
R. Mahendran, Preeti Birwal, and Megh R. Goyal (Eds.)

of spoilage organisms. However, the use of this technology can have some adverse effects on food products. Figure 15.1 illustrates the common limitations observed in the commercialization of cold plasma treatment.

The generation of various reactive oxygen species (ROS) and reactive nitrogen species (RNS) could alter the characteristics of food products [6, 44]. The formation of chemical species during CP treatment largely depends on the gases used during the process.

The limitations of CP treatment include sensorial (odor, flavor, color, and texture), germination, and food packaging surface changes. The odor of samples processed through plasma treatment is due to oxidative changes leading to rancidity [7]. The changes in flavor are observed due to the formation of secondary oxidation products from ROS having an impact on sensory properties [52]. Color degradation of plasma-treated samples is due to the reaction between the reactive species generated and certain chemical compounds responsible for maintaining the color of samples [1]. Textural changes take place at a higher treatment time, thus reducing the firmness of food samples [12]. Similarly, the increase in process time also harms the germination property as the cell wall ruptures and the germ layer comes in contact with reactive species resulting in reduced viability [47].

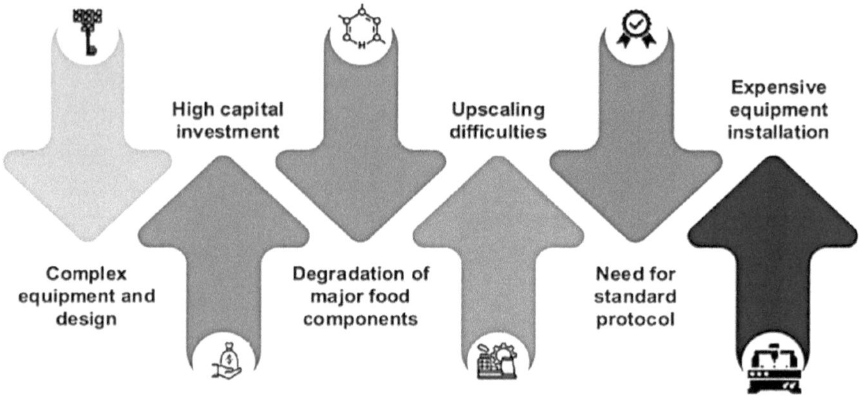

FIGURE 15.1 Limitations encountered in CP process.

In the case of food packaging materials, surface etching takes place which reduces the glass transition and melting temperature during processing [41]. Water processed using cold plasma treatment turns acidic and such water is not suitable for drinking purposes [29].

In terms of toxicology, still more research should be done to assess the mutagenicity of cold plasma treated samples. Some studies use insect models to evaluate the changes produced by plasma treatment [20]. *In vivo* and *in vitro* studies are mostly done to evaluate the mutagenicity of cold plasma-processed samples.

This chapter provides a brief overview of the detrimental effects of CP applications on foods.

15.2 LIMITATIONS OF CP IN FOOD APPLICATIONS

The limitations of CP treatment largely rely on the surface etching of food packaging material and structure changes of molecules present in food such as lipids, proteins, and bioactive compounds, which affect the sensorial characteristics such as color, texture, taste, and odor of food (Figure 15.2). The negative impact of CP treatment on foods is discussed in detail in the following sections.

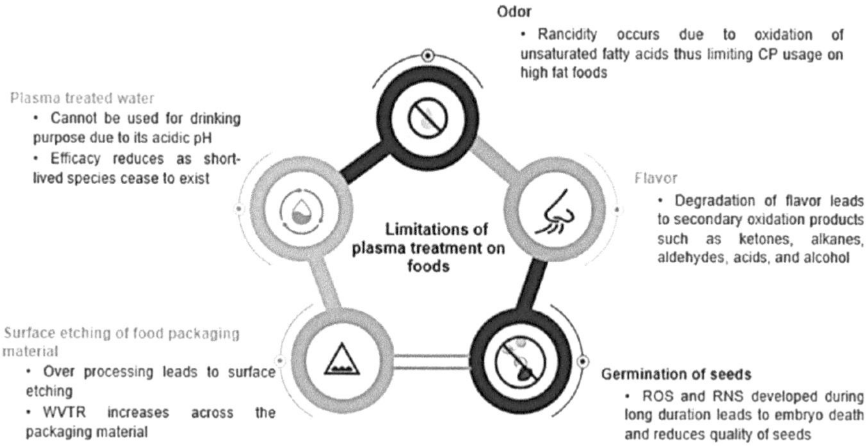

FIGURE 15.2 Impact of cold plasma treatment on foods.

15.2.1 IMPACT OF PLASMA TREATMENT ON THE SENSORY ASPECT

15.2.1.1 ODOR

CP processing of high-fat foods results in several disadvantages concerning its organoleptic properties such as loss of flavor and nutrition, which might lead to reduced consumer acceptance. A complex reaction takes place between the unsaturated fatty acids and the free radicals generated during plasma treatment leading to oxidation in the foods. Oxidation of lipids follows a spontaneous set of chain reactions, namely, initiation, propagation, and termination [18, 24].

Free electrons are generated from the free radicals produced during the CP process and could initiate the oxidation mechanism by reacting with the methyl group. The free electron combines with the hydrogen atom of the alkyl group of unsaturated fatty acid resulting in cleavage of the double bond, thus producing an alkyl free radical (Eqn. 1).

$$RH \rightarrow R^{*} + H \tag{1}$$

The free alkyl radical combined with oxygen molecule forms lipid peroxide radical (ROO*). This free radical reacts with the hydrogen of the alkyl group to form a stable carboxylic acid and a free methyl radical (Eqns. 2 and 3).

$$R^{*} + O_2 \rightarrow ROO* \tag{2}$$

$$ROO* + RH \rightarrow ROOH + R* \tag{3}$$

The propagation step continues to produce many alkyl free radicals. The termination step is initiated by the reaction of alkyl free radicals generated during the previous steps (Eqns. 4 to 6). The termination step of lipid oxidation is as follows:

$$R^{*} + R^{*} \rightarrow R - R \tag{4}$$

$$RO^{*} + R^{*} \rightarrow ROR \tag{5}$$

$$ROO* + R* \rightarrow ROOR \tag{6}$$

On exposure to CP, high-fat foods show a rapid increase in thiobarbituric acid reactive substances (TBARS) value owing to an increase in unsaturated fatty acids oxidation. TBARS value is an estimation of secondary by-products produced during oxidation, that is, hexanal malondialdehyde (MDA) [13].

Apart from producing such molecules, it also results in a more intense phase of oxidation during storage leading to reduced shelf life. Dried squids when treated using jet plasma result in the production of secondary oxidation products such as aldehydes, alcohols, furans, carbonyls, and hydrocarbons that exhibit volatile and non-volatile behavior [7]. These products are formed due to the reaction between lipid peroxide-free radicals (ROO*) and the ambient atmosphere. This results in the development of rancidity in treated samples.

Although CP results in the oxidation of high-lipid foods, under low-energy conditions, yet they prove to be the best in reducing the microbial species and preserving the quality characteristics. This was proved in a study where chub mackerel treated with a dielectric barrier discharge (DBD) (output voltage 130 kV) was consumable for up to 14 days without any change in sensory aspects [26]. Hence, process optimization is an important step to eliminate such interactions between high-lipid foods and plasma-generated reactive species.

15.2.1.2 FLAVOR

The changes in flavor are another common drawback of plasma treatment. The changes in flavor are identified by the perception of active volatile components generated on chewing. These compounds are elucidated by the taste receptors present on the tongue. During the CP process, secondary oxidation products (ketones, alkanes, aldehydes, acids, and alcohol) are produced that have a profound impact on the flavor of the sample. The production of secondary oxidation products is ascribed to the impact of CP treatment time and energy used. The formation of such products increases on enhancing the exposure time and energy.

The degradation of flavor in foods can be minimized until the food product can resist the changes made to it by plasma treatment time and energy. When the treatment time exceeds, degradation occurs. One such example is the CP treatment on a cheese slice that is subjected to treatment using CP with a mixture of helium and oxygen at 3.5 kV and 50 Hz. It was deduced from the study that the extension of process time beyond 10 min led to an increase in off-flavor generation. The secondary oxidation product formation occurs that depends on the composition of the sample under treatment and in turn results in a negative impact on the sensory property. Besides, a reduction in vitamin C of cucumber slices under plasma treatment was reported by Varilla [52] and Wang [53]. The degradation was concluded due to the effect of UV generation during CP treatment.

Thus, the changes in bioactive compounds and their degradation largely limit CP usage in flavored food products. DBD system can be used to minimize such effects in solid foods whereas a capillary tube set-up can be used for treating liquid foods to prevent nutrient degradation [34]. Therefore, the composition of the sample and the chemistry involved should be considered beforehand to minimize the negative effects observed on flavor during plasma treatment.

15.2.1.3 COLOR

The pigments (natural or synthetic) present in them and due to chemical reactions (enzymatic and non-enzymatic) mostly influence the color of food products. The sensory effects are largely dependent on the color of samples as it is one of the primary concerns regarding the appeal of foods. CP treatment has a profound impact on the color of samples as the reactive species produced result in a breakdown of compounds present in the sample, thus leading to color change [27]. The color change of foods after a particular treatment is evaluated based on ΔE values, which are calculated using the formula (Eqn. 7) where the difference in the color value of control and treated samples is substituted.

$$\Delta E = \sqrt{(L_2 - L_1)^2 + (a_2 - a_1)^2 + (b_2 - b_1)^2} \quad (7)$$

The color change mainly depends on the plasma conditions (working gas, power, time, input voltage, and distance between the sample and source) and type of sample (liquid or solid, whole or diced). At reduced exposure times in CP treatment, minute effects on color degradation of food samples are observed. Many studies signify the effects of CP processing on changes in the color of different foods. In an experiment on microbial reduction conducted by [10] on various juices (sour cherry nectar, tomato, apple, and orange), intensive color degradation was observed in apple juice. The change was deduced due to the effect of the isomerization of pigment present in them and oxidation reactions. The fruit's darkening was also attributed to moisture loss due to cutaneous wax melting on the surface of the fruit.

Similar to the above-discussed sample, walnut discoloration due to moisture changes during low-pressure cold plasma treatment was observed by [1]. CP also has an impact on the anthocyanin content of foods. The ROS generated during the treatment consist of active species such as ozone and hydroxyl radical that initiate oxidative cleavage of chromophores causing anthocyanin reduction. Plasma processing at the high voltage condition results in the reduction of anthocyanin content of blueberries [46]. Similar to the color pigments mentioned before, carotenoids are also broken down by plasma reactive species. Moreover, the produced oxidative species oxidize deoxymyoglobin/oxymyoglobin to metmyoglobin which in turn changes the color of meat. The reaction between hydrogen sulfide, oxygen, and the protein component, myoglobin results in the formation of sulphmyoglobin producing a green color in meat samples [28].

Adding to this, hydrogen peroxide generated during CP treatment reacts with myoglobin present in meat to form choleglobin, which also results in meat discoloration [16]. Compared to other foods, no significant color change was observed in fruits and vegetables. In a study on mandarins, the color change was insignificant due to the presence of a thick peel layer that resists plasma treatment [54].

The use of a corona discharge plasma jet harmed the sample brightness on exposure of more than 2 minutes. Optimization of process parameters is one of the crucial steps in reducing the negative effects of CP treatment.

15.2.1.4 TEXTURE

Cold plasma treatment helps in preserving the quality of food products without a high degree of deterioration. However, prolonged exposure to increase the efficacy of microbial inactivation also results in quality degradation having no effect after a particular decrease in log value of microbial reduction is achieved [25]. Ekezie et al. [13] demonstrated that CP processing increases the surface hydrophobicity and particle size of myofibrillar proteins present in meat sources [12].

On extending the exposure time, there occurs denaturation of protein. The effect of direct plasma treatment on wheat grains was studied by Los et al. [31], where a reduction in surface hydrophobicity was observed. It was also reported that 20 min of plasma treatment reduced the quality of grains [30]. In an experiment on CP jet treatment on blueberries, a reduction in firmness was reported at longer treatment times (120 s). Due to a decrease in firmness, the cell wall integrity is reduced, thereby reducing the anthocyanin pigments present in them [27]. The cell wall tends to break or fusion between the cells is identified when samples are treated for a longer time under plasma. Electroporation was observed in lettuce when treated for 10 min under an argon/oxygen combination. During such instances, nutrient loss occurs [5].

15.2.1.5 IMPACT OF PLASMA TREATMENT ON THE GERMINATION PROCESS

Germination viability and contaminating microorganisms can be eliminated from the food grains by using CP treatment. Though microbial decontamination of grains requires longer CP treatment, a longer duration of plasma interaction leads to negative effects on germination. On longer exposure to plasma treatment, the seeds tend to develop ROS and RNS which in the long run retards the quality of seed germination [31]. The embryo is present in a seed with layers of cell wall coatings and envelopes. Therefore, on treatment, surface etching takes place that leads to penetration of reactive species directly into the embryo and storage tissues causing adverse interactions [15].

The reactive species in combination with rapid water uptake lead to damage of genetic material and other membranes [4], initiate the fermentation process, and lead to embryo death [47]. The moisture present in the plant material along with the seed coat thickness determines the degree of damage that occurs on a seed. The ambient air-assisted CP process has the least negative effect on the DNA of pea seedlings. On the contrary, CP treatment with nitrogen as plasma gas produces the most negative effect on pea DNA and sprouting due to the formation of reactive oxygen and nitrogen species (RONS) [50].

$$O + NO + M \rightarrow NO_2 + M \quad (8)$$

$$NO + O_3 \rightarrow NO_2 + O_2 \quad (9)$$

$$N(D^2) + NO \rightarrow N_2O \quad (10)$$

$$NO_2 + O_3 \rightarrow NO_3 + O_2 \quad (11)$$

$$(resp.\ NO_2 + O + M \rightarrow NO_3 + M) \quad (12)$$

$$NO_3 + NO_2 \rightarrow N_2O_5 \quad (13)$$

Therefore, CP technology exhibits the improvement of surface decontamination and germination of various seeds and grains, but its potentially harmful effects need to be further investigated.

15.2.1.6 IMPACT OF PLASMA TREATMENT ON FOOD PACKAGING MATERIAL

CP treatment of food packaging surfaces leads to etching, degradation, and cross-linking which in turn alters the mechanical properties of the food packaging films [23]. Similarly, its usage in the decontamination of food packaging films was studied and several positive and negative changes were reported. Although CP treatment resulted in surface modification and enhancement of barrier properties of packaging film, roughness on the surface of films was also observed [41].

The etching of the film surface was identified due to overtime processing with CP. This leads to an increase in the surface area leading to an increase in water transfer rate due to concentration gradient. Such instances can negate the use of treated films to pack high water activity foods. Romani et al. [43] also concluded the appearance of yellowness on packaging films because of oxidation of proteins by the ROS generated on packaging films made of fish protein on exposure to glow discharge plasma. Thus, the transparency of treated films is also affected. Though such films aid in the packing of light-sensitive materials, consumer preference might decrease [41].

Apart from the above-mentioned effects, CP treatment also affects the thermal properties of biopolymer-based packaging films such as enthalpy, glass transition temperature, and melting temperature. In an experiment conducted by Pankaj et al. [40], the effect of DBD on sodium caseinate-based film was studied. A decrease in the glass transition temperature of the CP-exposed film was noticed in contrast to the control film (without plasma treatment). This decrease is attributed to the etching of the surface and the breakdown of chemical bonds leading to a chain scission reaction [42]. Similarly, a decrease in glass transition temperature, enthalpy, and melting temperature was observed in fish protein-based films on glow discharge plasma treatment compared to control films [41].

15.2.1.7 IMPACT OF PLASMA TREATMENT ON WATER

Notably, the pH level of the solution drastically changed from neutral to acidic range after plasma treatment [36]. The generation of HNO_3 and HNO_2 (Eqns. 6–11) during CP treatment results in changes in pH, thus inactivating bacterial cells [2, 6, 35].

$$O_2 \rightarrow O^{\cdot} + O^{\cdot} \quad 14$$

$$O^{\cdot} + O_2 \rightarrow O_3 \quad 15$$

$$e^- + N_2 \rightarrow N^* + N^* + e^- \quad 16$$

$$N_2^* + 2H_2O \rightarrow 2NO^{\cdot} + .4H_2 \quad 17$$

$$NO^{\cdot} + O^{\cdot} \rightarrow .NO_2 \quad 18$$

$$2NO_2 + H_2O \rightarrow HNO_2 + HNO_3 \quad 19$$

Sang et al. [44] reported that the efficacy of elimination of chemical contaminants is enhanced when the pH of the solution changes to alkaline from acidic nature. At low pH (acidic environment), the production of OH• and O_3 is reduced due to the consumption of electrons by H+ ions. However, at high pH (alkaline environment), the decomposition of H_2O_2 and O_3 results in OH• formation, thus improving the chemical contaminant removal efficiency [43].

The efficacy of plasma-treated water lasts for only one hour for microbial inactivation purposes. This is due to the short-lived reactive species that are generated during treatment [29]. The water treated using cold plasma is acidic due to the formation of nitric acid. Thus, this limits the consumption of plasma-treated water for drinking purposes.

15.3 TOXICOLOGICAL EFFECTS OF PLASMA TREATMENT

The use of CP in various sterilization systems results in the formation of ROS and RNS. Thus, the toxicology of such treatments should be taken into consideration during the design of such systems to reduce toxic exposure to workers and the environment surrounding the facility [14]. The chemistry induced in CP treatment of foods and the toxicity of such processes can be interpreted using FTIR, NMR, and chromatographic studies [45].

15.3.1 MUTAGENICITY OF PLASMA-TREATED SAMPLES

The biological safety of foods treated with plasma is crucial to examine prior to their consumption. The CP process could produce secondary metabolites on the surface of wheat grains, which is another important step to imply the safety of plasma-treated foods [32]. The cytotoxic activity of plasma-activated water (PAW) was studied on eukaryotic cells where the ROS generated was found to be one of the toxic species. The reaction of oxidation of sugar moieties to organic acids leads to the production of aldehydes possessing a mutagenic effect [9, 45].

A study was conducted on CP treatment of dairy industry waste where the waste discharge was treated using CP technology and the effect of treatment on the toxicity of aquatic habitats was evaluated [39]. They reported a decrease in toxic effects of wastewater, but, on increased exposure (48h), the sensitive organisms chosen for reference showed toxic effects.

Chen and Lin [6] experimented with the influence of plasma gas on deionized water for 30 minutes and found that the formed reactive species led to apoptosis [55]. In another study by Heslin et al. [23], the assessment of the cytotoxic and mutagenic potential of DBD plasma-treated lettuce in an *in vitro* mammalian cell (CHO-K1) model and in an *in vivo* larva model (*Galleria mellonella*). It was reported that a reduced spontaneous mutation and *in-vitro* cytotoxic effect whereas impactful *in-vivo* toxicity with around 10% with respect to the survival rate of the larva was observed in lettuce broth at 5 minutes treatment time [22].

Cold plasma studies are always conducted in a lab-scale environment and thus large-scale experimentation of such techniques is not reported to date.

15.3.2 IN-VIVO MUTAGENICITY STUDIES OF PLASMA-TREATED FOODS

Mutagenic assays are conducted on live organisms, and their physical and behavioral changes are monitored over time with constant feeding of plasma-treated food samples. Plasma-treated edible packaging films were tested for oral toxicity using rats (Sprague–Dawley) of both male and female varieties. Packaging film samples treated with CP were fed to rats and monitored for 14 days. It was reported that no changes in toxic levels or deaths were observed, but alterations in blood components such as creatinine, bilirubin, and hemoglobin were noted [20]. Mutagenic assessment is necessary to quantify the cytotoxic behavior of foods treated using CP [9]. These assays are conducted using various models, such as evaluating the fate of proteins, lipids, and microbial populations, or against living organisms such as rats or insect pests.

Food safety and maintenance of standards are crucial for products processed using any novel technique for large-scale processing of that particular technology. To evaluate the effect of plasma-treated samples, insect models were used, and their cells were taken for study by culturing in a medium containing the treated samples. Mutagenicity is determined by evaluating the absorbance of cell culture supernatant [32].

15.3.3 IMPACT OF PLASMA TREATMENT ON THE ENVIRONMENT

Cold plasma, when used in the process of decontaminating water or any processing equipment, leads to the production of harmful gases such as NO_X and traces of other hydrocarbons, which in a year is equal to the gases produced from the exhaust of an automobile in 10 minutes [40]. When carried out on a large scale, CP should consider regulatory approval for the release of toxic compounds, and its effect on the workforce in such a facility should be monitored [14]. At high voltage levels of processing, the ROS and RNS generated must be controlled, as the presence of such compounds in the environment affects nucleic acids and DNA [39].

15.4 CONSUMER PERSPECTIVE OF PLASMA-TREATED FOOD PRODUCTS

Novel non-thermal technologies and their potential in maintaining food safety are still new to consumers. Awareness should be created among them about the safety of such novel treatments. Thus, the plasma-treated samples must be proven safe according to standards. There is still a need for standards for plasma-treated foods. Consumers prefer products that are similar to control samples in terms of color and sensory attributes. Present-day consumers are more interested in the technology used than the health benefits it provides. Thus, successful scientific evidence regarded as safe is required from the processor's end to gain consumers' trust [21].

The main set of documents to get consumers' acceptance of a novel technology is regulatory approval, optimized process control, and scientific validation [38]. One main criterion that intrudes on consumer acceptance of plasma-treated foods is the ROS it generates during processing. The ROS toxic effects on food substrate and the environment are major issues that hinder wide acceptance [11]. Though CP processing of meat has an intense effect on microbial safety, the changes in color result in reduced consumer acceptance [53].

15.5 CHALLENGES IN THE COMMERCIALIZATION OF CP PROCESS FOR FOODS

Cold plasma treatment surfaces a major limitation in the large-scale application as most of the research work concerning cold plasma is done at a laboratory level. Scale-up and capital investment are major concerns when it comes to the application of cold plasma commercially [19]. Though ROS helps in the eradication of spoilage pathogens from the food sample, the pathway of reaction that takes place during plasma treatment is necessary for the large-scale utilization of such novel technology [19]. CP treatment requires extensive research on its safety during consumption [33].

Optimization of the CP process on an industrial scale is required for its development and commercialization. CP is a surface treatment and the microorganisms present in layers below the surface of the sample remain even after processing. Thus, treatment using such a technique for blocks of meat or fresh produce sample does not have effectiveness in terms of microbial safety [49].

Extensive research is required in plasma treatment for increased penetration into the food matrix [17]. A study conducted on the decontamination of spices using CP concluded that optimization of the process, safety studies, and economical assessment are prime concerns regarding its large-scale applicability [3]. Another disadvantage in using CP is the need for a uniform size and shape of the sample as the changes in sample parameters might have an impact on the plasma treatment time and power [8]. Siddique [49] reported that commercialization of plasma treatment can be considered in treatments where over time plasma treatment results in microbial inactivation and also retards any drastic quality changes [48].

The cost of operation using cold plasma technology relies on plasma gas and power supply. In the case of helium, nitrogen, oxygen, and atmospheric air, an operation of 1,000 h would have an operational cost of around \$636,000–\$9,096,000; \$9,000–\$72,000, \$18,000–\$144,000, and ~\$0, respectively [37].

The power consumption also depends on the size and distance between electrodes. Thus, it can be deduced from the above-mentioned expense that using gas mixtures such as helium, nitrogen, and oxygen leads to increased expenses on processing at a large scale compared to the usage of atmospheric air. Also, the setting up of the plasma system on a large scale has several issues concerned with its hardware compatibility. The major challenge is setting up proper hardware configurations to have a smooth functioning system [51].

Thus, large-scale utilization of this technology is cost-effective and requires process optimization, skilled manpower, and consumer awareness and acceptance of such products in the market.

15.6 SUMMARY

Cold plasma treatment is considered one of the most efficient novel non-thermal treatments. However, some limitations and toxic effects of using such technologies persist. Therefore, a lot of research is required in the areas of pilot scale development of plasma treatment plants, process, and product optimization studies, cost and economic evaluations, toxic compounds release, and their effect on the manpower working in the facility. Though plasma treatment inactivates microbes present in the sample, it provides surface treatment, and penetration is still not achieved. Thus, studies regarding the effect of plasma on thick layers of solid foods should be considered as the use of high voltage or reducing the distance between electrodes and sample leads to degradation of quality characteristics. Although plasma treatment has several advantages in terms of microbial inactivation and extension of shelf-life, yet no studies reported the safety of plasma-treated foods using *in-vivo* and *in-vitro* studies. Awareness among consumers regarding the safety of such food products should be done to use CP on a large scale.

KEYWORDS

- **cold plasma**
- **lipid oxidation**
- **mutagenicity**
- **reactive species**
- **sensory**
- **toxicology**

REFERENCES

1. Ahangari, M., Ramezan, Y., & Khani, M. R. (2021). Effect of low-pressure cold plasma treatment on microbial decontamination and physicochemical properties of dried walnut kernels (*Juglans regia* L.). *Journal of Food Process Engineering, 44*(1), Article e13593. https://doi.org/10.1111/jfpe.13593.
2. Bagheri, H., & Abbaszadeh, S. (2021). Microbial decontamination of spices using cold plasma. *Nutrition & Food Science Research, 8*(1), 53–59.
3. Balasubramanian, B., Pogozelski, W. K., & Tullius, T. D. (1998). DNA strand breaking by the hydroxyl radical is governed by the accessible surface areas of the hydrogen atoms of the DNA backbone. *Proceedings of the National Academy of Sciences of the United States of America, 95*(17), 9738–9743.
4. Bermúdez-Aguirre, D., Wemlinger, E., Pedrow, P., Barbosa-Cánovas, G., & Garcia-Perez, M. (2013). Effect of atmospheric pressure cold plasma (APCP) on the inactivation of *Escherichia coli* in fresh produce. *Food Control, 34*(1), 149–157.

5. Chandana, L., Sangeetha, C. J., Shashidhar, T., & Subrahmanyam, C. (2018). Non-thermal atmospheric pressure plasma jet for the bacterial inactivation in an aqueous medium. *Science of The Total Environment, 640-641,* 493–500.
6. Chen, Z., Lin, L., Cheng, X., Gjika, E., & Keidar, M. (2016). Effects of cold atmospheric plasma generated in deionized water in cell cancer therapy. *Plasma Processes and Polymers, 13*(12), 1151–1156.
7. Choi, S., Puligundla, P., & Mok, C. (2017). Effect of corona discharge plasma on microbial decontamination of dried squid shreds including physicochemical and sensory evaluation. *LWT, 75,* 323–328.
8. Coutinho, N. M., Silveira, M. R., Rocha, R. S., Moraes, J., & Ferreira, M. V. S. (2018). Cold plasma processing of milk and dairy products. *Trends in Food Science & Technology, 74,* 56–68.
9. Cubas, A. L. V., de Medeiros Machado, M., dos Santos, J. R., Zanco, J. J., & Ribeiro, D. H. B. (2019). Effect of chemical species generated by different geometries of air and argon non-thermal plasma reactors on bacteria inactivation in water. *Separation and Purification Technology, 222,* 68–74.
10. Cullen, P. J., Lalor, J., Scally, L., Boehm, D., & Milosavljević, V. (2018). Translation of plasma technology from the lab to the food industry. *Plasma Processes and Polymers, 15*(2), 1–29. https://doi.org/10.1002/ppap.201700085
11. Dasan, B. G., & Boyaci, I. H. (2018). Effect of cold atmospheric plasma on inactivation of *Escherichia coli* and physicochemical properties of apple, orange, tomato juices, and sour cherry nectar. *Food and Bioprocess Technology, 11*(2), 334–343.
12. Domonkos, M., Tichá, P., Trejbal, J., & Demo, P. (2021). Applications of cold atmospheric pressure plasma technology in medicine, agriculture and food industry. *Applied Sciences, 11*(11), 1–19. https://doi.org/10.3390/app11114809.
13. Ekezie, F. G. C., Cheng, J. H., & Sun, D. W. (2019). Effects of atmospheric pressure plasma jet on the conformation and physicochemical properties of myofibrillar proteins from king prawn (*Litopenaeus vannamei*). *Food Chemistry, 276,* 147–156.
14. Estévez, M. (2015). Oxidative damage to poultry: From farm to fork. *Poultry Science, 94*(6), 1368–1378.
15. Feizollahi, E., Misra, N. N., & Roopesh, M. S. (2020). Factors influencing the antimicrobial efficacy of dielectric barrier discharge (DBD) atmospheric cold plasma (ACP) in food processing applications. *Critical Reviews in Food Science and Nutrition, 61*(4), 666–689.
16. Frączek, J., Hebda, T., Ślipek, Z., & Kurpaska, S. (2005). Effect of seed coat thickness on seed hardness. *Canadian Biosystems Engineering / Le Genie des Biosystems au Canada, 47,* 1–5. Retrieved from: https://library.csbe-scgab.ca/docs/journal/47/c0503.pdf (accessed on 16 June 2024).
17. Fröhling, A., Durek, J., Schnabel, U., Ehlbeck, J., Bolling, J., & Schlüter, O. (2012). Indirect plasma treatment of fresh pork: Decontamination efficiency and effects on quality attributes. *Innovative Food Science & Emerging Technologies, 16,* 381–390.
18. Ganesan, A. R., Tiwari, U., Ezhilarasi, P. N., & Rajauria, G. (2021). Application of cold plasma on food matrices: A review on current and future prospects. *Journal of Food Processing and Preservation, 45*(1), Article e15070. https://doi.org/10.1111/jfpp.15070.
19. Gavahian, M., Chu, Y. H., Mousavi Khaneghah, A., Barba, F. J., & Misra, N. N. (2018). A critical analysis of the cold plasma induced lipid oxidation in foods. *Trends in Food Science & Technology, 77,* 32–41.

20. Gavahian, M., & Khaneghah, A. M. (2020). Cold plasma as a tool for the elimination of food contaminants: Recent advances and future trends. *Critical Reviews in Food Science and Nutrition, 60*(9), 1581–1592.
21. Han, S. H., Suh, H. J., Hong, K. B., Kim, S. Y., & Min, S. C. (2016). Oral toxicity of cold plasma-treated edible films for food coating. *Journal of Food Science, 81*(12), T3052–T3057. https://doi.org/10.1111/1750-3841.13523.
22. Hertwig, C., Meneses, N., & Mathys, A. (2018). Cold atmospheric pressure plasma and low energy electron beam as alternative nonthermal decontamination technologies for dry food surfaces: A review. *Trends in Food Science & Technology, 77,* 131–142. https://doi.org/10.1016/j.tifs.2018.05.012.
23. Heslin, C., Boehm, D., Gilmore, B. F., Megaw, J., & Bourke, P. (2020). Safety evaluation of plasma-treated lettuce broth using in vitro and in vivo toxicity models. *Journal of Physics D: Applied Physics, 53*(27), Article 275401. https://doi.org/10.1088/1361-6463/ab7ac8.
24. Hoque, M., McDonagh, C., Tiwari, B. K., Kerry, J. P., & Pathania, S. (2022). Effect of cold plasma treatment on the packaging properties of biopolymer-based films: A review. *Applied Sciences, 12*(3), Article 1346. https://doi.org/10.3390/app12031346.
25. Jadhav, H. B., & Annapure, U. (2021). Consequences of non-thermal cold plasma treatment on meat and dairy lipids – A review. *Future Foods, 4,* Article 100095. https://doi.org/10.1016/j.fufo.2021.100095.
26. Wang, J., Zhuang, H., & Lawrence, K. L. J. (2016). Disinfection of chicken fillets in packages with atmospheric cold plasma: Effects of treatment voltage and time. *International Journal of Laboratory Hematology, 38*(1), 42–49.
27. Chen, J., Wang, S. -z., Chen, J. -y., Chen, D. -z., & Deng, S. -g. (2007). Effect of cold plasma on maintaining the quality of chub mackerel (*Scomber japonicus*): Biochemical and sensory attributes. *Journal of Organizational Behavior, 28*(3), 303–325.
28. Lacombe, A., Niemira, B. A., Gurtler, J. B., Fan, X., & Sites, J. (2015). Atmospheric cold plasma inactivation of aerobic microorganisms on blueberries and effects on quality attributes. *Food Microbiology, 46,* 479–484.
29. Lee, H., Yong, H. I., Kim, H. J., Choe, W., & Yoo, S. J. (2016). Evaluation of the microbiological safety, quality changes, and genotoxicity of chicken breast treated with flexible thin-layer dielectric barrier discharge plasma. *Food Science & Biotechnology, 25*(4), 1189–1195.
30. Lim, J., Byeon, Y. S., Hong, E. J., Ryu, S., & Kim, S. B. (2021). Effect of post-discharge time of plasma-treated water (PTW) on microbial inactivation and quality of fresh-cut potatoes. *Journal of Food Processing and Preservation, 45*(5), 1–8.
31. Los, A., Ziuzina, D., Akkermans, S., Boehm, D., & Cullen, P. J. (2018). Improving microbiological safety and quality characteristics of wheat and barley by high voltage atmospheric cold plasma closed processing. *Food Research International, 106,* 509–521.
32. Los, A., Ziuzina, D., Boehm, D., Cullen, P. J., & Bourke, P. (2019). Investigation of mechanisms involved in germination enhancement of wheat (*Triticum aestivum*) by cold plasma: Effects on seed surface chemistry and characteristics. *Plasma Processes and Polymers, 16*(4), 1–12.
33. Los, A., Ziuzina, D., van Cleynenbreugel, R., Boehm, D., & Bourke, P. (2020). Assessing the biological safety of atmospheric cold plasma treated wheat using cell and insect models. *Foods, 9*(7), Article 898. https://doi.org/10.3390/foods9070898.

34. Mandal, R., Singh, A., & Singh, A. P. (2018). Recent developments in cold plasma decontamination technology in the food industry. *Trends in Food Science & Technology, 80,* 93–103. https://doi.org/10.1016/j.tifs.2018.09.001.
35. Munekata, P. E. S., Domínguez, R., Pateiro, M., & Lorenzo, J. M. (2020). Influence of plasma treatment on the polyphenols of food products – A review. *Foods, 9*(7), 1–14.
36. Murugesan, P., Moses, J. A., & Anandharamakrishnan, C. (2020). Performance of an atmospheric plasma discharge reactor for inactivation of *Enterococcus faecalis* and *Escherichia coli* in aqueous media. *Journal of Environmental Chemical Engineering, 8*(4), Article 103891. https://doi.org/10.1016/j.jece.2020.103891.
37. Murugesan, P., V. EM., Moses, J. A., & Anandharamakrishnan, C. (2020). Water decontamination using non-thermal plasma: Concepts, applications, and prospects. *Journal of Environmental Chemical Engineering, 8*(5), Article 104377. https://doi.org/10.1016/j.jece.2020.104377.
38. Niemira, B. A. (2012). Cold plasma decontamination of foods. *Annual Review of Food Science and Technology, 3*(1), 125–142. https://doi.org/10.1146/annurev-food-022811-101159.
39. Pan, Y., Cheng, J. H., & Sun, D. W. (2019). Cold plasma-mediated treatments for shelf-life extension of fresh produce: A review of recent research developments. *Comprehensive Reviews in Food Science and Food Safety, 18*(5), 1312–1326. https://doi.org/10.1111/1541-4337.12465.
40. Pankaj, S. K., Bueno-Ferrer, C., Misra, N. N., O'Neill, L., & Tiwari, B. K. (2014). Physicochemical characterization of plasma-treated sodium caseinate film. *Food Research International, 66,* 438–444.
41. Patange, A., Boehm, D., Giltrap, M., Lu, P., Cullen, P. J., & Bourke, P. (2018). Assessment of the disinfection capacity and eco-toxicological impact of atmospheric cold plasma for treatment of food industry effluents. *Science of The Total Environment, 631–632,* 298–307. https://doi.org/10.1016/j.scitotenv.2018.03.136.
42. Prysiazhnyi, V., Zaporojchenko, V., Kersten, H., & Černák, M. (2012). Influence of humidity on atmospheric pressure air plasma treatment of aluminium surfaces. *Applied Surface Science, 258*(14), 5467–5471. https://doi.org/10.1016/j.apsusc.2012.01.136.
43. Romani, V. P., Olsen, B., Pinto Collares, M., Meireles Oliveira, J. R., Prentice-Hernández, C., & Guimarães Martins, V. (2019). Improvement of fish protein films properties for food packaging through glow discharge plasma application. *Food Hydrocolloids, 87,* 970–976. https://doi.org/10.1016/j.foodhyd.2018.08.051.
44. Sang, W., Cui, J., Feng, Y., Mei, L., & Zhang, Q. (2019). Degradation of aniline in aqueous solution by dielectric barrier discharge plasma: Mechanism and degradation pathways. *Chemosphere, 223,* 416–424. https://doi.org/10.1016/j.chemosphere.2019.02.002.
45. Sarangapani, C., Misra, N. N., Milosavljevic, V., Bourke, P., O'Regan, F., & Cullen, P. J. (2016). Pesticide degradation in water using atmospheric air cold plasma. *Journal of Water Process Engineering, 9,* 225–232. https://doi.org/10.1016/j.jwpe.2015.12.006.
46. Sarangapani, C., Patange, A., Bourke, P., Keener, K., & Cullen, P. J. (2018). Recent advances in the application of cold plasma technology in foods. *Annual Review of Food Science and Technology, 9,* 609–629. https://doi.org/10.1146/annurev-food-030117-012403.
47. Sarangapani, C., Toole, G. O., Cullen, P. J., & Bourke, P. (2017). Atmospheric cold plasma dissipation efficiency of agrochemicals on blueberries. *Innovative Food Science & Emerging Technologies, 44,* 235–241. https://doi.org/10.1016/j.ifset.2017.10.015.

48. Schieber, M., & Chandel, N. S. (2014). ROS function in redox signaling and oxidative stress. *Current Biology, 24*(10), R453–R462. https://doi.org/10.1016/j.cub.2014.03.034.
49. Siddique, S. S., Hardy, G. E. S. J., & Bayliss, K. L. (2018). Cold plasma: A potential new method to manage postharvest diseases caused by fungal plant pathogens. *Plant Pathology, 67*(5), 1011–1021. https://doi.org/10.1111/ppa.12852.
50. Thirumdas, R., Sarangapani, C., & Annapure, U. S. (2015). Cold Plasma: A novel non-thermal technology for food processing. *Food Biophysics, 10*(1), 1–11. https://doi.org/10.1007/s11483-014-9363-8.
51. Tomeková, J., Kyzek, S., Medvecká, V., Gálová, E., & Zahoranová, A. (2020). Influence of cold atmospheric pressure plasma on pea seeds: DNA damage of seedlings and optical diagnostics of plasma. *Plasma Chemistry and Plasma Processing, 40*(6), 1571–1584. https://doi.org/10.1007/s11090-020-10157-3.
52. Varilla, C., Marcone, M., & Annor, G. A. (2020). Potential of cold plasma technology in ensuring the safety of foods and agricultural produce: A review. *Foods, 9*(10), Article 1435. https://doi.org/10.3390/foods9101435.
53. Wang, R. X., Nian, W. F., Wu, H. Y., Feng, H. Q., & Zhang, K. (2012). Atmospheric-pressure cold plasma treatment of contaminated fresh fruit and vegetable slices: inactivation and physiochemical properties evaluation. *The European Physical Journal D, 66*(10), 1–7. https://doi.org/10.1140/epjd/e2012-30369-6.
54. Yadav, B., Spinelli, A. C., Govindan, B. N., Tsui, Y. Y., McMullen, L. M., & Roopesh, M. S. (2019). Cold plasma treatment of ready-to-eat ham: Influence of process conditions and storage on inactivation of *Listeria innocua. Food Research International, 123,* 276–285. https://doi.org/10.1016/j.foodres.2019.04.013.
55. Yeon, M., Jo, S., & Min, S. C. (2017). Mandarin preservation by microwave-powered cold plasma treatment. *Innovative Food Science & Emerging Technologies, 39,* 25–32. https://doi.org/10.1016/j.ifset.2016.11.008.

CHAPTER 16

COLD PLASMA TECHNOLOGY: FUTURE AVENUES AND CHALLENGES

M. DHARINI and R. MAHENDRAN

ABSTRACT

Cold plasma is being used in numerous fields, including food processing. The applications of cold plasma in food processing include germination, disinfestation, disinfection, property modification, pesticide reduction, allergen reduction, and toxicity reduction. However, there are a few crucial challenges in applying cold plasma to food materials that must be tackled to commercialize this technology in the food processing sector. Thus, the chapter was aimed at addressing the challenges faced and the expertise to handle the challenges encountered. This chapter covers the advanced applications of cold plasma in food processing, the challenges faced, and possible avenues for eliminating the challenges.

16.1 INTRODUCTION

Cold plasma is a novel technology known for its non-thermal and non-residual approach to biological materials. The advent of low-temperature plasma is a remarkable invention for subjecting biological materials to plasma. Plasma is composed of numerous reactive species which, unlike other non-thermal technologies, act synergistically on the target material [15]. This synergistic effect of plasma reactive species on microbial cells disturbs their homeostasis; another major advantage of plasma in decontamination and disinfestation is that the microbes or insects cannot develop resistance against this technology, unlike other methods [2, 26].

Nonthermal Plasma Technology for Food and Food Products.
R. Mahendran, Preeti Birwal, and Megh R. Goyal (Eds.)

Plasma reactive species generated interact with gas or liquid; while the gas interaction produces hydroxyl and nitric oxide radicals, the liquid interaction produces long-lived reactive species such as hydrogen peroxides, nitrites, and nitrates [7]. Cold plasma has been actively researched in the field of food processing for numerous applications in the past two decades; however, it's still in the emerging stage. Though the application of plasma technology is approved and commercialized in other sectors such as medicine, textile, aerospace, and polymer sciences, it's not yet commercialized in the food sector. This is due to the presence of some major challenges in using plasma on food materials despite various applications of plasma in food processing.

This chapter highlights the current challenges faced in using cold plasma technology and possible prospective techniques to deal with the challenges. This chapter comprises a brief understanding of the recent advances in plasma in the food sector, challenges faced in treating food materials, and the futuristic approaches to tackle the challenges faced.

16.2 RECENT ADVANCES IN COLD PLASMA TECHNOLOGY IN THE FOOD SECTOR

The broad applications of cold plasma in food processing are preserving and maintaining the safety of food commodities, modifying biomolecules, and retaining bioactive compounds in food [17]. Other than this common modification, there are some recent and important applications of cold plasma in the field of food processing that are having a tremendous and prospective impact on the food processing community.

16.2.1 COLD PLASMA ON MANIPULATION OF PROTEINS

Apart from the reduction in microbial load, cold plasma reacts readily with food components such as protein, carbohydrate, and fat to oxidize them; oxidation tends to change the vital properties of the food material. Most of the applications like property modification, altering the enzyme activity, pesticide, and allergen reduction are mainly due to changes plasma imparts on food components. Modification of protein structure might include changes such as side-chain modification, protein backbone modification, fragmentation, crosslinking, and conformation changes [3, 52].

Enzymes and allergens are also a form of proteins that can also be altered by cold plasma to rise or decrease the activity [23, 54]. The increase in enzyme activity of foods can be both beneficial and detrimental based on the commodity. In fruits and vegetables, the enzymes such as peroxidase and lipoxygenase lead to an increase in enzymatic browning [33]. Modification of these enzyme structures using plasma reduces its activity. However, plasma technology can also be used to increase the activity of enzymes. The minimal processing conditions were reported to increase the enzyme activity whereas, the extreme processing condition can cause aggregation of amino acids which can completely inactivate the enzyme activity [20, 55].

Cold plasma can also manipulate the allergenicity in food as allergens are also proteins. It was noted that the utilization of cold plasma unfolded the complex epitopes and oxidized the linear epitopes to fragments, making them inaccessible for binding with the antibody [56].

A study on cold plasma's effect on peanuts and shrimp showed that plasma can reduce their allergenicity. Cold plasma treatment of peanut protein reduced the IgE binding capacity [56], while in the case of white shrimp, a gradual reduction of tropomyosin was observed [49]. However, in another study on cashew nuts, the allergenicity was not significantly altered after plasma processing. Further studies must be conducted on different plant-based foods, as plasma effects on foods vary from one another.

16.2.2 PESTICIDE AND TOXICITY REDUCTION BY COLD PLASMA

Cold plasma is also used for the degradation of pesticides into less toxic components [58]. Chaitanya et al. [45] studied the efficacy of atmospheric cold plasma on pesticide degradation in blueberries and the quality retention.

Sarangapani in 2016 [47] used dielectric barrier discharge (DBD)-based ambient pressure plasma for the degradation of certain pesticides in water. The degradation percentage achieved by them was 78%, 69%, and 57% for dichlorvos, malathion, and endosulfan, respectively. They found that the reactive species accountable for the decay of pesticides O_3, H_2O_2, and ^-OH. Plasma also aids in reducing the mycotoxins produced by fungi. The species present in plasma-like OH, O_3, NO, and NO_2 are responsible for the degradation of toxins [31]. These reactive species act on the chemical bonds of toxins and break their activity [58].

Apart from pesticides and mycotoxins, plasma can aid in eliminating toxins produced by industrial effluents. The food industrial effluents are rich in nutrients; thus, they can act as a medium for microbial growth.

Furthermore, they also contaminate groundwater and affect aquatic life forms. Cold plasma treatment for 24 hours in meat and dairy industry effluents was observed to decrease the toxicity in aquatic organisms [41].

16.2.3 EXTRACTION AND ANTIMICROBIAL ACTIVITY ENHANCEMENT

The surface morphology of biological materials can also be modified by cold plasma to enhance the extraction of the desired component from the subsequent layers. The possible reason for this application is due to the surface layer erosion by cold plasma [13, 43]. Plasma is known for increasing the antimicrobial activity when it is combined with essential oils [30]. The plasma application on any food commodity causes etching on the food surface, the consecutive essential oil application increases antimicrobial activity as it is easily absorbed inside the food material [10].

16.2.4 COLD PLASMA IN FOOD PACKAGING

Other than its direct application on food materials, plasma can also be used in generating plasma-modified packaging material for maintaining the quality of the product during storage. The polymer surfaces, in general, are hydrophobic and have low energy, plasma can be applied to change this characteristic of polymers by introducing a functional group [11]. The functionalization of food packaging surfaces can aid in improved printability, wettability, sealability, and adhesion properties; these properties are excellent in producing multilayer food packaging material for maintaining the freshness of the product [4]. Further, the change in surface properties also improves the water vapor barrier properties of the polymers. Other than polymers, plasma can also improve the properties of biodegradable packaging materials [35].

The impact of plasma species on the film surface aids in the formation of fissures, capillary pores, and cracks which in turn increase the water absorption capacity of the film [40]. Etching tends to reduce the water vapor transition and solubility of the fish protein film [44]. Plasma is also used for modifying the surface of encapsulated active packaging material for sustained release of active ingredients for a prolonged duration; the surface roughening effect of plasma plays a significant part in the release of active ingredients [9].

These are a few recent advances in cold plasma application in the food processing sector. The following section deals with the challenges that are being encountered in the utilization of plasma in foods.

16.3 CHALLENGES ENCOUNTERED

As mentioned earlier, the presence of reactive species is the major parameter responsible for the desired change in the food. However, the same reactive species can be harmful in terms of producing toxic substances when it interacts with food material. Most of the plasma studies involve decreasing the microbial load in foods; nevertheless, the effect of plasma on food components can also be detrimental. The following are the challenges encountered while subjecting food to plasma treatment.

16.3.1 CONTROLLING THE DOSAGE OF PLASMA AND INTERACTION WITH FOOD

Controllable generation of plasma and plasma reactive species specificity are the key factors that differ from equipment to equipment and also on the operating variables such as pressure, power, operating gas, and mode of application [32]. Optimization of the reactive species generation is a major challenge that needs to be addressed to reduce the concerns on oxidation of biomolecules such as lipids. Though various reactive species are present in plasma, key reactive species with high reactivity such as OH, H, O_3, NO, N_2 are depicted as the ones responsible for the targeted change in food. Even in plasma-activated liquids, the long-lived species were always mentioned to be the ones accountable for the application; rather, the short-lived reactive species also could have contributed to its application [21]. Thus, the lack of research on the infamous short-lived reactive species is also an important challenge that must be addressed [64].

Apart from controlling the dosage of plasma, understanding the exact chemical change in the food material has not yet been studied clearly. The constraints to obtaining this knowledge are the presence of different types of species in plasma and the variation in the chemical characteristics of food materials subjected to plasma [46]. Thus, the lack of standardization of plasma dose, i.e., the kind and quantity of reactive species generated and their interaction with the food material is a basic challenge faced by the food processing community. Asides from the chemical interaction of plasma on foods, the physical nature (physical external properties) of food such as the presence of cracks, folds, and thick and uneven surfaces raise doubt on uniform exposure of plasma. Thus, the surface of the food can also alter the efficiency of the processing; the microbes can hide between the folds and cracks of the food material thus it can lead to incomplete decontamination of the product using plasma [50].

16.3.2 PHYSIOCHEMICAL CHANGES IN FOOD

Apart from achieving selectivity of plasma, the major purpose of using plasma is to achieve the safety of the food material without altering the quality of food. Yet, the species in plasma are familiar with causing lipid and protein oxidation, which poses a real problem in treating fatty foods such as fish, milk, and meat [39]. The change in the biomolecules produces off-flavor and odors in food material; especially the fishy smell of foods after plasma treatment forms a major stimulus for rejection of food products [38]. This must also be considered when subjecting to high-fat foods because this could diminish the efficiency of plasma processing on foods [24].

Another detrimental change that can be observed while treating seafood and meat is the color change [18]. It has been described that the oxidation of myofibrillar protein by plasma species gives a green color to the meat, which is not acceptable for consumption [61]. When treating fruit surfaces for decontamination, visible marks or discoloration on the fruit surface were observed. It has been reported that the generation of localized streamers in the atmospheric DBD unit could be the reason for the appearance of marks on the soft skin surface [42]. Various studies have been conducted on plasma-assisted decontamination of foods that had physiochemical defects, followed by plasma treatment [15]. This review work revealed that apart from decontamination, plasma tends to have a negative impact on the physiochemical nature of foods, which must be taken into consideration.

16.3.3 FOOD LEGISLATION OR REGULATORY STATUS

The USFDA has not yet recognized cold plasma as a commercial technology for treating food materials [51]. Being a novel technology, regulatory guidelines are necessary for its application. However, to the best of our knowledge, there has not been any guideline or legislation from a government body framed for the use of plasma in food materials. Providing safety guidelines for both process and product parameters is essential for the futuristic commercialization of this technology in food processing [18].

The USFDA approval of ozone processing of foods as a direct additive to food gives a mirage of hope for the approval of plasma in the future. Nevertheless, unlike ozone technology where only one reactive species of O_3 in the gas is applied, the complex nature of plasma (which contains numerous reactive species) makes it hard to move towards the commercialization of this technology. The presence of short-life, metastable, and long-life reactive species in plasma makes it complicated to provide regulatory guidelines for the safety of the technology. If only long-lived species are present in plasma, it would be easy to frame regulations; however, the presence of metastable reactive species makes it hard to characterize the plasma and frame guidelines for treatment.

In contrast to air plasma, the generation of plasma inside the liquid food is less complicated in terms of proving the regulatory guidelines of plasma-activated liquid foods. This is because the reactive species in the liquid state are stable compared to the gas state; thus, the dosage of the reactive species, reaction chemistry, and their pathways can be easily analyzed in plasma-activated liquids. However, the high reactivity of reactive species in plasma-activated liquids questions the oxidative stability and the generation of toxicity when the species react with biomolecules present in the liquid. Thus, toxicology study is also required for commercialization and providing regulatory guidelines for plasma technology [36].

16.3.4 DESIGN AND SCALE UP- COMMERCIALIZATION OF TECHNOLOGY

There are various cold plasma equipment designed all over the world to treat food materials. However, the major problem faced is the cost of fabrication and operation. Low-pressure plasma equipment uses a vacuum pump and a completely sealed chamber to maintain low pressure; this adds up to the investment cost of the equipment [50]. Whereas in atmospheric pressure plasma, the cost is mainly based on the operating gas used and the cost of the safety measures needed while using high voltage [32].

These equipment are used for studies of cold plasma; nevertheless, almost all plasma-treated studies have been conducted on a lab-scale (smaller quantities of food were treated). So far, most of the studies were conducted for batch-type lab-scale equipment; only limited studies have been available on a large scale or continuous plasma processing of foods [14, 65]. Further, while considering the scale-up option for batch-type plasma equipment for commercial purposes [18], it also poses a major drawback because it might add up to the cost and it could be hard to scale up the equipment and get a similar effect on the food material [53]. Further, it is difficult to obtain uniformity of plasma application on the food material, which is a major factor that would question the efficiency of processing.

16.3.5 TOXICITY AND DOSAGE

The plasma reactive species are the sole agents for numerous applications of plasma on food materials. However, these reactive species tend to oxidize food components such as proteins, carbohydrates, and lipids to form harmful toxic substances such as aldehydes [12]. Apart from the toxicity generation, the presence of nitrogen reactive species such as nitrates and nitrites in plasma-treated liquid samples. Their concentration could increase in plasma-treated drinking water and fruit juices above the WHO standard recommendation of nitrites presence [37]. The complex nature of food and its interaction with various plasma reactive species makes it a complex process that has been vaguely investigated [34]. Thus, cyto and genotoxicity studies of plasma-treated food materials are currently required for proving the safety of using plasma on food materials.

In the past few years, scientists have started working on the toxicity studies of food, owing to the disapproval from USFDA for the commercialization of this technology for food processing. However, the studies done so far are not sufficient to understand the concentration of plasma, and exposure time that can be used for treatment without producing toxicity in the sample has not yet been reported. Control of plasma dosage to treat samples without producing any harmful effect on the human cells is the current and major problem faced in the field of plasma medicine. The following are the studies that were done so far on the toxicity.

Plasma-activated water (PAW) was subjected to a toxicity study by injecting 300 μl of water into the oral cavity of immunodeficient mice for 2 weeks (3 days a week) [60]. When mice injected with PAW generated at a maximum exposure time of 15 minutes were subjected to a toxicity study, there was no evidence of acute toxicity in the mice. Non-significant differences were reported in the organs and their function. Biological indicators such as neutrophils and mononuclear immune cells in blood were observed to increase in plasma-treated mice compared to the control. This implies that the plasma-activated water activated the immune cells in the immunodeficient mice. Although this study shows that PAW did not cause any toxicity development in mice, toxicity assessment studies of food materials other than water could be beneficial because oxidation occurs in the presence of food biomolecules, which could be harmful to living beings. However, in another study (in medicine) conducted on helium PAW treatment on breast and gastrointestinal cancer cells, it was noted that cell apoptosis increased with the elevation in plasma application duration and the quantities of reactive species present in PAW [8]. This indicates that PAW could be harmful when applied to living cells.

When the directly treated edible film was subjected to mice, it didn't have any toxic effect on mice for the continuous 14-day observation [22]. However, similar to the previous study, the blood components in mice were altered when subjected to plasma-treated edible film.

Similarly, genotoxicity studies on plasma treatment of meat breasts [27] and cold sliced chicken [1] also proved that plasma didn't have any toxic effect. In an *in vivo*, *in vitro* toxicity, and mutagenic study on plasma-treated lettuce samples, it was observed that there were less cytotoxic and mutagenic changes in the *in vitro* and mutational changes study, respectively, when subjected to mice cells [28]. However, in the *in vivo* study, when the 5 min and 10 min plasma-treated lettuce broth was injected into larvae, almost 93 and 50% of the larvae died, respectively after 24 hours. Whereas when untreated lettuce broth was injected only 10% of larvae died after 24 hours. In another study, plasma-treated wheat as a whole and wheat model medium was subjected to toxicity study using mammalian cells and insect *Tribolium castaneum* [29]. In this study, it was observed that the toxicity of wheat model medium treated with plasma on mammalian cells was more when compared to whole wheat grain (subjected in the form of powder). Prolonged exposure of 20 min and supplementation of 10% wheat medium or powder increased the toxicity. 96.3% of the cell growth was inhibited when plasma-treated (20 min) wheat medium (2%–10% concentration) was supplied to mammalian cells. Whereas when 20 min plasma-treated wheat grain model was subjected to mammalian cells, 17%, 22%, and 61% of mammalian cell growth was inhibited.

The morphological investigation of mammalian cells revealed that the cells supplied with plasma-treated wheat model medium developed a round structure rather than the epithelial shape of normal cell structure. However, the *T. castaneum* fed with whole wheat grain didn't show any biological changes in the insects. This study gives an idea that the type of material and their physical nature can also aid in increasing or decreasing the toxicity of the product. The subjection of liquid foods to plasma can be unraveled to have more impact than solid foods. The reason for this could be the increased reactivity of liquid foods in comparison to the solid foods where only the surface of the product gets exposed to plasma. Thus, the contradiction in toxicological studies could be owing to the types of equipment used, the power applied to the treatment samples, and also the nature of the samples used. These studies proved that plasma reactive species produce toxic substances in cells and mice which must be considered as a potential risk in using cold plasma for food application. More research must be conducted on the toxicological study of plasma-treated food such that it will be easy to determine the safe dosage of plasma that produces less toxicity in animals and humans [37].

The challenges mentioned above are real and are faced in the practical application of plasma on food materials. Despite getting regulatory body approval in the field of medicine, similar challenges such as local heating, incorporation of toxicity in non-targeted surfaces, and matrix degradation while trying to heal or disinfect a tissue surface are faced [19]. Thus, these challenges are not specific only to food processing; each sector deals with some kind of challenge while conducting experiments. This can be overcome with the infusion of technology and smart thinking. The study on the avenues by which these challenges were tackled in other sectors must be taken into account for the betterment. The following section deals with the future possibilities and the path by which we can lead to rectify these challenges.

16.4 FUTURE AVENUES OF PLASMA IN FOOD AND AGRICULTURE

For regulatory bodies to approve this technology, plasma parameters must be optimized such that one source of plasma (with specific gas, pressure, flow rate, and power) must be applied for one product and one application; it is the one doctor, one patient rule as explained by [5] for oncology. In the case of the food sector, the knowledge of plasma in different foods is still unknown. Therefore, to take the knowledge that we have on the food sector from the armature stage to the developed stage, the working scientists from Food Safety and Technology of the German Research Foundation (DFG), the Senate Commission on Food Safety started collecting information and works on plasma processing of foods [48]; they have started giving opinions and recommendations for conducting research and adopting plasma for food applications in the year 2013. They have provided a few recommendations for the studies to be conducted such as case by case assessment of the treated product, providing information on process parameters such as process gas, degree of ionization, treatment temperature, and time and geometry of the treatment, provide information on microbial, biological, and oxidative changes, and requirement of toxicity study. Based on their recommendations for plasma treatment, since then many studies have emerged on biochemical, microbial, and toxicological aspects [12].

Therefore, these kinds of regulatory guidelines from authorized bodies can be useful in the future for improving the standards of research in this area of food processing; this, in turn, can lead to a ray of light for the commercialization of the technology.

16.4.1 PROSPECTIVE APPROACH TO CONTROL PLASMA DOSAGE AND UNDERSTANDING PLASMA INTERACTION

The plasma dose depends on the type and quantity of species formed in plasma and interacts with food material. This dosage of reactive species depends on the operating gases used for generating plasma. Thus, a study on modulating the operating gas solely or in combination can aid in deciding the species generated in plasma [51].

The amounts of reactive species and interactions vary from food materials used, and their properties must be experimented separately and collected to reach a conclusion. However, only limited studies are available on the activity of plasma species with food materials. This could be due to the complexity and time-consuming nature of conducting experiments for various food materials and examining the effect of different dosages on the food. Furthermore, the need for utilizing high-end equipment for analysis is also not possible for everyone. Thus, to overcome these problems and reduce the cost of operation, the numerical computational fluid dynamics method can be adopted in the future. However, there is a shortage of studies on the numerical simulation of plasma on food materials. Yet, it could be the future of plasma in food processing as it has been in the field of medicine and polymer science. These numerical plasma studies in other sectors can be taken as a base for initiating numerical works on the plasma treatment of food. This could be beneficial both in saving time and money in the case of conducting trials. Recently, in one particular study, they used computational fluid dynamics to model the microbial inactivation of plasma in a bacterial culture, which is a head start for starting numerical modeling of plasma in foods. A floating type of DBD plasma was used for modeling and predicting the dynamics of plasma, such as species concentration and distribution on the target surface [51]. When compared with the experimental analysis of the microbial reduction, a 1 log difference in microbial inactivation was observed in the numerical simulation. They figured out that the difference could be due to the non-inclusion of UV rays in the numerical simulation.

While conducting computational fluid dynamics deals with the molecular level of simulation, plasma molecular dynamics deals with the atomic level of simulation. Plasma molecular dynamics could be the next step in the future after computational fluid dynamics. It is the finest approach that can be combined with experimental study to predict the species dynamics in plasma and its activity (of individual reactive species) with various food biomolecules [6].

The classic molecular dynamics simulation is based on the interatomic force field / the force potential which drives the activity of species with other species and biomolecules at the atomic level. In a molecular simulation conducted by [63] on plasma interaction with water, it was found that the O reactive species immediately react with water and form. OH, whereas, OH and HO_2 radicals in air plasma which were diffused into the water were found to continuously interact with water to reform OH and HO_2, respectively. Another study on plasma molecular dynamics simulated that the O atom contributed to 78% of C–O bond breakage, 38% of C-C bond breakage, and 26% of C–N bond breakage in the microbial cell [62]. These kinds of studies are quintessential and are in extreme need in the food processing sector, as it deals with the processing of complex food materials.

16.4.2 ATTAINING UNIFORMITY OF PLASMA EXPOSURE

It is a notable fact that plasma species are highly antibacterial in nature. Plasma reactive species act synergistically on the microbial cell to inactivate them. However, in the case of solid foods, there is a problem in attaining uniform distribution of plasma exposure on all food surfaces. In the case of plasma-assisted decontamination, the presence of folds, fissures, and sphericity of solid foods as discussed earlier in the challenges pose a major problem in producing improper decontamination of the product. To address this challenge, custom-designed plasma equipment must be developed for attaining even exposure of plasma on the surface and complete decontamination of the product [16].

A paper-based flexible DBD unit was designed and fabricated for treating food material to ensure uniform exposure [59]. The flexibility of the electrode makes it more adaptable for any food material. For producing this electrode, they used metalized sheets which were engraved in a form of the hexagonal honeycomb structure and attached to the adhesive for connecting it to the power supply. In this type of equipment, both volume and surface plasma discharge was obtained. In the subsequent study, modified the structure of the electrode in the form of cones (origami) to treat spherical fruits [57]. In this study, the electrodes were in direct contact with the food material and the whole area of the fruit was exposed to plasma evenly. It was observed that the microbial reduction in the tomato after treatment was effective. Thus, using flexible electrodes which could be altered based on the commodity of usage could be beneficial to obtaining uniform exposure of plasma.

16.4.3 LAB TO THE FOOD INDUSTRY

Though the scale-up of the lab-scale system to industrial-scale might cause problems such as cost and uniformity of treatment, a continuous array of plasma jets can be used for plasma treatment; this will not pose any questions on the efficiency of processing [37]. For the commercialization of technology, it might take several years; in the case of electron beam technology, it took 30 years for commercialization for using it in reducing air pollution [25]. Similarly, the commercialization of plasma technology in food processing might take years; but until then further studies on plasma processing of food and their interaction with food biomolecules must be conducted to understand the process which would lead us to the path of commercialization.

16.5 SUMMARY

Plasma is a promising novel technology that has actively been researched for its numerous potential applications in food technology. Nevertheless, the lack of complete understanding of the reactive species dosage and their effect on individual food products creates various challenges while treating food materials. This chapter covered the major challenges faced, which are the control of plasma dosage, interaction with food materials, lack of uniformity of plasma exposure, unavailability of regulatory guidelines and approval, and toxicity of the treated samples. These challenges can be overcome with continuous research on different food materials and various process conditions. However, conducting studies on all food materials and understanding their interaction could take decades, and the cost of conducting experiments could also become a challenge. Thus, numerical simulation of plasma reactive species on quantifying the reactive species generated at various processing conditions and studying its individual interaction with plasma molecular dynamics could be beneficial in addressing all the above-said challenges. Further, the advent of flexible plasma electrode in recent studies also proved that uniform exposure of plasma on whole fruits and other food commodities is possible. The challenges faced in applying cold plasma in food processing are not entirely new. Similar challenges have been faced in other sectors. However, the complexity of the food material used for treatment is exceptional. Only for the food sector, this could be addressed with the use of numerical simulations. Though the increase in studies on plasma application in food proves its effectiveness in maintaining food safety, future studies must be focused mainly on controlling the plasma dosage and assessment of plasma interaction and toxicity in the food material for optimization. This might pave the way for obtaining regulatory approval for the commercialization of this technology in the food processing sector.

KEYWORDS

- **cold plasma**
- **dielectric barrier discharge**
- **German Research Foundation**
- **plasma modeling**
- **plasma-activated water**
- **reactive species**
- **regulatory approval**
- **WHO**

REFERENCES

1. Aly, A. (2013). Comparison between gamma irradiation and plasma technology to improve the safety of cold sliced chicken. *African Journal of Food Science, 7*(12), 461–467.
2. Anbarasan, R., Boopathy, B., Stephen, J., & Radhakrishnan, M. (2022). Cold plasma disinfestation of *Callosobruchus maculatus* infested soybeans: Its subsequent impact on soymilk extraction yield and quality. *Journal of Food Process Engineering, 46*(2), Article e14246. https://doi.org/10.1111/jfpe.14246.
3. Anbarasan, R., Jaspin, S., Bhavadharini, B., Pare, A., Pandiselvam, R., & Mahendran, R. (2022). Chlorpyrifos pesticide reduction in soybean using cold plasma and ozone treatments. *LWT, 159,* Article 113193. https://doi.org/10.1016/j.lwt.2022.113193.
4. Anukiruthika, T., Sethupathy, P., Wilson, A., Kashampur, K., Moses, J. A., & Anandharamakrishnan, C. (2020). Multilayer packaging: Advances in preparation techniques and emerging food applications. *Comprehensive Reviews in Food Science and Food Safety, 19*(3), 1156–1186.
5. Bekeschus, S., Favia, P., Robert, E., & von Woedtke, T. (2019). White paper on plasma for medicine and hygiene: Future in plasma health sciences. *Plasma Processes and Polymers, 16*(1), 1–12.
6. Bogaerts, A., De Bie, C., Eckert, M., Georgieva, V., & Martens, T. (2010). Modeling of the plasma chemistry and plasma-surface interactions in reactive plasmas. *Pure and Applied Chemistry, 82*(6), 1283–1299.
7. Braný, D., Dvorská, D., Halašová, E., & Škovierová, H. (2020). Cold atmospheric plasma: A powerful tool for modern medicine. *International Journal of Molecular Sciences, 21*(8), Article 2932. https://doi.org/10.3390/ijms21082932.
8. Chen, Z., Lin, L., Cheng, X., Gjika, E., & Keidar, M. (2016). Effects of cold atmospheric plasma generated in deionized water in cell cancer therapy. *Plasma Processes and Polymers, 13*(12), 1151–1156.
9. Cui, H., Surendhiran, D., Li, C., & Lin, L. (2020). Biodegradable zein active film containing chitosan nanoparticle encapsulated with pomegranate peel extract for food packaging. *Food Packaging and Shelf Life, 24,* Article 100511. https://doi.org/10.1016/j.fpsl.2020.100511.
10. Cui, H., Wu, J., Li, C., & Lin, L. (2017). Promoting anti-listeria activity of lemongrass oil on pork loin by cold nitrogen plasma assist. *Journal of Food Safety, 37*(2), Article e12316. https://doi.org/10.1111/jfs.12316.
11. Cui, H., Yang, X., Abdel-Samie, M. A., & Lin, L. (2020). Cold plasma treated phlorotannin/Momordica charantia polysaccharide nanofiber for active food packaging. *Carbohydrate Polymers, 239,* Article 116214. https://doi.org/10.1016/j.carbpol.2020.116214.
12. Cullen, P. J., Lalor, J., Scally, L., Boehm, D., & Milosavljević, V. (2018). Translation of plasma technology from the lab to the food industry. *Plasma Processes and Polymers, 15*(2), 85–91. https://doi.org/10.1002/ppap.201700085.
13. Dakshayani, R., Paul, A., & Mahendran, R. (2021). Cold plasma-induced effects on bioactive constituents and antioxidant potential of lotus petal powder. *IEEE Transactions on Plasma Science, 49*(2), 507–512.

14. Dharini, M., Jaspin, S., & Mahendran, R. (2022). Optical emission spectra characterization of plasma discharge during the processing of milk and evaluation of its quality attributes. *IEEE Transactions on Plasma Science, 50*(4), 1128–1136.
15. Domonkos, M., Tichá, P., Trejbal, J., & Demo, P. (2021). Applications of cold atmospheric pressure plasma technology in medicine, agriculture and food industry. *Applied Sciences, 11*(11), Article 4809. https://doi.org/10.3390/app11114809.
16. Feizollahi, E., Misra, N. N., & Roopesh, M. S. (2020). Factors influencing the antimicrobial efficacy of dielectric barrier discharge (DBD) atmospheric cold plasma (ACP) in food processing applications. *Critical Reviews in Food Science and Nutrition, 61*(4), 666–689.
17. Ganesan, A. R., Tiwari, U., Ezhilarasi, P. N., & Rajauria, G. (2021). Application of cold plasma on food matrices: A review on current and future prospects. *Journal of Food Processing and Preservation, 45*(1), Article e15070. https://doi.org/10.1111/jfpp.15070.
18. Gavahian, M., Chu, Y. H., & Jo, C. (2019). Prospective applications of cold plasma for processing poultry products: Benefits, effects on quality attributes, and limitations. *Comprehensive Reviews in Food Science and Food Safety, 18*(4), 1292–1309.
19. Gilmore, B. F., Flynn, P. B., O'Brien, S., Hickok, N., Freeman, T., & Bourke, P. (2018). Cold plasmas for biofilm control: Opportunities and challenges. *Trends in Biotechnology, 36*(6), 627–638.
20. Gu, Y., Shi, W., Liu, R., Xing, Y., Yu, X., & Jiang, H. (2021). Cold plasma enzyme inactivation on dielectric properties and freshness quality in bananas. *Innovative Food Science & Emerging Technologies, 69,* Article 102649. https://doi.org/10.1016/j.ifset.2021.102649.
21. Guo, D., Liu, H., Zhou, L., Xie, J., & He, C. (2021). Plasma-activated water production and its application in agriculture. *Journal of the Science of Food and Agriculture, 101*(12), 4891–4899.
22. Han, S. H., Suh, H. J., Hong, K. B., Kim, S. Y., & Min, S. C. (2016). Oral toxicity of cold plasma-treated edible films for food coating. *Journal of Food Science, 81*(12), T3052–T3057.
23. Han, Y., Cheng, J., & Sun, D. (2019). Changes in activity, structure and morphology of horseradish peroxidase induced by cold plasma. *Food Chemistry, 301,* 1–9. Advance online publication. https://doi.org/10.1016/j.foodchem.2019.124240.
24. Hernández-Hernández, H. M., Moreno-Vilet, L., & Villanueva-Rodríguez, S. J. (2019). Current status of emerging food processing technologies in Latin America: Novel non-thermal processing. *Innovative Food Science & Emerging Technologies, 58,* Article 102233. https://doi.org/10.1016/j.ifset.2019.102233.
25. Kim, H. H. (2004). Nonthermal plasma processing for air-pollution control: A historical review, current issues, and future prospects. *Plasma Processes and Polymers, 1*(2), 91–110.
26. Lackmann, J.-W., Schneider, S., Edengeiser, E., Jarzina, F., & Brinckmann, S. (2013). Photons and particles emitted from cold atmospheric-pressure plasma inactivate bacteria and biomolecules independently and synergistically. *Journal of the Royal Society Interface, 10*(89), 591–597. https://doi.org/10.1098/rsif.2013.0591.
27. Lee, H., Yong, H. I., Kim, H. J., Choe, W., & Yoo, S. J. (2016). Evaluation of the microbiological safety, quality changes, and genotoxicity of chicken breast treated with flexible thin-layer dielectric barrier discharge plasma. *Food Science & Biotechnology, 25*(4), 1189–1195.

28. Liu, C., Springer, D., & Clifford, G. D. (2017). Performance of an open-source heart sound segmentation algorithm on eight independent databases. *Physiological Measurement, 38*(8), 1730–1745.
29. Los, A., Ziuzina, D., van Cleynenbreugel, R., Boehm, D., & Bourke, P. (2020). Assessing the biological safety of atmospheric cold plasma treated wheat using cell and insect models. *Foods, 9*(7), Article 898. https://doi.org/10.3390/foods9070898.
30. Manoharan, D., Radhakrishnan, M., Tiwari, B. K., & Schlüter, O. K. (2021). Plasma-assisted combination processes. In *Innovative Food Processing Technologies* (pp. 667–681). Elsevier. https://doi.org/10.1016/B978-0-12-815781-7.00007-X.
31. Mir, S. A., Dar, B. N., Shah, M. A., Sofi, S. A., & Hamdani, A. M. (2021). Application of new technologies in decontamination of mycotoxins in cereal grains: Challenges, and perspectives. *Food and Chemical Toxicology, 148,* Article 111978. https://doi.org/10.1016/j.fct.2021.111976.
32. Mir, S. A., Siddiqui, M. W., Dar, B. N., Shah, M. A., & Wani, M. H. (2020). Promising applications of cold plasma for microbial safety, chemical decontamination and quality enhancement in fruits. *Journal of Applied Microbiology, 129*(3), 474–485.
33. Misra, N. N., Pankaj, S. K., Segat, A., & Ishikawa, K. (2016). Cold plasma interactions with enzymes in foods and model systems. *Trends in Food Science & Technology, 55,* 39–47.
34. Mollakhalili-Meybodi, N., Yousefi, M., Nematollahi, A., & Khorshidian, N. (2021). Effect of atmospheric cold plasma treatment on technological and nutrition functionality of protein in foods. *European Food Research and Technology, 247*(7), 1579–1594.
35. Morent, R., De Geyter, N., Desmet, T., Dubruel, P., & Leys, C. (2011). Plasma surface modification of biodegradable polymers: A review. *Plasma Processes and Polymers, 8*(3), 171–190.
36. Múgica-Vidal, R., Sainz-García, E., Álvarez-Ordóñez, A., Prieto, M., & González-Raurich, M. (2019). Production of antibacterial coatings through atmospheric pressure plasma: A promising alternative for combatting biofilms in the food industry. *Food and Bioprocess Technology, 12*(8), 1251–1263.
37. Muhammad, A. I., Liao, X., Cullen, P. J., Liu, D., & Xiang, Q. (2018). Effects of nonthermal plasma technology on functional food components. *Comprehensive Reviews in Food Science and Food Safety, 17*(5), 1379–1394.
38. Olatunde, O. O., Benjakul, S., & Vongkamjan, K. (2019). Combined effects of high voltage cold atmospheric plasma and antioxidants on the qualities and shelf-life of Asian sea bass slices. *Innovative Food Science & Emerging Technologies, 54,* 113–122.
39. Olatunde, O. O., Shiekh, K. A., & Benjakul, S. (2021). Pros and cons of cold plasma technology as an alternative non-thermal processing technology in seafood industry. *Trends in Food Science & Technology, 111,* 617–627.
40. Pankaj, S. K., Bueno-Ferrer, C., Misra, N. N., Milosavljević, V., & O'Donnell, C. P. (2014). Applications of cold plasma technology in food packaging. *Trends in Food Science & Technology, 35*(1), 5–17.
41. Patange, A., Boehm, D., Giltrap, M., Lu, P., Cullen, P. J., & Bourke, P. (2018). Assessment of the disinfection capacity and eco-toxicological impact of atmospheric cold plasma for treatment of food industry effluents. *Science of The Total Environment, 631,* 298–307.
42. Perni, S., Liu, D. W., Shama, G., & Kong, M. G. (2008). Cold atmospheric plasma decontamination of the pericarps of fruit. *Journal of Food Protection, 71*(2), 302–308.

43. Rajan, A., Boopathy, B., Radhakrishnan, M., Rao, L., Schlüter, O. K., & Tiwari, B. K. (2023). Plasma processing: A sustainable technology in agri-food processing. *Sustainable Food Technology, 1*(1), 9–49.
44. Romani, V. P., Olsen, B., Pinto Collares, M., Meireles Oliveira, J. R., Prentice, C., & Guimarães Martins, V. (2019). Plasma technology as a tool to decrease the sensitivity to water of fish protein films for food packaging. *Food Hydrocolloids, 94,* 210–216.
45. Sarangapani, C., O'Toole, G., Cullen, P. J., & Bourke, P. (2017). Atmospheric cold plasma dissipation efficiency of agrochemicals on blueberries. *Innovative Food Science & Emerging Technologies, 44,* 235–241.
46. Sarangapani, C., Patange, A., Bourke, P., Keener, K., & Cullen, P. J. (2018). Recent advances in the application of cold plasma technology in foods. *Annual Review of Food Science and Technology, 9,* 609–629.
47. Sarangapani, C., Thirumdas, R., Devi, Y., Trimukhe, A., Deshmukh, R. R., & Annapure, U. S. (2016). Effect of low-pressure plasma on physico-chemical and functional properties of parboiled rice flour. *LWT – Food Science and Technology, 69,* 482–489.
48. Schlüter, O., Ehlbeck, J., Hertel, C., Habermeyer, M., & Roth, A. (2013). Opinion on the use of plasma processes for treatment of foods. *Molecular Nutrition & Food Research, 57*(5), 920–927.
49. Shriver, S. K. (2011). Effect of selected nonthermal processing methods on the allergen reactivity of Atlantic white shrimp (*Litopenaeus setiferus*). Unpublished doctoral dissertation, https://api.semanticscholar.org/CorpusID:104551313. Corpus ID: 104551313.
50. Šimončicová, J., Kryštofová, S., Medvecká, V., Ďurišová, K., & Kaliňáková, B. (2019). Technical applications of plasma treatments: Current state and perspectives. *Applied Microbiology and Biotechnology, 103*(13), 5117–5129.
51. Sonawane, S. K., T M., & Patil, S. (2020). Non-thermal plasma: An advanced technology for food industry. *Food Science and Technology International, 26*(8), 727–740.
52. Surowsky, B., Fischer, A., Schlueter, O., & Knorr, D. (2013). Cold plasma effects on enzyme activity in a model food system. *Innovative Food Science & Emerging Technologies, 19,* 146–152.
53. Thirumdas, R., Kadam, D., & Annapure, U. S. (2017). Cold plasma: An alternative technology for the starch modification. *Food Biophysics, 12*(1), 129–139.
54. Tolouie, H., Mohammadifar, M. A., Ghomi, H., & Hashemi, M. (2018). Cold atmospheric plasma manipulation of proteins in food systems. *Critical Reviews in Food Science and Nutrition, 58*(15), 2583–2597.
55. Ucar, Y., Ceylan, Z., Durmus, M., Tomar, O., & Cetinkaya, T. (2021). Application of cold plasma technology in the food industry and its combination with other emerging technologies. *Trends in Food Science & Technology, 114,* 355–371.
56. Venkataratnam, H., Sarangapani, C., Cahill, O., & Ryan, C. B. (2019). Effect of cold plasma treatment on the antigenicity of peanut allergen Ara h 1. *Innovative Food Science & Emerging Technologies, 52,* 368–375.
57. Wang, Q., Pal, R. K., Yen, H. W., Naik, S. P., & Orzeszko, M. K. (2022). Cold plasma from flexible and conformable paper-based electrodes for fresh produce sanitation: Evaluation of microbial inactivation and quality changes. *Food Control, 137,* Article 108915. https://doi.org/10.1016/j.foodcont.2022.108915.
58. Xiang, Q., Huangfu, L., Dong, S., Ma, Y., & Li, K. (2021). Feasibility of atmospheric cold plasma for the elimination of food hazards: Recent advances and future trends.

Critical Reviews in Food Science and Nutrition, 63(20), 4431–4449. Advance online publication. https://doi.org/10.1080/10408398.2021.2002257.

59. Xie, J., Chen, Q., Suresh, P., Roy, S., White, J. F., & Mazzeo, A. D. (2017). Paper-based plasma sanitizers. *Proceedings of the National Academy of Sciences of the United States of America, 114*(20), 5119–5124.
60. Xu, D., Cui, Q., Xu, Y., Wang, B., & Tian, M. (2018). Systemic study on the safety of immuno-deficient nude mice treated by atmospheric plasma-activated water. *Plasma Science and Technology, 20*(4), Article aa9842. https://doi.org/10.1088/2058-6272/aa9842.
61. Yong, H. I., Han, M., Kim, H. J., Suh, J. Y., & Jo, C. (2018). Mechanism underlying green discoloration of myoglobin induced by atmospheric pressure plasma. *Scientific Reports, 8*(1), Article 1–9. https://doi.org/10.1038/s41598-018-29767-6.
62. Yusupov, M., Neyts, E. C., Khalilov, U., Snoeckx, R., Van Duin, A. C. T., & Bogaerts, A. (2012). Atomic-scale simulations of reactive oxygen plasma species interacting with bacterial cell walls. *New Journal of Physics, 14*(9), Article 093043. https://doi.org/10.1088/1367-2630/14/9/093043.
63. Yusupov, M., Neyts, E. C., Simon, P., Berdiyorov, G., & Snoeckx, R. (2014). Reactive molecular dynamics simulations of oxygen species in a liquid water layer of interest for plasma medicine. *Journal of Physics D: Applied Physics, 47*(2), Article 025205. https://doi.org/10.1088/0022-3727/47/2/025205.
64. Zhang, K., Perussello, C. A., Milosavljević, V., Cullen, P. J., Sun, D. W., & Tiwari, B. K. (2019). Diagnostics of plasma reactive species and induced chemistry of plasma treated foods. *Critical Reviews in Food Science and Nutrition, 59*(5), 812–825. https://doi.org/10.1080/10408398.2018.1461458.
65. Ziuzina, D., Misra, N. N., Han, L., Cullen, P. J., & Moiseev, T. (2020). Investigation of a large gap cold plasma reactor for continuous in-package decontamination of fresh strawberries and spinach. *Innovative Food Science & Emerging Technologies, 59,* Article 102229. https://doi.org/10.1016/j.ifset.2019.102229.

INDEX

D

E

G

J

K

M

P

Q

R

T

Y

Z